Also Available from McGraw-Hill

Schaum's Outline Series in Mechanical Engineering

Most outlines include basic theory, definitions, and hundreds of solved problems and supplementary problems with answers.

Titles on the Current List Include:

Acoustics
Basic Equations of Engineering Science
Continuum Mechanics
Engineering Economics
Engineering Mechanics, 4th edition
Engineering Thermodynamics
Fluid Dynamics, 2d edition
Fluid Mechanics & Hydraulics, 2d edition
Heat Transfer
Lagrangian Dynamics
Machine Design

Mathematical Handbook of Formulas
  & Tables
Mechanical Vibrations
Operations Research
Statics & Mechanics of Materials
Statics & Strength of Materials
Strength of Materials, 2d edition
Theoretical Mechanics
Thermodynamics with Chemical
  Applications, 2d edition

Schaum's Solved Problems Books

Each title in this series is a complete and expert source of solved problems containing thousands of problems with worked out solutions.

Related Titles on the Current List Include:

3000 Solved Problems in Calculus
2500 Solved Problems in Differential Equations
2500 Solved Problems in Fluid Mechanics and Hydraulics
1000 Solved Problems in Heat Transfer
3000 Solved Problems in Linear Algebra
2000 Solved Problems in Mechanical Engineering Thermodynamics
2000 Solved Problems in Numerical Analysis
700 Solved Problems in Vector Mechanics for Engineers: Dynamics
800 Solved Problems in Vector Mechanics for Engineers: Statics

Available at your College Bookstore. A complete list of Schaum titles may be obtained by writing to: Schaum Division
McGraw-Hill, Inc.
Princeton Road, S-1
Hightstown, NJ 08520

CONVECTIVE HEAT AND MASS TRANSFER

Third Edition

W. M. Kays

Professor of Mechanical Engineering, Emeritus
Stanford University

M. E. Crawford

Professor of Mechanical Engineering
The University of Texas at Austin

McGraw-Hill, Inc.

New York   St. Louis   San Francisco   Auckland   Bogotá
Caracas   Lisbon   London   Madrid   Mexico   Milan   Montreal
New Delhi   Paris   San Juan   Singapore   Sydney   Tokyo   Toronto

This book was set in Times Roman.
The editors were John J. Corrigan and John M. Morriss;
the production supervisor was Kathryn Porzio.
The cover was designed by Joseph Gillians.
Project supervision was done by The Universities Press (Belfast) Ltd.
R. R. Donnelley & Sons Company was printer and binder.

CONVECTIVE HEAT AND MASS TRANSFER

1 2 3 4 5 6 7 8 9 0 DOC DOC 9 0 9 8 7 6 5 4 3

ISBN 0-07-033721-7

Library of Congress Cataloging-in-Publication Data

Kays, W. M. (William Morrow)
   Convective heat and mass transfer / W. M. Kays, M. E. Crawford.—
3rd ed.
      p.   cm.—(McGraw-Hill series in mechanical engineering)
   Includes bibliographical references and index.
   ISBN 0-07-033721-7
   1. Heat—Convection.   2. Mass transfer.   I. Crawford, M. E.
(Michael E.)   II. Title.   III. Series.
QC327.K37   1993
621.402′2—dc20
                           92-24670

ABOUT THE AUTHORS

W. M. Kays is Emeritus Professor of Mechanical Engineering at Stanford University. He received his Ph.D. degree in Mechanical Engineering in 1951 at Stanford, where he has spent all of his professional life, serving as Head of the Mechanical Engineering Department from 1961 to 1972 and Dean of Engineering from 1972 to 1984. He is a Member of the National Academy of Engineering and a Fellow of the American Society of Mechanical Engineering. Professor Kays has done extensive research on compact heat-exchanger surfaces and on heat transfer from the turbulent boundary layer. He is the author (with A. L. London) of *Compact Heat Exchangers*.

M. E. Crawford has, since 1980, been a member of the faculty at The University of Texas, where he teaches courses in the Thermal/Fluids Systems Group of the Mechanical Engineering Department. Prior to that, he was on the faculty at MIT. His doctoral work was carried out in the Mechanical Engineering Department at Stanford University. His teaching focuses on heat transfer and turbulence modeling. Dr Crawford's current research programs include development of simulation models of bypass transition, numerical and experimental studies of gas turbine film cooling, and liquid metal magnetohydrodynamic flows in packed beds. His research associated with development of the numerical simulation code STAN5 for application to high-performance gas turbine engines has earned him a international reputation in the gas turbine industry. This code, and its successor, TEXSTAN, continue to be used by major turbine engine companies in the United States, as well as many universities and research institutes, both in the USA and abroad. Dr Crawford is actively involved in consulting to industry and government, and he participates in short courses on numerical methods in fluid dynamics and gas turbine cooling. He has been a NATO–AGARD lecturer at the von Kármán Institute for Fluid Dynamics in Belgium. He is a member of the honor societies Pi Tau Sigma and Tau Beta Pi, and he is active in the American Society of Mechanical Engineering, the American Institute of Aeronautics and Astronautics, and the American Physical Society.

CONTENTS

Preface to the Third Edition   xi
Preface to the Second Edition   xv
Preface to the First Edition   xix
List of Symbols   xxii
Units Conversion Table   xxxiii

1   Introduction   1
2   Conservation Principles   5
3   Fluid Stresses and Flux Laws   10
4   The Differential Equations of the Laminar Boundary Layer   19
5   The Differential Equations of the Turbulent Boundary Layer 44
6   The Integral Equations of the Boundary Layer   62
7   Momentum Transfer: Laminar Flow inside Tubes   75
8   Momentum Transfer: The Laminar External Boundary Layer   88
9   Heat Transfer: Laminar Flow inside Tubes   108
10 Heat Transfer: The Laminar External Boundary Layer   159
11 Momentum Transfer: The Turbulent Boundary Layer   192
12 Momentum Transfer: Turbulent Flow in Tubes   244
13 Heat Transfer: The Turbulent Boundary Layer   255
14 Heat Transfer: Turbulent Flow inside Tubes   311
15 The Influence of Temperature-Dependent Fluid Properties   355
16 Convective Heat Transfer at High Velocities   370
17 Free-Convection Boundary Layers   396
18 Heat-Exchanger Analysis and Design   417
19 Compact Heat-Exchanger Surfaces   443
20 Mass Transfer: Formulation of a Simplified Theory   480
21 Mass Transfer: Some Solutions to the Conserved-Property
        Equation   503

22 Mass Transfer: Some Examples of Evaluation of the Driving Force   517

Appendixes   541
A   Property Values   541
B   Dimensions and Conversion to SI   559
C   Some Tables of Functions Useful in Boundary-Layer Analysis   563
D   Operations Implied by the ∇ Operator   566
E   Turbulent Boundary-Layer Benchmark Data   570

Indexes
   Author Index   591
   Subject Index   595

PREFACE TO THE
THIRD EDITION

The trends in the development of the science of convective heat transfer, as described in the Preface to the Second Edition, 1980, have continued unabated. The influence of the digital computer has become even more pervasive since the personal computer came upon the scene, and today students and engineers can carry out computations at home or at their desk that 20 years ago would would have required a large main-frame computer and a well-staffed computer center.

One must pose the same question as in the previous edition. Where does this leave the older analytic approach with its classical solutions for particular boundary conditions and its many approximate procedures for more general boundary conditions? And the answer to this question is the same: the classical approach has become *less* important, but it has still not lost its importance.

This new edition has been prepared in response to this answer, but also in recognition of the continually growing importance of computer-based finite-difference solutions and new mathematical models, especially in the calculation of turbulent boundary layers. However, this is not a book on numerical methods, but rather, as before, an introduction to boundary-layer theory, recognizing the increasing importance of computer-based solutions.

The old Chapter 4 on the differential equations of the boundary layer has been subdivided into two chapters, with all of the material on the turbulent boundary layer equations now in a new Chapter 5, and with a considerable expansion of that material.

The four chapters on the turbulent boundary layer and turbulent flow in tubes have been completely rewritten to reflect new methods and new experimental data. Most of the other chapters have been revised only moderately.

A major change has been the inclusion of two new chapters on heat-exchanger analysis and design (Chapter 18) and on compact heat-exchanger surfaces (Chapter 19). The rationale for this addition is that heat-exchanger analysis and design provides one of the most important applications of convective heat-transfer theory, and this is particularly true for heat exchangers involving gases. An introductory course in convective heat transfer *should* include a section on heat exchangers, and it is inconvenient to require students to have two textbooks.

McGraw-Hill and the authors would like to thank the following reviewers for their many helpful comments and suggestions: Ralph Grief, University of California–Berkeley; John Lienhard, Massachusetts Institute of Technology; Jack Lloyd, Michigan State University; Terry Simon, University of Minnesota; Brent Webb, Brigham Young University; and Ralph Webb, Pennsylvania State University.

The authors would like to express their thanks to Kiran Kimbell for her editorial assistance in preparing the third edition.

W. M. Kays
M. E. Crawford

COMPUTER SOFTWARE

Reference is made in various places in the text to the use of finite-difference procedures and computers to solve the boundary-layer equations, and indeed this is the only option available to make use of the turbulence models discussed in Chapters 11 and 13. A particular computer code was used to generate the solutions described in Chapters 11 and 13, but since many such codes exist, and most can use the models discussed in the text, the authors are reluctant to advocate any particular code. However, for those interested, the particular code used to generate the solutions described in Chapters 11 and 13, which is named STAN7, may be obtained without cost by sending a formatted $3\frac{1}{2}$ inch disk (double density) in a self-addressed and stamped disk-mailing envelope to:

Professor W. M. Kays
Department of Mechanical Engineering
Stanford University
Stanford, CA 94305

The program is designed for use on any Macintosh or DOS computer. Please indicate on the disk whether it is the Macintosh or DOS version that is desired. The code is in compiled form; the source code is not being offered.

This program solves the equations for an axisymmetric coordinate system for both laminar and turbulent boundary layers, and thus includes flow inside nozzles as well as flow over external surfaces. Free convection on a vertical surface is also included. For the turbulent boundary layer it uses either a mixing-length model or the $k\text{-}\varepsilon$ model, and the user has the option to change any of the constants appearing in the models. Fluid properties can be either constant or temperature-dependent. The properties of air and water are included with the program—property files for other fluids can be supplied by the user. Viscous energy dissipation is an option so that the high-velocity boundary layer can be solved. The program is extensively documented in a text file.

Chapters 18 and 19 describe a procedure for designing a heat exchanger. For a heat exchanger involving gases, where the pressure drop as well as the thermal performance is of importance, the design procedure is an iterative one. The procedure can be carried out by hand calculations, but it becomes very tedious and is far better carried out by computer. A second Macintosh computer disk contains a set of programs, PLATES, TUBES, CONDTUBE, and MATRIX, that will design a heat exchanger, or predict the performance of a heat exchanger, using any of the surface configurations described in Chapter 19. These programs will carry out design calculations in a few seconds that would otherwise take 20 minutes or longer. In calculations that also involve optimization of a heat exchanger with respect to a thermal system, hand calculation is virtually out of the question. To obtain these programs, send a second Macintosh formatted disk (double density) in a self-addressed stamped disk-mailing envelope to the same address.

This option is not available for DOS computers, although a DOS disk containing these programs can be purchased from Intercept Software, 3425 S. Bascom Ave., Campbell, CA 95008 [Phone: (408)-377-4870]. This disk contains data on all of the heat-transfer surfaces described in the book *Compact Heat Exchangers* by W. M. Kays and A. L. London (McGraw-Hill, New York, 1984).

A second boundary-layer finite-difference computer code, TEXSTAN, is available for those interested in a more extensive research tool for convective heat- and mass-transfer problems. This code is written in Fortran and is intended primarily for use on either "fast" PCs or UNIX-based workstations. It is a parabolic partial differential equation solver for equations of the form: Convection = Diffusion ± Source. Currently within the code the momentum sources include pressure gradient, free-convection body force, and a generalized body force. For the stagnation enthalpy equation (or temperature, if low speed), the sources include viscous dissipation, body force work, and a generalized volumetric source. For convective mass transfer, the appropriate source terms can be defined.

A large variety of flow geometries can be analyzed by TEXSTAN, including the external wall shear flow family of flat plates, airfoil-shaped surfaces, and bodies of revolution of the missile and nozzle shape. The internal wall shear flow family include circular pipes, planar ducts, and annular ducts, for both constant and converging and/or diverging cross-sectional areas, provided flow reversal does not occur.

Turbulence models in TEXSTAN include three levels of mean field closure via the eddy viscosity: Prandtl mixing length with Van Driest damping; a one-equation turbulence kinetic energy model; and numerous low-turbulence-Reynolds-number two-equation $k-\varepsilon$ models. The turbulent heat-flux closure is via a turbulent Prandlt (or Schmidt) number concept. Algebraic and full Reynolds stress models, along with turbulent heat-flux transport models, can easily be adapted to the code.

TEXSTAN can handle a variety of different boundary conditions. For the energy equation, the axial variation of either the wall heat flux or the wall enthalpy may be specified. The mass flux at a wall may also vary in the axial direction. For geometries with two different walls, such as annuli or planar ducts, asymmetric thermal and transpiration boundary conditions can be accommodated. For external wall shear flows, the free-steam velocity, rather than the pressure, is treated as a variable boundary condition, and the free-stream stagnation enthalpy is held constant. The user may specify the initial profiles of the dependent variables, or an automatic initial profile generator may be used.

Fluid properties in TEXSTAN may be treated as constant or variable. Constant fluid properties are user-supplied. Variable fluid properties are supplied through user-developed subroutines. The current routines include those for air at both low and elevated pressures, water, nitrogen, helium, and combustion products.

It is our intention to make available to the purchaser of this text an academic copy of TEXSTAN with the right to reproduce. Inquiries can be made to:

Professor Michael E. Crawford
Department of Mechanical Engineering
The University of Texas
Austin, TX 78712
(512)471-3107
crawford@eddy.me.utexas.edu

PREFACE TO THE SECOND EDITION

When the first edition of this book was being prepared in the early 1960s, the art and science of convective heat transfer were well into what might be called the second phase. The first phase was the period of almost exclusive reliance on experimental correlations of overall heat-transfer behavior, with the pertinent variables reduced to the nondimensional groupings so familiar to all heat-transfer engineers. In the second phase the effort was increasingly to develop mathematical models of the basic phenomena and then to deduce system behavior through mathematical reasoning, an effort which greatly expanded the ability of the analyst or designer to handle new and complex applications and which also enhanced understanding of the phenomena involved. The primary objective of the first edition was to bring together some of these analytic methods and results, to encourage their routine use by heat-transfer engineers, and perhaps also to encourage the development of a new breed of heat-transfer engineer with a somewhat different point of view.

Since publication of the first edition in 1966, two closely related developments in convective heat transfer are having a profound influence. The large-capacity digital computer, together with new and better finite-difference techniques, has largely removed the mathematical difficulties of handling boundary-layer flows, and this has been especially significant in the case of turbulent flows. When it was no longer necessary to make mathematical compromises, it became possible to focus attention on the basic transport mechanisms, and thus our knowledge and our ability to model the basic mechanisms have greatly improved. It is now possible and practicable to routinely calculate both laminar and turbulent boundary layers, and tube flows, with high precision for a very wide variety of conditions.

The question now is: Where does this leave the older analytic approach with its classical solutions for particular boundary conditions and its many approximate procedures for more general boundary conditions? The authors have two answers to this question. Intelligent use of computer-based methods requires an understanding of both the basic processes and, in at least a general way, the consequences of particular sets of conditions. This understanding is difficult to obtain when the computer is relied on exclusively. Equally important is the fact that a very high percentage of engineering heat-transfer problems do not require the high precision and detail generally available from a computer solution, but they are problems for which quick, low-cost answers are essential. For such problems the computer-based finite-difference solution is elegant, but overkill. The point at which overkill occurs is moving inexorably away from the classical methods, but in the authors' view the distance to go is still large, and in any case is going to vary greatly with local conditions. It is a question of engineering judgment; an engineer must optimize not only the system being designed but also his or her own expenditure of time.

In this second edition the authors have retained the basic objectives of the first edition, while at the same time modernizing it and shifting emphasis where appropriate, but they have also tried to provide a theoretical frame work for finite-difference methods. The relevant differential equations are developed, and simple turbulent transport models applicable to finite-difference procedures are discussed. Since the literature abounds with references to computer programs and model developments, the authors have tried mainly to refer to survey articles in the discussions.

As before, not everything can be covered. The topics chosen and the depth of coverage represents a personal judgment as to what is of first importance for a mechanical, aerospace, or nuclear engineering student at about the fifth-year level. A chapter on free convection has been added, and the discussion of the effects of surface roughness has been greatly expanded. Several chapters have been reorganized to completely separate laminar and turbulent flows, and the approach to turbulent transport processes has been drastically modified. Because of their continuing evolution, though, higher-order turbulence closure models are not discussed. Finally, it should be emphasized that only two-dimensional boundary-layer flows are treated, and only single-phase systems are considered.

Finally, the second author would like to express his gratitude to Professor Kays for the honor and privilege of being asked to coauthor the second edition. Professor Kays' style of teaching, both in the classroom and otherwise, and his approach to the subject of heat transfer will be forever with me. I would also like to express my indebtedness to Professor R. J. Moffat at Stanford, who taught me all I profess to know about experimental heat transfer and the art of written and oral communication. Lastly, I would like to express appreciation to two more colleagues, Professor A. L. London at Stanford and Professor J. L. Smith, Jr, at M.I.T., who have taught me the closely allied field of thermodynamics and the general methodology behind engineering problem solving.

W. M. Kays
M. E. Crawford

PREFACE TO THE FIRST EDITION

Prior to World War II, convective heat and mass transfer were largely empirical sciences, and engineering design was accomplished almost exclusively by the use of experimental data, generalized to some degree by dimensional analysis. During the past two decades great strides have been made in developing analytical methods of convection analysis, to the point where today experiment is assuming more its classical role of testing the validity of theoretical models. This is not to say that direct experimental data are not still of vital importance in engineering design, but there is no question that the area of complete dependence on direct experimental data has been greatly diminished. With this change our understanding of convection phenomena has been greatly enhanced, and we find ourselves in a position to handle, with confidence, problems for which experiment would be time-consuming and expensive. This book has been prepared as a response to this trend.

It is axiomatic that the engineering student must learn to reason from first principles so that she or he is not at a loss when faced by new problems. But time spent solving a complex problem from first principles is time wasted if the solution already exists. By their very nature analytic convection solutions often tend to be lengthy and difficult. Thus familiarity with, and an understanding of, some of the more important of the available analytic convective solutions should be an important part of the background of the heat-transfer engineer. One of the objectives of this book is to bring together in an easily usable form some of the many solutions to the boundary-layer equations. Although these are available in the heat-transfer literature, they are not always readily accessible to the practicing engineer, for whom time is an important consideration.

The author feels that a study of these solutions, in a logical sequence, also provides the best way for a student to develop an understanding of convective heat and mass transfer. Thus it is hoped that this book will serve both as a classroom text and as a useful reference book for the engineer.

This book is the outgrowth of a set of notes which the author has developed over the past ten years to supplement lectures in the "convection" portion of a one-year course in heat transfer for first-year graduate students. The students in the course have been largely mechanical, nuclear, and aeronautical engineers, interested in problems associated with thermal power systems and thermal environmental control.

It is assumed that the student has a typical undergraduate background in applied thermodynamics, fluid mechanics, and heat transfer. Heat transfer, although not mandatory, is usually of considerable help in orienting the student's thinking and establishing a sense of need for a deeper study of the subject. In particular, some familiarity with the commonly employed empirical methods of calculating convection heat-transfer rates is assumed, but only so that the student has an appreciation for the usefulness of a heat-transfer coefficient and some grasp of the basic physics of the convection process.

The choice of subject matter reflects quite frankly the author's own interests, and the depth to which each topic is pursued represents a compromise made necessary by what can be practicably accomplished in approximately one semester (or perhaps two quarters). It will found that the momentum boundary layer is heavily compressed, with only sufficient material presented to support the heat- and mass-transfer sections. The student desiring to concentrate heavily in boundary-layer theory will undoubtedly want to take a separate course on viscous fluid mechanics, for which adequate texts exist. And, for that matter, there is certainly a great deal more to convective heat and mass transfer than is presented here, not only in the topics considered but also in those not even mentioned. In the latter category the reader may miss such topics as natural convection, heat-exchanger theory, rotating surfaces, nonsteady flows, two-phase flows, boiling and condensation, non-Newtonian fluids, internally radiating gases, rarefied gases, magnetohydrodynamic flows, and coupling between heat and mass transfer. But this only suggests why second editions are usually bigger than first editions.

Finally, I would like to acknowledge my indebtedness to some of my colleagues, without whose assistance, conscious or otherwise, this book could never have been written. First, Professor A. L. London taught me all that I profess to know about teaching, introduced me to heat transfer, and has been a constant source of help and inspiration. Professor W. C. Reynolds has worked with me on some of the research that is summarized in the book, and substantial parts of it are the result of his work alone. Several months spent with Professor D. B. Spalding at Imperial College in London were a rare privilege, and his influence will be found throughout the book. But most specifically, Spalding's generalization of the convective mass-transfer problem forms the entire basis for the last three chapters. Although available in Spalding's many papers, it is hoped that its inclusion here will encourage its more extensive use. Lastly I would like to express appreciation to Mr. R. J. Moffat, who read the manuscript and made many helpful suggestions.

W. M. Kays

LIST OF SYMBOLS

English letter symbols

A	area, total surface area in a heat exchanger, m^2
A^+	Van Driest constant; see Eq. (11-25)
A_c	flow cross-sectional area, m^2
A_f	fin area, m^2
A_{fr}	frontal area of a heat exchanger, m^2
A_m	see Eq. (9-42)
A_w	wall area, m^2
a	thickness of separating plates in a heat exchanger, m
B	mass-transfer driving force; see Eq. (20-34)
B_f	transpiration parameter; see Eq. (11-44)
B_h	heat-transfer transpiration parameter; see Eq. (13-37)
Br'	Brinkman number, $V^2/(q_0''D)$; see Eq. (9-25)
b	plate spacing in a heat exchanger, m
b_f	transpiration parameter; see Eq. (11-50)
b_h	transpiration parameter; see Eq. (13-41)
C	surface curvature in flow direction, $1/R$, $1/m$
C	capacity rate, $\dot{m}c$, W/K
C_c	capacity rate on heat-exchanger cold-fluid side, $\dot{m}_c c_c$, W/K
C_D	drag coefficient
C_h	capacity rate on heat-exchanger hot-fluid side, $\dot{m}_h c_h$, W/K
C_n	see Eq. (9-36)

C_r	rotor capacity rate in a rotating periodic-flow heat exchanger, = (mass of rotor) × (specific heat of rotor) × (rev/s), W/K
Cr	$v/u_\infty R$, a curvature parameter
c	specific heat at constant pressure, J/(kg · K), m²/(s² · K)
c_f	local friction coefficient; see Eq. (6-10)
c_{f_m}	mean friction coefficient with respect to length
$\bar{c}_{f_{app}}$	apparent mean friction coefficient; Eq. (7-19)
c_{f0}	value of local friction coefficient with vanishingly small mass-transfer rate; see Eq. (11-47)
c_v	specific heat at constant volume, J/(kg · K), m²/(s² · K)
c_j	specific heat at constant pressure for component j of a mixture, J/(kg · K), m²/(s² · K)
D	inside diameter of a circular tube, m
D_h	hydraulic diameter, $D_h \; 4r_h = 4A_cL/A$, m
D_j	mass-diffusion coefficient for component j in a multi-component mixture, m²/s
d	outside diameter of a circular tube, m
$\mathscr{D}_{ij}$	mass diffusion coefficient for a binary (two-component) mixture, m²/s; note that $\mathscr{D}_{ij} = \mathscr{D}_{ji}$
E	friction power per unit of surface area, W/m²; see Eq. (19-4)
$\dot{E}$	rate of energy transfer by convection across a control surface, J/s, m² · kg/s³
e	internal thermal and chemical energy, J/kg, m²/s²
F	mass-flux ratio, $\dot{m}''/G_\infty$
F	resultant of all external forces acting on a control volume, N, kg · m/s²
G	mass flux, or mass velocity, ρu, kg/(m² · s); see Eqs. (2-2) and (2-3)
G	Clauser shape factor; see Eq. (11-43)
G_∞	mass velocity in the free-stream, ρu_∞, kg/(m² · s)
G	mass flux, or mass velocity vector, $\rho \mathbf{V}$, at any point in the stream, kg/(m² · s)
G_x, G_y, etc.	components of the mass flux vector, kg/m²
$\mathbf{G}_{\text{diff},j}$	mass flux of component j transported by diffusion, kg/m²; see Eq. (3-12)
G_n	see Eq. (9-38)
Gr_x	local Grashof number, $g\beta q_0'' x^3(t_0 - t_\infty)/v^2$

Gr_x^*	modified Grashof number, $g\beta q_0''x^4/k\nu^2$
Gr_D	Grashof number based on diameter
Gr_L	Grashof number based on L = surface area/surface perimeter
g	mass-transfer conductance, kg/(s · m²); see Eq. (20-33)
g_i	enthalpy conductance, kg/(s · m²); see Eq. (16-9)
g	acceleration due to gravity, m/s²
g^*	value of mass-transfer conductance for vanishingly small mass-transfer rate, kg/(s · m²); see Eq. (20-36)
H	boundary-layer shape factor; see Eq. (8-37)
H_0	"heat" of combustion, per unit of fuel mass, at a temperature t_0, J/kg, m²/s³
h	heat-transfer coefficient, or convection conductance, W/(m² · K), J/(s · m² · K), kg/(s³ · K); see Eq. (1-1)
h_k	roughness heat-transfer coefficient, W/(m² · K); see Eq. (13-46)
i	static enthalpy, and enthalpy of a mixture, $e + P/\rho$, J/kg, m²/s²
i'	instantaneous value of the fluctuating component of static enthalpy; J/kg, m²/s²; see Eq. (5-15)
$\bar{i}$	local time-mean static enthalpy, J/kg, m²/s²; see Eq. (5-15)
i_j	partial enthalpy of component j of a mixture, J/kg, m²/s
i^*	stagnation enthalpy, $i + \frac{1}{2}u^2$, J/kg, m²/s²
$\overline{i^*}$	local time-mean stagnation enthalpy, J/kg
i_R	reference enthalpy for evaluation of fluid properties, J/kg; see Eq. (16-42)
K	acceleration parameter; see Eq. (11-45)
K_p	equilibrium constant; see Chap. 22
k	instantaneous value of turbulence kinetic energy, $\frac{1}{2}u_i'u_i'$, m²/s²; see Eq. (5-39)
k'	fluctuating component of k, $k' = k - \bar{k}$, m²/s²
$\bar{k}$	local time-mean value of k, $\frac{1}{2}\overline{u_i'u_i'}$, m²/s²
k	thermal conductivity, W/(m · K), J/(s · m · K); see Eq. (3-8)
k_{eff}	$k + k_t$; see Eq. (16-7)
k_s	equivalent "sand grain" roughness, m
k_t	eddy or turbulent conductivity, W/(m · K), J/(s · m · K); see Eq. (13-3)

k_T	thermal-diffusion ratio; see Eq. (3-9)
L	flow length of a tube, m
Le_j	Lewis number, γ_j/Γ, $\mathrm{Pr}/\mathrm{Sc}_j$ (sometimes defined as the inverse of this)
l	length of a fin in a heat exchanger, m
l	mixing length, m; see Eq. (11-5)
l_t	turbulence length scale; see Eq. (11-31)
M	a blowing-rate parameter; see Eq. (13-43)
M	Mach number; see Eq. (16-32)
m	mass, kg
m	an exponent; see Eqs. (8-20) and (8-21)
m_j	mass concentration (mass fraction) of substance j in a mixture
$\dot{m}$	mass-flow rate, kg/s
$\dot{m}''$	total mass flux (mass-flow rate per unit area) at surface or phase interface, $\mathrm{kg}/(\mathrm{m}^2 \cdot \mathrm{s})$
$\dot{m}_j''$	mass flux of substance j at surface or phase interface, $\mathrm{kg}/(\mathrm{m}^2 \cdot \mathrm{s})$
$\dot{m}_j'''$	rate of creation of substance j, per unit volume, by chemical reaction, $\mathrm{kg}/(\mathrm{m}^3 \cdot \mathrm{s})$
$\mathfrak{M}$	molecular "weight", kg/kmol
n_α	mass fraction of element α in a mixture of compounds
$n_{\alpha,j}$	mass fraction of element α in a compound substance j
NTU	nondimensional size of a heat exchanger, AU/C
Nu	Nusselt number, hD/k, $4r_h h/k$, hD_h/k, xh/k
P	pressure, $\mathrm{N/m}^2$, Pa, $\mathrm{kg}/(\mathrm{m} \cdot \mathrm{s}^2)$
P_a	partial pressure of substance a in a gas mixture, $\mathrm{N/m}^2$
P'	fluctuating component of pressure, Pa, $\mathrm{N/m}^2$; see Eq. (5-14)
$\bar{P}$	local time-mean value of pressure, Pa, $\mathrm{N/m}^2$; see Eq. (5-14)
p	tube perimeter; see Eq. (14-18)
p^+	nondimensional pressure gradient (wall coordinates); see Eq. (11-13)
$\mathscr{P}$	conserved property of the second kind; see Eq. (20-21)
Pe	Péclet number, Re Pr
Pe_t	turbulent Péclet number; see Eq. (13-7)
Pr	Prandtl number, $\mu c/k$, μ/Γ, ν/α
$\mathrm{Pr}_{\mathrm{eff}}$	$\mu_{\mathrm{eff}}/(k_{\mathrm{eff}}/c)$; see Eq. (16-8)

Pr_t	turbulent Prandtl number; see Eq. (13-5)
q	heat, energy in transit by virtue of a temperature gradient, J, $m^2 \cdot kg/s^2$
$\dot{q}$	heat-transfer rate, J/s, W, $N \cdot m/s$, $m^2 \cdot kg/s^2$
$\dot{q}''$	heat-flux vector, heat-transfer rate per unit area, $J/(s \cdot m^2)$, W/m^2, kg/s^3
$\dot{q}_0''$	heat-flux, heat-transfer rate per unit area, at surface or phase interface, $J/(s \cdot m^2)$, W/m^2, kg/s^3
R	gas constant, $J/(kmol \cdot K)$; see App. B
R	radius of a body of revolution (see Fig. 6-4); radius of a cylinder or sphere (see Fig. 10-3), m
R	radius of curvature in flow direction, m
Ra	Rayleigh number, Gr Pr
R_n	see Eq. (9-36)
Ri	Richardson number, $(u/R)/(du/dR)$
r	radial distance in cylindrical or spherical coordinates, m
r	boundary-layer thickness ratio, Δ/δ
r	mass ratio of oxidant to fuel in a simple chemical reaction
r_c	recovery factor; see Eq. (16-22)
r_h	hydraulic radius, $A_c L/A$, m; see Eq. (7-17)
r_i	inner radius of annulus, m
r_o	outer radius of annulus, m
r_0	radius of a circular tube, m
r^+	nondimensional radial coordinate, r/r_0
r^*	r_o/r_i, radius ratio for annulus; see Fig. 9-6
R	heat-exchanger heat-transfer resistance, $1/U$, $m^2 \cdot K/W$
Re	Reynolds number, $4r_h G/\mu$, DG/μ, $xu_\infty\rho/\mu$, xu_∞/ν, $\delta_2 u_\infty/\nu$, $\Delta_2 u_\infty/\nu$, etc.
Re_k	roughness Reynolds number, $k_s u_T/\mu$
Re_t	Reynolds number of turbulence; see Eq. (11-32)
S	source function, thermal energy created per unit volume, $J/(s \cdot m^3)$, W/m^3
S_{ij}	strain-rate tensor, 1/s; see Eq. (5-3)
Sc_j	Schmidt number, μ/γ_j, $m/\rho D_h$
Sc_k	Schmidt number for diffusion of turbulence kinetic energy; see Eq. (11-38)
Sc_ε	Schmidt number for diffusion of turbulence dissipation; see Eq. (11-40)
St	Stanton number, h/Gc, $h/u_\infty\rho c$, etc.

St_0	value of local Stanton number with vanishingly small mass-transfer rate; see Eqs. (13-40) and (13-41)
St_k	roughness Stanton number; see Eq. (13-47)
s	entropy, $J/(kg \cdot K)$
T	absolute temperature, K
T	a boundary-layer shape factor; see Eq. (8-37)
T_R	reference temperature for evaluation of fluid properties, K; see Eqs. (15-15) and (16-40)
Tu	turbulence intensity; see Eq. (11-58)
t	temperature (either °C or K)
t'	fluctuating component of temperature, °C or K; see Eq. (5-14)
$\bar{t}$	local time-mean value of temperature; see Eq. (5-14)
$\overline{t'v'}$	apparent turbulent heat flux, $K \cdot m/s$
t_{aw}	adiabatic wall temperature, °C or K; see Eq. (16-22)
t_e	fluid temperature at entrance to a tube, °C or K
t_m	mixed mean fluid temperature, °C or K; see Eq. (9-5)
t_0	fluid temperature at surface or phase interface, °C or K
t_∞	temperature in the free stream at the outer edge of the boundary layer, °C or K
t^*	stagnation temperature, °C or K; see Eq. (16-1)
t_{db}	dry-bulb temperature, °C or K
t_{wb}	wet-bulb temperature, °C or K
t^+	nondimensional temperature in wall coordinates; see Eq. (13-10)
U	overall heat-transfer conductance in a heat exchanger, $W/(m^2 \cdot K)$; see Eq. (18-5)
u	velocity component in the x direction, m/s
u'	instantaneous value of the fluctuating component of velocity in the x direction, m/s; see Eq. (5–13)
$\bar{u}$	local time-mean velocity in the x direction, m/s; see Eqs. (5-20) and (5-13)
u_c	velocity at centerline of tube, m/s
$\overline{u'v'}\,\rho$	turbulent shear stress in a two-dimensional boundary layer, N/m^2
$\overline{u'X'}$	turbulent velocity–body force correlation, $N/(s \cdot m^2)$; see Eq. (5-38)
u_i, u_j	generalized velocity in tensor notation, m/s; see Eq. (3-7)
$\overline{u_j'u_i'}\,\rho$	turbulent stress tensor or Reynolds stress tensor, N/m^2

$\overline{u_j'i'}\,\rho$	enthalpy flux vector, $J/(s \cdot m^2)$
u^+	nondimensional velocity in wall coordinates; see Eq. (11-13); also nondimensional velocity in a tube, u/V
u_τ	"shear velocity," or "friction velocity," m/s; see definition preceding Eq. (11-13)
u_∞	velocity in the free stream at the outer edge of the boundary layer, m/s
V	volume, m^3
V	mean fluid velocity in a tube, m/s; see Eq. (7-3); also uniform velocity upstream of a blunt body, m/s
$\mathbf{V}$	velocity vector, m/s
$\mathbf{V}'$	fluctuating component of the velocity vector in a turbulent flow, m/s; see Fig. 5-1
$\bar{\mathbf{V}}$	local mean value of the velocity vector, m/s; see Fig. (5-1)
v	velocity component in the y direction, m/s
v'	instantaneous value of the fluctuating component of velocity in the y direction, m/s; see Eq. (5-13)
$\bar{v}$	local time-mean velocity in the y direction, m/s; see Eq. (5-13)
$\overline{v't'}\,\rho c$	turbulence heat flux in a two-dimensional boundary layer, $J/(s \cdot m^2)$; see Eq. (5-33)
v	fluid specific volume, m^3/kg
v_0	normal fluid velocity at a surface, m/s
v_0^+	nondimensional form of v_0 in wall coordinates; see Eq. (11-13)
v_r	velocity component in the r direction, m/s
W	mechanical or electrical work, J, $N \cdot m$, $m^2 \cdot kg/s^2$
$\dot{W}$	rate of doing mechanical work, J/s, W, $m^2 \cdot kg/s^3$
w	velocity component in the z direction, m/s
w'	instantaneous value of the fluctuating component of velocity in the z direction, m/s; see Eq. (5-13)
$\bar{w}$	local time-mean velocity in the z direction, m/s; see Eq. (5-13)
X	body force acting on a fluid in the x direction, per unit of volume, N/m^3, $kg/(s^2 \cdot m^2)$
X'	instantaneous value of the fluctuating component of body force in the x direction, N/m^3
X_i	body force acting in the x_i direction, N/m^3
X_a	mole fraction of component a in a mixture; see Eq. (22-1)

x	a spatial coordinate in cartesian and cylindrical systems; see App. D; also flow length in a tube, or distance measured along the surface of a body, m
$\mathbf{x}$	vector location (x, y, z) or (x_1, x_2, x_3)
x^+	nondimensional axial distance inside a tube, $(x/r_0)/(\text{Re Pr})$; also nondimensional form of x in wall coordinates; see Eq. (11-27)
x_i, x_j	generalized coordinates in tensor notation, m; see Eq. (3-7)
Y	upper-bound value of spatial coordinate y in control volume analysis, m
y	a spatial coordinate in a cartesian system, distance normal to the surface in a boundary layer, m
y^+	nondimensional distance from wall in wall coordinates; see Eq. (11-13)
z	a spatial coordinate in a cartesian system; also elevation with respect to a given datum in a gravity field, m

Greek letter symbols

α	molecular thermal diffusivity, $k/\rho c$, m^2/s
α	surface area per unit of volume in a heat exchanger, $1/\text{m}$
β	a pressure-gradient parameter; see Eq. (11-44)
β	wedge angle; see Fig. 8-2
β	volumetric coefficient of thermal expansion, K^{-1}
β	surface area of one side of heat exchanger per unit volume on that side
Γ	thermal diffusion coefficient, k/c, $\text{kg}/(\text{m} \cdot \text{s})$; see Eq. (3-10)
Γ	gamma function; see App. C
γ	ratio of specific heats, c/c_v
γ_j	mass-diffusion coefficient for substance j in a mixture, ρD_j, $\text{kg}/(\text{m} \cdot \text{s})$; see Eq. (3-13)
γ_m	see Eq. (9-42)
Δ	designates a difference when used as a prefix
Δ	thickness of a thermal boundary layer, m; for example, see Eq. (10-25)
Δ_2	enthalpy thickness of a thermal boundary layer, m; see Eq. (6-14)
Δ_4	conduction thickness of a thermal boundary layer, k/h, m; see Eq. (6-16)

δ	thickness of separating wall in a heat exchanger, m
δ	thickness of a momentum boundary layer, m; for example, see Eq. (8-32)
$\delta(\ )$	differential displacement of the variable $(\ )$
δ_1	displacement thickness of a momentum boundary layer, m; see Eq. (6-5)
δ_2	momentum thickness of a boundary layer, m; see Eq. (6-6)
δ_3	Clauser boundary-layer thickness, m; see Eq. (11-42)
δ_4	shear thickness of a boundary layer, m; Eq. (8-36)
δ_{99}	99 percent boundary-layer thickness, m; see Eq. (11-8)
δ_{ij}	Kronecker delta function; see footnote to Eq. (3-7)
δy_0	roughness displacement in mixing length; see Eq. (11-52)
ε	heat-exchanger effectiveness; see Eq. (18-2)
ε	turbulence dissipation rate, $N/(s \cdot m^2)$; see Eq. (5-38)
ε_H	eddy diffusivity for heat transfer, m^2/s; see Eq. (13-1)
ε_M	eddy diffusivity for momentum transfer, m^2/s; see Eq. (11-2)
ε_k	eddy diffusivity for the diffusion of turbulence kinetic energy, m^2/s; see Eq. (11-29)
ε_ε	eddy diffusivity for the diffusion of turbulence energy dissipation, m^2/s; see Eq. (11-40)
ζ	dependent variable in Blasius equation, $u/u_\infty = \zeta'(\eta)$; see Eq. (8-8)
η	similarity parameter; see Eq. (8-11); independent variable in Blasius and other similarity solution equations
η	film-cooling effectiveness; see Eq. (13-44)
η_f	heat-transfer effectiveness of a fin in a heat exchanger; see Eqs. (18-5) and (18-6)
η_0	overall surface effectiveness
θ	time, s
θ	$(t_0 - t)/(t_0 - t_c)$, nondimensional fluid temperature in a tube, a solution to Eq. (9-32)
θ	nondimensional temperature used in full-coverage film cooling; see Eq. (13-41)
θ	$(t_0 - t)/(t_0 - t_\infty)$, nondimensional fluid temperature in an external boundary layer, for the case of a step change in surface temperature
θ	nondimensional fluid temperature in a high-velocity boundary layer; see Eq. (16-18)
θ	a temperature difference, K; see Eq. (10-25)

θ	angular coordinate in a spherical system; see App. D
κ	mixing-length constant; see Eq. (11-7)
λ	a nondimensional boundary-layer parameter; see Eq. (8-39)
λ	a diffusion coefficient, Γ or γ_j; kg/(m · s); see Eq. (20-32)
λ	longitudinal conduction parameter in a heat exchanger; see Eq. (18-14)
λ	mixing-length constant; see Eq. (11-8)
λ_n	see Eq. (9-38)
μ	dynamic viscosity coefficient, N · s/m², Pa · s, kg/(m · s); see Eq. (3-1),
μ_t	eddy or turbulent viscosity, N · s/m²; see Eq. (11-4)
μ_{eff}	$\mu + \mu_t$; see Eq. (16-7)
v	kinematic viscosity, μ/ρ, m²/s
ξ	distance from beginning of tube or plate to point where heat transfer starts; also a dummy length variable; may be dimensional or nondimensional, according to context
ρ	fluid density, mass per unit volume, kg/m³
σ	Stefan–Boltzmann constant, W/(m² · K⁴), kg/(s³ · K⁴); see App. B
σ	ratio of free-flow to frontal area in a heat exchanger
σ	normal stress on an element of fluid, N/m², kg/(m · s²); see Eqs. (3-4)–(3-6)
σ_{ij}	fluid stress tensor, N/m²; see Eqs. (3-7), (5-1), and (5-6)
τ	shear stress, N/m², kg/(m · s²); see Eqs. (3-1)–(3-3)
τ_{ij}	viscous stress tensor, N/m²; see Eq. (5-1)
τ	nondimensional temperature in an external boundary layer, $(t_0 - t)/(t_0 - t_\infty)$, a solution to Eq. (10-4)
τ_0	shear stress evaluated at wall surface, N/m², kg/(m · s²)
ϕ	angular coordinate in cylindrical and spherical coordinate systems; see App. D
ϕ	dissipation function; see Eq. (4-32)
ϕ	relative humidity; see Eq. (22-6)
ψ	stream function; see Eq. (8-9), m²/s
ω	absolute humidity; see Eq. (22-5)
Ω_{ij}	a rotation tensor; see Eq. (5-4)

Subscripts

c	refers to colder-fluid side in a heat exchanger
c	evaluated at centerline of tube

0	evaluated within the boundary layer or considered phase but at the surface or wall
o	refers to the outer surface of an annulus
oo	outer surface of an annulus when outer surface alone is heated
∞	evaluated at the free-stream
D	evaluated based on diameter; see Eq. (12-2)
e	evaluated at the tube entrance
h	refers to hotter-fluid side in a heat exchanger
m	evaluated at the mixed mean state; also denotes other mean values
R	refers to reference temperature, or properties evaluated at reference temperature; radius of cylinder in Eqs. (10-22) and (10-23)
x	evaluated at a particular point along the surface
aw	evaluated at "adiabatic" wall state
L	evaluated at the surface but within the neighboring phase
T	evaluated at the transferred-substance state; see Eqs. (20-1)–(20-3)
j	refers to a substance j that is a component of a mixture
α	refers to a particular chemical element α in a compound and/or mixture
i	refers to the inner surface of an annulus
ii	inner surface of an annulus when inner surface alone is heated
CP	refers to constant-property solution with all fluid properties introduced at either mixed-mean or free-stream state
Ⓗ	refers to solution for constant heat rate per unit of tube length
Ⓣ	refers to solution for constant surface temperature

(The meanings of other subscripts should be apparent from the context of their use.)

Note

The dimensions of all quantities listed are given in standard SI base or derived units, or both. Some quantities are customarily expressed in decimal multiples or submultiples of these units, and care should be taken to note this fact in making arithmetic calculations. Typical examples are pressure, usually expressed in kilopascals, kPa, specific heat capacity, kJ/(kg · K), and enthalpy, MJ/kg or kJ/kg.

CHAPTER
1

INTRODUCTION

Among the tasks facing the engineer is the calculation of energy-transfer rates and mass-transfer rates at the interface between phases in a fluid system. Most often we are concerned with transfer at a solid–fluid interface where the fluid may be visualized as moving relative to a stationary solid surface, but there are also important applications where the interface is between a liquid and a gas.

If the fluids are everywhere at rest, the problem becomes one of either simple heat conduction where there are temperature gradients normal to the interface (which will be subsequently referred to as the *surface*) or simple mass diffusion where there are mass concentration gradients normal to the surface. However, if there is fluid motion, energy and mass are transported both by potential gradients (as in simple conduction) and by movement of the fluid itself. This complex of transport processes is usually referred to as *convection*. Thus the essential feature of a convective heat-transfer or a convective mass-transfer process is the transport of energy or mass to or from a surface by both molecular conduction processes and gross fluid movement.

Popular usage forces us to use the term *convection* somewhat loosely. We will speak of the *convective* terms in our differential

1

equations, as opposed to diffusive terms; here we refer to that *part* of the transport process attributable to the gross fluid motion alone.

If the fluid motion involved in the process is induced by some external means (pump, blower, wind, vehicle motion, etc.), the process is generally called *forced convection*. If the fluid motion arises from external force fields, such as gravity, acting on density gradients induced by the transport process itself, we usually call the process *natural convection*.

Engineering applications of convective heat and mass transfer are extremely varied. In a multifluid heat exchanger we are concerned solely with heat-transfer rates between the fluids and the solid surfaces of the heat exchanger separating the fluids. Calculation of the temperature of a cooled turbine blade or the throat of a rocket nozzle involves convective heat transfer alone, but if a fluid is injected through the surface (transpiration cooling), the problem is also a mass-transfer one. If the surface material is allowed to vaporize and/or burn to protect it from a high-temperature gas (ablation), we have another combination convective heat- and mass-transfer problem. The aerodynamic heating of high-speed aircraft is a convective heat-transfer process, but it also becomes a mass-transfer process when temperatures are so high that the gas dissociates, forming mass concentration gradients. The sling psychrometer presents a combination heat- and mass-transfer process. Combustion of a volatile fluid in air involves heat and mass transfer as well as chemical reaction within the transfer region.

Obviously the combination heat-transfer, mass-transfer, and chemical reaction problem is the most challenging of the convection problems. Nevertheless, the bulk of this book is devoted to convective heat transfer, for it will be shown that for many applications the mass-transfer and the combination problems lead (after suitable simplifying assumptions) to the same differential equations as simple heat transfer. Thus much of convective heat-transfer theory can be used directly in mass transfer, and it is perhaps simpler to develop the solutions on the heat-transfer framework than to employ a more general approach.

In simple convective heat transfer it is convenient in most cases to define a convection *heat-transfer conductance*, or *coefficient*, such that the heat flux at the surface is the product of the conductance and a temperature potential difference. Thus

$$\dot{q}_0'' = h(t_1 - t_2) \tag{1-1}$$

The conductance h is essentially a fluid mechanic property of the system, whereas the temperature difference is, of course, a thermodynamic quantity. The usefulness of Eq. (1-1) lies largely in the fact that in a great many technical applications $\dot{q}_0''$ is close to being directly proportional to $t_1 - t_2$, as the linearity of the applicable differential equation reveals. Nevertheless, numerous nonlinear problems are encountered where h itself is a function of the temperature difference. It is important to note that this does not destroy the validity of Eq. (1-1) as a definition of h, although it may well reduce the usefulness of the conductance concept.

In convective mass transfer we find it convenient to define a convective *mass-transfer conductance* such that the total mass flux at the surface is the product of the conductance g and a driving force B. Thus

$$\dot{m}'' = gB \qquad (1\text{-}2)$$

The conductance g is again essentially a fluid mechanic property of the system, whereas the driving force B is a thermodynamic property.† However, it is a somewhat more subtle property than the simple temperature difference of Eq. (1-1), and it has quite a different meaning depending on the nature of the problem. As a matter of fact, Eq. (1-1) can be *extracted* as a special case of Eq. (1-2), but consideration of the rather general nature of Eq. (1-2) is deferred until after simple convective heat-transfer theory is developed. The mass-transfer problem can also be nonlinear, which simply means that g becomes a function of B as well as the fluid mechanics of the system, but this again does not destroy the validity of Eq. (1-2).

In relatively simple mass-transfer applications where there is no chemical reaction involved, and where the mass concentrations of the transferred substance are small relative to 1, Eq. (1-2) will be shown to reduce to

$$\dot{m}'' = g(m_{j_1} - m_{j_2}) \qquad (1\text{-}3)$$

where the m_j are mass concentrations. Equation (1-3) has the same form as Eq. (1-2), and this fact has given rise to heat- and mass- transfer *analogies*. However, it is important to realize that Eq. (1-3) is but a special case of the more general Eq. (1-2).

Stated in the simplest possible way, the primary objective of this book is to develop methods to evaluate the functions h and g for a variety of engineering applications. Knowing h, we can evaluate $\dot{q}_0''$ given the temperature difference, or vice versa. For the mass-transfer, or combination, problem we also have to learn how to evaluate B, since B is not quite so simple as the temperature difference of the heat-transfer problem.

Almost all of the applications considered in this book are those for which the so-called *boundary-layer* approximations are valid. That is,

---

† The mass-transfer development here is based largely on the work of D. B. Spalding of the Imperial College of Science and Technology, London.

consideration will be restricted to a thin region near a phase boundary where most of the resistance to momentum, heat, and mass transfer resides. The vast majority of heat- and mass-transfer problems of engineering interest are subsumed under this restriction, but certainly by no means all. However, the solution of convection problems, usually three-dimensional, for which the boundary-layer approximations are not valid is beyond the scope of this book, although the basic differential equations for such problems are indeed developed.

In the solution of practically all heat- and mass-transfer convection problems, the corresponding fluid dynamic problem must first be solved, provided it is not completely coupled to the heat- and mass-transfer problems. Although this is not intended to be a book on viscous fluid dynamics, it will prove convenient to develop a considerable portion of viscous momentum boundary-layer theory to provide the foundation on which convection theory can be built. The criterion for selection of topics is simply that only those parts of momentum boundary-layer theory are developed that contribute directly to convection solutions to follow.

A final introductory note should be added about the manner of development of the various partial differential equations. In their most general form the equations are large and cumbersome, and the basic physics going into their development is often obscured by algebraic complexity. To provide a simple and clear development, we derive all differential equations in two dimensions and at the same time introduce the conventional boundary-layer approximations. Following this, in each case we extend the equations to three dimensions, dropping the boundary-layer approximations, and finally we render the equations in vector notation, and in some cases cartesian tensor notation. Nothing is really lost by this procedure, for the extensions will be perfectly clear and obvious, and much is gained in algebraic simplicity. And, of course, it is generally the boundary-layer equations that we solve later, although we have the more general equations at our disposal when we need them.

CHAPTER
2

CONSERVATION PRINCIPLES

The solution of a convection problem starts with the combination of an appropriate conservation principle and one or more flux laws. In this chapter we set down the ground rules that we will employ for application of the conservation principles and then introduce these principles and a suitable system of nomenclature.

THE CONTROL VOLUME

Consider a defined region in space across the boundaries of which mass, energy, and momentum may flow, within which changes of mass, energy, and momentum storage may take place, and on which external forces may act. Such a region is termed a *control volume,* and the boundary surfaces are called the *control surface.* The control volume comprises our region of interest in application of the various conservation principles discussed below. The volume may be finite in extent, or all its dimensions may be infinitesimal.

The complete definition of a control volume must include at least the implicit definition of some kind of coordinate system, for the control volume may be moving or stationary, and the coordinate system may be fixed on the control volume or elsewhere.

A control volume across whose surface no matter passes during the process in question is sometimes referred to as a *fixed-mass system*, or a *simple thermodynamic system*.

PRINCIPLE OF CONSERVATION OF MASS

Let the term *creation* have the connotation of *outflow* minus *inflow* plus *increase of storage*: then the principle of conservation or mass, when applied to a control volume (Fig. 2-1), may be expressed as

$$\text{Rate of creation of mass} = 0 \qquad (2\text{-}1)$$

If, in a flow system, the relative velocity normal to the control surface is designated as V, m/s, the total mass flux crossing the surface is

$$G = V\rho, \quad \text{kg/(s} \cdot \text{m}^2) \qquad (2\text{-}2)$$

where ρ is the fluid density, kg/m^3, and G the mass velocity or mass flux, kg/(s $\cdot$ m^2).

If G is constant over a cross-sectional area A, the total rate of mass flow across A is then

$$\dot{m} = AG = AV\rho, \quad \text{kg/s} \qquad (2\text{-}3)$$

THE MOMENTUM THEOREM

The *momentum* of an element of matter may be defined as a quantity equal to the product of the mass of that element and its velocity. Since velocity is a vector quantity, momentum is also a vector quantity. The momentum theorem may be expressed either in vector form or, using a cartesian coordinate system, as separate equations for the x, y, and z

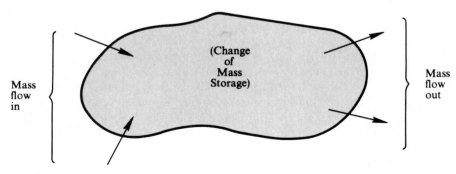

FIGURE 2-1
Control volume for application of conservation-of-mass principle.

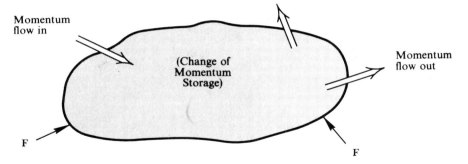

FIGURE 2-2
Control volume for application of momentum theorem.

directions, employing the components of momentum and external forces in each of these respective directions. The momentum theorem, when applied to a control volume, as in Fig. 2-2, may be expressed vectorially as

$$\text{Rate of creation of momentum} = \mathbf{F} \qquad (2\text{-}4)$$

with the following notation:

$m\mathbf{V}$ momentum

m mass, kg

$\mathbf{V}$ velocity, m/s, referred to an inertial or nonaccelerating coordinate system

$\dot{m}\mathbf{V}$ momentum rate across control surface

$\dot{m}$ rate of mass flow across control surface, kg/s

$\mathbf{F}$ resultant of all external forces acting on control surface or volume, N

Again the term *creation* has the same connotation as for Eq. (2-1), that is, *outflow minus inflow plus increase of storage.*

Application of Eq. (2-4) involves a summation of the momentum flux terms over the entire control surface to evaluate the outflow and inflow terms, a summation of rates of changes of momentum over the control volume to evaluate the increase of storage term, and a summation of the external forces over the surface and/or volume to evaluate the resultant external force.

Note that the velocities associated with the outflow and inflow terms are, in general, different from the velocities involved in the storage terms, and all must be referred to a nonaccelerating coordinate system.

PRINCIPLE OF CONSERVATION OF ENERGY

The principle of conservation of energy is a general expression of the first law of thermodynamics as applied to a control volume. Consider the energy that might cross the control surface and the changes of energy storage that might take place in the control volume shown in Fig. 2-3. The principle of conservation of energy can then be stated as

$$\text{Rate of creation of energy} = S \qquad (2\text{-}5)$$

Application of Eq. (2-5) involves the same kind of summation as required for Eqs. (2-1) and (2-4). Energy can cross the surface in the form of work or heat transfer, or it can cross as energy stored in any mass that flows across the surface. Some of the forms of energy frequently encountered in flow systems are as follows:

Energy that can cross a control surface

Mechanical
(or electrical) work W

Flow work   Work done by each unit mass of flowing fluid on the control volume, or by the control volume, as it flows across the control surface. In general, the flow work done by a unit mass is equal to the product of the normal fluid stress and the fluid specific volume, both evaluated at the control surface. If velocity gradients are not large, the normal stress is essentially the simple thermodynamic pressure, and the unit flow work becomes merely Pv, where v is the specific volume of the fluid.

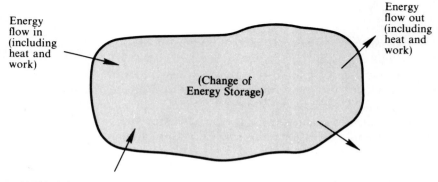

FIGURE 2-3
Control volume for application of conservation-of-energy principle.

Heat q, energy transferred by virtue of a temperature gradient

Energy stored in each unit mass as it crosses the control surface

Internal thermal energy e (including stored chemical energy)
Kinetic energy $\frac{1}{2}V^2$
Potential energy zg, where z is elevation with respect to some datum in a gravity field of constant strength g
Enthalpy i, a thermodynamic property, the sum $e + Pv$

Energy that can be involved in a change of energy storage within a control volume

Internal thermal energy e
Kinetic energy $\frac{1}{2}V^2$
Potential energy zg
Energy generation S

CHAPTER
3

FLUID
STRESSES
AND
FLUX
LAWS

In this chapter we set forth the equations we will use to evaluate the normal stresses and the shear stresses in a viscous fluid as well as the basic flux laws we will use to evaluate local heat flux and local mass diffusion flux. In Chapter 4 we combine these relations with the conservation principles of the previous chapter to develop a set of differential equations that must be satisfied in the region of interest.

VISCOUS FLUID STRESSES

Let us draw a rectangular box around a point in a viscous fluid flow field, as illustrated in the cartesian coordinate representation in Fig. 3-1. This two-dimensional box defines the stress nomenclature to be employed. Normal stresses are designated by σ and shear stresses by τ. The subscript convention follows that of elastic body mechanics. Notation is shown for the xy plane, with a similar scheme employed for the xz and yz planes.

For a Newtonian fluid (and only Newtonian fluids are considered here), the shear stresses are postulated to be directly proportional to the time rate of deformation of a fluid element, the proportionality factor

10

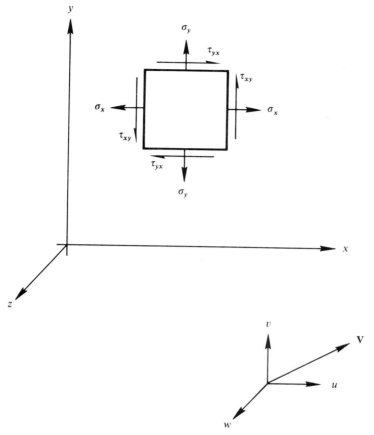

FIGURE 3-1
Stresses acting on element of fluid; components of fluid velocity.

being the viscosity coefficient μ. In cartesian coordinates, with u, v, w representing the components of velocity in the x, y, z directions, again as shown in Fig. 3-1, the shear stresses then become

$$\tau_{xy} = \tau_{yx} = \mu\left(\frac{\partial v}{\partial x} + \frac{\partial u}{\partial y}\right) \tag{3-1}$$

$$\tau_{yz} = \tau_{zy} = \mu\left(\frac{\partial w}{\partial y} + \frac{\partial v}{\partial z}\right) \tag{3-2}$$

$$\tau_{zx} = \tau_{xz} = \mu\left(\frac{\partial u}{\partial z} + \frac{\partial w}{\partial x}\right) \tag{3-3}$$

The equations for the normal stresses in a viscous fluid are simply postulated here without development or further argument. They can be derived for a simple gas by the kinetic theory of gases, but require certain critical assumptions for a more general development. In particular, the use of $-\frac{2}{3}\mu$ as the coefficient of the second term appears to be strictly limited to the low-density gas regime.[1]

$$\sigma_x = -P - \frac{2}{3}\mu\left(\frac{\partial u}{\partial x} + \frac{\partial v}{\partial y} + \frac{\partial w}{\partial z}\right) + 2\mu\frac{\partial u}{\partial x} \tag{3-4}$$

$$\sigma_y = -P - \frac{2}{3}\mu\left(\frac{\partial u}{\partial x} + \frac{\partial v}{\partial y} + \frac{\partial w}{\partial z}\right) + 2\mu\frac{\partial v}{\partial y} \tag{3-5}$$

$$\sigma_z = -P - \frac{2}{3}\mu\left(\frac{\partial u}{\partial x} + \frac{\partial v}{\partial y} + \frac{\partial w}{\partial z}\right) + 2\mu\frac{\partial w}{\partial z} \tag{3-6}$$

P is the simple thermodynamic pressure under these assumptions and, for the majority of applications, is the only term of real significance. For a *constant-density fluid* it is seen later that the second term in each equation is always zero and that only when there are very large velocity gradients in the direction of the stress (last term) does σ differ appreciably from P. Analysis of a normal shock wave is an example where all the terms are of significance.

Equations (3-1)–(3-6) can be compactly written using cartesian tensor notation† as

$$\sigma_{ij} = -\left(P + \frac{2}{3}\mu\frac{\partial u_k}{\partial x_k}\right)\delta_{ij} + \mu\left(\frac{\partial u_i}{\partial x_j} + \frac{\partial u_j}{\partial x_i}\right) \tag{3-7}$$

Some representative data on viscosity coefficients of fluids are given in App. A. The viscosity coefficient varies somewhat with the temperature of a fluid. Generally for a liquid it is a decreasing function of temperature, whereas for a gas it is an increasing function. μ has the dimensions Pa · s, or kg/(m · s).

---

† In using tensor notation the indices in the subscripts can have the values 1, 2, 3, and they have the following meaning: $x_1 = x$, $x_2 = y$, $x_3 = z$, and $u_1 = u$, $u_2 = v$, $u_3 = w$. Repeated indices in a term of an equation imply summation of the term over the three values of k (*summation convention*). The Kronecker delta is $\delta_{ij} = 0$, $i \neq j$, and $\delta_{ij} = 1$, $i = j$. In the symbol σ_{ij} the subscript i is the direction of the outward normal to the surface on which the stress acts, and the subscript j indicates the direction of the stress itself. By symmetry of the stress tensor, $\sigma_{12} = \sigma_{21} = \tau_{xy}$, $\sigma_{23} = \sigma_{32} = \tau_{yz}$, $\sigma_{31} = \sigma_{13} = \tau_{zx}$, $\sigma_{11} = \sigma_x$, $\sigma_{22} = \sigma_y$, $\sigma_{33} = \sigma_z$.

FOURIER'S LAW OF HEAT CONDUCTION

Fourier's law of heat conduction states that heat transfer by molecular interactions at any point in a solid or fluid is proportional in magnitude and coincident with the direction of the negative gradient of the temperature field. Conduction heat transfer is thus a vector quantity, and the basic Fourier equation for the heat flux vector is

$$\dot{\mathbf{q}}'' = -k\,\nabla t \qquad (3\text{-}8)$$

where k is the thermal conductivity of the conducting media and has the dimensions W/(m · K).

Fourier's law is the simplest form of a general energy flux law and is strictly applicable only when the system is uniform in all respects except for the temperature gradient, that is, there are no mass concentration gradients or gradients in other intensive properties.

Given a system with gradients of temperature, pressure, mass concentration, magnetic field strength, and so on, there is no a priori justification for ignoring the possibility that each of these gradients might contribute to the energy flux. The simplest expression that could describe this relationship would be a linear combination of terms, one for each of the existing potential gradients. Experience shows that there are, in fact, measurable "coupled" effects, such as energy fluxes due to mass concentration gradients, and that this more general form of the linear rate equation is necessary under certain conditions.

The relationships among the coefficients of this equation have been the subject of several investigations[2-4] in irreversible thermodynamics. These relationships fall into the domain of irreversible thermodynamics by the very nature of the diffusive transport process. One of the principal contributions of irreversible thermodynamics has been to show that, for example, if energy is transported because of a concentration gradient then mass will be transported because of a temperature gradient, and the coefficients of these two coupled interactions will be the same, provided that the driving potentials are appropriately chosen as a thermodynamically consistent set.

Since we are primarily concerned here with temperature gradients and mass concentration gradients, only the coupling between these effects is discussed. An explicit relation for the energy flux can be derived for a low-density gas from kinetic theory. The following equation for the thermal flux in a binary gas mixture (two components of the gas only) is given by Baron[5] and follows from the development by Chapman and Cowling:[6]

$$\dot{\mathbf{q}}'' = -k\,\nabla t + \left(i_1 + i_2 + \frac{RTk_T}{m_1 m_2 \mathfrak{M}}\right)(\mathbf{G}_{\text{diff},\,1}) \qquad (3\text{-}9)$$

where 1 and 2 refer to the two components of the mixture, i is the enthalpy, m is the mass concentration, $\mathfrak{M}$ is the mean molecular weight of the mixture, R is the universal gas constant, $\mathbf{G}_{\text{diff}, 1}$ is the mass diffusion flux of component 1, and k_T is a "thermal diffusion ratio." The effect incorporated in the last term is frequently known as the *diffusion-thermo*, or *Dufour*, effect.

Information on k_T is rather meager and largely derived from kinetic theory for gases, but calculations to establish the importance of this effect in typical boundary-layer flows have been made.[5,7] In this book, however, we assume that this effect is negligible in all the applications considered, although the transport of enthalpy by mass diffusion (first term in the first set of parentheses) is included where appropriate. With this exception, then, Fourier's law of heat conduction is assumed to represent the heat flux in all cases.

Some representative data on the thermal conductivity k of fluids are given in App. A. The thermal conductivity of liquids is relatively independent of temperature, but for gases it is an increasing function of temperature similar to the viscosity coefficient.

In some applications the thermal conductivity occurs naturally in combination with specific heat. In these cases it becomes convenient to define

$$\Gamma = k/c, \quad \text{kg/(s} \cdot \text{m)} \tag{3-10}$$

where c is the specific heat of the fluid, $J/(\text{kg} \cdot \text{K})$. (For gases c will be understood to be the specific heat at constant pressure c_p.) Thus if Γ is employed, Eq. (3-8) becomes

$$\dot{\mathbf{q}}'' = -\Gamma c \, \nabla t \tag{3-11}$$

FICK'S LAW OF DIFFUSION

Before introducing Fick's law, it is useful to discuss some definitions, and to make a distinction between mass transfer by *convection*, or bulk fluid movement, and mass transfer by *diffusion*, which is primarily caused by concentration gradients. We will define

$$\mathbf{G} = \text{total mass flux vector, kg/(s} \cdot \text{m}^2)$$

$$\mathbf{V} = \mathbf{G}/\rho$$

$$= \text{total velocity vector, m/s}$$

where ρ is the mixture density, kg/m^3.

These terms have the same meaning whether the fluid has one component, multiple components, or multiple components with concentration gradients.

If there are multiple components in the fluid, we will define the *total* mass flux vector for some component j as

$$\mathbf{G}_{\text{tot},\,j}, \quad \text{kg}/(\text{s} \cdot \text{m}^2)$$

We will then define the *convected* mass flux of component j as the flux of j attributable to bulk fluid movement, i.e.,

$$\mathbf{G}_{\text{conv},\,j} = m_j \mathbf{G}$$

where m_j is the mass concentration of component j, kilogram of j per kilogram of mixture. Also $m_j = \rho_j/\rho$.

Finally we will define the *diffusion* mass flux of component j as the difference, if any, between the total and convected fluxes:

$$\mathbf{G}_{\text{diff},\,j} = \mathbf{G}_{\text{tot},\,j} - \mathbf{G}_{\text{conv},\,j}$$

It follows from these definitions that

$$\sum_j m_j = 1$$

$$\sum_j \mathbf{G}_{\text{tot},\,j} = \mathbf{G}$$

$$\sum_j \mathbf{G}_{\text{conv},\,j} = \mathbf{G} \sum_j m_j = \mathbf{G}$$

and thus

$$\sum_j \mathbf{G}_{\text{diff},\,j} = 0$$

Note then that even if there is not bulk movement of the fluid, it is still possible to have diffusion taking place, but the vector sum of the diffusion of the various components of the mixture will be zero.

It is found that the rate of *diffusion* of any single component of a mixture of fluids, $\mathbf{G}_{\text{diff},\,j}$, is a function of the concentration gradient of that component, and is also a function of certain other potential gradients acting on the system, such as temperature and pressure gradients, and the concentration gradients of each of the other components. An exact equation can be developed from kinetic theory for the mass flux of one component in a multicomponent mixture of low density, simple gases,[8] although it is a rather complex expression.

A more restricted relation, which is exact for mass-diffusion in a binary mixture in the absence of other potential gradients, is Fick's law:

$$\mathbf{G}_{\text{diff},\,j} = -\rho D_j \nabla m_j \tag{3-12}$$

where D_j is a mass diffusion coefficient, m^2/s.

The corresponding energy relation, Fourier's law, was restricted to cases in which the system was completely described by the specification of one gradient, the temperature gradient. Similarly, the description of the mass-diffusion system must be fixed by ∇m_j alone before Fick's law is strictly applicable.

Most of the specific applications considered in this book involve only *binary* mixtures, so that Fick's law is applicable (in the absence of the Soret effect discussed later). However, we would like to develop the differential equations somewhat more generally, and we would like to consider at least some applications of diffusion in multicomponent systems. Knuth[9] shows that Fick's law describes accurately the diffusion of one component in a multicomponent mixture provided that the *binary* diffusion coefficients for each pair of components in the mixture are all the same and equal to D_j, that is,

$$D_j = \mathscr{D}_{12} = \ldots = \mathscr{D}_{ij}$$

where the $\mathscr{D}_{ij} = \mathscr{D}_{ji}$ are the diffusion coefficients for a *binary* mixture of i and j alone.

We use Fick's law exclusively, and the degree of approximation introduced in multicomponent applications then depends on the actual differences present in the $\mathscr{D}_{ij}$. Generally, binary diffusion coefficients are rather strongly dependent on relative molecular weight, and thus the molecular weight differences provide an approximate measure of the applicability of Fick's law to multicomponent systems. It should also be added that in a *turbulent* flow the simple transport model that we use also leads to approximate equality of the diffusion coefficients.

We find it convenient to write the coefficient in Fick's law in combination with the fluid density as follows:

$$\gamma_j = \rho D_j \tag{3-13}$$

Thus Eq. (3-12) becomes

$$\mathbf{G}_{\text{diff},\,j} = -\gamma_j \, \nabla m_j \tag{3-14}$$

Then γ_j is dimensionally similar to μ and Γ. Some representative data on γ_j for binary mixtures are given in App. A.

As discussed earlier, mass diffusion is a function not only of concentration gradients but also of the other potential gradients acting on the system. Here we are primarily concerned with the coupling effects of mass concentration and temperature gradients. Although pressure gradients and body forces, among others, can also induce mass diffusion, they are not generally important in the types of applications considered. For a gas explicit relations can be derived from kinetic theory. The following equation for the diffusion of component 1 in the binary mixture under the influence of both a concentration gradient and a temperature gradient is given by Baron[5] and follows from a development by Chapman and

Cowling:[6]

$$\mathbf{G}_{\text{diff},j} = -\rho \mathscr{D}_{12}\left(\nabla m_j + \frac{\mathfrak{M}_1 \mathfrak{M}_2}{\mathfrak{M}^2} k_T \nabla \ln T\right) \tag{3-15}$$

The effect incorporated in the last term is frequently known as the *thermo-diffusion*, or *Soret*, effect. Again, in this book we assume that this effect is negligible in the applications considered.

DIMENSIONLESS GROUPS OF TRANSPORT PROPERTIES

The shear-stress equations and the two flux laws can each be written in terms of one coefficient; and each coefficient is dimensionally the same as the viscosity. Thus we can form three dimensionless groups by simple ratios, which appear later as parameters in our differential equations.

Prandtl number $\qquad \Pr = \dfrac{\mu}{\Gamma} = \dfrac{\mu c}{k}$ $\qquad\qquad$ (3-16)

Schmidt number $\qquad \text{Sc}_j = \dfrac{\mu}{\gamma_j} = \dfrac{\mu}{\rho D_j}$ $\qquad\qquad$ (3-17)

Lewis number† $\qquad \text{Le}_j = \dfrac{\gamma_j}{\Gamma} = \dfrac{\Pr}{\text{Sc}_j}$ $\qquad\qquad$ (3-18)

TURBULENT-FLOW TRANSPORT COEFFICIENTS

The three transport coefficients defined above are all associated with molecular transport processes. In a turbulent flow these definitions are still valid, but they are involved in time-dependent terms in the differential equations. It proves mathematically convenient to postulate a model of the turbulent transport process that will lead to shear and flux laws similar in form to the molecular laws given. These, in turn, give rise to *turbulent* transport coefficients, which are dimensionally similar to the molecular coefficients but are functions of the dynamics of the flow field rather than being simple fluid properties. Consideration of the turbulent transport mechanism is deferred until later.

---

† The inverse, Γ/γ_j, is also frequently defined as the Lewis number.

REFERENCES

1. Rosenhead, L.: *Proc. Roy. Soc. (London)*, ser. A, vol. 226, p. 1, 1954.
2. DeGroot, S. R.: *Thermodynamics of Irreversible Processes,* Interscience, New York, 1952.
3. Keenan, J. H.: *Appl. Mech. Rev.,* vol. 49, 1955.
4. Onsager, L.: (I) *Phys. Rev.,* vol. 37, p. 405, 1931; (II) *Phys. Rev.,* vol. 38, p. 2265, 1931.
5. Baron, J. R.: *ARS J.,* vol. 32, pp. 1053–1059, 1962.
6. Chapman, S., and T. G. Cowling: *The Mathematical Theory of Non-Uniform Gases,* Cambridge University Press, London, 1961.
7. Tewfik, O. E.: *Int. J. Heat Mass Transfer,* vol. 7, pp. 409–421, 1964.
8. See, for example, Hirschfelder, J. O., C. F. Curtiss, and R. B. Bird: *Molecular Theory of Gases and Liquids,* Wiley, New York, 1954.
9. Knuth, E. L.: *Phys. Fluids,* vol. 2, p. 340, 1959.

CHAPTER

4

THE DIFFERENTIAL EQUATIONS OF THE LAMINAR BOUNDARY LAYER

In this chapter we develop the differential equations that must be satisfied in the flow field about a body when there is heat transfer and mass transfer between the fluid and the body. From the conservation-of-mass principle we develop two equations: the continuity equation and the mass-diffusion equation. From the momentum theorem we develop the momentum equation of the laminar boundary layer and the Navier–Stokes equations. From the conservation-of-energy principle we develop various forms of the energy equation of the laminar boundary layer and the general viscous energy equation.

THE CONCEPT OF THE BOUNDARY LAYER

A complete viscous fluid solution for flow about a body poses considerable mathematical difficulty for all but the most simple flow geometries. A great practical breakthrough was made when Prandtl discovered that for most applications the influence of viscosity is confined to an extremely thin region very close to the body and that the remainder of the flow field could to a good approximation be treated as inviscid, that is, it could be calculated by the methods of potential flow theory.

The thin region near the body surface, which is known as the boundary layer, lends itself to relatively simple analysis by the very fact of its thinness relative to the dimensions of the body. A fundamental assumption of the boundary-layer approximation is that the fluid immediately adjacent to the body surface is at rest relative to the body, an assumption that appears to be valid except for very low-pressure gases when the mean free path of the gas molecules is large relative to the body. Thus the hydrodynamic or *momentum boundary layer* may be defined as the region in which the fluid velocity changes from its free-stream, or potential flow, value to zero at the body surface (Fig. 4-1).

In reality there is no precise "thickness" to the boundary layer defined in this manner; and until we have developed a precise definition for boundary-layer thickness, we simply imply that the boundary-layer thickness is the distance in which *most* of the velocity change takes place.

If the boundary-layer thickness is very small relative to all other flow dimensions, it is apparent that, for the two-dimensional boundary-layer representation shown in Fig. 4-1, the following conditions must prevail within the boundary layer:

$$u \gg v$$

$$\frac{\partial u}{\partial y} \gg \frac{\partial u}{\partial x}, \quad \frac{\partial v}{\partial x}, \quad \frac{\partial v}{\partial y}$$

These conditions in turn simplify the application of Eq. (3-1) to the momentum theorem; and, although not so readily obvious, they also lead

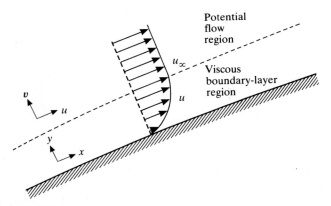

FIGURE 4-1
Momentum or viscous boundary layer on an external surface.

to a simplification of Eq. (3-4), namely,

$$\sigma_x \approx -P$$

and

$$\frac{\partial P}{\partial y} \approx 0, \qquad \frac{\partial P}{\partial x} \approx \frac{dP}{dx}$$

We develop the momentum equation of the boundary layer on this basis; then, after solving the equation for a particular case, we can show the conditions under which the boundary layer is truly "thin." We refer to the above conditions as the *boundary-layer approximations*.

Flow inside a tube is a form of boundary-layer problem in which, near the tube entrance, the boundary layer grows in much the same manner as over an external surface until its growth is stopped by symmetry at the centerline of the tube (Fig. 4-2). Thus the tube radius becomes the ultimate boundary-layer thickness.

When there is heat transfer or mass transfer between the fluid and the surface, it is also found that in most practical applications the major temperature and concentration changes occur in a region very close to the surface. This gives rise to the concept of the *thermal boundary layer* and the *concentration boundary layer,* and again the relative thinness of these boundary layers permits the introduction of boundary-layer approximations similar to those introduced for the momentum boundary layer, for example,

$$\frac{\partial t}{\partial y} \gg \frac{\partial t}{\partial x}$$

$$\frac{\partial m_j}{\partial y} \gg \frac{\partial m_j}{\partial x}$$

The thermal and concentration boundary layers are not necessarily the same thickness as the momentum boundary layer, and the conditions

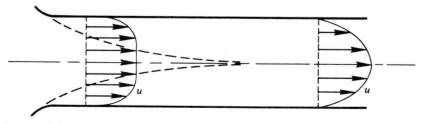

FIGURE 4-2
Development of a momentum boundary layer in the hydrodynamic entry region of a tube.

under which the boundary-layer approximations are valid are consequently different.

THE CONTINUITY EQUATIONS

Consider a two-dimensional flow in the xy plane, and cut out a control volume of infinitesimal dimensions (Fig. 4-3). Consider the total mass rate of flow across the control surface, and the rate of change of mass storage within, for a unit depth normal to the plane of the figure.

Let **G** represent mass flux (mass flow rate per unit of normal area) and G_x and G_y the x and y components, respectively. Then application of the principle of conservation of mass, Eq. (2-1), yields

$$\text{Outflow} = \left(G_y + \frac{\partial G_y}{\partial y}\,\delta y\right)\delta x + \left(G_x + \frac{\partial G_x}{\partial x}\,\delta x\right)\delta y$$

$$\text{Inflow} = G_x\,\delta y + G_y\,\delta x$$

$$\text{Increase of storage} = \frac{\partial \rho}{\partial \theta}\,\delta x\,\delta y$$

where θ is *time*. Then

$$\frac{\partial G_x}{\partial x} + \frac{\partial G_y}{\partial y} + \frac{\partial \rho}{\partial \theta} = 0 \tag{4-1}$$

which is the continuity equation for a two-dimensional flow in cartesian coordinates.

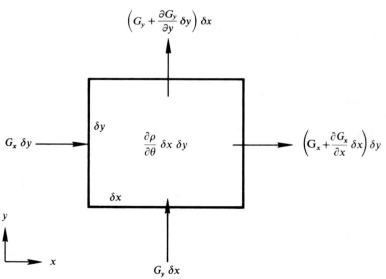

FIGURE 4-3
Control volume for development of the continuity equation.

The corresponding three-dimensional equation must then be

$$\frac{\partial G_x}{\partial x} + \frac{\partial G_y}{\partial y} + \frac{\partial G_z}{\partial z} + \frac{\partial \rho}{\partial \theta} = 0 \tag{4-2}$$

or, in vector notation,

$$\mathbf{\nabla} \cdot \mathbf{G} + \frac{\partial \rho}{\partial \theta} = 0 \tag{4-3}$$

The vector notation allows the coordinate system to remain unspecified. Appendix D summarizes the vector and scalar operations implied by the $\mathbf{\nabla}$ operator for cartesian, cylindrical, and spherical coordinate systems.

In cartesian tensor notation the continuity equation becomes

$$\frac{\partial}{\partial x_j}(\rho u_j) + \frac{\partial \rho}{\partial \theta} = 0 \tag{4-4}$$

For a constant-density flow Eq. (4-3) becomes

$$\mathbf{\nabla} \cdot \mathbf{G} = 0 = \mathbf{\nabla} \cdot \mathbf{V} \tag{4-5}$$

and Eq. (4-4) becomes

$$\frac{\partial u_j}{\partial x_j} = \frac{\partial u}{\partial x} + \frac{\partial v}{\partial y} + \frac{\partial w}{\partial z} = 0 \tag{4-6}$$

For constant-density flow in a *two-dimensional boundary layer* the continuity equation becomes simply

$$\frac{\partial u}{\partial x} + \frac{\partial v}{\partial y} = 0 \tag{4-7}$$

For flow in a *circular tube* where axial symmetry obtains it is convenient to employ a cylindrical coordinate system with nomenclature as defined in Fig. 4-4.

Then Eq. (4-3) reduces to

$$\frac{\partial(u\rho)}{\partial x} + \frac{1}{r}\frac{\partial}{\partial r}(\rho r v_r) + \frac{\partial \rho}{\partial \theta} = 0 \tag{4-8}$$

and, for constant-density flow, to

$$\frac{\partial u}{\partial x} + \frac{1}{r}\frac{\partial}{\partial r}(r v_r) = 0 \tag{4-9}$$

Note that the continuity equation does not involve any boundary-layer assumptions and thus is generally applicable.

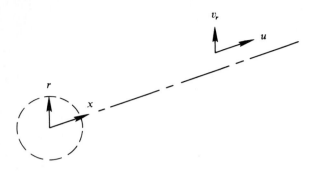

FIGURE 4-4
Coordinate system and velocity components for axisymmetric flow in a circular tube.

THE MOMENTUM EQUATIONS

The Steady-Flow Momentum Equation of the Boundary Layer

Consider steady flow along a semi-infinite two-dimensional surface of small curvature with a free-stream velocity u_∞. Let x be measured along the surface and y normal to the surface. The requirement of zero tangential velocity at the surface results in the development of a *momentum boundary layer*, defined as the region in which u changes from 0 to u_∞ (Fig. 4-5).

Cut out an infinitesimal stationary control volume of unit depth within the boundary layer, and consider the external forces acting on this control volume in the x direction and the x momentum fluxes crossing the control surface (Fig. 4-6). (For brevity let us omit body forces such as gravity, although these can simply be added if desired.)

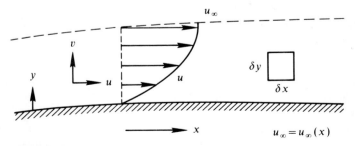

FIGURE 4-5
Coordinate system, velocity components, and control volume for development of the momentum differential equation of the boundry layer.

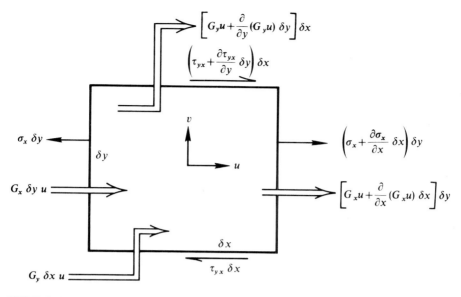

FIGURE 4-6
Control volume, momentum fluxes, and external forces for development of the steady-flow momentum differential equation of the boundary layer.

Note that the stipulation of steady flow implies a laminar flow, since a turbulent flow is by its nature unsteady.

Applying the momentum theorem, in the x direction, we have

$$\text{Outflow of momentum} = \left[G_x u + \frac{\partial}{\partial x} (G_x u) \, \delta x \right] \delta y$$

$$+ \left[G_y u + \frac{\partial}{\partial y} (G_y u) \, \delta y \right] \delta x$$

$$\text{Inflow of momentum} = (G_x \, \delta y) u + (G_y \, \delta x) u$$

$$\text{Increase of momentum storage} = 0$$

$$\text{External forces} = - \sigma_x \, \delta y - \tau_{yx} \, \delta x + \left(\sigma_x + \frac{\partial \sigma_x}{\partial x} \, \delta x \right) \delta y$$

$$+ \left(\tau_{yx} + \frac{\partial \tau_{yx}}{\partial y} \, \delta y \right) \delta x$$

Combining with Eq. (2-4) and simplifying,

$$\frac{\partial}{\partial x} (G_x u) + \frac{\partial}{\partial y} (G_y u) = \frac{\partial \sigma_x}{\partial x} + \frac{\partial \tau_{yx}}{\partial y}$$

Expanding the left-hand terms,

$$G_x \frac{\partial u}{\partial x} + u \frac{\partial G_x}{\partial x} + G_y \frac{\partial u}{\partial y} + u \frac{\partial G_y}{\partial y} = \frac{\partial \sigma_x}{\partial x} + \frac{\partial \tau_{yx}}{\partial y}$$

By the continuity equation for a steady flow,

$$\frac{\partial G_x}{\partial x} + \frac{\partial G_y}{\partial y} = 0$$

Thus

$$G_x \frac{\partial u}{\partial x} + G_y \frac{\partial u}{\partial y} = \frac{\partial \sigma_x}{\partial x} + \frac{\partial \tau_{yx}}{\partial y}$$

Next, employing Eqs. (3-1) and (3-4) and invoking the boundary-layer approximations, we get

$$\frac{\partial \sigma_x}{\partial x} = -\frac{dP}{dx}$$

$$\frac{\partial \tau_{yx}}{\partial y} = \frac{\partial}{\partial y} \left(\mu \frac{\partial u}{\partial y} \right)$$

Also note that

$$G_x = \rho u, \qquad G_y = \rho v$$

With these substitutions, we have the *momentum equation of the boundary layer*:

$$\rho u \frac{\partial u}{\partial x} + \rho v \frac{\partial u}{\partial y} + \frac{dP}{dx} = \frac{\partial}{\partial y} \left(\mu \frac{\partial u}{\partial y} \right) \tag{4-10}$$

Note that this equation is valid for variable properties ρ and μ.

For *axisymmetric flow in a circular tube* we can develop a similar boundary-layer equation in cylindrical coordinates by identical methods:

$$\rho u \frac{\partial u}{\partial x} + \rho v_r \frac{\partial u}{\partial r} + \frac{dP}{dx} = \frac{1}{r} \frac{\partial}{\partial r} \left(r\mu \frac{\partial u}{\partial r} \right) \tag{4-11}$$

The Navier–Stokes Equations

Now, suppose we apply the momentum theorem in precisely the same manner to a three-dimensional flow, without the simplifying assumptions of the boundary layer but including changes with time and including external body forces. Employing a cartesian coordinate system and applying the momentum theorem in the x direction, we obtain

$$\rho\left(\frac{\partial u}{\partial \theta} + u\frac{\partial u}{\partial x} + v\frac{\partial u}{\partial y} + w\frac{\partial u}{\partial z}\right) + \frac{\partial P}{\partial x} = \frac{\partial}{\partial x}\left\{\mu\left[2\frac{\partial u}{\partial x} - \frac{2}{3}\left(\frac{\partial u}{\partial x} + \frac{\partial v}{\partial y} + \frac{\partial w}{\partial z}\right)\right]\right\}$$

$$+ \frac{\partial}{\partial y}\left[\mu\left(\frac{\partial u}{\partial y} + \frac{\partial v}{\partial x}\right)\right]$$

$$+ \frac{\partial}{\partial z}\left[\mu\left(\frac{\partial w}{\partial x} + \frac{\partial u}{\partial z}\right)\right] + X \qquad (4\text{-}12)$$

Similar equations are obtained for the y and z directions.

The origin of the various terms in (4-12) can readily be seen by comparison with (4-10) and the stress equations (3-1)–(3-6). X is the body force in the x direction, N/m^3.

These three quations can be written more compactly in cartesian tensor notation:[1]

$$\rho\left(\frac{\partial u_i}{\partial \theta} + u_j\frac{\partial u_i}{\partial x_j}\right) = \frac{\partial \sigma_{ji}}{\partial x_j} + X_i \qquad (4\text{-}13)$$

The stress tensor σ_{ij} is defined in Eq. (3-7).

For *constant-property flow* a considerably simpler formulation can be obtained:

$$\rho\left(\frac{\partial u}{\partial \theta} + u\frac{\partial u}{\partial x} + v\frac{\partial u}{\partial y} + w\frac{\partial u}{\partial z}\right) + \frac{\partial P}{\partial x} = \mu\left(\frac{\partial^2 u}{\partial x^2} + \frac{\partial^2 u}{\partial y^2} + \frac{\partial^2 u}{\partial z^2}\right) + X \qquad (4\text{-}14)$$

or, more briefly,

$$\rho\frac{Du}{D\theta} + \frac{\partial P}{\partial x} = \mu \nabla^2 u + X \qquad (4\text{-}15)$$

where D represents the *substantial derivative*, defined as indicated.

Again, analogous equations can be written for the y and z directions; or, abandoning a coordinate system, the constant-properties equations can be put in vector notation as

$$\rho\frac{D\mathbf{V}}{D\theta} + \nabla P = \mu \nabla^2\mathbf{V} + \mathbf{F} \qquad (4\text{-}16)$$

In tensor notation the constant-properties equations become

$$\rho\left(\frac{\partial u_i}{\partial \theta} + u_j\frac{\partial u_i}{\partial x_j}\right) + \frac{\partial P}{\partial x_i} = \mu\frac{\partial^2 u_i}{\partial x_j\,\partial x_j} + X_i \qquad (4\text{-}17)$$

It should be noted here that the momentum equation of the two-dimensional boundary layer, Eq. (4-10), can readily be derived from the Navier–Stokes equations, and indeed this is the procedure preferred by many authors rather than deriving the boundary-layer equation directly after making the boundary-layer assumptions. This procedure involves an assumption that the boundary layer is "thin," after which an order-of-magnitude analysis of the various terms in the Navier–Stokes equations is made. This analysis leads to a conclusion that the criterion for a "thin" boundary layer on a flat plate with constant free-stream velocity is that the Reynolds number based on distance from the leading edge x, $xu_\infty\rho/\mu = \mathrm{Re}_x$, must be large relative to unity. Further details on the order-of-magnitude analysis may be found in, for example, Schlichting[2] or Streeter.[3]

THE MASS-DIFFUSION EQUATIONS

Steady-Flow Mass-Diffusion Equation of the Boundary Layer

Consider steady flow along the surface of a body, and let there be a transfer of mass to or from the fluid at the surface. If the transferred substance contains different chamical species than the free-stream fluid, concentration gradients will be established. In general, there may be any number of chemical components in the resulting mixture, and, as suggested by Fick's law, Eq. (3-14), each will tend to diffuse in the direction of the negative of its own concentration gradient (but see comments on the applicability of Fick's law to multicomponent diffusion in Chap. 3). Additionally, various components may combine chemically to form new species, so that creation and destruction of various components of the mixture may occur at any point, resulting in concentration gradients. Thus, if there is chemical reaction, mass diffusion can occur even with no mass transfer at the body surface. The principle of conservation of mass requires a continuity equation for each species: and the mass-diffusion equation is the species continuity equation.

Let us consider some particular component of the mixture, to be designated by a subscript j. At the body surface this component has a particular concentration, possibly determined by the conditions of thermodynamic equilibrium at the surface. In the free stream the concentration of j is presumably some known value, frequently zero. Then the region in which the concentration of j changes from its body-surface value to its free-stream value is defined as the *concentration boundary layer* and may be represented as in Fig. 4-5, but with m_j replacing u and with m_j having a finite value at the surface, rather than zero as for u.

Let us cut out a stationary infinitesimal control volume of unit depth within this boundary layer and examine the ways in which component j can cross the control surface (Fig. 4-7) in steady flow.

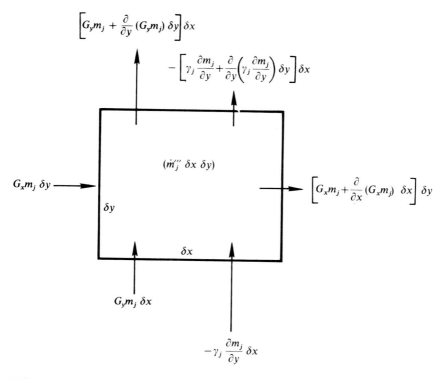

FIGURE 4-7
Control volume and mass-transfer terms for development of the steady-flow mass-diffusion differential equation of the boundary layer.

Across each side, component j is carried by the movement of the entire stream (convected transfer), and these terms are then the product of the total mass flow rate and the concentration of component j, m_j, kg of component j per kg of total mixture. Transport of component j also takes place quite independently by diffusion. We assume that Fick's law of diffusion holds. However, here we invoke the boundary-layer approximations, assuming that the only concentration gradient of significance is in the y direction, normal to the surface.

To combine the terms on Fig. 4-7, we need to modify our conservation-of-mass principle, Eq. (2-1), to account for the creation term. Within the control volume, component j may be created (or destroyed) by chemical reaction. We designate the rate of creation of j per unit volume by $\dot{m}_j'''$, kg of j per m^3. Note that

$$\sum_j \dot{m}_j''' = 0$$

Thus

$$\text{Rate of creation of component } j = \dot{m}_j'''$$

Then

$$\text{Outflow of } j = \left[G_x m_j + \frac{\partial}{\partial x} (G_x m_j) \, \delta x \right] \delta y + \left[G_y m_j + \frac{\partial}{\partial y} (G_y m_j) \, \delta y \right] \delta x$$
$$- \left[\gamma_j \frac{\partial m_j}{\partial y} + \frac{\partial}{\partial y} \left(\gamma_j \frac{\partial m_j}{\partial y} \right) \delta y \right] \delta x$$

$$\text{Inflow of } j = G_x m_j \, \delta y + G_y m_j \, \delta x - \gamma_j \frac{\partial m_j}{\partial y} \, \delta x$$

$$\text{Increase of storage of } j = 0 \quad \text{(steady flow)}$$

Substituting and simplifying,

$$\frac{\partial}{\partial x} (G_x m_j) + \frac{\partial}{\partial y} (G_y m_j) - \frac{\partial}{\partial y} \left(\gamma_j \frac{\partial m_j}{\partial y} \right) = \dot{m}_j'''$$

Expanding the left-hand terms, we have

$$G_x \frac{\partial m_j}{\partial x} + m_j \frac{\partial G_x}{\partial x} + G_y \frac{\partial m_j}{\partial y} + m_j \frac{\partial G_y}{\partial y} - \frac{\partial}{\partial y} \left(\gamma_j \frac{\partial m_j}{\partial y} \right) = \dot{m}_j'''$$

But, by the continuity equation for a steady flow,

$$\frac{\partial G_x}{\partial x} + \frac{\partial G_y}{\partial y} = 0$$

Thus

$$G_x \frac{\partial m_j}{\partial x} + G_y \frac{\partial m_j}{\partial y} - \frac{\partial}{\partial y} \left(\gamma_j \frac{\partial m_j}{\partial y} \right) = \dot{m}_j''' \tag{4-18}$$

or

$$u\rho \frac{\partial m_j}{\partial x} + v\rho \frac{\partial m_j}{\partial y} - \frac{\partial}{\partial y} \left(\gamma_j \frac{\partial m_j}{\partial y} \right) = \dot{m}_j''' \tag{4-19}$$

If component j is *chemically inert* in the system, that is, there is *no chemical reaction*, then $\dot{m}_j''' = 0$, and (4-18) becomes

$$G_x \frac{\partial m_j}{\partial x} + G_y \frac{\partial m_j}{\partial y} - \frac{\partial}{\partial y} \left(\gamma_j \frac{\partial m_j}{\partial y} \right) = 0 \tag{4-20}$$

For *axisymmetric flow in a circular tube with axisymmetric diffusion* an identical method leads to

$$u\rho \frac{\partial m_j}{\partial x} + v_r \rho \frac{\partial m_j}{\partial r} - \frac{1}{r} \frac{\partial}{\partial r} \left(\gamma_j r \frac{\partial m_j}{\partial r} \right) = 0 \tag{4-21}$$

General Equation for Steady-Flow Diffusion

If we discard the boundary-layer approximations and consider the complete steady-flow convection–diffusion problem in three dimensions, we need only add one more convection term and two additional diffusion terms. However, in this case the cartesian coordinate representation becomes cumbersome. A neater formulation of the equations can be obtained by using vector notation. Thus the more general counterparts of Eqs. (4-18) and (4-20) become, respectively,

$$\mathbf{G} \cdot \nabla m_j - \nabla \cdot \gamma_j \, \nabla m_j = \dot{m}_j''' \tag{4-22}$$

$$\mathbf{G} \cdot \nabla m_j - \nabla \cdot \gamma_j \, \nabla m_j = 0 \tag{4-23}$$

Diffusion of Chemical Element α

It is sometimes more convenient, in considering diffusion with chemical reaction, to follow a particular chemical element rather than a compound. The advantage obviously lies in the fact that whereas a compound may be created or destroyed at any point in the flow field, an element is considered indestructible.

Define the mass concentration of an element α as follows. Let

n_α = kg of element α per kg of total mixture

$n_{\alpha,j}$ = kg of element α per kg of substance j

Note that $n_{\alpha,j}$ is not a variable but is simply fixed by the composition of substance j.

It then follows from the definitions that

$$n_\alpha = \sum_j n_{\alpha,j} m_j$$

$$\sum_\alpha n_\alpha = 1$$

Conservation of element α requires that

$$\sum_j n_{\alpha,j} \dot{m}_j''' = 0$$

We now use this expression, together with Eq. (4-18), to develop a boundary-layer diffusion equation for α. Multiply Eq. (4-18) through by $n_{\alpha,j}$:

$$n_{\alpha,j} G_x \frac{\partial m_j}{\partial x} + n_{\alpha,j} G_y \frac{\partial m_j}{\partial y} - n_{\alpha,j} \frac{\partial}{\partial y}\left(\gamma_j \frac{\partial m_j}{\partial y}\right) = n_{\alpha,j} \dot{m}_j'''$$

Summing over j yields

$$G_x \sum_j n_{\alpha,j} \frac{\partial m_j}{\partial x} + G_y \sum_j n_{\alpha,j} \frac{\partial m_j}{\partial y} - \sum_j n_{\alpha,j} \frac{\partial}{\partial y}\left(\gamma_j \frac{\partial m_j}{\partial y}\right) = \sum_j n_{\alpha,j} \dot{m}_j''' = 0$$

But since $n_{\alpha,j}$ is not a function of x or y, we can write

$$G_x \frac{\partial}{\partial x} \sum_j n_{\alpha,j} m_j + G_y \frac{\partial}{\partial y} \sum_j n_{\alpha,j} m_j - \frac{\partial}{\partial y} \sum_j \frac{\partial}{\partial y} (n_{\alpha,j} m_j) = 0$$

We now substitute the result that

$$\sum_j n_{\alpha,j} m_j = n_\alpha$$

and obtain the final form

$$G_x \frac{\partial n_\alpha}{\partial x} + G_y \frac{\partial n_\alpha}{\partial y} - \frac{\partial}{\partial y} \sum_j \gamma_j \frac{\partial}{\partial y} (n_{\alpha,j} m_j) = 0 \qquad (4\text{-}24)$$

Note that Eq. (4-24) is very similar to (4-20); and, in fact, if the γ_j are all the same, it will become identical, which is in anticipation of a simplification to be introduced later.

If we now discard the boundary-layer approximations and consider the complete steady-flow diffusion problem in three dimensions, it follows directly that

$$\mathbf{G} \cdot \nabla n_\alpha - \nabla \cdot \left(\sum_j \gamma_j n_{\alpha,j} \nabla m_j \right) = 0 \qquad (4\text{-}25)$$

which is comparable to Eq. (4-23).

THE ENERGY EQUATIONS

The Steady-Flow Energy Equation of the Boundary Layer

Consider steady flow along a semi-infinite surface with a free-stream velocity u_∞ and a free-stream temperature t_∞. Let the surface temperature be t_0, which may vary along the surface. There may or may not be mass transfer from the surface, and there may or may not be concentration gradients throughout the flow field. Chemical reaction may be taking place in the flow field.

In general, there will be heat exchange between the surface and the fluid. Thermodynamic equilibrium requires that the fluid just adjacent to the surface be at t_0, and the region in which the temperature changes from t_0 to t_∞ is called the *thermal boundary layer* (Fig. 4-8).

Let us cut out an infinitesimal stationary control volume of unit depth and examine the various rates of energy transfer across the control surface (Fig. 4-9). Several forms of energy transfer are present. Fluid crossing the boundary will carry its own internal thermal and chemical energy, as well as its kinetic energy and potential energy with respect to external force fields, and will do flow work on the control volume. Heat

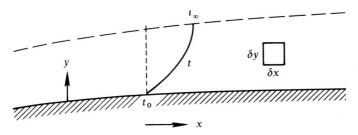

FIGURE 4-8
The thermal boundary layer.

will be transferred as a result of the temperature gradients at the surface. If there are concentration gradients, there will be mass diffusion, and each component of the fluid mixture will carry its own energy. Work will be done on the control volume, and its surfaces, by shear forces.

These various energy-transfer streams are indicated symbolically on Fig. 4-9. Note that the boundary-layer approximation has been invoked so that velocity, temperature, and concentration gradients in the y direction only are considered.

Before evaluating the various energy-transfer terms on Fig. 4-9, let us examine in a little more detail the enthalpy i that will appear in all the

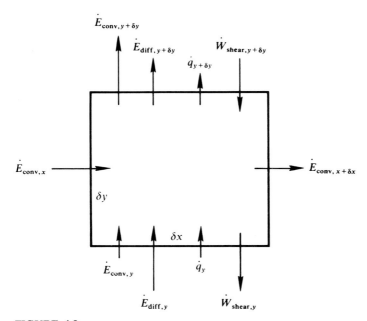

FIGURE 4-9
Control volume and energy-transfer terms for development of the steady-flow energy differential equation of the boundary layer.

convection terms and in the mass-diffusion terms. As a part of the boundary-layer approximation, it is assumed that the gradient of the normal stress is simply the gradient of the thermodynamic pressure. Thus the flow work done by each element of fluid as it crosses the control surface is merely Pv or $\cdot P/\rho$. The mixture enthalpy $i = e + P/\rho$ is then the sum of the internal energy and the flow work of the fluid, per unit mass. For substances that can react chemically, the internal energy includes the stored chemical energy. This point is developed in detail later. The distinction between i and i_j should next be noted: i without subscript refers to the entire mixture, J/kg, of mixture; i_j is the *partial enthalpy* of component j, J/kg of component j. Thus

$$\sum_j m_j i_j = i$$

The partial enthalpy i_j is a function of chemical composition, temperature, and pressure, although frequently we assume perfect gas behavior and hence neglect the pressure dependence. We have more to say later about the evaluation of i for particular fluid systems.

Making the assumptions (in addition to the boundary-layer approximations) of *steady flow, no internal heat generation, and no work done by external fields*, the energy-transfer terms become

$$\dot{E}_{\text{conv}, x} = G_x \, \delta y \, (i + \tfrac{1}{2}u^2) \quad \text{(enthalpy plus kinetic energy,}$$
$$\text{assuming } u^2 \gg v^2)$$

$$\dot{E}_{\text{conv}, x + \delta x} = \delta y \left\{ G_x(i + \tfrac{1}{2}u^2) + \frac{\partial}{\partial x} [G_x(i + \tfrac{1}{2}u^2)] \, \delta x \right\}$$

$$\dot{E}_{\text{conv}, y} = G_y \, \delta x \, (i + \tfrac{1}{2}u^2)$$

$$\dot{E}_{\text{conv}, y + \delta y} = \delta x \left\{ G_y(i + \tfrac{1}{2}u^2) + \frac{\partial}{\partial y} [G_y(i + \tfrac{1}{2}u^2)] \, \delta y \right\}$$

$$\dot{E}_{\text{diff}, y} = -\left(\sum_j \gamma_j \frac{\partial m_j}{\partial y} i_j \right) \delta x \quad \text{(sum of diffusion rate of each}$$
$$\text{component times enthalpy of that}$$
$$\text{component, neglecting Soret effect)}$$

$$\dot{E}_{\text{diff}, y + \delta y} = -\left[\sum_j \gamma_j \frac{\partial m_j}{\partial y} i_j + \frac{\partial}{\partial y} \left(\sum_j \gamma_j \frac{\partial m_j}{\partial y} i_j \right) \delta y \right] \delta x$$

$$\dot{q}_y = -k \left(\frac{\partial t}{\partial y} \right) \delta x \quad \text{(conduction heat transfer, neglecting Dufour effect)}$$

$$\dot{q}_{y + \delta y} = -\left[k \left(\frac{\partial t}{\partial y} \right) + \frac{\partial}{\partial y} \left(k \frac{\partial t}{\partial y} \right) \delta y \right] \delta x$$

$$\dot{W}_{\text{shear}, y} = (\tau_{yx} u) \, \delta x \quad \text{(shear force times velocity)}$$

$$\dot{W}_{\text{shear}, y + \delta y} = \left[\tau_{yx} u + \frac{\partial}{\partial y} (\tau_{yx} u) \, \delta y \right] \delta x$$

Combining the above by the principle of conservation of energy, Eq. (2-5), and simplifying, we obtain

$$\frac{\partial}{\partial x}[G_x(i + \tfrac{1}{2}u^2)] + \frac{\partial}{\partial y}[G_y(i + \tfrac{1}{2}u^2)] - \frac{\partial}{\partial y}\left(k\frac{\partial t}{\partial y}\right)$$

$$-\frac{\partial}{\partial y}\left(\sum_j \gamma_j \frac{\partial m_j}{\partial y}i_j\right) - \frac{\partial}{\partial y}(\tau_{yx}u) = 0$$

Now expand the first two terms in this equation:

$$G_x\frac{\partial}{\partial x}(i + \tfrac{1}{2}u^2) + (i + \tfrac{1}{2}u^2)\frac{\partial G_x}{\partial x} + G_y\frac{\partial}{\partial y}(i + \tfrac{1}{2}u^2)$$

$$+ (i + \tfrac{1}{2}u^2)\frac{\partial G_y}{\partial y} - \frac{\partial}{\partial y}\left(k\frac{\partial t}{\partial y}\right) - \frac{\partial}{\partial y}\left(\sum_j \gamma_j \frac{\partial m_j}{\partial y}i_j\right) - \frac{\partial}{\partial y}(\tau_{yx}u) = 0$$

But, since by the continuity equation for a steady flow $\partial G_x/\partial x + \partial G_y/\partial y = 0$, the second and fourth terms drop out. We then have

$$G_x\frac{\partial}{\partial x}(i + \tfrac{1}{2}u^2) + G_y\frac{\partial}{\partial y}(i + \tfrac{1}{2}u^2) - \frac{\partial}{\partial y}\left(k\frac{\partial t}{\partial y}\right)$$

$$-\frac{\partial}{\partial y}\left(\sum_j \gamma_j \frac{\partial m_j}{\partial y}i_j\right) - \frac{\partial}{\partial y}(\tau_{yx}u) = 0 \quad (4\text{-}26)$$

Equation (4-26) is useful in its own right, and we use it in a later chapter. However, at present we combine it with the momentum equation to get a different form.

First take the term containing τ_{yx} and expand it:

$$\frac{\partial}{\partial y}(\tau_{yx}u) = \tau_{yx}\frac{\partial u}{\partial y} + u\frac{\partial \tau_{yx}}{\partial y}$$

From Eq. (3-1), under the boundary-layer approximations,

$$\tau_{yx} = \mu\frac{\partial u}{\partial y}$$

Substituting, we get

$$\frac{\partial}{\partial y}(\tau_{yx}u) = \mu\left(\frac{\partial u}{\partial y}\right)^2 + u\frac{\partial}{\partial y}\left(\mu\frac{\partial u}{\partial y}\right)$$

To obtain an expression for the last term on the right-hand side, transpose the momentum equation, Eq. (4-10), and multiply through by u:

$$u\frac{\partial}{\partial y}\left(\mu\frac{\partial u}{\partial y}\right) = \rho u^2\frac{\partial u}{\partial x} + \rho uv\frac{\partial u}{\partial y} + u\frac{dP}{dx}$$

Now, combine the last two equations, substitute the result into Eq. (4-26), carry out the differentiation of the first two terms of Eq. (4-26), substitute $u\rho = G_x$ and $v\rho = G_y$, and the result can readily be simplified to

$$\rho u \frac{\partial i}{\partial x} + \rho v \frac{\partial i}{\partial y} - \frac{\partial}{\partial y}\left(k\frac{\partial t}{\partial y}\right) - \frac{\partial}{\partial y}\left(\sum_j \gamma_j \frac{\partial m_j}{\partial y} i_j\right) - \mu\left(\frac{\partial u}{\partial y}\right)^2 - u\frac{dP}{dx} = 0$$

(4–27)

This is the steady-flow energy equation of the boundary layer, including variable fluid properties, mass diffusion and chemical reaction, and viscous energy dissipation.

Both Eqs. (4-26) and (4-27) are perfectly valid energy equations of the boundary layer. The distinction is that the terms in Eq. (4-26) represent conservation of the sum of both mechanical and thermal energy (the First Law of Thermodynamics), while Eq. (4-27) is a statement of conservation of thermal energy alone and thus includes a thermal energy *creation* term (the term involving viscosity). The fact that Eq. (4-27) is in reality a *thermal* energy equation is obscured by the use of the enthalpy i, which contains a mechanical energy component. However, this component, the flow work, is actually offset by the pressure-gradient term.

It should be noted that in developing the equation we neglected *potential energy* due to *external fields* for the sake of brevity. However, had we includes a force field in the x direction and then included the corresponding body force in the momentum equation that we combined with the energy equation, we would have found that the effects of the external fields cancel. Thus Eq. (4-27) is valid in the presence of external force fields. On the other hand, a body force X (which might be a gravity force) would result in an additional term, $-uX$, on the left-hand side of Eq. (4-26).

Note that we have also neglected any internal heat sources, such as might be produced by the resistance offered to an electric current. Such effects, which are lumped into a source function S, W/m^3, in the more general energy equation given below, can be subtracted from the left-hand sides of Eqs. (4-26) and (4-27) if desired. Note, however, that conversion of chemical energy to thermal energy has already been included in the enthalpy terms by the definition of the internal energy.

The corresponding equation for *axisymmetric flow in a circular tube with axisymmetric heating* can be developed in an identical manner to yield

$$\rho u \frac{\partial i}{\partial x} + \rho v_r \frac{\partial i}{\partial r} - \frac{1}{r}\frac{\partial}{\partial r}\left(kr\frac{\partial t}{\partial r}\right) - \frac{1}{r}\frac{\partial}{\partial r}\left(r\sum_j \gamma_j \frac{\partial m_j}{\partial y} i_j\right) - \mu\left(\frac{\partial u}{\partial r}\right)^2 - u\frac{dP}{dx} = 0$$

(4-28)

A little more insight into the significance of the mass-diffusion term in the energy equation and the net mass flux G can be gained if we reexamine the mass crossing the lower surface of the control volume shown in Fig. 4-7. It should be borne in mind here that we are discussing mixtures of fluids rather than single-component fluids; hence when we speak of the velocity of the mixture, the term *velocity* is ambiguous. There is no velocity that is entirely appropriate; what is meant by mixture velocity is the mass-weighted average of the velocities of the various components. Some of the components are moving faster, and some slower, than this average.

According to Fig. 4-7, the total flow of component j crossing the lower surface is

$$G_y m_j \, \delta x - \gamma_j \frac{\partial m_j}{\partial y} \, \delta x$$

The total flow of *all* components crossing this surface must be the summation of this quantity over j; *but, by definition of G_y,* this is also $G_y \, \delta x$. Thus

$$G_y \, \delta x = G_y \, \delta x \sum_j m_j - \delta x \sum_j \gamma_j \frac{\partial m_j}{\partial y}$$

But

$$\sum_j m_j = 1$$

Therefore

$$G_y = G_y - \sum_j \gamma_j \frac{\partial m_j}{\partial y}$$

or

$$\sum_j \gamma_j \frac{\partial m_j}{\partial y} = 0 \qquad (4\text{-}29)$$

This result simply means that diffusion of a certain mass of one or more components in the positive y direction is always accompanied by diffusion of an equal mass of other components in the negative y direction. This obviously must be the case, for it is not possible to have a concentration gradient of one component without corresponding opposite concentration gradients of the other components. Note again that G is the net mass flux, regardless of whether diffusion is taking place.

If one now examines the mass-diffusion term in Eq. (4-26) in the light of conclusion (4-29), it is apparent that this term is of importance only when the enthalpy of the various components differs; if each component has the same enthalpy then the i_j factors out and the summation goes to zero. This fact permits some later simplifications to the energy equation.

The General Viscous Energy Equation

Now let us consider the general three-dimensional problem in cartesian coordinates, including nonsteady flow with a source function. In so doing, let us use the complete expressions for the normal stresses, Eqs. (3-4)–(3-6), when evaluating the work done on the control volume as each unit of mass crosses the surface, rather than merely the thermo-dynamic pressure as before. Let us also use the complete shear-stress equations, Eqs. (3-1)–(3-3). This complicates the problem algebraically to a considerable degree, but in principle the procedure is no different from that used to develop Eq. (4-26), and from there to develop Eq. (4-27). In the final step, analogous to the development of Eq. (4-27), the Navier–Stokes equations are combined with the energy equation. The resulting equation can be simplified to a form in which the various terms are arranged in the same order as in Eq. (4-27), so that the origin of each term can be readily appreciated.

$$\left(\rho u \frac{\partial i}{\partial x} + \rho v \frac{\partial i}{\partial y} + \rho w \frac{\partial i}{\partial z} + \rho \frac{\partial i}{\partial \theta} \right)$$

$$- \left[\frac{\partial}{\partial x} \left(k \frac{\partial t}{\partial x} \right) + \frac{\partial}{\partial y} \left(k \frac{\partial t}{\partial y} \right) + \frac{\partial}{\partial z} \left(k \frac{\partial t}{\partial z} \right) \right]$$

$$- \left[\frac{\partial}{\partial x} \left(\sum_j \gamma_j \frac{\partial m_j}{\partial x} i_j \right) + \frac{\partial}{\partial y} \left(\sum_j \gamma_j \frac{\partial m_j}{\partial y} i_j \right) + \frac{\partial}{\partial z} \left(\sum_j \gamma_j \frac{\partial m_j}{\partial z} i_j \right) \right]$$

$$- \mu \phi - \left(u \frac{\partial P}{\partial x} + v \frac{\partial P}{\partial y} + w \frac{\partial P}{\partial z} + \frac{\partial P}{\partial \theta} \right) - S = 0 \qquad (4\text{-}30)$$

where ϕ is the *dissipation function*:

$$\phi = 2 \left[\left(\frac{\partial u}{\partial x} \right)^2 + \left(\frac{\partial v}{\partial y} \right)^2 + \left(\frac{\partial w}{\partial z} \right)^2 \right] + \left[\left(\frac{\partial u}{\partial y} + \frac{\partial v}{\partial x} \right)^2 + \left(\frac{\partial v}{\partial z} + \frac{\partial w}{\partial y} \right)^2 \right.$$

$$+ \left. \left(\frac{\partial w}{\partial x} + \frac{\partial u}{\partial z} \right)^2 \right] - \frac{2}{3} \left[\frac{\partial u}{\partial x} + \frac{\partial v}{\partial y} + \frac{\partial w}{\partial z} \right]^2$$

The terms in the first and last brackets of the dissipation function come from the viscous part of the normal stresses, whereas the second bracketed terms come from the shear stresses. Note that in the boundary-layer approximation only 1 of these 12 terms is retained.

Equation (4-30) can be written more compactly by using vector notation and the substantial derivative, thus avoiding reference to any coordinate system:

$$\rho \frac{Di}{D\theta} - \mathbf{\nabla} \cdot k \, \mathbf{\nabla} t - \mathbf{\nabla} \cdot \left(\sum_j \gamma_j i_j \, \mathbf{\nabla} m_j \right) - \mu \phi - \frac{DP}{D\theta} - S = 0 \qquad (4\text{-}31)$$

In cartesian tensor notation the general viscous energy equation becomes

$$\rho \frac{Di}{D\theta} - \frac{\partial}{\partial x_l} \left(k \frac{\partial t}{\partial x_l} \right) - \frac{\partial}{\partial x_l} \left(\sum_j \gamma_j i_j \frac{\partial m_j}{\partial x_l} \right) - \mu \phi - \frac{DP}{D\theta} - S = 0 \qquad (4\text{-}32)$$

where

$$\phi = \left(\frac{\partial u_i}{\partial x_j} + \frac{\partial u_j}{\partial x_i} - \frac{2}{3} \frac{\partial u_l}{\partial x_l} \delta_{ij} \right) \frac{\partial u_i}{\partial x_j}$$

The most general coordinate system for the continuity, Navier–Stokes, and viscous energy equations is that of curvilinear orthogonal coordinates; see Back.[4]

Various Particularizations of the Energy Equation

We have actually already considered some particular forms of the energy equation, having directly derived Eq. (4-27) for steady flow in a two-dimensional boundary layer and, by implication, Eq. (4-28) for steady axisymmetric flow and heating in a circular tube. Since we are going to be concerned with the energy equation throughout this book, it is worthwhile at this time to assemble some of the other forms of this equation that are employed later.

If we restrict consideration for the moment to *steady flow,* the time dependence of Eq. (4-31) can be dropped, and the $\rho(Di/D\theta)$ term can then be expressed in terms of the mass-flux vector. If we further limit consideration to *no source function* and to *low velocities* such that the dissipation function may be neglected (and this approximation is discussed further later) and then *neglect the pressure-gradient term* (which is often justfiable), we obtain

$$\mathbf{G} \cdot \mathbf{\nabla} i - \mathbf{\nabla} \cdot k \, \mathbf{\nabla} t - \mathbf{\nabla} \cdot \left(\sum_j \gamma_j i_j \, \mathbf{\nabla} m_j \right) = 0 \qquad (4\text{-}33)$$

and for the case of *no mass concentration gradients*

$$\mathbf{G} \cdot \mathbf{\nabla} i - \mathbf{\nabla} \cdot k \, \mathbf{\nabla} t = 0 \qquad (4\text{-}34)$$

which is similar in form to (4-23) and (4-25).

For flow in a *circular tube,* or a circular tube annulus, we are concerned with problems where the flow is hydrodynamically axisym-

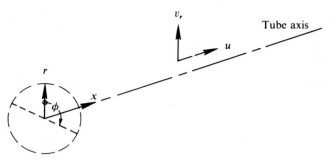

FIGURE 4-10
Coordinate system for flow in a circular tube.

metric but where the heat transfer may not be axisymmetric. By employing the cylindrical coordinate system shown on Fig. 4-10, Eq. (4-34) becomes

$$u\rho \frac{\partial i}{\partial x} + v_r\rho \frac{\partial i}{\partial r} - \left[\frac{1}{r}\frac{\partial}{\partial r}\left(rk\frac{\partial t}{\partial r}\right) + \frac{1}{r^2}\frac{\partial}{\partial \phi}\left(k\frac{\partial t}{\partial \phi}\right) + \frac{\partial}{\partial x}\left(k\frac{\partial t}{\partial x}\right)\right] = 0 \quad (4\text{-}35)$$

Note that this equation contains a conduction term in the x direction, which was neglected under the boundary-layer assumption in the development of Eq. (4-28).

For flow in *rectangular* or *triangular tubes* a cartesian coordinate system becomes convenient (Fig. 4-11), and the corresponding equation is then

$$u\rho \frac{\partial i}{\partial x} + v\rho \frac{\partial i}{\partial y} + w\rho \frac{\partial i}{\partial z} - \left[\frac{\partial}{\partial x}\left(k\frac{\partial t}{\partial x}\right) + \frac{\partial}{\partial y}\left(k\frac{\partial t}{\partial y}\right) + \frac{\partial}{\partial z}\left(k\frac{\partial t}{\partial z}\right)\right] = 0$$

$$(4\text{-}36)$$

Obviously there are a great variety of particular forms of the energy equation, depending on the idealizations that can be made in particular

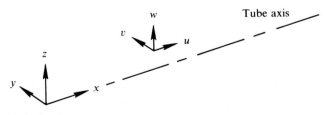

FIGURE 4-11
Coordinate system and corresponding velocity components for flow in a rectangular tube.

applications. Equations (4-30)–(4-32) provide the basic relations from which these forms can be developed.

In addition to the system geometry and the idealizations, one other factor affects the form of the energy equation: the fluid. The applicable *thermodynamic equation of state* has a rather considerable influence on the form that the energy equation ultimately takes, for the enthalpy i must be related to the other fluid properties. For example, let us go back to the boundary-layer energy equation (4-27) and consider the case of a perfect gas. Then

$$di = c\,dt$$

where c is understood to represent the constant-pressure specific heat.

Let us omit mass diffusion for brevity and note that for a gas the dP/dx term can generally be shown to be negligible for cases where the stagnation enthalpy equation does not have to be used. Then Eq. (4-27) becomes

$$u\rho c\,\frac{\partial t}{\partial x} + v\rho c\,\frac{\partial t}{\partial y} - \frac{\partial}{\partial y}\left(k\,\frac{\partial t}{\partial y}\right) - \mu\left(\frac{\partial u}{\partial y}\right)^2 = 0 \qquad (4\text{-}37)$$

On the other hand, for an *incompressible liquid,* $\rho = $ constant, and

$$di = c\,dt + \frac{1}{\rho}\,dP$$

Substituting in Eq. (4-27) and again omitting mass diffusion but *not* the pressure-gradient term, we obtain

$$\rho u\left(c\,\frac{\partial t}{\partial x} + \frac{1}{\rho}\frac{dP}{dx}\right) + \rho v\left(c\,\frac{\partial t}{\partial y} + 0\right) - \frac{\partial}{\partial y}\left(k\,\frac{\partial t}{\partial y}\right) - \mu\left(\frac{\partial u}{\partial y}\right)^2 - u\,\frac{dP}{dx} = 0$$

This time the pressure-gradient term cancels without any assumptions about its influence, and we again obtain Eq. (4-37).

If the thermal conductivity k can be treated as constant, which is a quite good assumption for liquids and a reasonable assumption for a gas with small temperature gradients, Eq. (4-37) becomes

$$u\,\frac{\partial t}{\partial x} + v\,\frac{\partial t}{\partial y} - \frac{k}{\rho c}\frac{\partial^2 t}{\partial y^2} - \frac{\mu}{\rho c}\left(\frac{\partial u}{\partial y}\right)^2 = 0$$

Defining $\alpha = k/\rho c$, the *thermal diffusivity* of the fluid, and introducing the Prandtl number, $\mathrm{Pr} = \mu c/k$, we obtain

$$u\,\frac{\partial t}{\partial x} + v\,\frac{\partial t}{\partial y} - \alpha\left[\frac{\partial^2 t}{\partial y^2} + \frac{\mathrm{Pr}}{c}\left(\frac{\partial u}{\partial y}\right)^2\right] = 0 \qquad (4\text{-}38)$$

Here we see the dissipation function (the last term) depends not only on the velocity but also on the Prandtl number. Viscous energy dissipation can be important for high-Prandtl-number fluids (oils, for example), even for rather moderate velocities and velocity gradients. For gases, on the other hand, where the Prandtl number is near 1, we shall see later that the velocity must approach the speed of sound before this term is significant.

Under conditions where viscous energy dissipation can be neglected, Eq. (4-38) finally becomes

$$ u\frac{\partial t}{\partial x} + v\frac{\partial t}{\partial y} - \alpha\frac{\partial^2 t}{\partial y^2} = 0 \qquad (4\text{-}39) $$

and integration of this equation, or its equivalent in other coordinate systems, for various boundary conditions comprises a substantial part of some of the later chapters.

PROBLEMS

4-1. Consider steady flow of a constant-property fluid in a long duct formed by two parallel planes. Consider a point sufficiently far removed from the duct entrance that the y component of velocity is zero and the flow is entirely in the x direction. Write the Navier–Stokes equations for both the x and y directions. What can you deduce about the pressure gradients?

4-2. Show that for a constant-property viscous fluid flowing at extremely low velocities, so that all inertia terms may be neglected, and in which there are no body forces, the Navier–Stokes equations may be reduced to $\nabla^2 P = 0$.

4-3. Consider steady axisymmetric flow in the entrance region of a pipe, that is, in the region in which the velocity profile is developing. Starting with Eq. (4-11) and assuming that along the centerline the influence of viscosity is negligible, derive a relation between centerline axial velocity and pressure.

4-4. Consider flow in the eccentric annulus of a journal bearing in which there is no axial flow. Deduce the applicable laminar boundary-layer equations (continuity, momentum, energy) for a constant-density fluid, with uniform composition, in an appropriate coordinate system.

4-5. Deduce a set of boundary-layer differential equations (continuity, momentum, energy) for steady flow of a constant-property fluid without body forces, and with negligible viscous dissipation, in a coordinate system suitable for analysis of the boundary layer on the surface of a rotating disk.

4-6. Starting with the general viscous energy equation, show by a succession of steps how and why it reduces to the classic heat-conduction equation for a solid, and finally to the Laplace equation.

4-7. Derive Eq. (4-18) using Eq. (4-1) and the definitions from the Fick's law section of Chap. 3.

4-8 Derive the conservation laws for axisymmetric flow in a pipe using control volume principles similar to that developed in the text for cartesian coordinate flow. Assume steady flow, steady state, and constant properties. For the momentum equation neglect body forces. For the energy equation neglect body force work and the energy source. Also, for the energy equation use the thermal equation of state for ideal gases or incompressible liquids, and assume the Mach number is small for the case of the ideal gas, but do not neglect viscous dissipation. Assume boundary-layer flow, but do not neglect axial conduction.

4-9. Derive the constant-property energy equation (4-39) starting with Eq. (4-26). Be sure to state all assumptions made.

REFERENCES

1. Deissler, R. G.: *Am. J. Phys.*, vol. 44, pp. 1128–1130, 1976.
2. Schlichting, H.: *Boundary Layer Theory*, 6th ed., pp. 117–121, McGraw-Hill, New York, 1968.
3. Streeter, V. L.: *Fluid Dynamics*, pp. 242–243, McGraw-Hill, New York, 1948.
4. Back, L. H.: TR 32-1332, Jet Propulsion Laboratory, California Institute of Technology, Pasadena, September 15, 1968. (See also *Handbook of Tables for Applied Engineering Science*, 2d ed., pp. 496–501, Chemical Rubber Company Press, Cleveland, Ohio, 1973.)

CHAPTER
5

THE DIFFERENTIAL EQUATIONS OF THE TURBULENT BOUNDARY LAYER

In this chapter we develop the differential equations applicable to turbulent boundary-layer flows, often called the *Reynolds-averaged equations*. These evolve from the same basic equations developed for laminar flows, namely the conservation-of-mass equation, the Navier–Stokes equations, and the conservation-of-energy equation. However, they differ in basic form by the addition of the time-dependent term, reflecting the fact that turbulent flows are inherently unsteady. For a large class of turbulent flows the unsteady equations can be greatly simplified through a time-averaging process to yield equations similar to those for steady laminar flows. The consequence of this reformulation is the introduction of new unknowns within the equations, called the turbulent or Reynolds stresses for momentum and the turbulent heat fluxes for energy.

The chapter begins by presenting the variables and differential equations applicable to a time-dependent turbulent flow. The concepts of Reynolds decomposition and time-averaging are then developed, followed by formulation of the variable-property time-averaged turbulent flow equations, along with their two-dimensional boundary-layer forms. Kinetic energy equations for the mean flow and turbulence field are then developed, along with discussion of the dissipation transport equation. The turbulence kinetic energy and dissipation transport equations form the basis for the $k-\varepsilon$ two-equation turbulence model as an alternative to the conventional Prandtl mixing-length model of turbulence.

MOMENTUM AND THERMODYNAMIC VARIABLES

The momentum variables in the turbulent equations are identical in typestyle to their laminar counterparts, but are considered to represent instantaneous quantities as a function of position (x, y, z) and time (θ). When the same variables appear with overbars they represent mean values as a function of position only, and when the variables appear primed they represent time-dependent fluctuations about their respective mean values. The momentum variables are generalized using index notation, since development of the equations for turbulent flow is more easily carried out in this way, and the vector concepts likewise are generalized using tensors. Most of the thermodynamic variables remain simple because of their scalar nature.

The cartesian coordinate directions (x, y, z) in index notation become (x_1, x_2, x_3), and the associated velocity components (u, v, w) become (u_1, u_2, u_3). Commonly used momentum variables include the x_i-direction mass flux ρu_i, the x_i-direction momentum per unit mass u_i, and the body force per unit volume X_i. Thermodynamic and energy variables associated with the turbulent flow equations include the pressure P, temperature t, enthalpy i, internal energy e, density ρ, and specific heat c. Molecular transport coefficients include the dynamic viscosity μ and the thermal conductivity k. The kinetic energy per unit mass is $\frac{1}{2}u_i u_i$, and S is a volumetric energy source term.

NEWTONIAN STRESS AND FOURIER HEAT-FLUX MODELS

The Newtonian stress model for turbulent flow is of the same form as Eq. (3-7) for laminar flow, but the terms are functions of both space and time. It is a second-order tensor of the form

$$\sigma_{ij} = -P\delta_{ij} + \tau_{ij} \tag{5-1}$$

where δ_{ij} is the Kronecker delta, a unit second-order tensor, and the viscous stress tensor τ_{ij} reflects the deformations caused by spatial velocity gradients $\partial u_i/\partial x_j$. These gradients are a form of second-order tensor, which can be decomposed into the sum of a symmetric tensor S_{ij}

and an antisymmetric tensor Ω_{ij}:

$$\frac{\partial u_i}{\partial x_j} = S_{ij} + \Omega_{ij} \qquad (5\text{-}2)$$

Here S_{ij} is the strain-rate tensor, related to the deformation of the fluid, while the tensor Ω_{ij} relates to the rotation of the fluid. They are defined as

$$S_{ij} = \frac{1}{2}\left(\frac{\partial u_i}{\partial x_j} + \frac{\partial u_j}{\partial x_i}\right) \qquad (5\text{-}3)$$

and

$$\Omega_{ij} = \frac{1}{2}\left(\frac{\partial u_i}{\partial x_j} - \frac{\partial u_j}{\partial x_i}\right) \qquad (5\text{-}4)$$

Based on the assumptions that the viscous stress field is linearly proportional to the strain-rate field, is symmetric, and obeys the Stokes hypothesis, the viscous stress tensor becomes

$$\tau_{ij} = 2\mu S_{ij} - \tfrac{2}{3}\mu S_{kk}\delta_{ij} \qquad (5\text{-}5)$$

and Eq. (5-1) becomes

$$\sigma_{ij} = -P\delta_{ij} + 2\mu S_{ij} - \tfrac{2}{3}\mu S_{kk}\delta_{ij} \qquad (5\text{-}6)$$

The Fourier heat-flux model is of a form similar to the vector equation (3-8), but $\dot{q}''$ is temporarily replaced by $\dot{q}/A$ to avoid confusion with the primed quantities. For heat flux in the x_i direction

$$\left(\frac{\dot{q}}{A}\right)_i = -k\frac{\partial t}{\partial x_i} \qquad (5\text{-}7)$$

INSTANTANEOUS EQUATIONS OF TURBULENCE

The turbulence equations are similar to the laminar flow equations of Chap. 4, and in some instances they are identical, although perhaps rearranged. The time-derivative term in each equation reflects the unsteady nature of the turbulent flow.

Conservation of mass is identical with Eq. (4-4):

$$\frac{\partial \rho}{\partial \theta} + \frac{\partial}{\partial x_j}(\rho u_j) = 0 \qquad (5\text{-}8)$$

The *x_i-momentum equation* follows Eq. (4-17), and becomes

$$\frac{\partial}{\partial \theta}(\rho u_i) + \frac{\partial}{\partial x_j}(\rho u_j u_i) = \frac{\partial}{\partial x_j}(\sigma_{ji}) + X_i \qquad (5\text{-}9)$$

or, with the stress model (5-1),

$$\frac{\partial}{\partial \theta}(\rho u_i) + \frac{\partial}{\partial x_j}(\rho u_j u_i) = \frac{\partial}{\partial x_j}(-P\delta_{ij} + \tau_{ji}) + X_i \qquad (5\text{-}10)$$

The *conservation of energy* for a single species can be written using either stagnation enthalpy or total energy, i.e., internal energy plus kinetic energy. The former is typically used in convective heat transfer. The latter allows insight into the various forms of energy transfer between the mean flow field and the turbulence field, as well as transfer between mechanical and thermal energy forms, which eventually leads to an increase in entropy. Both forms of the equation evolve from each other by separating the pressure part of the stress tensor to form a flow work term. The *stagnation enthalpy equation* form is similar to that developed in Eq. (4-26):

$$\frac{\partial}{\partial \theta}\left[\rho\left(i^* - \frac{P}{\rho}\right)\right] + \frac{\partial}{\partial x_j}(\rho u_j i^*) = \frac{\partial}{\partial x_j}\left(-\frac{\dot{q}}{A}\right)_j + \frac{\partial}{\partial x_j}(u_i \tau_{ji}) + u_i X_i + S$$

$$(5\text{-}11)$$

where

$$i^* = e + \frac{P}{\rho} + \tfrac{1}{2}u_i u_i = i + \tfrac{1}{2}u_i u_i$$

Development of a *mechanical energy equation* is carried out by multiplying the x_i-momentum equation by the velocity component u_i and invoking the conservation of mass on the convective terms to yield

$$\frac{\partial}{\partial \theta}\left(\rho\frac{u_i u_i}{2}\right) + \frac{\partial}{\partial x_j}\left(\rho u_j \frac{u_i u_i}{2}\right) = u_i \frac{\partial}{\partial x_j}(\sigma_{ji}) + u_i X_i \qquad (5\text{-}12)$$

A *thermal energy equation* is developed in a manner similar to that carried out to develop the static enthalpy equation (4-27). It is a split equation, whereby the kinetic energy component is removed from the total energy equation by subtracting the mechanical energy equation (5-12).

REYNOLDS DECOMPOSITION

The time dependence of a turbulent flow is the result of relatively small-scale vorticity in the flow; virtually every fluid particle is a part of a so-called turbulent eddy or structure, of which there are a very large number in a typical flow. A mechanical or optical experimental measurement at a point within a turbulent flow reveals a velocity that fluctuates in a random manner around a steady time-independent velocity. Any attempt to completely describe the velocity field, either through measurements or numerical simulations, as a function of time is difficult to impossible, except for simple flows at relatively low Reynolds numbers. On the other hand, because the fluctuating velocity components for a majority of turbulent flows tend to be small relative to the mean velocity, it appears practicable to adopt a statistical approach to describe the effects of the time-dependent components of the flow, and thereby treat the flow as if it were steady.

Figure 5-1 shows a way of describing the velocity in a turbulent flow, reduced for clarity to a two-dimensional flow. $\bar{\mathbf{V}}$ is a vector representing the steady velocity component, and $\mathbf{V}'$ is the fluctuating component, which varies with time but which must be zero when averaged over time, in keeping with the statistical approach. The total instantaneous velocity is $\mathbf{V}$. The figure shows a cartesian decompositon of the instantaneous velocity into components in the x direction, or x_1 direction, and in the y direction, or x_2 direction. The x component is composed of the steady quantity $\bar{u}$ and its fluctuation quantity u', and the y component is composed of the steady quantity $\bar{v}$ and its fluctuation quantity v'. For two-dimensional turbulent flows experiments show that there exists a fluctuation quantity w' in the z direction, or x_3 direction, although $\bar{w}$ is zero. The instantaneous velocity components u_i for $i = 1, 2, 3$ can be expressed as

$$u = \bar{u} + u', \qquad v = \bar{v} + v', \qquad w = \bar{w} + w' \qquad (5\text{-}13)$$

These are referred to as the (Osborne) Reynolds decompositions. Likewise, scalar quantities such as pressure and temperature also fluctuate, and they are expressed in terms of decompositions as

$$P = \bar{P} + P', \qquad t = \bar{t} + t' \qquad (5\text{-}14)$$

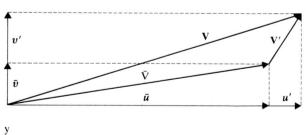

FIGURE 5-1
Velocity components for a turbulent flow (shown in two dimensions only).

The enthalpy terms need special consideration. Both the stagnation enthalpy and the static enthalpy have Reynolds decompositions similar to that for temperature:

$$i^* = \overline{i^*} + i^{*\prime}, \qquad i = \overline{i} + i' \tag{5-15}$$

Substituting decompositions for i and u_i into the definition $i^* = i + \frac{1}{2}u_iu_i$ and comparing with the decomposition for i^* leads to the following definitions for mean and fluctuating components:

$$\overline{i^*} = \overline{i} + \frac{1}{2}\bar{u}_i\bar{u}_i, \qquad i^{*\prime} = i' + \bar{u}_iu_i' + \frac{1}{2}u_i'u_i' \tag{5-16}$$

If the definition of $\overline{i^*}$ is to include the component $\frac{1}{2}\overline{u_i'u_i'}$, which is the mean value of the turbulence kinetic energy, then this component must appear in the definition of $i^{*\prime}$ with opposite sign.

Themodynamic and thermophysical properties such as density, dynamic viscosity, specific heat, and thermal conductivity can also fluctuate in response to the pressure and temperature fields. Arguments for neglecting the density fluctuations are based primarily on arguments related to the Mach number of the turbulence, as discussed in Hinze[2] or Cebeci and Bradshaw.[3] For buoyancy-driven flows, or flows with large absolute temperature differences, the fluctuating density field cannot be ignored. For the treatment of turbulence in this book, however, we will assume that fluctuations of all properties are small, and thus all properties will appear without an overbar to signify the lack of consideration of their fluctuations.

Decompositions involving velocity gradients and temperature gradients are straightforward. However, it becomes convenient to define the strain rate tensor, stress tensor, and viscous stress tensor decompositions

$$S_{ij} = \bar{S}_{ij} + s_{ij}', \qquad \sigma_{ij} = \bar{\sigma}_{ij} + \sigma_{ij}', \qquad \tau_{ij} = \bar{\tau}_{ij} + \tau_{ij}' \tag{5-17}$$

Body force and source terms have the decompositions

$$X_i = \bar{X}_i + X_i', \qquad S = \bar{S} + S' \tag{5-18}$$

TIME-AVERAGING AND TURBULENCE STATISTICS

If we assume that the flow has a measurable continuous time history, we can define a *time-averaging operator* to create statistics of the flow. Consider some variable or property of the flow, $f(\mathbf{x}, \theta)$. The statistical mean value of that variable or property at $\mathbf{x}$, where $\mathbf{x}$ is the vector location (x_1, x_2, x_3), is

$$\bar{f}(\mathbf{x}) = \lim_{\theta \to \infty} \frac{1}{\theta} \int_{\theta_0}^{\theta_0 + \theta} f(\mathbf{x}, \theta)\, d\theta \tag{5-19}$$

The variable $\bar{f}(\mathbf{x})$ exists if the limit converges, and it is stationary if it does not depend on the start time of the averaging process, θ_0. An alternative to Eq. (5-19) would be to consider ensemble-averaging, whereby a large number of measurements or samples of f are collected and averaged.

Working with the velocity component u, we can form its mean value at a location $\mathbf{x}$ by averaging its instantaneous variation over a long time:

$$\bar{u} = \frac{1}{\theta} \int_{\theta_0}^{\theta_0 + \theta} u(\mathbf{x}, \theta) \, d\theta \qquad (5\text{-}20)$$

From the definition of the Reynolds decomposition we require that the fluctuations about the mean be zero, i.e.,

$$\overline{u'} = \frac{1}{\theta} \int_{\theta_0}^{\theta_0 + \theta} [u(\mathbf{x}, \theta) - \bar{u}] \, d\theta = \frac{1}{\theta} \int_{\theta_0}^{\theta_0 + \theta} u' \, d\theta$$

and

$$\overline{u'} = 0$$

The variance, or second moment, of u' at a location $\mathbf{x}$ is obtained by changing the integrand in Eq. (5-19) to $[u(\mathbf{x}, \theta) - \bar{u}]^2$ and carrying out the time-averaging process:

$$\overline{u'u'} = \overline{u'^2} = \overline{u_1'u_1'} = \frac{1}{\theta} \int_{\theta_0}^{\theta_0 + \theta} [u(\mathbf{x}, \theta) - \bar{u}]^2 \, d\theta$$

The double-velocity correlation of the components u and v at a single point $\mathbf{x}$ in the flow is similar in concept to the covariance in statistics. It is obtained by carrying out the time-averaging process:

$$\overline{u'v'} = \overline{u_1'u_2'} = \frac{1}{\theta} \int_{\theta_0}^{\theta_0 + \theta} [u(\mathbf{x}, \theta) - \bar{u}][v(\mathbf{x}, \theta) - \bar{v}] \, d\theta$$

The variance for any of the velocity components u_i at a point will be a second-order tensor. Likewise, the double-velocity correlation between velocity components u_i and u_j at a point will also be a second-order tensor. Triple single-point correlations are formed using any of the three components, yielding a third-order tensor. Similarly, double- and triple-correlation statistics involving one or more velocity components and a thermodynamic variable can also be formed.

With these operations defined, we can state some simple averaging rules, which are readily developed by simple substitution:

$$\bar{\bar{u}} = \bar{u} \quad \text{(mean rule)}, \qquad \overline{u + v} = \bar{u} + \bar{v} \quad \text{(sum rule)}, \qquad \overline{u'} = 0, \qquad \overline{\bar{u}u'} = 0$$

$$\overline{uv} = \overline{(\bar{u} + u')(\bar{v} + v')} = \overline{\bar{u}\bar{v} + \bar{u}v' + \bar{v}u' + u'v'} = \overline{\bar{u}\bar{v}} + \overline{u'v'} = \bar{u}\bar{v} + \overline{u'v'}$$

$$\overline{u^2} = \overline{(\bar{u} + u')(\bar{u} + u')} = \overline{\bar{u}\bar{u} + \bar{u}u' + \bar{u}u' + u'u'} = \overline{\bar{u}\bar{u}} + \overline{u'u'} = \bar{u}\bar{u} + \overline{u'u'}$$

Assuming that the turbulence field is such that mean quantities are independent of the start of the averaging process, integration and differentiation can be interchanged:

$$\frac{\overline{\partial u}}{\partial x} = \frac{\partial \bar{u}}{\partial x} \quad \text{(derivative rule)}$$

These examples and rules for averaging are extended in a straightforward way to the variables written using indices.

REYNOLDS-AVERAGED TRANSPORT EQUATIONS OF TURBULENCE

Development of the Reynolds-averaged equations to describe turbulent flows that have a statistically steady mean is achieved by substituting the various velocity and scalar Reynolds decompositions into the instantaneous equations and operating on the equations with the time-averaging operator, Eq. (5-19). Equations for the fluctuating components can then be developed by subtracting the mean equations from their instantaneous forms. Transport equations for the various turbulence statistics are obtained by manipulating of the fluctuating component equations and then time-averaging the residual equations. These become the transport equations for various turbulence quantities such as turbulence kinetic energy, dissipation, Reynolds stresses, and turbulent heat-flux components.

CONSERVATION OF MASS

This equation is obtained by first substituting the Reynolds decomposition velocity terms into the instantaneous conservation-of-mass equation (5-8):

$$\frac{\partial \rho}{\partial \theta} + \frac{\partial}{\partial x}[\rho(\bar{u} + u')] + \frac{\partial}{\partial y}[\rho(\bar{v} + v')] + \frac{\partial}{\partial z}[\rho(\bar{w} + w')] = 0$$

which, upon separating components, becomes

$$\frac{\partial \rho}{\partial \theta} + \frac{\partial}{\partial x}(\rho \bar{u}) + \frac{\partial}{\partial x}(\rho u') + \frac{\partial}{\partial y}(\rho \bar{v}) + \frac{\partial}{\partial y}(\rho v') + \frac{\partial}{\partial z}(\rho \bar{w}) + \frac{\partial}{\partial z}(\rho w') = 0$$

Now, the entire equation is averaged over time using Eq. (5-19), the sum rule is applied to break up the averaging into individual terms, the derivative rule is applied to each term, and, on dropping the time term, the resulting equation becomes

$$\frac{\partial}{\partial x}(\overline{\rho \bar{u}}) + \frac{\partial}{\partial x}(\overline{\rho u'}) + \frac{\partial}{\partial y}(\overline{\rho \bar{v}}) + \frac{\partial}{\partial y}(\overline{\rho v'}) + \frac{\partial}{\partial z}(\overline{\rho \bar{w}}) + \frac{\partial}{\partial z}(\overline{\rho w'}) = 0$$

Applying the mean rule for time-averages of mean quantities, and noting that the time-average of the product of a mean quantity and a fluctuating quantity is zero, the resulting Reynolds-averaged conservation-of-mass equation is

$$\frac{\partial}{\partial x}(\rho\bar{u}) + \frac{\partial}{\partial y}(\rho\bar{v}) + \frac{\partial}{\partial z}(\rho\bar{w}) = 0 \qquad (5\text{-}21)$$

or, in index notation,

$$\frac{\partial}{\partial x_j}(\rho\bar{u}_j) = 0 \qquad (5\text{-}22)$$

Note that for constant-density turbulent flow the density term can be removed from the spatial-derivative term, with the resulting equation being the same as Eq. (4-6). Also, for constant-density flow, an equation for the fluctuating velocity components is obtained by substracting the Reynolds-averaged form from the instantaneous form, ignoring the time variation of density, to yield

$$\frac{\partial}{\partial x_j}(\rho u_j') = 0 \qquad (5\text{-}23)$$

CONSERVATION OF MASS FOR THE TURBULENT BOUNDARY LAYER

If we now restrict consideration to the simple two-dimensional turbulent boundary layer with constant density, the conservation-of-mass equation becomes

$$\frac{\partial\bar{u}}{\partial x} + \frac{\partial\bar{v}}{\partial y} = 0 \qquad (5\text{-}24)$$

MOMENTUM TRANSPORT EQUATION IN THE x_i-DIRECTION

Development of this equation proceeds by first substituting the Reynolds decompositions into the instantaneous x_i-direction momentum equation (5-9), time-averaging, and then applying the derivative rule:

$$\frac{\partial}{\partial x_j}[\overline{\rho(\bar{u}_j + u_j')(\bar{u}_i + u_i')}] = \frac{\partial}{\partial x_j}(\bar{\sigma}_{ji} + \overline{\sigma_{ji}'}) + \overline{X}_i + \overline{X_i'}$$

Again, recall that density does not carry an overbar, reminding us that we are not considering a density decomposition. Should fluctuations in density need to be considered, we would have to mass- or Favre-average rather than time-average.[1] Application of the time-averaging rules leads to the following form of the x_i-momentum equation:

$$\frac{\partial}{\partial x_j}(\rho \bar{u}_j \bar{u}_i + \rho \overline{u_j' u_i'}) = \frac{\partial}{\partial x_j}(\bar{\sigma}_{ji}) + \bar{X}_i \qquad (5\text{-}25)$$

Several variations of the x_i-momentum equation must be developed to assist in formation of other turbulence equations and in turbulent transport models. The first variation follows from decomposition of the stress tensor into the pressure term and the viscous stress tensor, coupled with movement of the new convective term to the right-hand side of the equation:

$$\frac{\partial}{\partial x_j}(\rho \bar{u}_j \bar{u}_i) = -\frac{\partial \bar{P}}{\partial x_i} + \frac{\partial}{\partial x_j}(\bar{\tau}_{ji} - \rho \overline{u_j' u_i'}) + \bar{X}_i \qquad (5\text{-}26)$$

where the pressure term has been converted using the relationship $\partial(-\bar{P}\delta_{ij})/\partial x_j = -\partial \bar{P}/\partial x_i$.

This form of the momentum equation gives rise to an analogy between the viscous stress tensor and the term $-\rho \overline{u_j' u_i'}$, *a turbulent stress tensor*, although in the turbulence literature this tensor often appears without the minus sign. It is a set of nine terms: $j = 1, 2, 3$ for each of the $i = 1, 2, 3$, or x-, y-, z-direction momentum equations, although symmetry arguments will reduce this to a set of six terms. While the $\rho \overline{u_i' u_j'}$ are derived mathematically as a consequence of time-averaging the x_i-momentum transport equation for turbulence, they are easily quantified by experimental techniques. They are highly nonlinear and important, reflecting the inherently time-dependent character of the turbulent flow. With the minus sign dropped, the term $\rho \overline{u_j' u_i'}$ is called the *Reynolds stress tensor*, and often given the second-order tensor symbol R_{ij}.

MOMENTUM EQUATION FOR THE TURBULENT BOUNDARY LAYER

If we now restrict Eq. (5-26) to the x-direction, the turbulent stress term for $i = 1$ becomes

$$\frac{\partial}{\partial x_j}(-\rho \overline{u_j' u_{i=1}'}) = \frac{\partial}{\partial x_1}(-\rho \overline{u_1' u_1'}) + \frac{\partial}{\partial x_2}(-\rho \overline{u_2' u_1'}) + \frac{\partial}{\partial x_3}(-\rho \overline{u_3' u_1'})$$

$$= \frac{\partial}{\partial x}(-\rho \overline{u' u'}) + \frac{\partial}{\partial y}(-\rho \overline{v' u'}) + \frac{\partial}{\partial z}(-\rho \overline{w' u'})$$

In the two-dimensional boundary layer the $\overline{w' u'}$ term is found from experimental evidence to be essentially zero, although w' exists for all turbulent flows due to the nature of the fluctuating vorticity. The variance $\overline{w'^2}$ is an important quantity, and is part of the turbulence kinetic energy of the flow. The term $\overline{u' u'} = \overline{u'^2}$ is likewise an important quantity and part of the turbulence kinetic energy of the flow, although for two-dimensional boundary-layer flows its streamwise gradient is found to be negligible except in the vicinity of the stagnation point or near boundary-layer separation. It is eliminated in keeping with the boundary-layer approximations. Likewise, the $\bar{\tau}_{ji}$ tensor for $i = 1$ is reduced to a single component τ_{yx}, reflecting the requirement to only consider y-direction gradients in the boundary-layer approximations.

For two-dimensional "steady" turbulent boundary-layer flows the x_i-momentum equation (5-26) for $i = 1$ is reduced using the boundary-layer approximations, and combined with the conservation-of-mass equation (5-22) to form the standard variable-property form

$$\rho \bar{u} \frac{\partial \bar{u}}{\partial x} + \rho \bar{v} \frac{\partial \bar{u}}{\partial y} = -\frac{d\bar{P}}{dx} + \frac{\partial}{\partial y} (\tau - \overline{\rho u'v'}) + \bar{X} \tag{5-27}$$

For constant-density flow with negligible body forces the boundary-layer momentum equation is recast using Newton's law of viscosity to become

$$\bar{u} \frac{\partial \bar{u}}{\partial x} + \bar{v} \frac{\partial \bar{u}}{\partial y} = -\frac{1}{\rho} \frac{d\bar{P}}{dx} + \frac{\partial}{\partial y} \left(v \frac{\partial \bar{u}}{\partial y} - \overline{u'v'} \right) \tag{5-28}$$

STAGNATION ENTHALPY TRANSPORT EQUATION

Development of this equation proceeds by first substituting the various Reynolds decompositions into the instantaneous stagnation enthalpy equation (5-11), time-averaging, and then applying the derivative rule:

$$\frac{\partial}{\partial x_j} \overline{[\rho(\bar{u}_j + u_j')(i^* + i^{*\prime})]} = \frac{\partial}{\partial x_j} \left[\overline{\left(-\frac{\dot{q}}{A} \right)_j} + \overline{\left(-\frac{\dot{q}}{A} \right)_j'} \right]$$

$$+ \frac{\partial}{\partial x_j} \overline{[(\bar{u}_i + u_i')(\bar{\tau}_{ji} + \tau_{ji}')]}$$

$$+ \overline{(\bar{u}_i + u_i')(\bar{X}_i + X')} + \overline{(\bar{S} + S')}$$

Application of the time-averaging rules leads to the following form of the stagnation enthalpy equation:

$$\frac{\partial}{\partial x_j} (\rho \overline{\bar{u}_j i^*} + \rho \overline{u_j' i^{*\prime}}) = \frac{\partial}{\partial x_j} \overline{\left(-\frac{\dot{q}}{A} \right)_j} + \frac{\partial}{\partial x_j} (\bar{u}_i \bar{\tau}_{ji} + \overline{u_i' \tau_{ji}'})$$

$$+ \bar{u}_i \bar{X}_i + \overline{u_i' X'} + \bar{S} \tag{5-29}$$

The new convective transport term as a result of the time-averaging can be expanded using the definition of $i^{*\prime}$ to yield

$$\overline{\rho u_j' i^{*\prime}} = \overline{\rho u_j' i'} + \rho \overline{\bar{u}_i u_j' u_i'} + \rho \overline{u_j' \frac{u_i' u_i'}{2}} \tag{5.30}$$

Substitution of the expanded convective term into the time-averaged equation and rearrangement leads to the general form of the stagnation enthalpy equation:

$$\frac{\partial}{\partial x_j}(\rho \overline{u_j} \, \overline{i^*}) = \frac{\partial}{\partial x_j}\left[\left(-\frac{\dot{q}}{A}\right)_j - \overline{\rho u_j' i'}\right] + \frac{\partial}{\partial x_j}[\bar{u}_i(\bar{\tau}_{ji} - \overline{\rho u_j' u_i'})]$$

$$+ \frac{\partial}{\partial x_j}\left[\overline{u_i'\left(\tau_{ji}' + \rho \frac{u_j' u_i'}{2}\right)}\right] + \bar{u}_i \bar{X}_i + \overline{u_i' X'} + \bar{S} \qquad (5\text{-}31)$$

This form of the stagnation enthalpy equation gives rise to an analogy between the molecular heat-flux vector and the term $-\overline{\rho u_j' i'}$, with the latter being considered as a turbulent *enthalpy flux vector*. Like its counterpart $-\overline{\rho u_i' u_j'}$, it is mathematically derived as a consequence of making the stagnation enthalpy transport equation for turbulence a time-independent equation. The other new terms include the various work and dissipation terms, along with the $\overline{u_i' X'}$ body force term and volumetric source term.

ENERGY EQUATIONS FOR THE TURBULENT BOUNDARY LAYER

This equation is developed in a manner similar to the boundary-layer momentum equation. We first consider the turbulent enthalpy-flux term. As with the momentum equation, the z-component term is found from experimental evidence to be negligible, due principally to there being no gradients in mean value of energy in the z-direction. The x-direction gradient is neglected primarily because of the boundary-layer approximations, although it may be important for low-Pr turbulent flows with significant wall temperature gradients in the flow direction. The stagnation enthalpy equation (5-31) is reduced by applying the boundary-layer approximations, combined with the conservation of mass equation (5-22), and neglecting the higher-order fluctuation correlation, to yield

$$\rho \bar{u} \frac{\partial \overline{i^*}}{\partial x} + \rho \bar{v} \frac{\partial \overline{i^*}}{\partial y} = \frac{\partial}{\partial y}(-\dot{q}'' - \overline{\rho v' i'}) + \frac{\partial}{\partial y}[\bar{u}(\tau - \overline{\rho u' v'})]$$

$$+ \bar{u}\bar{X} + \overline{u' X'} + \bar{S} \qquad (5\text{-}32)$$

where the molecular heat-flux term symbol has been replaced with the conventional double-prime notation to indicate a flux or per unit area.

For two-dimensional boundary-layer flow with low velocities, no fluctuating or mean work terms, no body forces or sources, and with $de = c \, dT$, Eq. (5-32) reduces to

$$\rho c \bar{u} \frac{\partial \bar{t}}{\partial x} + \rho c \bar{v} \frac{\partial \bar{t}}{\partial y} = \frac{\partial}{\partial y}(-\dot{q}'' - \rho c \overline{v' t'}) \qquad (5\text{-}33)$$

and for constant-density flow with constant specific heat the two-dimensional boundary-layer energy equation for turbulent flow with Fourier's law for molecular heat flux becomes

$$\bar{u}\frac{\partial \bar{t}}{\partial x} + \bar{v}\frac{\partial \bar{t}}{\partial y} = \frac{\partial}{\partial y}\left(\alpha\frac{\partial \bar{t}}{\partial y} - \overline{v't'}\right) \tag{5-34}$$

MECHANICAL ENERGY TRANSPORT EQUATION

This equation considers the transport of the mean kinetic energy per unit mass of the flow, and it is derived by multiplying the time-averaged x_i-direction momentum-transport equation (5-26) by $\bar{u}_i$ and applying the conservation-of-mass equation (5-22) to the convective term to yield

$$\frac{\partial}{\partial x_j}\left(\rho\bar{u}_j\frac{\bar{u}_i\bar{u}_i}{2}\right) = \bar{u}_i\frac{\partial \bar{P}}{\partial x_i} + \bar{u}_i\frac{\partial}{\partial x_j}(\bar{\tau}_{ji} - \rho\overline{u'_j u'_i}) + \bar{u}_i\bar{X}_i$$

Application of the chain rule to the stress terms leads to the final form of the mean mechanical energy equation:

$$\frac{\partial}{\partial x_j}\left(\rho\bar{u}_j\frac{\bar{u}_i\bar{u}_i}{2}\right) = \frac{\partial(\bar{u}_i\bar{P})}{\partial x_i} - \bar{P}\frac{\partial \bar{u}_i}{\partial x_i} + \frac{\partial}{\partial x_j}(\bar{u}_i\bar{\tau}_{ji}) - \bar{\tau}_{ji}\frac{\partial \bar{u}_i}{\partial x_j}$$
$$+ \frac{\partial}{\partial x_j}[\bar{u}_i(-\rho\overline{u'_j u'_i})] - (-\rho\overline{u'_j u'_i})\frac{\partial \bar{u}_i}{\partial x_j} + \bar{u}_i\bar{X}_i \tag{5-35}$$

The major significance of this rearranged form (5-35) is that we can identify stress–velocity gradient products, which usually represent source or sink terms in conversion or energy from one form to another, and velocity–stress products, which represent work rate or power terms.

TURBULENCE KINETIC ENERGY TRANSPORT EQUATION

This equation considers the turbulence kinetic energy per unit mass of the flow, and it is derived as follows. First a *transport equation for the fluctuating velocity* is created by inserting the various Reynolds decompositions into the instantaneous x_i-direction momentum equation (5-9) and then subtracting from it the time-averaged momentum equation (5-25) and ignoring the time term:

$$\frac{\partial}{\partial x_j}(\rho\bar{u}_j u'_i + \rho u'_j \bar{u}_i + \rho u'_j u'_i) = \frac{\partial}{\partial x_j}(\sigma'_{ji}) - \frac{\partial}{\partial x_j}(-\rho\overline{u'_j u'_i}) + X'_i \tag{5-36}$$

Second, the transport equation for the fluctuating velocity is multiplied by u_i':

$$u_i' \frac{\partial}{\partial x_j} (\rho \bar{u}_j u_i' + \rho u_j' \bar{u}_i + \rho u_j' u_i') = u_i' \frac{\partial}{\partial x_j} (\sigma_{ji}') - u_i' \frac{\partial}{\partial x_j} (-\overline{\rho u_j' u_i'}) + u_i' X_i'$$

and the resulting three convective terms are manipulated via chain-rule operations and the conservation-of-mass equations (5-22) and (5-23):

$$u_i' \frac{\partial}{\partial x_j} (\rho \bar{u}_j u_i') = \frac{\partial}{\partial x_j} \left(\rho \bar{u}_j \frac{u_i' u_i'}{2} \right), \qquad u_i' \frac{\partial}{\partial x_j} (\rho u_j' \bar{u}_i) = u_i' \rho u_j' \frac{\partial \bar{u}_i}{\partial x_j}$$

$$u_i' \frac{\partial}{\partial x_j} (\rho u_j' u_i') = \frac{\partial}{\partial x_j} \left(\rho u_j' \frac{u_i' u_i'}{2} \right)$$

Substitution of these rearranged terms and time-averaging leads to the following form of the turbulence kinetic energy equation:

$$\frac{\partial}{\partial x_j} \left(\overline{\rho \bar{u}_j \frac{u_i' u_i'}{2}} + \overline{\rho u_j' \frac{u_i' u_i'}{2}} \right) + \overline{\rho u_j' u_i' \frac{\partial \bar{u}_i}{\partial x_j}} = \overline{u_i' \frac{\partial}{\partial x_j} (\sigma_{ji}')} + \overline{u_i' X_i'}$$

The chain rule is applied to the fluctuating velocity–fluctuating stress-gradient correlation, and it is expanded using the Reynolds decomposition for the fluctuating stress tensor. Then applying the fluctuating velocity conservation-of-mass equation (5-23) with constant density to eliminate the fluctuating pressure–fluctuating velocity-gradient term yields

$$\overline{u_i' \frac{\partial}{\partial x_j} (\sigma_{ji}')} = \frac{\partial}{\partial x_j} \overline{(u_i' \sigma_{ji}')} - \overline{\sigma_{ji}' \frac{\partial u_i'}{\partial x_j}}$$

$$= \frac{\partial}{\partial x_j} (-\overline{u_i' P' \delta_{ji}} + \overline{u_i' \tau_{ji}'}) - \overline{\tau_{ji}' \frac{\partial u_i'}{\partial x_j}}$$

Substitution of this expansion into the turbulence kinetic energy equation leads to the following form:

$$\frac{\partial}{\partial x_j} \left(\overline{\rho \bar{u}_j \frac{u_i' u_i'}{2}} \right) = \frac{\partial}{\partial x_j} \left(-\overline{\rho u_j' \frac{u_i' u_i'}{2}} - \overline{u_j' P'} + \overline{u_i' \tau_{ji}'} \right)$$

$$+ (-\overline{\rho u_j' u_i'}) \frac{\partial \bar{u}_i}{\partial x_j} - \overline{\tau_{ji}' \frac{\partial u_i'}{\partial x_j}} + \overline{u_i' X_i'} \qquad (5\text{-}37)$$

In this equation the fluctuating viscous stress tensor–fluctuating velocity-gradient correlation can be rewritten in terms of the fluctuating strain-rate tensor s_{ij}' (recall that τ_{ij}' is symmetric) for the case where the compressibility term is neglected. To convert the correlation to its positive-definite form, note that the stress–velocity gradient correlation is the same with interchange of the i and j indices, allowing it to be added and divided by two, to create the second fluctuating stress tensor. The following equation results:

$$\overline{\tau_{ij}' \frac{\partial u_i'}{\partial x_j}} = 2\mu \overline{s_{ij}' \frac{\partial u_i'}{\partial x_j}} = 2\mu \frac{1}{2} \overline{\left(s_{ij}' \frac{\partial u_i'}{\partial x_j} + s_{ji}' \frac{\partial u_j'}{\partial x_i} \right)} = 2\mu \overline{s_{ij}' s_{ij}'}$$

This form shows that the fluctuating viscous stress tensor–fluctuating velocity-gradient correlation is positive-definite preceded by a minus sign, signifying a negative source or a sink term for the turbulence kinetic energy. It represents the rate at which the viscous stress performs deformation work against the fluctuating strain rate, and is dissipative in nature.

A common manipulation of the transport equation for turbulence kinetic energy is carried out by combining the gradient of the fluctuating velocity–fluctuating viscous stress correlation, which represents the rate of work by the fluctuating viscous stresses of the turbulent flow, with the fluctuating viscous stress tensor–fluctuating velocity-gradient correlation. Manipulation of these two terms leads to

$$\frac{\partial}{\partial x_j} \overline{(u_i' \tau_{ji}')} - \overline{\tau_{ji}' \frac{\partial u_i'}{\partial x_j}} = \rho \left\{ \frac{\partial}{\partial x_j} \mu \left[\frac{\partial}{\partial x_j} \overline{\left(\frac{u_i' u_i'}{2} \right)} \right] - \rho \nu \overline{\frac{\partial u_i'}{\partial x_j} \frac{\partial u_i'}{\partial x_j}} \right\}$$

where the second tensor is called the turbulence dissipation rate ε:

$$\varepsilon = \nu \overline{\frac{\partial u_i'}{\partial x_j} \frac{\partial u_i'}{\partial x_j}}$$

With this manipulation, Eq. (5-37) becomes

$$\frac{\partial}{\partial x_j} \left(\rho \bar{u}_j \overline{\frac{u_i' u_i'}{2}} \right) = \frac{\partial}{\partial x_j} \left[-\rho \overline{u_j' \frac{u_i' u_i'}{2}} - \overline{u_j' P'} + \mu \frac{\partial}{\partial x_j} \overline{\left(\frac{u_i' u_i'}{2} \right)} \right]$$

$$+ (-\rho \overline{u_j' u_i'}) \frac{\partial \bar{u}_i}{\partial x_j} - \rho \varepsilon + \overline{u_i' X_i'} \qquad (5\text{-}38)$$

The turbulence kinetic energy is commonly defined as one of the following quantities:

$$\bar{k} = \tfrac{1}{2} \overline{u_i' u_i'} = \tfrac{1}{2} \overline{(u_1' u_1' + u_2' u_2' + u_3' u_3')} = \tfrac{1}{2} \overline{(u'u' + v'v' + w'w')}$$

This definition is consistent in notation with the other Reynolds decompositions, in that a decomposition represents a mean value (with overbar symbol) and a fluctuating value (with prime symbol). As such, we define the instantaneous quantity $k = \tfrac{1}{2} u_i' u_i'$ and the mean value of that quantity as $\bar{k} = \tfrac{1}{2} \overline{u_i' u_i'}$. With these definitions, k' is the difference between the instantaneous and fluctuating quantities.

The structure of the $\bar{k}$-transport equation (5-38) takes a form similar to the other transport equations:

$$\text{Convection } (\bar{k}) = \text{Diffusion } (\bar{k}) \pm \text{Source } (\bar{k}) \qquad (5\text{-}39)$$

Comparison of this with Eq. (5-38) suggests that the $\partial/\partial x_j$ term to the right of the equal sign is composed of three diffusional quantities, although the first two terms are really energy-transfer terms representing diffusion of $\bar{k}$ by the turbulence itself and turbulent flow work (multiply and divide by ρ), while the third term represents viscous self-diffusion of $\bar{k}$. The last three terms of (5-38) represent sources and/or sinks of $\bar{k}$. The Reynolds stress–mean velocity-gradient product $(-\rho\overline{u_j'u_i'})\,\partial\bar{u}_i/\partial x_j$ is a source for $\bar{k}$, and it is generally positive, especially for boundary-layer flows. This term appears in the mean mechanical energy equation (5-35) with a negative sign, indicating that it is a transfer of energy from the mean field to the turbulence field. As indicated above, $\rho\varepsilon$ represents a dissipation of $\bar{k}$, being positive-definite with a minus sign. It will appear in a thermal energy equation with a positive sign, showing it to be a source of thermal energy in the energy-transfer process. The $\overline{u_i'X_i'}$ velocity–body force correlation is likewise a source of $\bar{k}$.

$\bar{k}$ EQUATION FOR THE TURBULENT BOUNDARY LAYER

Development of this equation primarily requires elimination of the x-direction gradient by applying the boundary-layer approximations and elimination of the z-direction components and gradients by restriction to two-dimensional flow. The $\bar{k}$-transport equation (5-38), combined with the conservation-of-mass equation (5-22), becomes, neglecting body forces,

$$\rho\bar{u}\frac{\partial\bar{k}}{\partial x} + \rho\bar{v}\frac{\partial\bar{k}}{\partial y} = \frac{\partial}{\partial y}\left[-\rho\overline{v'k'} - \overline{v'P'} + \mu\left(\frac{\partial\bar{k}}{\partial y}\right)\right]$$

$$+ (-\rho\overline{u'v'})\frac{\partial\bar{u}}{\partial y} - \rho\varepsilon \tag{5-40}$$

DISSIPATION TRANSPORT EQUATION

The derivation of the dissipation rate equation is quite tedious, involving differentiation of the instantaneous transport equation for the velocity fluctuation (5-36) with respect to x_l, multiplication of the resulting equation by the term $\nu\,\partial u_i'/\partial x_l$, then time-averaging. Implicit in the derivation is that the kinematic viscosity does not have spatial variations. The resulting transport equation contains a multitude of new fluctuation correlations, none of which explicitly appear in any other transport equation, and whose values are practically impossible to experimentally quantify. The general form of the resulting equation[4] is

$$\frac{\partial}{\partial x_j}(\rho \bar{u}_j \varepsilon) = \frac{\partial}{\partial x_j}[\rho(T_\varepsilon + \Pi_\varepsilon + D_\varepsilon)] + \rho(P_\varepsilon - \varepsilon_\varepsilon) \qquad (5\text{-}41)$$

This form is adapted from Mansour, Kim, and Moin,[4] except that the three diffusion-like terms are written explicitly to reflect their diffusive nature.

This equation is quite similar in form to Eq. (5-38) for the turbulence energy. In Eq. (5-38) the three-part diffusion term is equivalent to the $T_\varepsilon + \Pi_\varepsilon + D_\varepsilon$ term, where T_ε represents transfer of ε due to the turbulence itself, Π_ε represents transfer by pressure fluctuations, and D_ε represents viscous diffusion of ε. The P_ε term is a collection of four terms representing production of ε by the mean and the turbulence fields, and ε_ε represents the viscous dissipation of ε.

ε EQUATION FOR THE TURBULENT BOUNDARY LAYER

This equation evolves by elimination of the x-direction gradient because of the boundary-layer approximations and elimination of the z-direction components and gradients by the restriction of two-dimensional flow. The ε-transport equation (5-41), combined with the conservation-of-mass equation (5-22), becomes

$$\rho \bar{u}\frac{\partial \varepsilon}{\partial x} + \rho \bar{v}\frac{\partial \varepsilon}{\partial y} = \frac{\partial}{\partial y}[\rho(T_\varepsilon + \Pi_\varepsilon + D_\varepsilon)] + \rho(P_\varepsilon - \varepsilon_\varepsilon) \qquad (5\text{-}42)$$

PROBLEMS

5-1. Using appropriate assumptions, reduce Eq. (5-9) to compare it with Eq. (4-17).

5-2. Convert Eq. (5-11) to Eq. (4-32) using the definition of the substantial (or total) derivative.

5-3. Carry out the necessary algebra to show that Eqs. (5-16) represent the appropriate decompositions for the stagnation enthalpy.

5-4. Using the averaging rules, develop the conservation-of-mass equation (5-21), including all the intermediate steps.

5-5. Reduce Eq. (5-26) to the boundary-layer equation (5-28) using appropriate assumptions.

5-6. Derive the stagnation enthalpy equation (5-31), and reduce it to its low-velocity, constant-property boundary-layer form given by Eq. (5-34).

5-7. Carry out the derivation of the turbulence kinetic energy equation (5-38).

5-8. The construction of the boundary-layer equations for momentum and energy can be considered using the formulation of Eq. (5-39). Recast the laminar boundary-layer equations for momentum, Eq. (4-10), and energy Eq. (4-39), and their turbulent counterparts. Eqs. (5-28) and (5-34), into the form given by Eq. (5-39) to identify the convection, diffusion, and source terms for each equation.

REFERENCES

1. Rubesin, M. W., and W. C. Rose: NASA TM X-62248, Ames Research Center, Moffett Field, California, 1973.
2. Hinze, J. O.: *Turbulence,* 2d ed., p. 22, McGraw-Hill, New York, 1975.
3. Cebeci, T., and P. Bradshaw: *Physical and Computational Aspects of Convective Heat Transfer,* pp. 48–53, Springer-Verlag, New York, 1984.
4. Mansour, N. N., J. Kim, and P. Moin: *AIAA J.,* vol. 27, pp. 1068–1073, 1989.

CHAPTER

6

THE INTEGRAL EQUATIONS OF THE BOUNDARY LAYER

Particular solutions to the boundary-layer differential equations can be obtained, at least for a laminar flow, for almost any boundary condition. Exact analytic solutions have been obtained only for certain classes of problem, however, and numerical methods are frequently necessary for the more general problem. When the calculation procedure becomes tedious, it is often worthwhile to look for approximate methods of solution, such as integral methods. The integral equations of the boundary layer provide the basis for a number of approximate procedures, but are, in themselves, exact, at least within the boundary-layer approximation. The approximate nature of the integral solutions arises from the manner in which they are generally employed.

First we develop the momentum integral equation of the boundary layer in a fairly general manner, so that it can be employed for flow inside and over axisymmetric bodies as well as along plane surfaces. Then we use a similar procedure to develop an energy integral equation. In so doing, we also take the opportunity to introduce and define some useful integral parameters: the displacement thickness, the momentum thickness, and the enthalpy thickness.

THE MOMENTUM INTEGRAL EQUATION

Consider a body of revolution over which there is axisymmetric flow. Assume that a thin momentum boundary layer is developing on this surface, with the various dimensions and parameters defined as in Fig. 6-1.

Let us consider a stationary control volume fixed to the body surface. Take the length of the control volume in the x direction (along the surface) to be infinitesimal, δx, but let the height (normal to the surface) be a finite distance Y.

The distance Y may extend to a point within the boundary layer, or it may extend effectively beyond the boundary layer; in either case the assumption of a "thin" boundary layer allows the stipulation that $Y \ll R$, where R is the radius of revolution (transverse radius of curvature) of the body. Thus the surface could also be the inside of a circular-section nozzle, with R measured from a centerline in the main fluid stream. In either case, as R is made indefinitely large, we approach a two-dimensional surface, that is, a plane. Presumably we can express $R = R(x)$.

Let u_∞ designate the free-stream velocity in the x direction just *outside* the boundary layer. If the boundary layer is "thin," u_∞ will be approximately equal to the velocity at the surface, as computed by a potential flow analysis for the body. Presumably $u_\infty = u_\infty(x)$.

We assume that there is fluid flow through the body surface by injection, suction, or other forms of mass transfer. Using a subscript zero

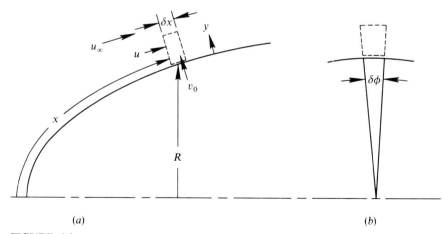

(a)   (b)

FIGURE 6-1
Coordinate system and location of the control volume for development of the momentum integral equation of the boundary layer on a body of revolution.

to designate fluid conditions at the interface, the velocity of the injected fluid is v_0 in the y direction.

By employing a control volume that is infinitesimal in the x direction but finite in the y direction, we intend to develop a differential equation that has, in effect, already been integrated in the y direction. We do so by simply applying the principles of conservation of mass and momentum to the control volume shown in Fig. 6-2 for steady flow. Note that $R \gg Y$.

Applying Eq. (2-1), we have

$$\left[R \, \delta\phi \, \delta x \, v_Y \rho_Y + R \, \delta\phi \int_0^Y u\rho \, dy + \delta\phi \frac{d}{dx} \left(R \int_0^Y u\rho \, dy \right) \delta x \right]$$

$$- \left(R \, \delta\phi \, \delta x \, v_0 \rho_0 + R \, \delta\phi \int_0^Y u\rho \, dy \right) = 0$$

from which

$$v_Y \rho_Y = v_0 \rho_0 - \frac{1}{R} \frac{d}{dx} \left(R \int_0^Y u\rho \, dy \right) \tag{6-1}$$

We then apply Eq. (2-4) to the control volume in the same manner,

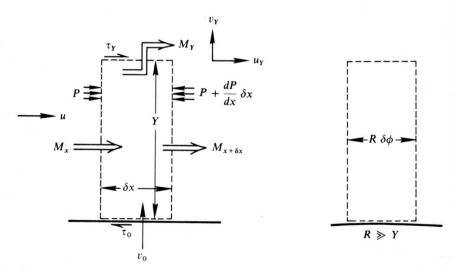

FIGURE 6-2
Control volume for development of the momentum integral equation of the boundary layer.

considering momentum flux and forces in the x direction only. Thus

$$M_x = R \, \delta\phi \int_0^Y \rho u^2 \, dy$$

$$M_{x+\delta x} = R \, \delta\phi \int_0^Y \rho u^2 \, dy + \delta\phi \frac{d}{dx}\left(R \int_0^Y \rho u^2 \, dy\right) \delta x$$

$$M_Y = R \, \delta\phi \, \delta x \, v_Y \rho_Y u_Y$$

(Note that M_Y is the product of the mass flow across the top of the control volume and the x component of velocity of that flow.)

Combining with the pressure and shear force terms in Fig. 6-2, and simplifying, we obtain

$$\tau_Y - \tau_0 - Y \frac{dP}{dx} = \frac{1}{R} \frac{d}{dx}\left(R \int_0^Y \rho u^2 \, dy\right) + \rho_Y v_Y u_Y$$

Now combine with Eq. (6-1), and at the same time let Y become sufficiently large that $u_Y = u_\infty$ and $\tau_Y = 0$; in other words, let Y be larger than the effective boundary-layer thickness:

$$-\tau_0 = \frac{1}{R}\frac{d}{dx}\left(R \int_0^Y \rho u^2 \, dy\right) - \frac{u_\infty}{R}\frac{d}{dx}\left(R \int_0^Y \rho u \, dy\right) + \rho_0 v_0 u_\infty + \int_0^Y \frac{dP}{dx} \, dy$$

$$(6\text{-}2)$$

Since Y extends past the boundary layer, the velocity u_∞ is related to the pressure P by inviscid flow behavior. Accordingly, we can take advantage of the Bernoulli equation for inviscid flow along a streamline to relate the pressure-gradient term to the free-stream velocity gradient:

$$\frac{dP}{\rho_\infty} = -d\left(\frac{u_\infty^2}{2}\right)$$

from which

$$\frac{dP}{dx} = -\rho_\infty u_\infty \frac{du_\infty}{dx} \qquad (6\text{-}3)$$

Then Eq. (6-2) becomes

$$-\tau_0 = \frac{1}{R}\frac{d}{dx}\left(R \int_0^Y \rho u^2 \, dy\right) - \frac{u_\infty}{R}\frac{d}{dx}\left(R \int_0^Y \rho u \, dy\right)$$

$$+ \rho_0 v_0 u_\infty + \int_0^Y \left(-\rho_\infty u_\infty \frac{du_\infty}{dx}\right) dy \qquad (6\text{-}4)$$

Equation (6-4) is the momentum integral equation of the boundary layer, developed under conditions where the fluid density may be variable, and applicable for flow over or inside an axisymmetric body where the boundary-layer thickness is much less than the local radius of curvature. For a two-dimensional body the radius R obviously drops out.

After defining and developing the boundary-layer thickness parameters below, we will be able to write more compact forms of the
momentum integral equation.

THE DISPLACEMENT AND MOMENTUM THICKNESS

Mathematically the boundary layer extends indefinitely in the y direction,
although practically speaking the major velocity change takes place, in
the usual case, very close to the body surface. Nevertheless, there is
generally some ambiguity in speaking of the boundary-layer "thickness"
if what is meant is the region in which velocity changes from u_∞ to 0. We
find it both convenient and far more useful to define some boundary-
layer thickness parameters that are completely unambiguous. These are
called the *displacement thickness* δ_1 and the *momentum thickness* δ_2.

Consider a two-dimensional flow along a flat surface (Fig. 6-3) such
that a boundary layer grows, starting at $x = 0$.

Take a control surface $ABCD$ and consider the following integral
statements:

$$\text{Mass flow across } AB = \int_0^Y u_\infty \rho_\infty \, dy$$

$$\text{Mass flow across } CD = \int_0^Y u\rho \, dy$$

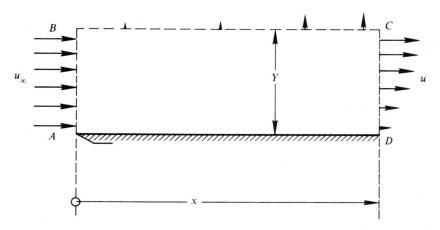

FIGURE 6-3
Control volume for development of the concept of displacement thickness and momentum
thickness of the boundary layer.

Thus, by conservation of mass,

$$\text{Mass flow across } BC = \int_0^Y u_\infty \rho_\infty \, dy - \int_0^Y u\rho \, dy$$

$$= u_\infty \rho_\infty \int_0^Y \left(1 - \frac{u\rho}{u_\infty \rho_\infty}\right) dy$$

Now let $Y \to \infty$ so that the entire boundary layer is enclosed; let us *define* δ_1 such that

$$\delta_1 u_\infty \rho_\infty = u_\infty \rho_\infty \int_0^\infty \left(1 - \frac{u\rho}{u_\infty \rho_\infty}\right) dy$$

$$\delta_1 = \int_0^\infty \left(1 - \frac{u\rho}{u_\infty \rho_\infty}\right) dy \qquad (6\text{-}5)$$

Thus δ_1 is a measure of the *displacement* of the main stream resulting from the presence of the flat plate and its boundary layer. Similarly,

$$x\text{-momentum flux across } AB = \int_0^Y \rho_\infty u_\infty^2 \, dy$$

$$x\text{-momentum flux across } CD = \int_0^Y \rho u^2 \, dy$$

If Y is sufficiently large that u is essentially u_∞ at Y then

$$x\text{-momentum flux across } BC = u_\infty \left[u_\infty \rho_\infty \int_0^Y \left(1 - \frac{u\rho}{u_\infty \rho_\infty}\right) dy \right]$$

Then the net rate of loss of momentum in the control volume is

$$\int_0^Y \rho_\infty u_\infty^2 \, dy - \int_0^Y \rho u^2 \, dy - u_\infty^2 \rho_\infty \int_0^Y \left(1 - \frac{u\rho}{u_\infty \rho_\infty}\right) dy = \int_0^Y \rho u(u_\infty - u) \, dy$$

Again let $Y \to \infty$ and then define δ_2 such that

$$\delta_2 u_\infty^2 \rho_\infty = \int_0^\infty \rho u(u_\infty - u) \, dy$$

$$\delta_2 = \int_0^\infty \frac{\rho u}{\rho_\infty u_\infty} \left(1 - \frac{u}{u_\infty}\right) dy \qquad (6\text{-}6)$$

Thus the *momentum thickness* is a measure of the momentum flux decrement caused by the boundary layer, which, according to the momentum theorem, is proportional to the drag of the plate.†

---

† Note that if $v_0 = 0$, or if R varies with x, then δ_1 and δ_2 no longer have the simple physical meaning implied in this development. Nevertheless, the defining equations (6-5) and (6-6) remain valid and unambiguous. The reader is referred to Hokenson[1] for boundary-layer thickness parameter definitions that are more general.

ALTERNATIVE FORMS OF THE MOMENTUM INTEGRAL EQUATION

Examination of Eq. (6-4) reveals that the integrals are essentially the same as appear in the definitions of δ_1 and δ_2, provided we change the upper limit in Eqs. (6-5) and (6-6) from ∞ to Y. This is valid since we have already assumed in the development of Eq. (6-4) that Y is sufficiently large that $u_Y = u_\infty$. Thus we can evidently replace the integrals in Eq. (6-4) and express the momentum integral equation in terms of the boundary-layer thickness parameters. After rearranging and simplifying, we obtain

$$\frac{\tau_0}{\rho_\infty u_\infty^2} + \frac{\rho_0 v_0}{\rho_\infty u_\infty} = \frac{d\delta_2}{dx} + \delta_2\left[\left(2 + \frac{\delta_1}{\delta_2}\right)\frac{1}{u_\infty}\frac{du_\infty}{dx} + \frac{1}{\rho_\infty}\frac{d\rho_\infty}{dx} + \frac{1}{R}\frac{dR}{dx}\right] \quad (6-7)$$

Equation (6-7) is simply an ordinary differential equation for δ_2 as a function of x; and this equation, although in a sense exact, also forms the basis for many approximate boundary-layer solutions. The approximations generally arise as the result of some assumptions about the relationship between τ_0 and δ_2, the thickness ratio δ_1/δ_2 (frequently called the *shape factor*), or an assumed function for the velocity profile u/u_∞.

Various particularizations of Eq. (6-7) can be obtained by dropping suitable terms. For example, for constant-density flow along a two-dimensional surface $(R \to \infty)$ with no suction or blowing $(v_0 = 0)$, we obtain the simpler form

$$\frac{\tau_0}{\rho_\infty u_\infty^2} = \frac{d\delta_2}{dx} + \delta_2\left[\left(2 + \frac{\delta_1}{\delta_2}\right)\frac{1}{u_\infty}\frac{du_\infty}{dx}\right] \quad (6-8)$$

A still more restricted form results if there is no pressure gradient, so that u_∞ is a constant:

$$\frac{\tau_0}{\rho_\infty u_\infty^2} = \frac{d\delta_2}{dx} \quad (6-9)$$

Later we will find it convenient to define a local friction coefficient c_f that is dimensionless and given by

$$c_f = \frac{\tau_0}{\frac{1}{2}\rho_\infty u_\infty^2} \quad (6-10)$$

Substituting in Eq. (6-9),

$$\frac{c_f}{2} = \frac{d\delta_2}{dx} \quad (6-11)$$

THE ENERGY INTEGRAL EQUATION

Integral methods for solution of the thermal boundary layer can be used to obtain approximate solutions in much the same manner as for the momentum boundary layer. The energy integral equation of the boundary layer is developed in a rather general manner applicable to a compressible, high-velocity fluid, with or without suction or injection, including viscous heat generation and chemical reaction, and for a body of revolution as well as a flat plate.

Consider a body of revolution over which there is steady axial fluid flow (Fig. 6-4). Let the surface temperature of the body vary in the direction of flow in any arbitrary manner, and let there be fluid flow through the body surface at a velocity (normal to the free-stream direction) that can vary in an arbitrary manner along the surface. The fluid flowing through the wall can, in general, be the same as or different from the free-stream fluid. There are no variations of any kind in the ϕ direction, and the *stagnation enthalpy* of the free-stream fluid is everywhere constant.

Consider the steady-state energy-transfer rates across the surfaces of a control volume in the boundary layer (Fig. 6-5). Let the control volume be of infinitesimal extent in the flow direction, but of finite height Y in the direction normal to the surface; and let it be fixed to the surface at $y = 0$. Assume that velocity and temperature gradients in the x direction are negligible relative to the corresponding gradients in the y direction (the boundary-layer approximation), and ignore the work done on the control volume by the normal *viscous* stresses. Convection will be the only energy transport mechanism in the x direction, while in the y

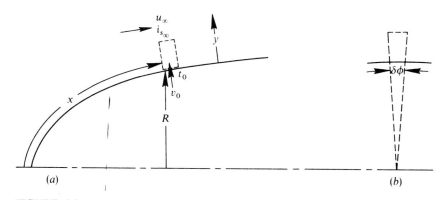

FIGURE 6-4
Coordinate system and location of the control volume for development of the energy integral equation of the boundary layer.

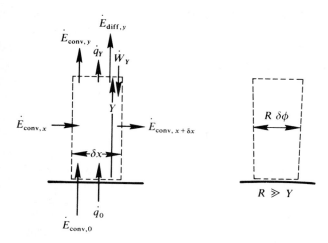

FIGURE 6-5
Control volume for development of the energy integral equation of the boundary layer.

direction there will be heat transfer and shear-work transfer as well as convection.

Application of the conservation-of-energy principle, Eq. (2-5), to the control volume for steady flow yields

$$(\dot{E}_{\text{conv}, x+\delta x} + \dot{E}_{\text{conv}, Y} + \dot{E}_{\text{diff}, Y} + \dot{q}_Y) - (\dot{E}_{\text{conv}, x} + \dot{E}_{\text{conv}, 0} + \dot{q}_0 + \dot{W}_Y) = 0$$

$$(6\text{-}12)$$

Each of the convection terms is the product of the mass flow and of the stagnation enthalpy (thermal and chemical enthalpy plus kinetic energy). It simplifies the algebra if we let the *enthalpy datum* be the *free-stream stagnation state*, which has already been designated as a constant. Thus

$$i_s = (i + \tfrac{1}{2}u^2) - (i_\infty + \tfrac{1}{2}u_\infty^2)$$

and

$$i_{s,\infty} = 0$$

We now write expressions for each of the terms in Eq. (6-12), based on the appropriate flux laws:

$$\dot{E}_{\text{conv}, x} = R\,\delta\phi \int_0^Y \rho u i_s \, dy$$

$$\dot{E}_{\text{conv}, x+\delta x} = R\,\delta\phi \int_0^Y \rho u i_s \, dy + \delta\phi \frac{d}{dx}\left(R \int_0^Y \rho u i_s \, dy\right) dx$$

$$\dot{E}_{\text{conv}, 0} = R\,\delta\phi\,\delta x\,\rho_0 v_0 i_{s,0}$$

$$\dot{q}_0 = R\,\delta\phi\,\delta x\,\dot{q}_0''$$

Now let Y be sufficiently large that all properties at Y are free-stream properties; that is, Y is larger than the boundary-layer thickness. Under these conditions,

$$\dot{E}_{\text{conv}, Y} = R \, \delta\phi \, \delta x \, \rho_Y v_Y i_{s, Y} = 0$$

$$\dot{E}_{\text{diff}, Y} = 0$$

$$\dot{q}_Y = 0, \qquad \dot{W}_Y = 0$$

Substituting in Eq. (6-12) and rearranging, we have

$$\dot{q}_0'' = \frac{1}{R} \frac{d}{dx} \left(R \int_0^Y \rho u i_s \, dy \right) - \rho_0 v_0 i_{s,0} \qquad (6\text{-}13)$$

Equation (6-13) is the energy integral equation of the boundary layer. After developing the thermal boundary-layer thickness parameters below, we will be able to write more compact forms of the energy integral equation.

THE ENTHALPY AND CONDUCTION THICKNESSES

The problem of defining a thickness of the thermal boundary layer is similar to that posed by the momentum boundary layer. Again we can define integral thickness parameters in an unambiguous way, and we make use of two—the *enthalpy thickness* and the *conduction thickness*. The physical significance of these parameters should be apparent merely from their definitions. The enthalpy thickness is defined as

$$\Delta_2 = \int_0^\infty \frac{\rho u i_s}{\rho_\infty u_\infty i_{s,0}} \, dy \qquad (6\text{-}14)$$

For a low-velocity *constant-property flow* of a perfect gas with no chemical reaction, Δ_2 can be expressed in terms of temperature as follows:

$$i_s = c(t - t_\infty)$$

$$\Delta_2 = \int_0^\infty \frac{u}{u_\infty} \left(\frac{t - t_\infty}{t_0 - t_\infty} \right) dy \qquad (6\text{-}15)$$

Equation (6-15) is also applicable to a constant-property liquid to a good approximation.

A *conduction thickness* Δ_4 may be defined as

$$\Delta_4 = \frac{k(t_0 - t_\infty)}{\dot{q}_0''}$$

where k is the thermal conductivity of the fluid.

Later we will want to define a *convection conductance,* or *heat-transfer coefficient,* as

$$h = \frac{\dot{q}_0''}{t_0 - t_\infty}$$

Thus

$$\Delta_4 = k/h \qquad (6\text{-}16)$$

The physical significance of both the enthalpy thickness and the conduction thickness may be appreciated by reference to Fig. 6-6.

ALTERNATIVE FORMS OF THE ENERGY INTEGRAL EQUATION

Making use of the definition of the enthalpy thickness of the boundary layer, we can now replace the integral in Eq. (6-13), provided only that we let $Y \to \infty$. Substituting, and then expanding the derivative and rearranging, we obtain

$$\frac{\dot{q}_0''}{\rho_\infty u_\infty i_{s,0}} + \frac{\rho_0 v_0}{\rho_\infty u_\infty} = \frac{d\Delta_2}{dx} + \Delta_2 \left(\frac{1}{u_\infty} \frac{du_\infty}{dx} + \frac{1}{\rho_\infty} \frac{d\rho_\infty}{dx} + \frac{1}{R} \frac{dR}{dx} + \frac{1}{i_{s,0}} \frac{di_{s,0}}{dx} \right) \qquad (6\text{-}17)$$

Note the similarity in form to Eq. (6-7). When it is expressed in this manner, the significance of the various terms is apparent.

Let us now examine a more restricted form of the energy integral equation, for example, low-velocity gas flow, no chemical reaction, $v_0 = 0$, and constant fluid properties across the boundary layer. Then Δ_2

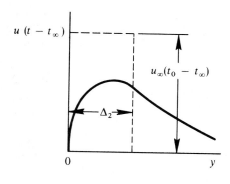
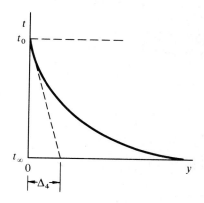

FIGURE 6-6
Graphical illustrations of the meaning of the enthalpy thickness and the conduction thickness of the boundary layer.

takes on the form given by Eq. (6-15), and we obtain

$$\frac{\dot{q}_0''}{\rho_\infty u_\infty c(t_0 - t_\infty)} = \frac{d\Delta_2}{dx} + \Delta_2 \left[\frac{1}{u_\infty} \frac{du_\infty}{dx} + \frac{1}{\rho_\infty} \frac{d\rho_\infty}{dx} + \frac{1}{R} \frac{dR}{dx} + \frac{1}{t_0 - t_\infty} \frac{d(t_0 - t_\infty)}{dx} \right]$$

(6-18)

A more compact expression of the same equation is

$$\frac{\dot{q}_0''}{c} = \frac{1}{R} \frac{d}{dx} [\Delta_2 R \rho_\infty u_\infty (t_0 - t_\infty)]$$

(6-19)

Note that in each of these equations the product $\rho_\infty u_\infty$ always appears, rather than ρ_∞ and u_∞ independently. Thus we frequently find it convenient to employ the mass velocity or mass flux, $G_\infty = \rho_\infty u_\infty$; Eq. (6-19) then becomes

$$\frac{\dot{q}_0''}{c} = \frac{1}{R} \frac{d}{dx} [\Delta_2 R G_\infty (t_0 - t_\infty)]$$

(6-20)

and Eqs. (6-17) and (6-18) can be similarly simplified. In internal flows, such as flows through nozzles, G_∞ is an especially convenient parameter, for it is merely the mass flow rate divided by the flow cross-sectional area [see Eq. (2-3)]. In such problems it is then unnecessary to evaluate separately the density and velocity. Note that this simplification is not available to us for the *momentum* integral equation (6-7), which means that to solve momentum problems we must separately evaluate free-stream velocity and density at all points along the surface.

Let us now see to what form the energy integral equation reduces for the simplest conceivable boundary-layer problem. Consider flow over a flat plate, $R \to \infty$, with constant pressure and free-stream velocity $du_\infty/dx = 0$, constant properties $d\rho_\infty/dx = 0$, and constant fluid-to-surface temperature difference $d(t_0 - t_\infty)/dx = 0$. Equation (6-18) then becomes

$$\frac{\dot{q}_0''}{\rho u_\infty c(t_0 - t_\infty)} = \frac{d\Delta_2}{dx}$$

(6-21)

If we now define a local heat-transfer coefficient

$$h = \frac{\dot{q}_0''}{t_0 - t_\infty}$$

(6-22)

then

$$\frac{h}{\rho u_\infty c} = \frac{d\Delta_2}{dx}$$

The group of variables on the left-hand side is nondimensional and is called the *local Stanton number* St. Thus

$$St = \frac{h}{\rho u_\infty c} = \frac{h}{G_\infty c} \tag{6-23}$$

and

$$St = \frac{d\Delta_2}{dx} \tag{6-24}$$

Note the analogous relation from the momentum integral equation (6-11).

It should also be apparent that an integral boundary-layer equation for the diffusion of some component j in a fluid mixture can be developed in the same manner.

PROBLEMS

6-1. Develop a momentum integral equation for steady flow without blowing or suction for use in the entry region of a circular tube. Note that Eq. (6-4) is not applicable for this case because it has been assumed that the boundary-layer thickness is small relative to the body radius R; in the present case the boundary layer ultimately grows to the centerline of the tube.

6-2. Develop the corresponding energy integral equation for Prob. 6-1.

6-3. Develop a boundary-layer integral equation for the diffusion of component j in a multicomponent mixture.

6-4. Derive the momentum integral equation (6-8) and the energy integral equation (6-21).

REFERENCE

1. Hokenson, G. L.: *AIAA J.*, vol. 15, pp. 597–600, 1977.

CHAPTER 7

MOMENTUM
TRANSFER:
LAMINAR
FLOW
INSIDE
TUBES

In this chapter we develop some of the very simplest solutions for the velocity distribution for steady laminar flow inside smooth tubes. Our primary concern is circular tubes, although some additional cross-sectional shapes are also considered.

FULLY DEVELOPED LAMINAR FLOW IN CIRCULAR TUBES

Consider the steady laminar flow of a viscous fluid inside a circular tube, as shown in Fig. 7-1. Let the fluid enter with a uniform velocity over the flow cross section. As the fluid moves down the tube a boundary layer of low-velocity fluid forms and grows on the wall surface because the fluid immediately adjacent to the wall must have zero velocity.

A particular and simplifying feature of viscous flow inside cylindrical tubes is the fact that the boundary layer must meet itself at the tube centerline, and the velocity distribution then establishes a fixed pattern that is invariant thereafter. We refer to the *hydrodynamic entry length* as that part of the tube in which the momentum boundary layer grows and the velocity distribution changes with length. We speak of the *fully*

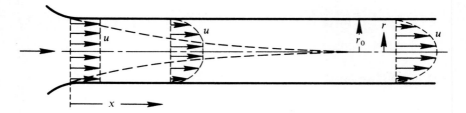

FIGURE 7-1
Development of the velocity profile in the hydrodynamic entry region of a pipe.

developed velocity profile as the fixed velocity distribution in the fully developed region. It should be added that we are assuming in this discussion that the fluid properties, including density, are not changing along the length of the tube.

Without yet worrying about how long the hydrodynamic entry length must be in order for a fully developed velocity profile to obtain, let us evaluate the fully developed velocity distribution for a laminar flow with constant viscosity. The applicable equation of motion must evidently be the momentum equation for axisymmetric flow in a circular tube, Eq. (4-11). However, by definition of a fully developed velocity profile, it is apparent that $v_r = 0$ and $\partial u / \partial x = 0$, and u is a function of r alone. Thus Eq. (4-11) becomes

$$\frac{\mu}{r}\frac{d}{dr}\left(r\frac{du}{dr}\right) = \frac{dP}{dx} \tag{7-1}$$

Since the pressure is independent of r, Eq. (7-1) can be integrated directly twice with respect to r to yield the desired velocity function. Applying the boundary conditions

$$\frac{du}{dr} = 0 \quad \text{at } r = 0$$

$$u = 0 \quad \text{at } r = r_0$$

we readily obtain

$$u = \frac{r_0^2}{4\mu}\left(-\frac{dP}{dx}\right)\left(1 - \frac{r^2}{r_0^2}\right) \tag{7-2}$$

Equation (7-2) is the familiar parabolic law. However, it proves more useful to express the velocity in terms of a *mean velocity* V rather than the pressure gradient. If we designate the flow cross-sectional area of a tube as A_c, the mass rate of flow across an elemental segment of that area dA_c is, by Eq. (2-3),

$$d\dot{m} = u\rho \, dA_c$$

Then the total mass flow rate through the tube is

$$\dot{m} = \int_{A_c} u\rho \, dA_c$$

Let us define a mean velocity V such that

$$\dot{m} = A_c V\rho, \qquad V = \frac{\dot{m}}{A_c\rho} \tag{7-3}$$

Then

$$V = \frac{1}{A_c\rho} \int_{A_c} u\rho \, dA_c \tag{7-4}$$

or, since the density is constant,

$$V = \frac{1}{A_c} \int_{A_c} u \, dA_c \tag{7-5}$$

For axisymmetric flow in a circular tube $dA_c = 2\pi r \, dr$ and $A_c = \pi r_0^2$. Thus

$$V = \frac{2}{r_0^2} \int_0^{r_0} ur \, dr \tag{7-6}$$

If we now substitute Eq. (7-2) into Eq. (7-6) and integrate, we obtain

$$V = \frac{r_0^2}{8\mu} \left(-\frac{dP}{dx} \right) \tag{7-7}$$

Equation (7-7), together with (7-3), can be used directly to calculate pressure drop. We can also combine (7-7) with (7-2) to obtain a simpler expression for the local velocity:

$$u = 2V \left(1 - \frac{r^2}{r_0^2} \right) \tag{7-8}$$

The shear stress at the wall surface can be evaluated from the gradient of the velocity profile at the wall. From Eq. (3-1),

$$\tau_0 = \mu \left(\frac{\partial u}{\partial r} \right)_{r=r_0} = \mu \left[2V \left(-\frac{2r_0}{r_0^2} \right) \right] = -\frac{4V\mu}{r_0} \tag{7-9}$$

To provide consistency with procedures to be used later, it is worth noting an alternative procedure to evaluate shear stress. Consider a stationary control volume as shown in Fig. 7-2. Let us apply the momentum theorem, Eq. (2-4), in the x direction, noting that, because of the fully developed nature of the flow, there is no net change in

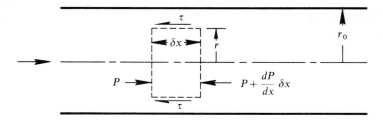

FIGURE 7-2
Control volume for analyzing fully developed flow in a pipe.

momentum flux. Thus

$$0 = P\pi r^2 - \left(P + \frac{dP}{dx}\,\delta x\right)\pi r^2 - \tau 2\pi r\,\delta x$$

$$\tau = \frac{r}{2}\left(-\frac{dP}{dx}\right) \tag{7-10}$$

and

$$\tau_0 = \frac{r_0}{2}\left(-\frac{dP}{dx}\right) \tag{7-11}$$

Equations (7-10) and (7-11) are equally applicable to a fully developed *turbulent* flow, as long as it is understood that τ refers to an apparent shear stress that is the linear combination of the viscous stress and the apparent turbulent shear stress.

Also,

$$\frac{\tau}{\tau_0} = \frac{r}{r_0} \tag{7-12}$$

Note, then, that in a fully developed pipe flow, whether laminar or turbulent, the apparent shear stress varies *linearly* from a maximum at the wall surface to zero at the pipe or tube centerline (Fig. 7-3).

Finally, Eq. (7-11) can be combined with Eq. (7-7), and we again obtain Eq. (7-9).

We can express the wall shear stress in terms of a nondimensional friction coefficient c_f defined as in Eq. (6-10). Let us base the definition arbitrarily on the mean velocity. Thus

$$\tau_0 = c_f \frac{\rho V^2}{2} \tag{7-13}$$

Then, employing (7-9) and considering the absolute value of the shear stress, to preserve the fact that wall shear is always opposite to the flow,

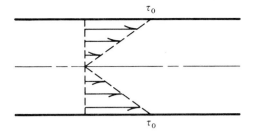

FIGURE 7-3
Shear-stress distribution for fully developed flow in a pipe.

we get

$$c_f = \frac{4V\mu/r_0}{\rho V^2/2} = \frac{8\mu}{r_0 \rho V} = \frac{16}{2r_0 \rho V/\mu}$$

We note for the fully developed velocity profile that c_f, the local friction coefficient, is independent of x. The nondimensional group of variables in the denominator is the *Reynolds number* Re. Thus

$$\text{Re} = \frac{2r_0 \rho V}{\mu} = \frac{DV\rho}{\mu} = \frac{DG}{\mu} \tag{7-14}$$

where $D = 2r_0$, the pipe diameter, and $G = \dot{m}/A_c$, the mean mass velocity. Thus

$$c_f = \frac{16}{\text{Re}} \tag{7-15}\dagger$$

We have dealt with this most elementary of internal flow problems in excessive detail perhaps, but it has provided us with an opportunity to define a number of terms that we use later.

FULLY DEVELOPED LAMINAR FLOW IN OTHER CROSS-SECTIONAL SHAPE TUBES

Laminar velocity profile solutions have been obtained for the fully developed flow case for a large variety of flow cross-sectional shapes. The applicable equation of motion for steady, constant-property, fully de-

---

† Integration of Eq. (7–11) and substitution of the friction coefficient definition results in the pressure-drop equation $\Delta P = 4c_f(L/D)(\rho V^2/2)$. The product $4c_f$ is f, the conventional friction factor used in fluid mechanics in conjunction with the Moody diagram. Then Eq. (7-15) becomes $f = 64/\text{Re}$.

veloped flow with no body forces, and with x the flow direction coordinate, can be readily deduced from the Navier–Stokes equation (4-15). Thus

$$\mu \, \nabla^2 u = \frac{dP}{dx} \qquad (7\text{-}16)$$

By assuming dP/dx to be constant over the flow cross section, this equation has been solved by various procedures, including numerically, for various shapes of tube. In most cases the shear stress will vary around

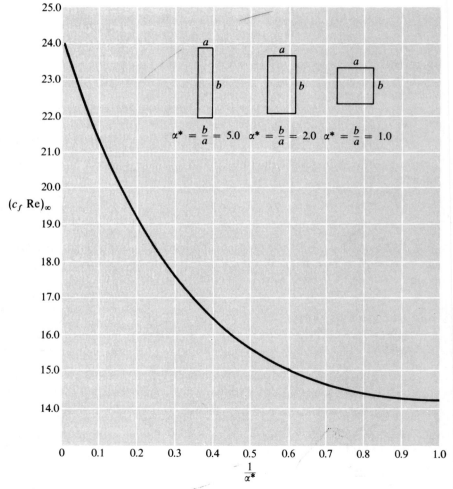

FIGURE 7-4
Friction coefficients for fully developed laminar flow in rectangular tubes.

the periphery of the tube; but if a mean shear stress with respect to peripheral area is defined (and this is the stress needed to calculate pressure drop), a friction coefficient can be defined in terms of Eq. (7-13).

On Fig. 7-4 the fully developed friction coefficients for the family of rectangular tubes,[1] extending from the square tube to flow between parallel planes, are plotted. Figure 7-5 gives similar results for flow between concentric annuli.[2]

For flow through an *equilateral triangular tube*

$$c_f \, \text{Re} = 13.33$$

The Reynolds number in all these results is defined as

$$\text{Re} = \frac{4r_h G}{\mu} \tag{7-17}$$

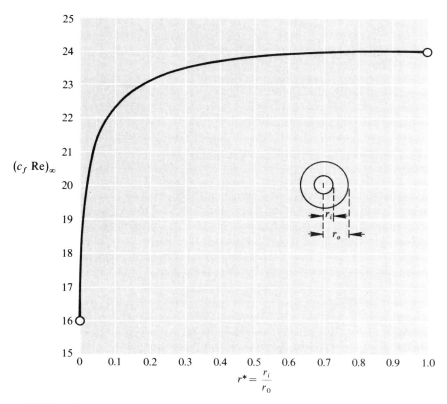

FIGURE 7-5
Friction coefficients for fully developed laminar flow in circular-tube annuli.

where

$$r_h = \frac{\text{cross-sectional area}}{\text{perimeter}} \quad \text{(the ``hydraulic radius''\dag)}$$

$$= A_c L / A$$

and where

A_c = cross-sectional area

L = tube length

A = total tube surface area in length L

G = mean mass velocity, $\dot{m}/A_c$

It has been found by experiment that if the above definition of Reynolds number is employed, laminar flow is obtained for flow inside a round tube as long as the Reynolds number is less than about 2300, and this criterion appears to be a good approximation for smooth tubes regardless of tube cross-sectional shape. Above this Reynolds number, the flow becomes unstable to small disturbances, and a transition to a turbulent type of flow generally occurs, although a fully established turbulent flow may not occur until the Reynolds number reaches about 10,000.

Fully developed flow solutions for a great many other tube shapes may be found in Shah and London[3] and in Kakac, Shah, and Aung.[4] All can be presented in precisely the same way as used here.

THE LAMINAR HYDRODYNAMIC ENTRY LENGTH

The preceding discussion has been concerned with the velocity distribution and friction coefficient at points far removed from the tube entrance where fully developed conditions obtain. The complete hydrodynamic solution for a tube must include some kind of entry length, as in Fig. 7-1. For a *circular tube* with axisymmetric, constant-property flow the differential equation of motion becomes the momentum equation (4-11),

---

† The choice of the hydraulic radius as the length dimension in the Reynolds number is actually a purely arbitrary one, since any length that characterizes the size of the flow passage would suffice as long as it is defined. However, we find that for a turbulent flow the hydraulic radius does lead to approximate correlation between the behavior of tubes of various cross-sectional shape, and we use the hydraulic radius for both laminar and turbulent flow to provide a consistent treatment and to avoid confusion. Note that $4r_h = D_h$, the hydraulic diameter; for circular tubes the tube diameter and hydraulic diameter are equivalent.

with μ constant:

$$\frac{\mu}{r}\frac{\partial}{\partial r}\left(r\frac{\partial u}{\partial r}\right) = \rho u\frac{\partial u}{\partial x} + \rho v_r\frac{\partial u}{\partial r} + \frac{dP}{\partial x} \qquad (7\text{-}18)$$

The simplest entry condition would be a uniform velocity at the tube entrance, $x = 0$, as shown in Fig. 7-1. Even for a sharp-cornered or abrupt-contraction entrance, the velocity profile develops in much the same manner, although the behavior for the first few diameters from the entrance is somewhat different.

Equation (7-18) incorporates the boundary-layer assumptions and thus would not be expected to be valid very close to the tube entrance. According to Shah and London, if the full Navier–Stokes equations are employed, it is found that for $\mathrm{Re} < 400$ and $(x/D)/\mathrm{Re} < 0.005$, Eq. (7-18) will lead to error in the computed velocity profiles. The solutions discussed below are thus only accurately applicable beyond this region.

The entry-length initial and boundary conditions of interest, then, are

$$\left.\begin{aligned} u &= V \\ v_r &= 0 \end{aligned}\right\} \quad \text{at } x = 0$$

$$\left.\begin{aligned} u &= 0 \\ v_r &= 0 \end{aligned}\right\} \quad \text{at } r = r_0$$

$$\frac{\partial u}{\partial r} = 0 \quad \text{at } r = 0$$

Hornbeck[10] has solved this problem *numerically* without any simplifying assumptions (other than the boundary layer assumptions). Figure 7-6 shows some of Hornbeck's results for the axial velocity profiles, and here it can be seen how the velocity profile develops and approaches the fully developed condition.

Approximate solutions to this problem can be obtained by solving the linearized momentum equation, such as described by Langhaar[5] or Sparrow, Lin, and Lundgren.[6] Friction coefficients from Langhaar's solution are plotted on Fig. 7-7, in the form $c_f\,\mathrm{Re} = f[\mathrm{Re}/(x/D)]$.

Three friction coefficients are indicated: c_f, c_{f_m}, and $\bar{c}_{f_{\mathrm{app}}}$. The *local* friction coefficient is described as c_f and is based on the actual local wall shear stress at x, using Eq. (7-13). In computing the pressure drop in a tube, the integrated mean wall shear stress from $x = 0$ to the point of interest is of more utility than the local shear stress. The mean friction coefficient from $x = 0$ to x is described as c_{f_m}. Part of the pressure drop in the entrance region of a tube is attributable to an increase in the total fluid momentum flux, which is associated with the development of the velocity profile. Pressure-drop calculations in this region must consider

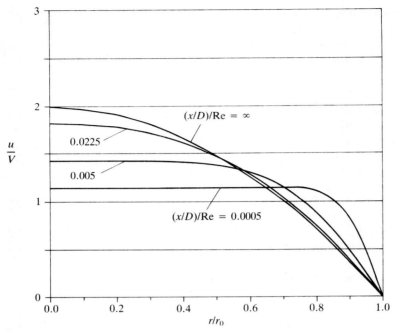

FIGURE 7-6
Axial velocity distribution in the hydrodynamic entrance region of a circular tube.

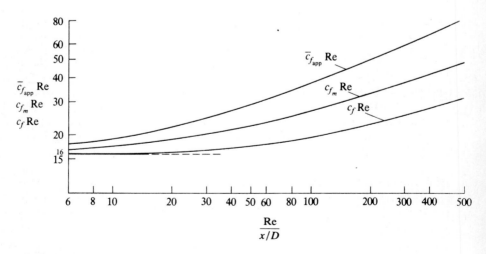

FIGURE 7-7
Friction coefficients for laminar flow in the hydrodynamic entry length of a circular tube (Langhaar[5]).

the variation in momentum flux as well as the effects of surface shear forces. The combined effects of surface shear and momentum flux have been incorporated in a single *apparent* mean friction coefficient $\bar{c}_{f_{app}}$. The pressure drop from 0 to x can then be evaluated from

$$\Delta P = 4\bar{c}_{f_{app}} \frac{\rho V^2}{2} \frac{x}{D} \qquad (7\text{-}19)$$

The approach of c_f toward $16/\text{Re}$ in Fig. 7-7 is a measure of the development of the velocity profile toward a fully developed profile. Note that c_f has reached within about two percent of its ultimate magnitude at $\text{Re}/(x/D) = 20$. Thus, a good approximate figure for the length of tube necessary for the development of the laminar velocity profile is

$$\frac{x}{D} = \frac{\text{Re}}{20} \qquad (7\text{-}20)$$

Solutions based on the same procedure as that used by Langhaar, but for flow in concentric-tube annuli—including flow between parallel planes—have been obtained by Heaton, Reynolds, and Kays.[7] A general method for computing velocity profiles and pressure drop in the entrance region of ducts of arbitrary cross section is given by Fleming and Sparrow.[8]

A considerable number of additional solutions for both fully developed flow and entry-length flow are presented in Shah and London,[3] and a correlation for entry-length flow is given in Shah.[9]

PROBLEMS

7-1. Consider steady, laminar, constant-property flow in a duct formed by two parallel planes. Let the velocity be uniform at the duct entrance. Calculate the development of the velocity profile in the entry length, using the momentum integral equation (6-4), and an assumption that the velocity profile may be approximated by a constant-velocity segment across the center portion of the duct and by simple parabolas in the growing boundary layer adjacent to the walls. Note that the mean velocity, Eq. (7-3), must be a constant, and thus Eq. (7-4) must be satisfied. [Some help in this problem may be obtained from an examination of the development of Eq. (8-35) in the next chapter.] Evaluate the hydrodynamic entry length and compare with Eq. 7-20).

7-2. Starting with the momentum theorem, develop an equation for the pressure drop for steady flow of a constant-property fluid in a tube of constant cross-sectional shape as a function of the friction coefficient, mean velocity, and tube length. Start with a control volume that is of infinitesimal dimension in the flow direction but that extends across the entire flow section. Then reconsider the problem when fluid density varies in some known manner along the tube but can be considered as effectively constant over the flow cross section. Discuss the implications of the latter assumption.

7-3. Consider fully developed laminar flow of a constant-property fluid in a circular tube. At a particular flow cross section calculate the total axial momentum flux by integration over the entire cross section. Compare this with the momentum flux evaluated by multiplying the mass flow rate times the mean velocity. Explain the difference, then discuss the implications for the last part of Prob. 7-2.

7-4. Consider steady, laminar, constant-property, fully developed flow between parallel planes. Derive expressions for the velocity profile, mean velocity, and friction coefficient as a function of Reynolds number.

7-5. Two air tanks are connected by two parallel circular tubes, one having an inside diameter of 1 cm and the other an inside diameter of 0.5 cm. The tubes are 2 m long. One of the tanks has a higher pressure than the other, and air flows through the two tubes at a combined rate of 0.00013 kg/s. The air is initially at 1 atm pressure and 16°C. Assuming that fluid properties remain constant and that the entrance and exit pressure losses are negligible, calculate the pressure differences between the two tanks.

7-6. A particular heat exchanger is built of parallel plates, which serve to separate the two fluids, and parallel continuous fins, which extend between the plates so as to form rectangular flow passages. For one of the fluids the plate separation is 1 cm and the nominal fin separation is 2 mm. However, manufacturing tolerance uncertainties lead to the possibility of a 10 percent variation in the fin separation. Consider the extreme case where a 10 percent oversize passage is adjacent to a 10 percent undersize passage. Let the flow be laminar and the passages sufficiently long that an assumption of fully developed flow throughout is reasonable. For a fixed pressure drop how does the flow rate differ for these two passages, and how does it compare to what it would be if the tolerance were zero.?

7-7. Develop the analysis that leads to the linear shear stress distributon described by Eq. (7-12).

7-8. Using the methodology developed in the text for a circular pipe, develop the fully developed mean velocity profile and fully developed friction coefficient for the flow between parallel planes. Compare your friction result with Figs. 7-5 and 7-5.

7-9. Repeat Prob. 7-8 for an annulus with radius ratio r^*. Compare your velocity-profile result with Eq. (9-26) and your friction result with Fig. 7-5 and Shah and London.[3]

7-10. The apparent fricton factor $\bar{c}_{f_{app}}$ reflects pressure-drop contributions from wall friction and momentum-flux change as the fluid accelerates from its flat entrance profile to its fully developed profile. Integrate the momentum equation (7-18) over the interval from $r = 0$ to r_0 and from $x = 0$ to x, assuming constant properties, to obtain a relation for the pressure drop ΔP in terms of c_{f_m} and the momentum-flux change, and convert the solution to Eq. (7-19). The result leads to the definition of $\bar{c}_{f_{app}}$.

7-11. *Computer analysis of laminar entry flow in a circular pipe:* calculate this flow and plot the fully developed velocity profile. Plot and compare the development with $\mathrm{Re}/(x/D)$ of the product $\bar{c}_{f_{\mathrm{app}}}$ Re with Fig. 7-7. Construct a similar form of plot for the development of the ratio of the flow maximum velocity to the mean velocity. Determine the entry length. You can choose how to set up the problem in terms of a suitable choice of Reynolds number, dimensions, and fluid properties (constant). For the initial conditon let the velocity profile be flat.

7-12 *Computer analysis of laminar entry flow between parallel planes and in a circular-tube annulus with* $r^* = 0.1$: follow the instructions in Prob. 7-11. Where appropriate, compare your results with this text and with Shah and London.[3]

REFERENCES

1. Kays, W. M., and S. H. Clark: TR no. 17, Department of Mechanical Engineering, Stanford University, Stanford, California, August 15, 1953.
2. Lundberg, R. E., W. C. Reynolds, and W. M. Kays: NASA TN D-1972, Washington, August 1963.
3. Shah, R. K., and A. L. London: "Laminar Flow Forced Convection in Ducts," *Advances in Heat Transfer,* Academic Press, New York, 1978.
4. Kakac, S., R. K. Shah, and W. Aung: *Handbook of Single-Phase Convective Heat Transfer,* John Wiley, New York, 1987.
5. Langhaar, H. L.: *J. Appl. Mech.,* vol. 9, 1942.
6. Sparrow, E. M., S. H. Lin, and T. S. Lundgren: *Phys. Fluids,* vol. 7, pp. 338–347, 1964.
7. Heaton, H. S., W. C. Reynolds, and W. M. Kays: *Int. J. Heat Mass Transfer,* vol. 7, p. 763, 1964.
8. Fleming, D. P., and E. M. Sparrow: *J. Heat Transfer,* vol. 91, pp. 345–354, 1969.
9. Shah, R. K.: *J. Fluids Engng,* vol. 100, pp. 177–179, 1978.
10. Hornbeck, R. W.: *Appl. Sci. Res., Ser. A,* vol. 13, pp. 224–232, 1964.

CHAPTER
8

MOMENTUM TRANSFER: THE LAMINAR EXTERNAL BOUNDARY LAYER

The development of the laminar momentum boundary layer for a constant-property fluid flowing over an external surface is considered in this chapter. We use the term *external flow,* as opposed to *internal flow,* to characterize flow along a surface on which the boundary-layer thickness is small relative to the distance to any other surface. Thus the viscous flow region is bounded by a solid surface on one side only, and is bounded on the other side by an inviscid potential flow. In an internal flow (flow in a pipe) we are usually concerned with the viscous effects on the entire region between boundary surfaces, although the hydrodynamic entry region of a pipe has the characteristics of an external flow.

First we consider the family of *similarity* solutions to the momentum equation of the boundary layer for a steady laminar flow. We treat the concept of similarity solutions to a partial differential equation in some detail, since this forms the basis for the majority of useful laminar boundary-layer solutions, including the thermal and concentration boundary layers. Nonsimilar boundary layers are then discussed, and *local similarity* and *local nonsimilarity* solutions are described.

Then we develop two approximate laminar boundary-layer solutions, employing the momentum integral equation, including a relatively simple procedure for handling a flow in a pressure gradient.

The momentum boundary-layer theory presented only outlines and highlights this large area of study, and the reader is referred to the more comprehensive treatises on the subject, such as Schlichting[1] or Moore[2] for further details.

SIMILARITY SOLUTIONS: THE LAMINAR INCOMPRESSIBLE BOUNDARY LAYER WITH CONSTANT PROPERTIES AND CONSTANT FREE-STREAM VELOCITY

Consider steady flow without turbulence over a semi-infinite flat plate aligned with the flow (Fig. 8-1). Let the free-stream velocity u_∞ be constant, and let all fluid properties be constant. The physical requirement of zero velocity at the surface must result in the development of a momentum boundary layer, starting at the leading edge of the plate, as shown in Fig. 8-1. The applicable equations of motion must be the momentum equation† (4-10), together with the continuity equation (4-7). Since u_∞ is stipulated as constant, dP/dx must be zero according to Eq. (6-3). With viscosity constant, and introducing $\nu = \mu/\rho$, our equations then become

$$\nu \frac{\partial^2 u}{\partial y^2} = u \frac{\partial u}{\partial x} + v \frac{\partial u}{\partial y} \tag{8-1}$$

$$\frac{\partial u}{\partial x} + \frac{\partial v}{\partial y} = 0 \tag{8-2}$$

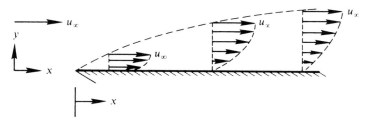

FIGURE 8-1
Development of a laminar boundary layer on a flat plate with constant free-strain velocity.

---

† Many authors prefer to develop the boundary-layer momentum equation from the more general Navier–Stokes equation (4-14), employing an order-of-magnitude argument, which, in turn, shows that Eq. (8-1) is valid only for $\text{Re}_x = (u_\infty \rho x/\mu) \gg 1$.

The boundary conditions are

$$\left. \begin{array}{c} u = 0 \\ v = 0 \end{array} \right\} \quad \text{at } y = 0$$

$$u \to u_\infty \quad \text{at } y \to \infty$$

$$u = u_\infty \quad \text{at } x = 0$$

The shapes that the velocity profiles (Fig. 8-1) must take in order to satisfy the boundary conditions suggest that they may possibly be geometrically similar, differing only by a stretching factor on the y coordinate, such a factor being some function of the distance x along the plate. It frequently occurs that partial differential equations have solutions that involve this kind of similarity; if a differential equation has a family of similar solutions, these are usually much easier to obtain than other solutions.

The statement that the velocity profiles at all the x positions are geometrically similar, differing only by a multiplying factor, is equivalent to the statement that

$$u = f[y \cdot g(x)] \qquad (8\text{-}3)$$

Let us now see if an assumption that the solution can be expressed in this form truly leads to such a solution. Let

$$\eta = y \cdot g(x) \qquad (8\text{-}4)$$

Then

$$u = f(\eta) \qquad (8\text{-}5)$$

By adopting the nomenclature $f' = df/d\eta$ and $g' = dg/dx$, Eq. (8-3) can be substituted into the boundary-layer equations (8-1) and (8-2) as follows:

$$\frac{\partial u}{\partial x} = \frac{\partial f}{\partial x} = \frac{\partial f}{\partial \eta} \frac{\partial \eta}{\partial x} = f'yg'$$

$$\frac{\partial u}{\partial y} = \frac{\partial f}{\partial y} = \frac{\partial f}{\partial \eta} \frac{\partial \eta}{\partial y} = f'g$$

$$\frac{\partial^2 u}{\partial y^2} = \frac{\partial^2 f}{\partial y^2} = \frac{\partial}{\partial y}\left(\frac{\partial f}{\partial y}\right) = \frac{\partial}{\partial \eta}\left(\frac{\partial f}{\partial y}\right)\frac{\partial \eta}{\partial y} = f''g^2$$

Substituting these derivatives into Eqs. (8-1) and (8-2) yields

$$vf''g^2 = ff'yg' + vf'g$$

$$f'yg' + \frac{\partial v}{\partial y} = 0$$

Combining these equations to eliminate v, and then rearranging to separate variables, we obtain

$$\frac{1}{f}\frac{d}{d\eta}\left(\frac{f''}{f'}\right) = \frac{1}{v}\frac{g'}{g^3} \tag{8-6}$$

Since one side is a function of η only and the other side is a function of x only, and since x and η are independent, then each side must be a constant; that is, we have successfully separated variables and reduced the problem to the solution of two ordinary differential equations. This result immediately suggests that a similarity solution does indeed exist.

Consider first the right-hand side, and set it equal to an arbitrary constant, $-k$:

$$\frac{1}{v}\frac{g'}{g^3} = -k$$

Solving,

$$\frac{dg}{g^3} = -kv\,dx, \qquad -\frac{1}{2g^2} = -kvx + C$$

At $y = \infty$, $u = u_\infty$; but at $x = 0$ a boundary condition is that $u = u_\infty$ even at $y = 0$. Thus $g(0)$ must be infinite. It thus follows that $C = 0$. Then

$$g = \frac{1}{\sqrt{2kvx}}, \qquad \eta = yg = \frac{y}{\sqrt{2kvx}} \tag{8-7}$$

Note the implications of this result. Evidently the velocity function is of the form

$$u = f\left(\frac{y}{\sqrt{x}}\right)$$

This always turns out to be the case for those solutions of the boundary-layer equations where similar velocity profiles are obtained, and $y/\sqrt{x}$ is sometimes referred to as a similarity parameter.

Let us now consider the left-hand side of Eq. (8-6):

$$\frac{1}{f}\frac{d}{d\eta}\left(\frac{f''}{f'}\right) = -k$$

$$d\left(\frac{f''}{f'}\right) = -kf\,d\eta$$

$$\frac{f''}{f'} = -k\int f\,d\eta + C$$

To evaluate the constant of integration, note that

$$\left.\begin{array}{l} \eta = 0 \\ u = 0 \\ v = 0 \\ f = 0 \end{array}\right\} \quad \text{at } y = 0$$

and, from Eq. (8-1),

$$\frac{\partial^2 u}{\partial y^2} = 0 = \frac{\partial^2 f}{\partial \eta^2} = f'' \quad (\text{if } v_0 = 0)$$

Thus $C = 0$, and

$$\frac{f''}{f'} = -k \int_0^\eta f \, d\eta$$

Let us now define a *nondimensional* velocity in terms of a *derivative* of a function of η, the derivative being for the purpose of eliminating the integral in the above equation. Let

$$\zeta'(\eta) = \frac{u}{u_\infty} = \frac{f}{u_\infty}$$

Then

$$f = u_\infty \zeta'$$
$$f' = u_\infty \zeta''$$
$$f'' = u_\infty \zeta'''$$

and

$$\frac{f''}{f'} = \frac{\zeta'''}{\zeta''}$$

Substituting,

$$\frac{\zeta'''}{\zeta''} = -k \int_0^\eta u_\infty \frac{d\zeta}{d\eta} d\eta = -ku_\infty \zeta$$

$$\zeta''' + ku_\infty \zeta \zeta'' = 0$$

Since k is an arbitrary constant but ku_∞ must be nondimensional, let

$$ku_\infty = \tfrac{1}{2}$$

Thus

$$\zeta''' + \tfrac{1}{2}\zeta\zeta'' = 0 \tag{8-8}$$

where

$$\eta = \frac{y}{\sqrt{vx/u_\infty}} \qquad \zeta'(\eta) = \frac{u}{u_\infty}$$

Equation (8-8) is the Blasius equation, an ordinary differential equation for which the boundary conditions must be

$$u = 0 \quad \text{at} \ y = 0$$

$$v = 0 \quad \text{at} \ y = 0$$

$$u \to u_\infty \quad \text{at} \ y \to \infty$$

Then

$$\zeta'(0) = 0, \qquad \zeta'(\infty) = 1$$

Since

$$f''(0) = 0, \qquad \zeta'''(0) = 0 \quad \text{(provided that } v_0 = 0\text{)}$$

Then, from Eq. (8-8),

$$\zeta(0)\zeta''(0) = 0$$

It is not possible that all the derivatives are zero at the wall [in fact, $\zeta''(0) = 0$ would correspond to zero shear stress at the wall], so we conclude that

$$\zeta(0) = 0$$

Equation (8-8) can also be developed by introducing a stream function ψ, although this is not necessary. The stream function is simply a device for replacing the u and v components of velocity by a single function. If we define $\psi(x, y)$ by

$$u = \frac{\partial \psi}{\partial y}, \qquad v = -\frac{\partial \psi}{\partial x} \tag{8-9}$$

we find that such a definition satisfies the continuity equation (8-2). It can readily be shown that ψ is directly related to ζ; in fact,

$$\psi = \sqrt{vxu_\infty} \ \zeta \tag{8-10}$$

As before, $\zeta = \zeta(\eta)$, and

$$\eta = \frac{y}{\sqrt{vx/u_\infty}} \tag{8-11}$$

Note that, by combining these equations, $u = u_\infty \zeta'(\eta)$.

With these definitions, the Blasius equation is readily developed by just substituting Eq. (8-9) into Eq. (8-1) and then replacing ψ by ζ through Eq. (8-10).

Equation (8-8) was first solved numerically (and undoubtedly by hand) by Blasius,[3] but it can be readily solved now by standard computer-based numerical procedures. A brief summary of typical results is given in Table 8-1.

TABLE 8-1
Solutions to the laminar constant-property boundary layer with an impermeable wall and $u_\infty =$ constant

η	ζ	ζ'	ζ''
0	0	0	0.3321
0.2	0.00664	0.06641	0.3320
0.4	0.02656	0.13277	0.3315
0.6	0.05974	0.19894	
0.8	0.10611	0.26471	
1.0	0.16557	0.32979	
1.2	0.23795	0.39378	
1.4	0.32298	0.45627	
1.6	0.42032	0.51676	
1.8	0.52952	0.57477	
2.0	0.65003	0.62977	
2.2	0.78120	0.68132	
2.4	0.92230	0.72899	
2.6	1.07252	0.77246	
2.8	1.23099	0.81152	
3.0	1.39682	0.84605	
3.2	1.56911	0.87609	
3.4	1.74696	0.90177	
3.6	1.92954	0.92333	
3.8	2.11605	0.94112	
4.0	2.30576	0.95552	
4.2	2.49806	0.96696	
4.4	2.69238	0.97587	
4.6	2.88826	0.98269	
4.8	3.08534	0.98779	
5.0	3.28329	0.99155	

For higher values of η, $\zeta = \eta - 1.72$.

The velocity profiles are given by $\zeta' = u/u_\infty$. We can also evaluate a friction coefficient based on free-stream velocity, using the defining equation (6-10):

$$c_f = \frac{\tau_0}{\frac{1}{2}\rho u_\infty^2} \tag{8-12}$$

The wall shear stress τ_0 can then be evaluated from (3-1), invoking the boundary-layer approximations:

$$\tau_0 = \mu \left(\frac{\partial u}{\partial y} \right)_0 = \mu \left[f' \left(\frac{\partial \eta}{\partial y} \right) \right]_0 = \mu \sqrt{u_\infty^3/vx} \; \zeta''(0)$$

but $\zeta''(0) = 0.332$ and thus

$$c_f = \frac{0.664}{\sqrt{u_\infty x/\nu}} = \frac{0.664}{\mathrm{Re}_x^{1/2}}$$
(8-13)

where we define

$$\mathrm{Re}_x = \frac{u_\infty x}{\nu} = \frac{u_\infty \rho x}{\mu} = \frac{G_\infty x}{\mu}$$
(8-14)

We term c_f the *local* friction coefficient since it is used to evaluate the *local* wall shear stress. We may also be interested in the total shear force from 0 to x, in which case we would like to know the *mean* shear stress from 0 to x. We can define a mean friction coefficient on this basis; it is easy to show that this is simply the average of c_f with respect to x. Performing the integration, the mean friction coefficient c_{f_m} becomes

$$c_{f_m} = \frac{1.328}{\mathrm{Re}_x^{1/2}}$$
(8-15)

By use of Eq. (6-11), c_f can also be expressed in terms of the *momentum thickness* Reynolds number

$$c_f/2 = 0.2204/\mathrm{Re}_{\delta_2}$$
(8-16)

The previously defined boundary-layer thickness parameters δ_1 and δ_2, Eqs. (6-5) and (6-6), can now be readily evaluated from the data of Table 8-1 (note that in performing the integrations $dy = \sqrt{\nu x/u_\infty}\, d\eta$). Thus

$$\delta_1 = 1.72\sqrt{\nu x/u_\infty} \quad \text{(displacement thickness)} \tag{8-17}$$
$$\delta_2 = 0.664\sqrt{\nu x/u_\infty} \quad \text{(momentum thickness)} \tag{8-18}$$

Note also that δ_2 can be perhaps more easily evaluated by integration of Eq. (6-11), obtained from the momentum integral equation.

Our original simplifying assumption in developing the boundary-layer equations was that the boundary layer remains "thin," so that $\partial u/\partial y$ is always very much greater than any other velocity gradient. This assumption would seem to be valid as long as $\delta_1 \ll x$. From Eq. (8-17)

$$\frac{\delta_1}{x} = \frac{1.72\sqrt{\nu x/u_\infty}}{x} = \frac{1.72}{\mathrm{Re}_x^{1/2}}$$
(8-19)

Thus the boundary-layer approximation would appear to be valid as long as $\mathrm{Re}_x \gg 1$. Thus, for example, our boundary-layer approximation would not be valid in the region very close to the leading edge of the plate.

SIMILARITY SOLUTIONS FOR THE LAMINAR INCOMPRESSIBLE BOUNDARY LAYER FOR $u_\infty = Cx^m$

Similar solutions to the boundary-layer equations can be obtained if the free-stream velocity varies as

$$u_\infty = Cx^m \tag{8-20}$$

Before examining this family of solutions, let us consider what physical problems are approximated by such a velocity variation. Note the family of two-dimensional potential flows shown in Fig. 8-2.

These flows are very easily solved by potential flow theory, and it is found that the velocity along the surface varies according to Eq. (8-20) in each case; in fact, it is found that

$$m = \frac{\beta/\pi}{2 - \beta/\pi} = \frac{x}{u_\infty} \frac{du_\infty}{dx} \tag{8-21}$$

The boundary-layer solutions corresponding to these potential flows are frequently called *wedge-flow* solutions for obvious reasons. They are also frequently called the Falkner and Skan flows after the authors who first published their boundary-layer solutions.[4] The previously obtained solution for constant free-stream velocity is a special case of these flows,

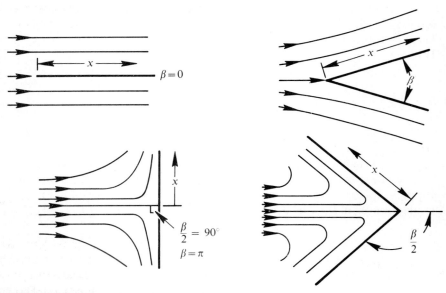

FIGURE 8-2
The family of "wedge" flows.

$\beta = 0$, and the family includes the two-dimensional stagnation point $\beta = \pi$.

For constant-property, two-dimensional flow the boundary-layer *momentum equation* (4-10) becomes

$$v \frac{\partial^2 u}{\partial y^2} = u \frac{\partial u}{\partial x} + v \frac{\partial u}{\partial y} + \frac{1}{\rho} \frac{dP}{dx} \tag{8-22}$$

The pressure-gradient term can be expressed in terms of the velocity via the Bernoulli equation, as was done in Eq. (6-3), which is valid in the free stream where a potential flow presumably exists. Combining Eq. (8-20) with Eq. (6-3), we get

$$\frac{dP}{dx} = -\rho C x^m C m x^{m-1} = -\frac{\rho u_\infty^2 m}{x}$$

Thus

$$v \frac{\partial^2 u}{\partial y^2} = u \frac{\partial u}{\partial x} + v \frac{\partial u}{\partial y} - \frac{u_\infty^2 m}{x} \tag{8-23}$$

In addition to Eq. (8-23), the continuity equation (8-2) is still applicable.

We now seek similarity solutions in the same manner as before. Actually the same similarity parameter is applicable; the simplest procedure is to substitute Eq. (8-9) into Eq. (8-23) and then transform to (ζ, η) using Eqs. (8-10) and (8-11). An ordinary differential equation results, so that similarity solutions must exist:

$$\zeta''' + \tfrac{1}{2}(m+1)\zeta\zeta'' + m(1 - \zeta'^2) = 0 \tag{8-24}$$

The boundary conditions are the same as for the previous problem:

$$\zeta(0) = 0, \quad \zeta'(0) = 0, \quad \zeta'(\infty) = 1$$

[Note that the constant-u_∞ problem, Eq. (8-8), is merely a special case of this more general problem.]

Equation (8-24) can be solved numerically for the indicated boundary conditions. For $m > 0$, the solutions are unique. For $m < 0$, two families of solutions exist for a limited range of negative β, corresponding either to a decelerating mainstream to the point of incipient separation or to flows in a laminar boundary layer after separation. A partial summary[4] of the results for $m > 0$ and for the $m < 0$ decelerating flows is given in Table 8-2. A friction coefficient can be evaluated as before:

$$c_f = \frac{2\zeta''(0)}{\mathrm{Re}_x^{1/2}} \tag{8-25}$$

TABLE 8-2
Solutions to the laminar constant-property boundary layer with an impermeable wall and $u_\infty = Cx^m$

β	m	$\zeta''(0)$
2π	∞	∞
π	1.0	1.233 Stagnation
1.57	0.333	0.759
0.627	0.111	0.510
0	0	0.332 Flat plate
-0.314	-0.0476	0.220
-0.625	-0.0904	0 Separation

We make more use of the wedge solutions later when we consider heat transfer.

SIMILARITY SOLUTIONS FOR THE LAMINAR INCOMPRESSIBLE BOUNDARY LAYER FOR $v_0 \neq 0$

A boundary condition employed in the examples considered so far has been that the velocity normal to the surface at the surface is zero, that is, $v_0 = 0$. A number of interesting applications occur in which there is mass transfer, and hence a normal velocity component, at the surface. Evaporation from or condensation onto a surface is one example. Another is that of a porous surface through which a fluid (possibly the same fluid as the main stream) is being forced in order to protect the surface from an extremely high-temperature main stream. Another might be the case of a porous surface through which part of the main stream is being sucked, possibly to prevent boundary-layer separation because of an adverse pressure gradient.

The applicable differential equation is still Eq. (8-22), and we are merely changing a boundary condition. The general problem where v_0 varies in any arbitrary manner along the surface can probably be solved only by direct numerical integration of Eq. (8-22). However, let us examine the conditions under which similarity solutions exist. This merely involves examination of the possible boundary conditions for Eq. (8-23) to see if a change of boundary conditions allows v_0 to be nonzero, and, if so, how v_0 must vary along the surface.

From Eqs. (8-9), (8-20), and (8-20),

$$v = -\frac{\partial \psi}{\partial x}, \qquad \psi = \sqrt{vxu_\infty}\, \zeta(\eta) \qquad u_\infty = Cx^m$$

Combining these,

$$v = -\zeta(\eta)\frac{m+1}{2}x^{(m-1)/2}\sqrt{Cv} - Cy\frac{m-1}{2}\zeta'(\eta)x^{m-1}$$

But $\eta = 0$ and $v = v_0$ at $y = 0$. Thus

$$v_0 = -\zeta(0)\frac{m-1}{2}x^{(m-1)/2}\sqrt{Cv} \tag{8-26}$$

We now let $C = u_\infty/x^m$, and obtain

$$v_0 = -\frac{m+1}{2}\zeta(0)\frac{u_\infty}{\sqrt{u_\infty x/v}} \tag{8-27}$$

and

$$\zeta(0) = -\frac{2}{m+1}\frac{v_0}{u_\infty}\sqrt{\frac{u_\infty x}{v}} = -\frac{2}{m+1}\frac{v_0}{u_\infty}\text{Re}_x^{1/2} \tag{8-28}$$

Thus $\zeta(0) = $ constant, as a boundary condition to Eq. (8-24), gives us a family of similarity solutions for $v_0 \neq 0$. However, for $\zeta(0)$ to be a constant, v_0 must vary along the surface in the particular manner prescribed by Eq. (8-26); that is,

$$v_0 \propto x^{(m-1)/2} \tag{8-29}$$

Note that only for the stagnation point $m = 1$ can v_0 be a constant. for $u_\infty = $ constant, $m = 0$ and v_0 must vary as $1/\sqrt{x}$ in order for similar velocity profiles to obtain.

The group of variables

$$\frac{v_0}{u_\infty}\text{Re}_x^{1/2} \tag{8-30}$$

is frequently referred to as the "blowing" (or "suction") parameter. Equation (8-24) has been solved numerically for various values of this parameter, as well as various values of m. Some of the results for $m = 0$ are given in Table 8-3, from which the local friction coefficient may be evaluated from Eq. (8-25). Further results may be found in Ref. 5. However, the major area of usefulness of these solutions arises in connection with heat- and mass-transfer solutions to be considered later, where the complete $\zeta(\eta)$ is required for each case.

NONSIMILAR MOMENTUM BOUNDARY LAYERS

In the preceding sections we transformed the momentum equation and boundary conditions using η, a similarity variable, to eliminate the x

TABLE 8-3
Solutions to the laminar constant-property boundary layer with blowing or suction at the wall and u_∞ = constant

$\dfrac{v_0}{u_\infty} \mathrm{Re}_x^{1/2}$	$\zeta''(0)$	
−2.500	2.590	
−0.750	0.945	Suction
−0.250	0.523	
0	0.332	
+0.250	0.165	
+0.375	0.094	Blowing
+0.500	0.036	(Separation)
+0.585	0	

dependence from the equation and boundary conditions. We found that the transformation was successful, provided that u_∞ was constant or u_∞ and/or v_0 had power-law variations with x. For arbitrary u_∞, v_0 constant, or arbitrary transverse surface curvature, the boundary-layer flow will be nonsimilar.

Several classes of solution methods exist for solving the nonsimilar laminar boundary-layer problem. Perhaps the most exact solution is a numerical solution to the governing partial differential equations by finite-difference or finite-element methods. A second class comprises the approximate boundary-layer solutions employing the momentum integral equation. This class is discussed in the sections following this one.

The third class of approximate solutions employs the concept of local similarity and local nonsimilarity. Consider the general transformation of the independent variables in the momentum equation and boundary conditions from (x, y) to (ξ, η). If either ξ or derivatives with respect to ξ remain after transformation, similarity solutions will not exist.† However, *local similarity* solutions can be obtained by dropping the ξ derivatives from the transformed momentum equation and boundary conditions and retaining ξ as a parameter. The resulting solution is generally valid if ξ or the discarded derivatives are small. *Local*

---

† It is possible that another set of similarity variables will result in similarity. The free-parameter method described by Hansen[6] is a useful method for searching for similarity variables.

nonsimilarity solutions carry the essence of the local similarity solution a step further by retaining the ξ derivatives in the solution procedure. Sparrow, Quack, and Boerner[7] develop a procedure to generate an additional equation and boundary conditions for $\partial \zeta / \partial \xi$ and solve it and the momentum equation simultaneously. A solution procedure using asymptotic expansions is described by Dewey and Gross.[8] With both the local similar and local nonsimilar methods, the solutions are independent of the upstream history of the flow.

AN APPROXIMATE LAMINAR BOUNDARY-LAYER SOLUTION FOR CONSTANT FREE-STREAM VELOCITY DEVELOPED FROM THE MOMENTUM INTEGRAL EQUATION

To illustrate the power and simplicity of the momentum integral equations as a device for approximate solution of the boundary layer, let us again consider the laminar constant-property boundary layer with constant free-stream velocity. The applicable form of the momentum integral equation is Eq. (6-9). The wall shear stress can be expressed in terms of the velocity gradient at the wall, and thus Eq. (6-9) becomes

$$\frac{\nu}{u_\infty^2} \left(\frac{\partial u}{\partial y} \right)_0 = \frac{d\delta_2}{dx} \tag{8-31}$$

We now must propose some kind of velocity profile in the boundary layer so that we can develop an expression for the momentum thickness δ_2. The power of the method lies in the fact that the solution is not highly dependent on the shape of the velocity profile. Previous experience with laminar flow in a tube might suggest that a simple parabola would be satisfactory; and, as a matter of fact, a quite good solution can be obtained with a simple parabola. A better solution is obtained, however, if the boundary-layer differential equation (8-1) is examined and it is noted that the second derivative of u with respect to y must go to zero at the wall. A simple parabola does not have this property, but a cubic parabola does. Consider the cubic parabola

$$u = ay + by^3$$

A parabolic velocity profile implies that the boundary layer has a definite, finite thickness, rather than extending indefinitely in the y direction as in the exact solution, and this artificiality must be accepted in the integral methods. Suppose we let this distance be a boundary-layer thickness δ. Then we can introduce the boundary conditions that $\partial u / \partial y = 0$ and

$u = u_\infty$ when $y = \delta$. The constants a and b can then be evaluated, yielding

$$\frac{u}{u_\infty} = \frac{3}{2}\frac{y}{\delta} - \frac{1}{2}\left(\frac{y}{\delta}\right)^3 \tag{8-32}$$

Substitution of (8-32) into the integral definition of δ_2, Eq. (6-6), yields an expression for δ_2. This expression is, in turn, differentiated with respect to x and substituted in place of the right-hand side of Eq. (8-31). The left-hand side is obtained directly from Eq. (8-32), with the final result

$$\delta\, d\delta = \frac{140}{13}\frac{\nu}{u_\infty}\, dx$$

This equation is readily integrated for the boundary condition $\delta = 0$ at $x = 0$ (for example, the boundary layer originates at the beginning of the plate), yielding

$$\delta = 4.64\sqrt{\nu x / u_\infty} \tag{8-33}$$

Equation (8-33) gives the boundary-layer thickness (artificial though it may be) as a function of x. The wall shear stress can next be evaluated by substituting Eq. (8-33) into Eq. (8-32), which gives the complete velocity field, and then evaluating the gradient at the wall. Next the friction coefficient, Eq. (6-10), is introduced, and the final result for the local friction coefficient is

$$c_f = \frac{0.646}{\mathrm{Re}_x^{1/2}} \tag{8-34}$$

This result is within three percent of the exact Blasius solution, Eq. (8-13), and is illustrative of the power of the method. We use this result later in connection with a heat-transfer solution, where it proves more convenient than the exact solution.

AN APPROXIMATE LAMINAR BOUNDARY-LAYER SOLUTION FOR ARBITRARILY VARYING FREE-STREAM VELOCITY OVER A BODY OF REVOLUTION

A somewhat different use of the momentum integral equation in an approximate solution is illustrated by the following simple and powerful procedure.

Consider Eq. (6-7) with density constant and no injection or suction:

$$\frac{\tau_0}{\rho u_\infty^2} = \frac{d\delta_2}{dx} + \delta_2\left(2 + \frac{\delta_1}{\delta_2}\right)\frac{1}{u_\infty}\frac{du_\infty}{dx} + \frac{\delta_2}{R}\frac{dR}{dx} \tag{8-35}$$

To simplify the algebra, let us introduce still another boundary-layer thickness: the *shear thickness*

$$\delta_4 = \frac{\mu u_\infty}{\tau_0} = \frac{u_\infty}{(\partial u / \partial y)_0} \tag{8-36}$$

As a further simplification, let

$$T = \frac{\delta_2}{\delta_4}, \qquad H = \frac{\delta_1}{\delta_2} \tag{8-37}$$

Substituting these definitions into Eq. (8-35) and then rearranging, we get

$$\frac{u_\infty}{v} \frac{d\delta_2^2}{dx} + \frac{2\delta_2^2 u_\infty}{v} \frac{dR}{R dx} = 2\left[T - (2 + H) \frac{\delta_2^2}{v} \frac{du_\infty}{dx} \right] \tag{8-38}$$

Let us define a local parameter:

$$\lambda = \frac{\delta_2^2}{v} \frac{du_\infty}{dx} \tag{8-39}$$

Let us now make the *assumption* that the boundary-layer velocity profile depends only on local conditions and, specifically, that

$$T = T(\lambda), \qquad H = H(\lambda)$$

A dimensional analysis of the variables on which the momentum thickness must depend would also suggest that the λ group is a significant parameter. The right-hand side of (8-38) is, then, entirely a function of λ; that is,

$$\frac{u_\infty}{v} \frac{d\delta_2^2}{dx} + \frac{2\delta_2^2 u_\infty}{v} \frac{dR}{R dx} = F(\lambda) \tag{8-40}$$

where

$$F(\lambda) = 2[T - (2 + H)\lambda]$$

Attempts to find a satisfactory form for $F(\lambda)$ have followed various courses. Examination of various *exact* solutions indicates that the fundamental assumption is valid to a good degree of approximation; for example, a plot of F versus λ from the wedge solutions and from various other particular solutions yields curves that are quite close together, especially for accelerating flows. Thwaites[9] made such a comparison and further showed that the function could be quite well approximated by a linear relation, which greatly simplifies the algebra:

$$F = a - b\lambda = 0.44 - 5.68\lambda \tag{8-41}$$

A more exact but somewhat less convenient procedure is suggested by Spalding.[10]

Substituting Eq. (8-41) into Eq. (8-40), rearranging, and combining the derivatives, we get

$$d(R^2 \delta_2^2 u_\infty^b) = a v u_\infty^{b-1} R^2 \, dx$$

Integrating between 0 and x yields

$$R^2 \delta_2^2 u_\infty^b - R_0^2 \delta_{2,0}^2 u_{\infty,0}^b = a v \int_0^x u_\infty^{b-1} R^2 \, dx$$

Take $x = 0$ as the leading edge of the body; one of the following is also zero there: R_0, $\delta_{2,0}$, or $u_{\infty,0}$. Thus

$$R^2 \delta_2^2 u_\infty^b = a v \int_0^x u_\infty^{b-1} R^2 \, dx$$

Taking the given values of a and b and solving for δ_2, we find

$$\delta_2 = \frac{0.664 v^{0.5}}{R u_\infty^{2.84}} \left(\int_0^x u_\infty^{4.68} R^2 \, dx \right)^{0.5} \tag{8-42}$$

Solution of (8-42) then involves only a simple integration procedure that depends on the given variation of u_∞ and R with x. Once we have the momentum thickness δ_2, the shear thickness and the local wall shear stress can be evaluated from Table 8-4 based on the wedge solutions.

More exact procedures for calculating the laminar boundary layer may be found in many places in the literature, and the appropriate one should be used where precise results are required. Most, however, require tedious numerical calculations, and for many engineering purposes Eq. (8-42) will suffice.

TABLE 8-4
Functions for use with Eq. (8-42)

$\lambda = \dfrac{\delta_2^2}{v}\dfrac{du_\infty}{dx}$	$T = \dfrac{\delta_2}{\delta_4} = \dfrac{\delta_2 \tau_0}{\mu u_\infty}$	$H = \dfrac{\delta_1}{\delta_2}$
−0.082	0	3.70 Separation
−0.080	0.039	3.58
−0.070	0.089	3.17
−0.060	0.113	2.99
−0.040	0.153	2.81
−0.024	0.182	2.71
0	0.220	2.60 Flat plate
0.016	0.244	2.55
0.048	0.291	2.44
0.080	0.333	2.34 Stagnation (approx.)
0.12	0.382	2.23
0.25	0.500	2.00

PROBLEMS

8-1. Air at 26°C and 1 atm pressure flows normally to a 5 cm diameter circular cylinder at a velocity of 9 m/s. It can be shown from potential flow theory that in the vicinity of the forward stagnation point for flow normal to a cylinder the velocity along the surface, u_∞ (which is the velocity just outside of any boundary layer), is given by

$$u_\infty = \frac{4Vx}{D}$$

where V is the oncoming normal velocity, x is the distance along the surface measured from the stagnation point, and D is the diameter of cylinder. Calculate the displacement thickness of the boundary layer at the stagnation point, and discuss the significance of the result.

8-2. Derive Eq. (8-21) from potential flow theory.

8-3. Solve the laminar boundary layer for constant free-stream velocity, using the momentum integral equation and an assumption that the velocity profile may be approximated by

$$\frac{u}{u_\infty} = \sin\frac{\pi y}{2\delta}$$

Evaluate the momentum thickness, displacement thickness, and friction coefficient, and compare with the exact solution.

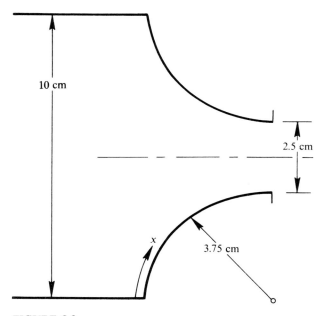

10 cm

2.5 cm

x

3.75 cm

FIGURE 8-3

8-4. Redevelop Eq. (8-42) for the case where density and dynamic viscosity are functions of x.

8-5. Air emerges from the axisymmetric nozzle shown in Fig. 8-3 at a centerline velocity of $10 \, \text{m/s}$ at 1 atm pressure and 21°C. Assuming an essentially constant-density and constant-temperature process (and this is ensured by the low velocity), calculate the displacement thickness of the boundary layer at the nozzle throat, assuming that the free-stream velocity along the inner surface of the nozzle varies linearly with distance, starting with $u_\infty = 0$ at the sharp corner. Calculate the air mass flow rate through the nozzle and the overall pressure drop through the nozzle. On the basis of these results, discuss the concept of a nozzle "discharge coefficient." What would be the discharge coefficient of this nozzle? If you were to define a Reynolds number based on throat diameter and mean velocity, how would the discharge coefficient vary with Reynolds number?

8-6. Derive Eq. (8-24) in a manner similar to that used in the development of Eq. (8-8) for zero pressure gradient.

8-7. *Computer analysis of the laminar momentum boundary layer over a flat plate with zero pressure gradient and constant properties:* Calculate the boundary layer flow and compare the resuls with the similarity solution for development in the streamwise direction of such quantities as the boundary-layer thickness, displacement thickness, and momentum thickness. Evaluate the concept of boundary-layer similarity. Compare the friction coefficient results based on x Reynolds number and momentum-thickness Reynolds number with the results in the text. Examine several consecutive velocity profiles to evaluate the momentum-thickness distribution to assess the validity of the momentum integral Eq.(6-7). Feel free to investigate any other attribute of the boundary-layer flow. You can choose how to set up the problem in terms of a suitable choice of initial x Reynolds number, dimensions, and fluid properties (constant). For the initial conditon, let the velocity profile be constructed from Table 8-1.

8-8. *Computer analysis of the laminar momentum boundary layer over a flat plate with a pressure gradient based on* $u_\infty = Cx^m$ *and constant properties:* Calculate the boundary layer flow for $m = 0.111$ and compare the results with the similarity solution for $m = 0.111$ and $m = 0.0$. Evaluate the concept of boundary-layer similarity. Compare the friction coefficient results based on x Reynolds number with the results in the text. Examine several consecutive velocity profiles to evaluate the momentum-thickness distribution to assess the validity of the momentum integral, Eq. (6-7). Feel free to investigate any other attribute of the boundary layer flow. You can choose how to set up the problem in terms of a suitable choice of initial x Reynolds number, dimensions, and fluid properties (constant). For the initial condition let the velocity profile be constructed from one of several methods: a Runge–Kutta type of solution to Eq. (8-24), or an estimated profile based on Prob. 8-3 and a suitable estimate of δ from the open literature.

REFERENCES

1. Schlichting, H.: *Boundary Layer Theory*, 6th ed., McGraw-Hill, New York, 1968.
2. Moore, F. K. (ed.): *Theory of Laminar Flows,* vol. 13: *High Speed Aerodynamics and Jet Propulsion,* Princeton University Press, Princeton, NJ 1964.
3. Blasius, H.: *Z. (Angew) Math. Phys.,* vol. 56, 1908.
4. Falkner, V. M., and S. W. Skan: *Phil. Mag.,* vol. 12, p. 865, 1931.
5. NASA TN 3151, Washington, September 1954.
6. Hansen, A. G.: *Similarity Analysis of Boundary Value Problems in Engineering,* Prentice-Hall, Englewood Cliffs, NJ, 1964.
7. Sparrow, E. M., H. Quack, and C. J. Boerner, *AIAA J.,* vol. 8, pp. 1936–1942, 1970.
8. Dewey, C. F., Jr, and J. F. Gross: *Advances in Heat Transfer,* vol. 4, pp. 317–446. Academic Press, New York, 1964.
9. Thwaites, H.: *Aero. Q.,* vol. 1, p. 245, 1949.
10. Spalding, D. B.: *J. Fluid Mech.,* vol. 4, p. 22, 1958.

CHAPTER
9

HEAT TRANSFER: LAMINAR FLOW INSIDE TUBES

In this chapter we consider heat exchange between a fluid and a solid surface, when the fluid is flowing steadily with laminar motion inside a smooth tube. We assume that all body forces are negligible and that the fluid is forced through the tube by some external means, unrelated to the temperature field in the fluid. All the laminar flow solutions considered in this chapter are based on the idealization of constant fluid properties. The influence of temperature-dependent properties is discussed in Chap. 15.

Since the major technical applications of these results are in the analysis and design of heat exchangers, it would be well at this point to discuss the distinction between heat-exchanger theory and the convection theory with which we are concerned here. Heat-exchanger theory concerns the total heat-transfer rates and terminal fluid temperatures in multifluid heat exchangers, the influence of the flow arrangement of the two or more fluids, and the effect of total surface area on these heat-transfer rates and temperatures. The application of heat-exchanger theory depends upon prior knowledge of the local heat-transfer conductance between the fluid and the tube wall surface, Eq. (1-1), for the particular flow situation existing within the heat-exchanger tubes. This conductance, or heat-transfer coefficient, may be obtained experimentally, or in some cases it may be deduced by analytic means. In this study we are concerned primarily with the analytic evaluation of the conductance under various conditions. The results of the evaluation can then be used, *together* with heat-exchanger theory, for the analysis of complete heat exchangers. Some of our results have significant application outside the realm of heat-exchanger theory, and we will also discover situations where the conductance concept loses its useful significance. In these cases it is more convenient to work directly with temperatures and heat-transfer rates than to employ a conductance.

We start with a consideration of the region far removed from the entrance to the tube, where both a fully developed velocity profile and a fully developed temperature profile obtain. Here we consider tubes of various flow cross-section shape: a circular tube, circular-tube annuli, rectangular tubes, and triangular tubes. We consider heating (or cooling) from the two surfaces of an annulus and the effects of a peripheral heat-flux variation around a tube.

Next we consider a class of problems in which the velocity profile is fully developed and remains fixed while the temperature profile develops. The fluid temperature upstream of some point is assumed to be uniform and equal to the surface temperature, there being no heat transfer in this region. Following this point, heat transfer takes place: then we are concerned with the development of the temperature profile. These *thermal-energy-length* solutions are considered in some detail for the circular tube, and results are also presented for rectangular tubes and concentric circular-tube annuli. A method is then developed whereby the thermal-entry-length solutions for constant surface temperature and constant heat flux can be used to solve for the temperature distribution resulting from any arbitrary axial distribution of surface temperature or heat flux. Finally, some results are presented for the combined hydrodynamic- and thermal-entry length, that is, where both the velocity and fluid temperatures are uniform at the tube entrance.

THE ENERGY DIFFERENTIAL EQUATIONS FOR FLOW THROUGH A CIRCULAR TUBE

We have already developed the applicable differential energy equation for flow in a circular tube under conditions of steady laminar flow with low velocities, no concentration gradients, negligible pressure gradient, and no source functions, Eq. (4-35). Let us restrict the problem to constant properties and introduce, for a perfect gas, the enthalpy relation

$$di = c \, dt$$

Equation (4-35) then becomes

$$upc\frac{\partial t}{\partial x} + v_r\rho c\frac{\partial t}{\partial r} - \left[\frac{k}{r}\frac{\partial}{\partial r}\left(r\frac{\partial t}{\partial r}\right) + \frac{k}{r^2}\frac{\partial^2 t}{\partial \phi^2} + k\frac{\partial^2 t}{\partial x^2}\right] = 0 \qquad (9\text{-}1)$$

[As noted in the discussion following Eq. (4-37), this equation is also valid for an incompressible liquid with no assumption regarding the pressure-gradient term such as was necessary in the development of Eq. (4-35).]

Let us now further restrict the problem to symmetric heat transfer $(\partial^2 t/\partial \phi^2 = 0)$ and hydrodynamically fully developed flow $(v_r = 0)$. Equation (9-1) then becomes, after transposition,

$$\frac{1}{r}\frac{\partial}{\partial r}\left(r\frac{\partial t}{\partial r}\right) + \frac{\partial^2 t}{\partial x^2} = u\left(\frac{\rho c}{k}\right)\frac{\partial t}{\partial x} \qquad (9\text{-}2)$$

Let $\alpha = k/\rho c$, the *molecular thermal diffusivity* of the fluid. Thus

$$\frac{1}{r}\frac{\partial}{\partial r}\left(r\frac{\partial t}{\partial r}\right) + \frac{\partial^2 t}{\partial x^2} = \frac{u}{\alpha}\frac{\partial t}{\partial x} \qquad (9\text{-}3)$$

If we can neglect axial conduction relative to radial conduction (and we investigate the conditions under which this is permissible later) then $\partial^2 t/\partial x^2 = 0$, and we finally obtain the energy equation for laminar flow in a circular tube, solutions to which comprise the major part of this chapter:

$$\frac{1}{r}\frac{\partial}{\partial r}\left(r\frac{\partial t}{\partial r}\right) = \frac{u}{\alpha}\frac{\partial t}{\partial x} \qquad (9\text{-}4)$$

THE CIRCULAR TUBE WITH FULLY DEVELOPED VELOCITY AND TEMPERATURE PROFILES

The existence of a fully developed temperature profile at points far removed from the entrance to the tube is a little more difficult to visualize than the fully established velocity profile of Chap. 7. However, under a few possible heating conditions we can visualize the possibility of there existing a nondimensional temperature profile that is invariant with tube length, and under these conditions a solution to the energy equation can be effected very easily.

Criterion for a Fully Developed Temperature Profile

The term *fully developed temperature profile* implies that there exists, under certain conditions, a generalized temperature profile that is invariant with tube length, as in Fig. 9-1.

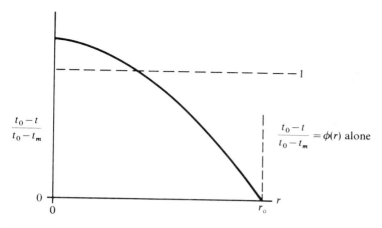

FIGURE 9-1
A fully developed temperature profile.

For convenience t_m is defined as the *mixed mean fluid temperature.* This temperature is also sometimes referred to as the mass-averaged temperature, bulk fluid temperature, or mixing-cup temperature: it is the temperature that characterizes the average thermal energy state of the fluid. We have

$$\text{Axial convected thermal energy rate} = \dot{m}ct_m = (A_cV\rho)ct_m$$

$$= \int_{A_c} u\rho ct \, dA_c$$

Then

$$t_m = \frac{1}{A_cV} \int_{A_c} ut \, dA_c \tag{9-5}$$

[Compare with the definition of the mean velocity, Eq. (7-5).]

If the nondimensional temperature is defined in terms of t_m, the mixed mean temperature, and t_0, the wall temperature, and if the nondimensional temperature profile is invariant in the flow direction, we can write the following for the conditions at the wall:

$$\left[\frac{\partial}{\partial r}\left(\frac{t_0 - t}{t_0 - t_m}\right)\right]_{r=r_0} = \text{const} = -\frac{(\partial t/\partial r)_{r=r_0}}{t_0 - t_m}$$

Let us now introduce a *convection conductance,* or *heat-transfer coefficient h,* defined as in Eq. (6-22), but using the mixed mean temperature t_m; then

$$\dot{q}_0'' = h(t_0 - t_m) \tag{9-6}$$

Note that this conductance is defined with reference to the mixed mean temperature; this is quite arbitrary, but convenient.

From the definition of thermal conductivity, the following must apply at the wall surface (and this must be the case whether the flow is laminar or turbulent):

$$\dot{q}_0'' = -k\left(\frac{\partial t}{\partial r}\right)_{r=r_0}$$

Combining these three equations yields

$$\frac{\dot{q}_0''/k}{\dot{q}_0''/h} = \frac{h}{k} = \text{const}$$

Thus

$$h = \text{const}, \quad \text{invariant with } x$$

Experimentally this is often noted to be at least approximately true at points removed from the tube entrance for certain particular boundary conditions. Thus the nondimensional temperature profile is commonly observed to be invariant with tube length; and, in fact, most heat-exchanger analysis is based on a constant conductance h.

The statement that this profile is invariant with x can be expressed as

$$\frac{\partial}{\partial x}\left(\frac{t_0 - t}{t_0 - t_m}\right) = 0$$

Differentiating and solving for $\partial t/\partial x$,

$$\frac{\partial t}{\partial x} = \frac{dt_0}{dx} - \frac{t_0 - t}{t_0 - t_m}\frac{dt_0}{dx} + \frac{t_0 - t}{t_0 - t_m}\frac{dt_m}{dx} \tag{9-7}$$

This expression can now be substituted into the right-hand side of the energy differential equation (9-4). If we do so, we see that there are at least two boundary conditions (and these are special cases of a more general boundary condition, which we shall examine later) for which it will be possible to integrate the energy equation directly with respect to r, that is, treat it as an ordinary differential equation. Let us first examine some possible conditions.

Consider the case of *constant heat rate* Ⓗ per unit of tube length. Technically, constant-heat-rate problems arise in a number of situations: electric resistance heating, radiant heating, nuclear heating, and in counterflow heat exchangers when the fluid capacity rates† are the same.

---

† Product of mass flow rate and specific heat of fluid.

This is, then, a rather important boundary condition. Writing the convection rate equation,

$$\dot{q}_0'' = h(t_0 - t_m) = \text{const}$$

If h is a constant then

$$t_0 - t_m = \text{const}$$

from which

$$\frac{dt_0}{dx} - \frac{dt_m}{dx} = 0, \qquad \frac{dt_0}{dx} = \frac{dt_m}{dx}$$

Thus, substituting in Eq. (9-7),

$$\frac{\partial t}{\partial x} = \frac{dt_0}{dx} = \frac{dt_m}{dx} \tag{9-8}$$

It also follows that dt_m/dx is a constant, as a simple energy balance reveals.

Next consider the case where the *surface temperature* Ⓣ is *constant.* This is another very common convection application, and it occurs in such heat exchangers as evaporators, condensers, and, in fact, any heat exchanger where one fluid has a very much higher capacity rate than the other. For constant surface temperature,

$$\frac{dt_0}{dx} = 0$$

and Eq. (9-7) reduces to

$$\frac{\partial t}{\partial x} = \frac{t_0 - t}{t_0 - t_m} \frac{dt_m}{dx} \tag{9-9}$$

For these two conditions, *constant heat rate* and *constant surface temperature,* Eqs. (9-8) and (9-9) can be substituted into the energy equation (9-4) to yield the following two differential equations:

Constant heat rate, Ⓗ

$$\frac{1}{r}\frac{\partial}{\partial r}\left(r\frac{\partial t}{\partial r}\right) = \frac{u}{\alpha}\left(\frac{dt_m}{dx}\right)$$

Constant surface temperature, Ⓣ

$$\frac{1}{r}\frac{\partial}{\partial r}\left(r\frac{\partial t}{\partial r}\right) = \frac{u}{\alpha}\frac{t_0 - t}{t_0 - t_m}\left(\frac{dt_m}{dx}\right)$$

The corresponding temperature variations with tube length for these two conditions are shown in Fig. 9-2.

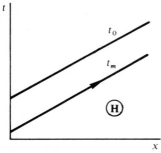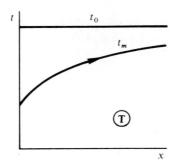

FIGURE 9-2

In both cases the independent variables (x, r) have been separated so that we can hold x constant and integrate directly with respect to r and determine the temperature profile.

The constant-surface-temperature and constant-heat-rate boundary conditions are shown by Sparrow and Patankar[1] to be special cases of a more general exponential heat-flux boundary condition. But these two do cover the usual extremes met in heat-exchanger design, and are thus of great technical importance. It must be remembered, however, that we are still restricted to a long tube; near the tube entrance a fully developed temperature profile will not exist.

Constant-Heat-Rate Solution Ⓗ

The equation to be solved is

$$\frac{1}{r}\frac{\partial}{\partial r}\left(r\frac{\partial t}{\partial x}\right) = \frac{u}{\alpha}\left(\frac{dt_m}{dx}\right) \tag{9-10}$$

The applicable boundary conditions are

$$t = t_0 \quad \text{at } r = r_0$$

$$\frac{\partial t}{\partial r} = 0 \quad \text{at } r = 0$$

Substituting the parabolic velocity profile, Eq. (7-8), for u yields

$$\frac{1}{r}\frac{\partial}{\partial r}\left(r\frac{\partial t}{\partial r}\right) = \frac{2V}{\alpha}\left(1 - \frac{r^2}{r_0^2}\right)\frac{dt_m}{dx}$$

This equation can now be directly integrated twice with respect to r, and the two constants of integration can be evaluated from the boundary conditions:

$$t = t_0 - \frac{2V}{\alpha}\frac{dt_m}{dx}\left(\frac{3r_0^2}{16} + \frac{r^4}{16r_0^2} - \frac{r^2}{4}\right) \tag{9-11}$$

Equation (9-11) is then the desired temperature profile. With the temperature profile established, the mixed mean temperature can be evaluated next. From Eq. (9-5),

$$t_m = \frac{2}{r_0^2 V} \int_0^{r_0} utr\, dr \tag{9-12}$$

If Eqs. (9-8) and (9-11) are substituted for u and t, this equation can be readily integrated to yield

$$t_m = t_0 - \frac{11}{96}\frac{2V}{\alpha}\frac{dt_m}{dx} r_0^2 \tag{9-13}$$

At this point the conductance h can be evaluated by use of its defining equation (9-6):

$$\dot{q}_0'' = h(t_0 - t_m) = h\frac{11}{96}\frac{2V}{\alpha}\frac{dt_m}{dx} r_0^2 \tag{9-14}$$

The surface heat flux must yet be evaluated, and this can be done in one of two ways. The temperature profile, Eq. (9-11), can be differentiated to give the temperature gradient at the wall surface, from which $\dot{q}_0''$ can be evaluated from the conduction rate equation. A still simpler method is to make an energy balance on a differential control volume, as shown in Fig. 9-3.

Applying the conservation-of-energy principle,† Eq. (2-5), and solving for the wall surface heat flux, we have

$$\dot{q}_0'' = \frac{r_0 V\rho c}{2}\frac{dt_m}{dx} \tag{9-15}$$

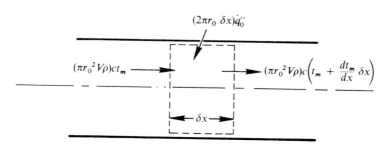

FIGURE 9-3
Control volume for an energy flow in a circular tube.

---

† Note in this case that any conversion of mechanical to thermal energy is neglected, so that conservation of thermal energy only is considered.

Combining (9-14) and (9-15), and solving for h,

$$h = \frac{48}{11}\frac{k}{D} = 4.364\frac{k}{D} \tag{9-16}$$

where $D = 2r_0$.

Note the simple form of the solution as soon as we introduce the conductance. Since all temperatures cancel, we see that $\dot{q}_0''$ is in this case truly proportional to the temperature difference, which is the reason for introducing the convection rate equation and defining a conductance. The rate equation and the conductance are purely conveniences that are extremely useful in presenting the solution in a compact form and in using the solution to calculate heat-transfer rates. Later we run into situations where the conductance does not introduce any simplification and therefore loses its useful significance, and in those cases we dispense with it.

Note that the conductance h depends only on k and D, and is independent of V, ρ, c, and so on. This is only true for laminar flow, however, and only for the special case of fully developed velocity and temperature profiles. This independence results from the fact that the energy transport is a purely molecular conduction problem under these conditions.

Equation (9-16) is most often expressed in nondimensional form as follows:

$$hD/k = 4.364$$

or

$$\mathrm{Nu} = 4.364 \tag{9-17}$$

where $\mathrm{Nu} = hD/k$ is called the *Nusselt number*.

This solution can be represented in a different way that allows an interesting comparison. By dimensional reasoning,

$$\mathrm{Nu} = \mathrm{St}\,\mathrm{Pr}\,\mathrm{Re}$$

where

$\mathrm{St} = h/V\rho c$ or h/Gc,	the Stanton number
$\mathrm{Pr} = \mu c/k$,	the Prandtl number
$\mathrm{Re} = DV\rho/\mu$ or DG/μ,	the Reynolds number

Note that the Nusselt number and the Stanton number are simply two different nondimensional versions of the heat-transfer coefficient. Both are extensively used, but the choice is usually a personal one; one can always be evaluated from the other. The Prandtl number was previously defined in Chap. 3.

Thus

$$\text{St Pr Re} = 4.364$$

$$\text{St Pr} = \frac{4.364}{\text{Re}} \tag{9-18}$$

Note the similarity in form to the corresponding friction coefficient, Eq. (7-15). The Prandtl number is merely a fluid property parameter, so it is evident that a fixed relation exists between the Stanton number and the friction factor. It is found later in fully developed turbulent flow that the Stanton number is again proportional (approximately) to the friction coefficient.

Constant-Surface-Temperature Solution Ⓣ

The equation to be solved is

$$\frac{1}{r}\frac{\partial}{\partial r}\left(r\frac{\partial t}{\partial r}\right) = \frac{u}{\alpha}\frac{t_0 - t}{t_0 - t_m}\frac{dt_m}{dx} \tag{9-19}$$

The applicable boundary conditions are

$$t = t_0 \quad \text{at} \quad r = r_0$$

$$\frac{\partial t}{\partial r} = 0 \quad \text{at} \quad r = 0$$

The parabolic velocity profile, Eq. (7-8), is substituted for u, as before, yielding

$$\frac{1}{r}\frac{\partial}{\partial r}\left(r\frac{\partial t}{\partial r}\right) = \frac{2V}{\alpha}\left(1 - \frac{r^2}{r_0^2}\right)\frac{t_0 - t}{t_0 - t_m}\frac{dt_m}{dx}$$

This problem has been solved by Bhatti,[7] as reported by Kakac, Shah, and Aung.[8] The solution is in the form of an infinite series:

$$\frac{t_0 - t}{t_0 - t_m} = \sum_{n=0}^{\infty} C_{2n}\left(\frac{r}{r_0}\right)^{2n} \tag{9-20}$$

where the coefficients C_{2n} are given by

$$C_0 = 1, \quad C_2 = -\tfrac{1}{4}\lambda_0^2 = -1.828397$$

$$C_{2n} = \frac{\lambda_0^2}{(2n)^2}(C_{2n-4} - C_{2n-2})$$

$$\lambda_0 = 2.704364$$

The Nusselt number corresponding to the temperature distribution of Eq. (9-20) can be shown to be

$$\text{Nu} = \tfrac{1}{2}\lambda_0^2 = 3.657 \tag{9-21}$$

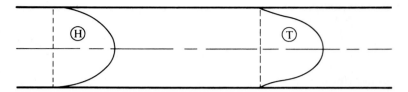

FIGURE 9-4
Fully developed temperature profiles for constant heat rate and constant surface temperature.

This result is 16 percent less than the solution for constant heat rate. The difference in conductance can be explained simply by the slightly different shape of the generalized temperature profile, Fig. 9-4.

It is thus seen that the convection conductance depends to some degree on the type of the surface temperature variation. More is seen of this effect later.

Exponentially Varying Heat-Flux Solution

The two cases just considered, where a fully developed temperature profile is physically possible at some distance from the tube entrance, are special cases of a more general boundary condition where the heat flux varies *exponentially* along the tube:

$$q'' = q_0'' \exp\left(\tfrac{1}{2}mx^+\right)$$

where m is a constant exponent that can assume positive or negative values. The case $m = 0$ corresponds to the constant-heat-rate case, and $m = -2\lambda_0^2 = -14.63$ corresponds to the constant-surface-temperature case. Shah and London[5] have developed the following correlation equation, which fits the exact solution within 3 percent over the range:

$$-51.36 < m < 100$$

$$\text{Nu} = 4.3573 + 0.0424\,m - 2.8368 \times 10^{-4}\,m^2 + 3.6250 \times 10^{-6}\,m^3$$
$$- 7.6497 \times 10^{-8}\,m^4 + 9.1222 \times 10^{-10}\,m^5 - 3.8446 \times 10^{-12}\,m^6$$

$$(9\text{-}22)$$

Effect of Peripheral Heat-Flux Variation

In the two examples already considered, both the heat flux and the surface temperature were uniform around the tube periphery. Frequently the heat flux is unevenly distributed around the tube (a bank of tubes radiantly heated on one side, for example), which leads to hot spots on the tube surface. If the tube wall material is thick and has a sufficiently high conductivity, peripheral conduction in the wall may alleviate this problem, but for thin-walled tubes the difficulty can be acute.

The problem of laminar flow in a circular tube with fully developed, constant-heat-rate conditions axially, but with any arbitrary peripheral variation of heat flux, has been treated by Reynolds.[2,3]

For the applicable differential energy equation, we go back to Eq. (9-1) but again set $v_r = 0$ and $\partial^2 t / \partial x^2 = 0$. This yields the equation for fully developed, constant heat rate:

$$\frac{1}{r}\frac{\partial}{\partial r}\left(r\frac{\partial t}{\partial r}\right) - \frac{1}{r^2}\frac{\partial^2 t}{\partial \phi^2} = \frac{u}{\alpha}\frac{\partial t}{\partial x} = \frac{u}{\alpha}\frac{dt_m}{dx}$$

Of particular interest is the rather simple result obtained for a cosine heat-flux variation, since many nonuniform heating problems can be at least approximated by a cosine variation (Fig. 9-5). For this case $\dot{q}_0''(\phi) = \dot{q}_a''(1 + b \cos \phi)$, and the local Nusselt number turns out to be

$$\text{Nu}(\phi) = \frac{1 + b \cos \phi}{\frac{11}{48} + \frac{1}{2}b \cos \phi} \qquad (9\text{-}23)$$

If $b = 0$, the Nusselt number becomes independent of ϕ and equal to 4.364, as would be expected.

Depending on the magnitude of b, the Nusselt number can vary in very strange ways. For example, if $b = 0.458$, it goes to infinity at $\phi = \pi$. This is a result of the definition of the Nusselt number, which is constructed from a conductance based on the difference between the local surface temperature and the mixed mean fluid temperature; that is,

$$\text{Nu} = hD/k$$

$$\dot{q}_0'' = h(t_0 - t_m)$$

or

$$t_0 - t_m = \dot{q}_0''/h$$

An infinite conductance merely means that the surface temperature is the same as the mixed mean temperature, and does not imply an infinite heat flux. In constant-heat-rate problems (constant heat rate per unit of

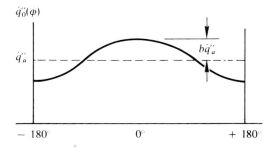

$\dot{q}_0'(\varphi)$

$\dot{q}_a''$

$b\dot{q}_a''$

$-180°$ $0°$ $+180°$ **FIGURE 9-5**

length) the mixed mean temperature of the fluid can always be determined at any point along the tube, by an energy balance; the heat-transfer problem is merely to determine the difference between the mean fluid temperature and the surface temperature.

The Effect of Fluid Axial Conduction

In the development of Eq. (9-4) the axial fluid conduction term $\partial^2 t/\partial x^2$ was neglected. For the fully developed *constant-heat-rate* case this term is always 0, so the Nusselt number is not affected by axial conduction. But for the *constant-surface-temperature* case axial conduction can be of significance at low values of Re and/or Pr. In fact, the significant parameter is the *Péclet number* Pe = Re Pr. This problem has been investigated by Michelsen and Villadsen,[9] and the following equations are recommended:

$$Nu = \begin{cases} 4.180654 - 0.183460 \, Pe & \text{for } Pe < 1.5 \\ 3.656794 + 4.487/Pe^2 & \text{for } Pe > 5 \end{cases} \qquad (9\text{-}24)$$

As can be seen, the effect of axial fluid conduction is totally negligible for Pe = 100, and is quite small even for Pe = 10. In practical applications axial conduction is frequently of considerable significance for laminar flow of liquid metals, which have very low Prandtl numbers. For gases axial conduction can be of importance only at extremely low Reynolds numbers, and for most liquids it is seldom of significance.

Effects of Viscous Dissipation and Internal Heat Sources

In the development of Eq. (9-10) for fully developed, constant heat rate, the influence of both viscous energy dissipation and any internal sources of heat was neglected. The two terms that are introduced by these effects can be seen in Eq. (4-31), namely $\mu\phi$ for viscous dissipation and S for a volume heat source. These terms can be carried through in the development, and will then appear in Eq. (9-10) as simply additive on the right-hand side. For fully developed conditions $\mu\phi$ becomes $\mu(\partial u/\partial r)^2$, and is thus independent of x. If S is held constant, it too is then also independent of x. In either case the temperature profile can still be fully developed and independent of x in a nondimensional sense. Equation (9-10) can be integrated directly as before, with a closed-form algebraic result, and a simple algebraic equation results for the Nusselt number. Kakac, Shah, and Aung[8] present the following results:

$$Nu = 192/(44 + 3\lambda) \qquad (9\text{-}25)$$

where

$$\lambda = SD/q_0'' + 64 \, Br'$$

and
$$\text{Br}' = V^2/q_0''D$$

Br' is known as the *Brinkman number* and is the characteristic non-dimensional parameter for viscous dissipation for heat-flux-specified problems.

Note that Eq. (9-25) reduces to Eq. (9-17) in the absence of both viscous dissipation and internal heat sources.

Viscous dissipation is generally negligible for low-Prandtl-number fluids unless velocity is extremely high, but can become very important for high-Prandtl-number fluids such as oils, even at very moderate velocities.

THE CONCENTRIC CIRCULAR-TUBE ANNULUS WITH FULLY DEVELOPED VELOCITY AND TEMPERATURE PROFILES, ASYMMETRIC HEATING

Heat transfer to a fluid flowing in a circular-tube annulus is a particularly interesting problem, and one of considerable technical importance because either or both of the surfaces can be heated independently. If we restrict consideration, for the moment, to points far removed from the entrance to the tube (as in the previously discussed circular-tube problems), there are again heating conditions of technical interest for which a fully developed temperature profile is possible—constant heat rate per unit of length but with the heat flux on each surface independently specified, and constant but independently specified temperatures on each surface. This problem in all its variations has been completely solved[4] (including the thermal-entry-length problem), but we consider here only the constant-heat-rate case. (In principle, the case of exponentially varying heat flux will lead to fully developed temperature profiles, but this more general case has apparently not been solved.)

The annulus system and the nomenclature to be employed are shown in Fig. 9-6.

The energy differential equation that is applicable to the constant-axial-heat-rate case is exactly the same as for the circular-tube, constant-heat-rate problem, that is, Eq. (9-10). The velocity profile must first be determined, and this can be done by integration of Eq. (7-1):

$$\frac{u}{V} = \frac{2}{M}\left[1 - \left(\frac{r}{r_o}\right)^2 + B \ln\frac{r}{r_o}\right] \tag{9-26}$$

where

$$B = \frac{r^{*2} - 1}{\ln r^*}$$

$$M = 1 + r^{*2} - B$$

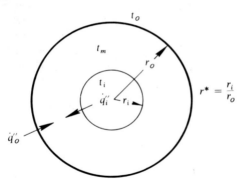

FIGURE 9-6
Nomenclature for flow in a circular-tube annulus.

and r^* is defined in Fig. 9-6. Then the energy equation becomes

$$\frac{1}{r}\frac{\partial}{\partial r}\left(r\frac{\partial t}{\partial r}\right) = \frac{2V}{\alpha M}\left[1 - \left(\frac{r}{r_o}\right)^2 + B\ln\frac{r}{r_o}\right]\frac{dt_m}{dx} \qquad (9\text{-}27)$$

As was the case for the circular tube, this equation may be directly integrated without difficulty. The heat flux at each of the two surfaces specifies the temperature gradient at the surfaces, which provides the necessary boundary conditions. However, it is not necessary to solve the equation for each particular application; the linearity of the energy equation suggests that superposition methods may be employed to build solutions for asymmetric heating by merely adding other solutions. Superposition methods are based on the fact that any number of solutions to a homogeneous and linear differential equation may be added and the result will still satisfy the differential equation. It remains only to add solutions in such a way as to satisfy the desired boundary conditions. In this problem only two fundamental solutions are needed: (1) the outer wall heated with the inner insulated and (2) the inner wall heated with the outer insulated. The algebra of this development is lengthy and is not repeated here; only the final form is given.

Denote the heat fluxes on the inner and outer tubes by $\dot{q}_i''$ and $\dot{q}_o''$, respectively, positive when heat flows *into* the fluid. The heat-transfer coefficients on the inner and outer tubes are defined as follows:

$$\dot{q}_i'' = h_i(t_i - t_m)$$
$$\dot{q}_o'' = h_o(t_o - t_m)$$

We define the inner- and outer-surface Nusselt numbers as

$$\text{Nu}_i = h_i D_h/k$$
$$\text{Nu}_o = h_o D_h/k$$

where $D_h = 4r_h = 2(r_o - r_i)$ for an annulus.

Next, let Nu_{ii} be defined as the inner-tube Nusselt number when the inner tube alone is heated, and let Nu_{oo} be the outer-tube Nusselt number when the outer tube alone is heated; that is, these are the fundamental solutions. Define θ_i^* and θ_o^* as influence coefficients, which may be evaluated from the fundamental solutions. The Nusselt numbers on the inner and outer tubes for any heat-flux ratio can then be evaluated from

$$Nu_i = \frac{Nu_{ii}}{1 - (\dot{q}_o''/\dot{q}_i'')\theta_i^*} \qquad (9\text{-}28)$$

$$Nu_o = \frac{Nu_{oo}}{1 - (\dot{q}_i''/\dot{q}_o'')\theta_o^*} \qquad (9\text{-}29)$$

The complete set of functions for the annulus for fully developed laminar flow is given in Table 9-1.

It is of interest to examine how the Nusselt number can vary with asymmetric heating. Take, for example, the case of flow between parallel planes, which is one of the limiting cases of the annulus: $r^* = 1$. Equations (9-28) and (9-29) become identical in this case, so let us refer to sides 1 and 2 rather than the inner and outer surfaces. Then

$$Nu_1 = \frac{5.385}{1 - 0.346\dot{q}_2''/\dot{q}_1''}$$

If $\dot{q}_2'' = 0$, $Nu_1 = 5.385$, which is, of course, the solution for only one side heated. If both sides are heated equally, $\dot{q}_2''/\dot{q}_1'' = 1$, and then $Nu_1 = Nu_2 = 8.23$. Now note what happens if

$$0.346\dot{q}_2''/\dot{q}_1'' = 1$$

TABLE 9-1
Circular-tube annulus solutions for constant heat rate and fully developed velocity and temperature profiles

r^*	Nu_{ii}	Nu_{oo}	θ_i^*	θ_o^*
0	∞	4.364	∞	0
0.05	17.81	4.792	2.18	0.0294
0.10	11.91	4.834	1.383	0.0562
0.20	8.499	4.883	0.905	0.1041
0.40	6.583	4.979	0.603	0.1823
0.60	5.912	5.099	0.473	0.2455
0.80	5.58	5.24	0.401	0.299
1.00	5.385	5.385	0.346	0.346

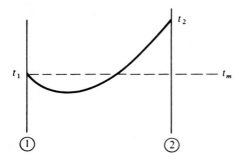

FIGURE 9-7
Temperature distribution resulting from unequal heating on the two sides of a flat duct.

In this case

$$\mathrm{Nu}_1 = \infty$$

and

$$\dot{q}_2''/\dot{q}_1'' = 2.9$$

This does not imply an infinite heat-transfer rate but merely that $t_1 - t_m = 0$ because of the temperature profile shown in Fig. 9-7.

If $\dot{q}_2''/\dot{q}_1'' > 2.9$, the Nusselt number, and thus the conductance, becomes negative. This fact does not destroy either the validity or the usefulness of the Nusselt number, as long as we understand the physical reason.

SOLUTIONS FOR TUBES OF NONCIRCULAR CROSS SECTION WITH FULLY DEVELOPED VELOCITY AND TEMPERATURE PROFILES

Rectangular and Triangular Tubes

Solutions have been obtained for rectangular and triangular tubes in much the same manner as for the circular tubes. The applicable energy differential equation may be deduced from Eq. (4-36):

$$\nabla^2 t = \frac{\partial^2 t}{\partial y^2} + \frac{\partial^2 t}{\partial z^2} = \frac{u}{\alpha}\frac{\partial t}{\partial x}$$

The velocity u is obtained from solutions of the corresponding momentum equation (7-16).

Some of these results, obtained by numerical integration,[5] are listed in Table 9-2. For both the constant-heat-rate case and the constant-surface-temperature case the temperature is assumed to be constant around the tube periphery, and the conductance used in calculating Nu is an average with respect to peripheral area. For tube walls that have finite conductivity in the peripheral and or axial direction Nu is different from those given in Table 9-2. These Nusselt number variations and those for

TABLE 9-2
Nusselt numbers for fully developed velocity and temperature profiles in tubes of various cross sections

Cross-sectional shape	b/a	Nu_H†	Nu_T
circle		4.364	3.66
square	1.0	3.61	2.98
rectangle	1.43	3.73	3.08
rectangle	2.0	4.12	3.39
rectangle	3.0	4.79	3.96
rectangle	4.0	5.33	4.44
rectangle	8.0	6.49	5.60
parallel plates	∞	8.235	7.54
parallel plate (one insulated)		5.385	4.86
triangle		3.11	2.49

† The constant-heat-rate solutions are based on constant *axial* heat rate, but with constant *temperature* around the tube periphery. Nusselt numbers are averages with respect to tube periphery.

various other boundary conditions are discussed in detail in Shah and London[5] and summarized by the same authors in Ref. 6 (see also Ref. 8).

In all cases the Nusselt number is defined as hD_h/k, where

$$D_h = 4 \times (\text{flow area perimeter})$$

This definition of hydraulic diameter is used consistently throughout this book.

An interesting feature of the solutions for shapes with sharp corners, like the square, is the fact that the local heat-transfer conductance varies around the periphery and approaches zero at the corners; this means that the heat flux goes to zero at the corners.

Other Tube Cross-Sectional Shapes

The number of possibly interesting shapes of tubes for technical applications is virtually infinite. For each of these there are a considerable number of possible heating boundary conditions that will lead to fully developed temperature profiles, and thereby to constant Nusselt numbers. A large number of solutions of one type or another have been developed for various of these cases. No attempt will be made here to consider all of these possibilities. Shah and London[5] tabulate a number of these solutions, and also the *Handbook of Single-Phase Convective Heat Transfer*[8] is a useful source of data.

CIRCULAR-TUBE THERMAL-ENTRY-LENGTH SOLUTIONS

So far we have considered only problems where we have fully established velocity and temperature profiles and in which the conductance h is constant with length. Next we consider some cases where the temperature of the fluid is uniform over the flow cross section at the point where heat transfer begins, but the velocity profile is already fully established and invariant. Under these conditions we find that the conductance varies along the length of the tube. We restrict consideration to laminar, incompressible, low-velocity flow.

Strictly speaking, these solutions apply rigorously only when a hydrodynamic starting length is provided so that the velocity profile is fully developed before heat transfer starts, a condition rarely encountered in technical applications. Such solutions are, however, excellent approximations for fluids whose Prandtl numbers are high relative to 1. It is well, therefore, to discuss briefly the significance of the Prandtl number in such problems.

The Prandtl number is a nondimensional group of fluid transport properties:

$$\text{Pr} = \frac{\mu c}{k}$$

If the numerator and denominator are multiplied by density, the Prandtl number can be written as

$$\text{Pr} = \frac{\nu}{\alpha} = \frac{\text{kinematic viscosity}}{\text{thermal diffusivity}}$$

The kinematic viscosity is a diffusivity for momentum, or for velocity, in the same sense that the thermal diffusivity is a diffusivity for heat, or for temperature. (*Diffusivity* is defined as the rate at which a particular effect is diffused through the medium.) If the Prandtl number is 1 then heat and momentum are diffused through the fluid at the same rates: if the velocity and temperature are both uniform at the entrance to a tube, then the velocity and temperature profiles develop together. Later, in consideration of external boundary layers, it is seen that Pr = 1 leads to great simplifications for this reason. If the Prandtl number is greater than 1, it must follow that the velocity profile develops more rapidly than the temperature profile. Actually if the Prandtl number is greater than about 5, the velocity profile leads the temperature profile sufficiently that a solution based on an already fully developed velocity profile will apply quite accurately even though there is no hydrodynamic starting length.

Of course, the opposite also holds: and for a fluid with a Prandtl number less than 1 the temperature profile develops more rapidly than the velocity profile.

The Prandtl number for any particular fluid generally varies somewhat with temperature, but only over a limited range. Figure 9-8 shows the Prandtl number spectrum. Note that the area of applicability of the solutions we are about to consider is limited to $Pr \geqslant 5$. Of course, if a hydrodynamic starting length is actually provided, the solutions are applicable for any fluid. The necessary length of tube for development of the laminar velocity profile is discussed in connection with Eq. (7-20).

Uniform Surface Temperature

The applicable differential energy equation is again Eq. (9-3):

$$\frac{\partial^2 t}{\partial r^2} + \frac{1}{r}\frac{\partial t}{\partial r} = \frac{u}{\alpha}\frac{\partial t}{\partial x} - \frac{\partial^2 t}{\partial x^2} \qquad (9\text{-}30)$$

It is convenient to put this equation in nondimensional form before discussing its solution. The form of the solution and the pertinent nondimensional parameters can be deduced from the differential equation itself. Let us introduce the following dimensionless variables:

$$\theta = \frac{t_0 - t}{t_0 - t_e}$$

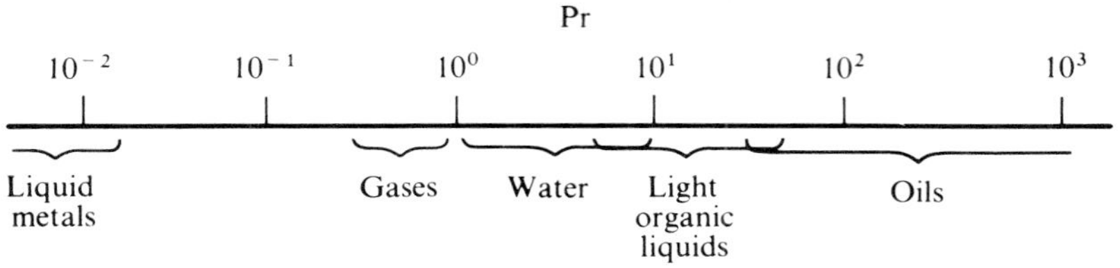

FIGURE 9-8
Prandtl number spectrum of fluids.

where t_0 is the constant surface temperature and t_e is the uniform entering fluid temperature;

$$r^+ = \frac{r}{r_0}, \qquad u^+ = \frac{u}{V}, \qquad x^+ = \frac{x/r_0}{\mathrm{Re}\,\mathrm{Pr}}$$

where x is the axial distance from the point where heat transfer starts.

The particular form of x^+ is chosen so that the parameters will be absorbed within the variables (at least after axial conduction is neglected). At first it may not be apparent that such a choice for x^+ will accomplish this, but if x/r_0 alone had been chosen for the nondimensional length, $\mathrm{Re}\,\mathrm{Pr}$ would appear in the normalized equation and it would be easily seen that this group can be absorbed within the definition of the nondimensional length.

If we now form the indicated derivatives and substitute into Eq. (9-30), we obtain

$$\frac{\partial^2 \theta}{\partial r^{+2}} + \frac{1}{r^+}\frac{\partial \theta}{\partial r^+} = \frac{u^+}{2}\frac{\partial \theta}{\partial x^+} - \frac{1}{(\mathrm{Re}\,\mathrm{Pr})^2}\frac{\partial^2 \theta}{\partial x^{+2}} \tag{9-31}$$

The last term in this equation takes into account heat conduction in the axial direction. Whether this term can be neglected apparently depends on the magnitude of $\mathrm{Re}\,\mathrm{Pr}$, large values of this parameter suppressing the term. The differential equation alone is insufficient to tell how large this parameter must be in order to neglect axial conduction, but it was seen in connection with the fully developed temperature profile problem that the effect of axial conduction is quite small for $\mathrm{Re}\,\mathrm{Pr} > 10$, and completely negligible for $\mathrm{Re}\,\mathrm{Pr} > 100$. For present purposes this term will be neglected.

For hydrodynamically fully developed laminar flow the parabolic velocity profile, Eq. (7-8), is applicable:

$$u = 2V\left(1 - \frac{r^2}{r_0^2}\right), \qquad u^+ = 2(1 - r^{+2})$$

Substituting in Eq. (9-31), the final form of the differential equation is obtained:

$$\frac{\partial^2 \theta}{\partial r^{+2}} + \frac{1}{r^+}\frac{\partial \theta}{\partial r^+} = (1 - r^{+2})\frac{\partial \theta}{\partial x^+} \tag{9-32}$$

The solution sought is then

$$\theta = \theta(x^+, r^+)$$

for the boundary conditions

$$\theta(0, r^+) = 1, \qquad \theta(x^+, 1) = 0, \qquad \left(\frac{\partial \theta}{\partial r^+}\right)_{x^+,\, r^+=0} = 0$$

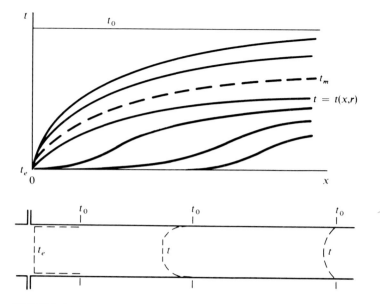

FIGURE 9-9
Development of the temperature profile in the thermal-entry region of a pipe.

Graphically the constant-surface-temperature, thermal-entry-length solution will appear as in Fig. 9-9. Once we have obtained this solution, we can calculate the heat flux at any x^+ from the slope of the temperature profile at the wall:

$$\dot{q}_0''(x^+) = -k\left(\frac{\partial t}{\partial r}\right)_{r=r_0} = k\frac{t_0 - t_e}{r_0}\left(\frac{\partial \theta}{\partial r^+}\right)_{r^+=1}$$

Next, we can determine the local mixed mean temperature of the fluid as a function of x^+ by either of two methods: (1) by integrating over the flow cross section, employing Eq. (9-12), or (2) by integrating $\dot{q}_0''(x^+)$ from $x^+ = 0$ to x^+ and applying an energy balance to define the mixed mean temperature

$$\theta_m(x^+) = \frac{t_0 - t_m}{t_0 - t_e}$$

Then we can define and evaluate a local convection conductance, or heat-transfer coefficient, h_x and a local Nusselt number, which is simply a nondimensional conductance:

$$\dot{q}_0''(x^+) = h_x(t_m - t_0) = -h_x(t_0 - t_e)\theta_m(x^+)$$

$$h_x = \frac{-\dot{q}_0''(x^+)}{(t_0 - t_e)\theta_m(x^+)} \tag{9-33}$$

$$\mathrm{Nu}_x = \frac{2r_0 h_x}{k} = -\frac{2r_0 \dot{q}_0''(x^+)}{k(t_0 - t_e)\theta_m(x^+)} = \frac{-2}{\theta_m}\left(\frac{\partial\theta}{\partial r^+}\right)_{r^+=1}$$

Thus we see that

$$\mathrm{Nu}_x = f(x^+) = f\left(\frac{x/r_0}{\mathrm{Re\ Pr}}\right)$$

Note that, by normalizing the differential equation, the dimensionless parameters necessary to describe the solution are derived before the latter is obtained.

The boundary conditions are such that the temperature gradient at the wall is infinite at $x^+ = 0$; therefore $\mathrm{Nu}_x = \infty$ at $x^+ = 0$. At points far down the tube we would expect the Nusselt number to approach the fully established condition previously discussed. The complete solution could be expected to appear as in Fig. 9-10.

Often in heat-exchanger analysis a mean conductance with respect to tube length, and thus a mean Nusselt number, is of more utility than the local conductance and Nusselt number:

$$\mathrm{Nu}_m = \frac{1}{x^+}\int_0^{x^+} \mathrm{Nu}_x\, dx^+ \tag{9-34}$$

It is generally simpler to employ heat-exchanger theory and work directly with the mixed mean temperature than to integrate Eq. (9-34). The solution for a simple heat exchanger with constant surface temperature

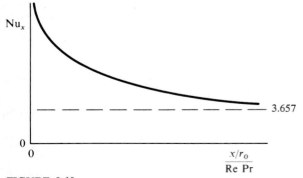

FIGURE 9-10
Variation of local Nusselt number in the thermal-entry region of a tube with constant surface temperature.

can be expressed as

$$\frac{t_m - t_e}{t_0 - t_e} = 1 - \exp\left(-\frac{2\pi r_0}{\dot{m}c}\int_0^x h_x\, dx\right)$$

$$= 1 - \exp\left(-2\int_0^{x^+} \mathrm{Nu}_x\, dx^+\right) = 1 - \theta_m$$

Combining and solving for Nu_m, we get

$$\mathrm{Nu}_m = \frac{1}{2x^+}\ln\frac{1}{\theta_m} \tag{9-35}$$

Let us now examine the final solution to Eq. (9-32) for the indicated boundary conditions. This equation is linear and homogeneous, and such a partial differential equation can always be solved by the method of separation of variables. Let us assume that the solution to (9-32) can be represented by the product

$$\theta = R(r^+) \cdot X(x^+)$$

Then substitution into Eq. (9-32) results in two ordinary differential equations:

$$X' + \lambda^2 X = 0$$

$$R'' + \frac{1}{r^+}R' + \lambda^2 R(1 - r^{+2}) = 0$$

We use $-\lambda^2$ as the separation constant, and the primes refer to differentiation with respect to the independent variables.

The first of these equations is satisfied by a simple exponential function

$$X = C\exp\left(-\lambda^2 x^+\right)$$

Solutions to the second that satisfy the boundary conditions can be found for an infinite number of suitable values of λ, and they have been obtained by both numerical and approximate means.[10] This is an equation of the Sturm–Liouville type, and it has been studied extensively. The final solution takes the form

$$\theta(x^+, r^+) = \sum_{n=0}^{\infty} C_n R_n(r^+)\exp\left(-\lambda_n^2 x^+\right) \tag{9-36}$$

where the λ_n are the eigenvalues, the R_n are the corresponding eigenfunctions, and the C_n are constants.

From (9-36), the heat flux at the wall surface can be readily evaluated from the derivative at the wall, as indicated previously:

$$\dot{q}_0''(x^+) = \frac{2k}{r_0}(t_0 - t_e)\sum_{n=0}^{\infty} G_n\exp\left(-\lambda_n^2 x^+\right) \tag{9-37}$$

where $G_n = -\frac{1}{2}C_n R'_n(1)$ for brevity. Next, the mixed mean temperature θ_m can be evaluated after integration of Eq. (9-37) from $x = 0$ to x^+:

$$\theta_m = 8 \sum_{n=0}^{\infty} \frac{G_n}{\lambda_n^2} \exp\left(-\lambda_n^2 x^+\right) \tag{9-38}$$

The local and mean Nusselt numbers are then readily evaluated from Eqs. (9-33) and (9-35):

$$\mathrm{Nu}_x = \frac{\displaystyle\sum_{n=0}^{\infty} G_n \exp\left(-\lambda_n^2 x^+\right)}{2 \displaystyle\sum_{n=0}^{\infty} (G_n/\lambda_n^2) \exp\left(-\lambda_n^2 x^+\right)} \tag{9-39}$$

$$\mathrm{Nu}_m = \frac{1}{2x^+} \ln\left[\frac{1}{8 \displaystyle\sum_{n=0}^{\infty} (G_n/\lambda_n^2) \exp\left(-\lambda_n^2 x^+\right)}\right] \tag{9-40}$$

The constants and eigenvalues for these infinite series, as presented by Sellars, Tribus, and Klein,[10] are given in Table 9-3. The complete results of the constant-surface-temperature solution are presented in Table 9-4.

For large x^+ the series become increasingly more convergent, until finally, for $x^+ > 0.1$, only the first term is of significance. Then Eq. (9-39) becomes

$$\mathrm{Nu}_x = \frac{G_0 \exp\left(-\lambda_0^2 x^+\right)}{2(G_0/\lambda_0^2) \exp\left(-\lambda_0^2 x^+\right)} = \frac{1}{2}\lambda_0^2 = 3.657 \tag{9-41}$$

TABLE 9-3
Infinite-series-solution functions for the circular tube; constant surface temperature; thermal-entry length

n	λ_n^2	G_n
0	7.313	0.749
1	44.61	0.544
2	113.9	0.463
3	215.2	0.415
4	348.6	0.383

For $n > 2$, $\lambda_n = 4n + \frac{8}{3}$; $G_n = 1.01276\lambda_n^{-1/3}$.

TABLE 9-4
Nusselt numbers and mean temperature for the circular tube; constant surface temperature; thermal-entry length

x^+	Nu_x	Nu_m	θ_m
0	∞	∞	1.000
0.001	12.80	19.29	0.962
0.004	8.03	12.09	0.908
0.01	6.00	8.92	0.837
0.04	4.17	5.81	0.628
0.08	3.77	4.86	0.459
0.10	3.71	4.64	0.396
0.20	3.66	4.15	0.190
∞	3.66	3.66	0.0

This is, of course, the Nusselt number for a fully developed temperature profile, as discussed earlier. Thus the thermal-entry length must be approximately $x^+ = 0.1$:

$$x^+ = 0.1 = \frac{x/r_0}{Re\ Pr} = \frac{2x/D}{Re\ Pr}$$

Thus

$$(x/D)_{\text{fully dev}} \approx 0.05\ Re\ Pr$$

For example, for air with $Pr = 0.7$, flowing with $Re = 500$,

$$(x/D)_{\text{fully dev}} \approx 17.5$$

On the other hand, for an oil with $Pr = 100$, flowing with $Re = 500$,

$$(x/D)_{\text{fully dev}} \approx 2500$$

In an oil heat exchanger it is very rare indeed that anything approaching a fully developed temperature profile is attained; the fully developed profile solutions previously discussed are of very little utility in high-Prandtl-number applications.

Let us consider application of this solution to an oil heat exchanger for which the Prandtl number is 100, the Reynolds number is 500, and each of the tubes in the heat exchanger is 100 diameters long (there may be several passes, but the fluid completely mixes between passes and a new thermal-entry length must develop in each). If the oil is being heated or cooled by an essentially constant-temperature fluid, this solution should provide the proper conductance for the oil side of the heat exchanger. In a problem of this sort the mean conductance is required for use in the heat-exchanger analysis. Under these conditions,

$$x^+ = \frac{2(100)}{500 \times 100} = 0.004$$

Thus, from Table 9-4, $Nu_m = 12.09$. Note the very large error that would have been obtained if the fully developed solution had been used.

One note of caution should be mentioned here. All the laminar flow solutions considered so far have been based on constant fluid properties. For liquids where the viscosity varies markedly with temperature (and oils are the worst offenders in this regard) considerable error may be introduced unless a correction is made for the variation of viscosity over the flow cross section. This subject is discussed in more detail in Chap. 15.

Uniform Heat Flux

The uniform-heat-flux, thermal-entry-length problem for laminar flow in a circular tube is very similar to the constant-surface-temperature problem. The same energy equation (9-32), must be solved, with the difference being that the constant-surface-temperature boundary condition is replaced by a constant temperature gradient at the surface. Also, the definition of θ is replaced by

$$\theta = \frac{t_e - t}{\dot{q}_0'' D / k}$$

The method of separation of variables and Sturm–Liouville theory have been employed[11] to obtain an eigenvalue solution, which is presented

TABLE 9-5
Infinite-series-solution functions for the circular tube; constant heat rate; thermal-entry length

m	λ_m^2	A_m
1	25.68	7.630×10^{-3}
2	83.86	2.053×10^{-3}
3	174.2	0.903×10^{-3}
4	296.5	0.491×10^{-3}
5	450.9	0.307×10^{-3}

For larger m, $\gamma_m = 4m + \frac{4}{3}$; $A_m = 0.428 \gamma_m^{-7/3}$

TABLE 9-6
Nusselt numbers for the circular tube; constant heat rate; thermal-entry length

x^+	Nu_x
0	∞
0.002	12.00
0.004	9.93
0.010	7.49
0.020	6.14
0.040	5.19
0.100	4.51
∞	4.36

here only in terms of the local Nusselt number:

$$Nu_x = \left[\frac{1}{Nu_\infty} - \frac{1}{2} \sum_{m=1}^{\infty} \frac{\exp(-\gamma_m^2 x^+)}{A_m \gamma_m^4} \right]^{-1} \tag{9-42}$$

The necessary eigenvalues and constants are given in Table 9-5.

$Nu_\infty = \frac{48}{11}$ or 4.364, as previously evaluated, and a complete tabulation of Nu, is given in Table 9-6.

In a constant-heat-flux problem the mixed mean temperature of the fluid varies linearly with x^+ and can be established by simple energy balance methods. The purpose of the heat-transfer solution is to relate the surface temperature to the mean fluid temperature through the definition of the conductance, that is,

$$t_0 - t_m = \frac{\dot{q}_0''}{h_x} = \frac{\dot{q}_0'' D}{Nu_x k}$$

Thus the mean fluid and the wall surface temperatures plotted as functions of distance along the tube appear as in Fig. 9-11.

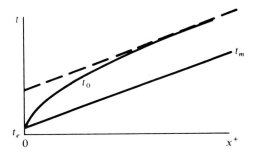

FIGURE 9-11
Temperature variations in the thermal-entry region of a tube with constant heat rate per unit of tube length.

THERMAL-ENTRY-LENGTH SOLUTIONS FOR THE RECTANGULAR TUBE AND ANNULUS

Laminar flow, thermal-entry-length solutions will be presented for two families of tube cross-sectional geometries. Entry-length solutions for other geometries and boundary conditions, as well as further discussion, can be found in Shah and London,[5] and in Kakac, Shah, and Aung.[8]

For the rectangular tube and annulus a fully developed laminar velocity profile, with all its implications, is assumed and the fluid properties are treated as constant. The problem is basically an extension of the circular-tube problem, but it is more difficult because of the more complex geometry. In the case of the annulus the differential equation to be solved is identical to the equation for the circular tube; only the boundary conditions differ. The entire family of circular-tube annuli from the single circular tube to the case of two parallel planes has been solved in a very general way for every possible heat-flux and surface-temperature combination on the two surfaces involved. For the rectangular tube there are three independent space variables in place of two; otherwise, essentially the same energy equation must be solved.

Only the final results of these analyses are presented. In the case of the annulus there are so many boundary condition possibilities that space permits presentation of only the most important.

The Rectangular Tube

The only member of the rectangle family that has been handled in a complete way is the case of flow between parallel planes,[4,10,12] for the obvious reason that this is geometrically a simpler case than even the circular tube. Only the case of both surfaces at the same constant temperature is presented here: some other boundary conditions are considered in the next section, where the parallel-planes geometry forms one of the limits of the circular-tube annulus problem.

Any constant-surface-temperature solution can be presented in the form of Eqs. (9-38)–(9-40) if we define x^+ in some consistent manner. For the solutions to be presented here let

$$x^+ = \frac{2(x/D_h)}{\text{Re Pr}}$$

where D_h is the hydraulic diameter defined as

$$D_h = \frac{4 \times \text{cross-sectional area}}{\text{perimeter}}$$

TABLE 9-7
Infinite-series-solution functions for the parallel-planes system; both surfaces at constant temperature; thermal-entry length

n	λ_n^2	G_n
0	15.09	1.717
1	171.3	1.139
2	498	0.952

For higher n, $\lambda_n = 16\sqrt{\frac{1}{3}}\,n + \frac{20}{3}\sqrt{\frac{1}{3}}$; $G_n = 2.68\lambda_n^{-1/3}$.

(For the flow between parallel planes D_h becomes twice the plate spacing.) The eigenvalues and constants for the parallel-planes system are given in Table 9-7. [These have been modified from the original references to make them usable directly in Eqs. (9-38)–(9-40).]

The square tube has been studied,[13] and only the first three eigenvalues and constants are available. However, even this truncated part of the series is quite sufficient for $x^+ > 0.01$. These functions are given in Table 9-8.

Thermal-entry-length solutions for the entire family of rectangular tubes have been obtained by numerical integration[14] for uniform surface temperature. An abstract of these results is given in Table 9-9. Here b/a is the ratio of the long side of the rectangle to the short side, as in Table 9-2. (The results given for the square, $b/a = 1.0$, have been calculated from the eigenvalues and constants given in Table 9-8, while those for $b/a = \infty$ have been calculated from the eigenvalues and constants given in

TABLE 9-8
Infinite-series-solution functions for the square tube; constant surface temperature; thermal-entry length

n	λ_n^2	G_n
0	5.96	0.598
1	35.64	0.462
2	78.9	0.138

TABLE 9-9
Nusselt numbers for the rectangular-tube family; constant surface temperature; thermal-entry length

	Nu_x					Nu_m				
	b/a					b/a				
x^+	1	2	4	6	∞	1	2	4	6	∞
0	∞	∞	∞	∞	∞	∞	∞	∞	∞	∞
0.01	4.55	5.72	6.57	7.02	8,52	8.63	8.58	9.47	10.01	11.63
0.02	4.12	4.72	5.55	6.07	7.75	6.48	6.84	7.71	8.17	9.83
0.05	3.46	3.85	4.87	5.48	7.55	4.83	5.24	6.16	6.70	8.48
0.10	3.10	3.54	4.65	5.34	7.55	4.04	4.46	5.44	6.04	8.02
0.20	2.99	3.43	4.53	5.24	7.55	3.53	3.95	5.00	5.66	7.78
∞	2.98	3.39	4.51	5.22	7.55	2.98	3.39	4.51	5.22	7.55

Table 9-7.) The accuracy of the numerical calculations is difficult to determine, but some inconsistencies are noted in Table 9-9.

The thermal-entry-length Nusselt numbers for the rectangular tube family for constant-heat-rate have been computed numerically[14] and are presented in Table 9-10.

The Concentric Circular-Tube Annulus

The concentric circular-tube annulus presents the possibility of a large variety of combinations of surface temperature and heat flux. This

TABLE 9-10
Nusselt numbers for the rectangular-tube family; constant-heat-rate axially; constant surface temperature peripherally; thermal-entry length

	Nu_x				
	b/a				
x^+	1	2	0.333	4	∞
0	∞	∞	∞	∞	∞
0.01	7.10	7.46	8.02	8.44	
0.02	5.69	6.05	6.57	7.00	8.80
0.05	4.45	4.84	5.39	5.87	
0.10	3.91	4.38	5.00	5.62	8.25
0.20	3.71	4.22	4.85	5.45	
∞	3.60	4.11	4.77	5.35	8.235

problem in all its possible variations has been solved exactly.[4] The most interesting as well as the most common problems that arise in annulus heat transfer are generally those where the heat fluxes on the two surfaces are specified. To save space, the eigenvalues and functions are omitted and only the Nusselt numbers and influence coefficients are presented as functions of x^+ (see Table 9-11).

TABLE 9-11
Nusselt numbers and influence coefficients for the circular-tube-annulus family; constant heat rate; thermal-entry length

$r^* = r_i/r_o$	x^+	Nu_{ii}	Nu_{oo}	θ_i^*	θ_o^*
0.05	0.002	33.2	13.4	0.1265	0.00255
	0.01	24.2	7.99	0.460	0.00760
	0.02	21.5	6.58	0.817	0.0125
	0.10	18.1	4.92	2.13	0.0278
	0.20	17.8	4.80	2.17	0.0293
	∞	17.8	4.79	2.18	0.0294
0.10	0.002	25.1	13.5	0.0914	0.00491
	0.01	17.1	8.08	0.311	0.0147
	0.02	14.9	6.65	0.540	0.241
	0.10	12.1	4.96	1.296	0.0531
	0.20	11.9	4.84	1.38	0.0560
	∞	11.9	4.83	1.38	0.0562
0.25	0.002	18.9	13.8	0.0605	0.01104
	0.01	12.1	8.28	0.194	0.0328
	0.02	10.2	6.80	0.325	0.0540
	0.10	7.94	5.04	0.746	0.118
	0.20	7.76	4.91	0.789	0.125
	∞	7.75	4.90	0.793	0.125
0.50	0.002	16.4	14.2	0.0437	0.0189
	0.01	10.1	8.55	0.1347	0.0570
	0.02	8.43	7.03	0.224	0.0934
	0.10	6.35	5.19	0.498	0.204
	0.20	6.19	5.05	0.526	0.215
	∞	6.18	5.04	0.528	0.216
1.00 (parallel planes)	0.0005	23.5	23.5	0.01175	0.01175
	0.005	11.2	11.2	0.0560	0.0560
	0.02	7.49	7.49	0.1491	0.1491
	0.10	5.55	5.55	0.327	0.327
	0.25	5.39	5.39	0.346	0.346
	∞	5.38	5.38	0.346	0.346

For these solutions x^+ is defined† as

$$x^+ = \frac{2(x/D_h)}{\text{Re Pr}}$$

and

$$D_h = 2(r_o - r_i)$$

Where the heat flux on each surface is constant along the length of the tube (invariant with x^+), these results are directly applicable to Eqs. (9-28) and (9-29). The use of these results for the case where the heat flux varies with length, as in a nuclear reactor, for example, is discussed in the next section.

THE EFFECT OF AXIAL VARIATION OF THE SURFACE TEMPERATURE WITH HYDRODYNAMICALLY FULLY DEVELOPED FLOW

All the convection solutions considered so far have been based on either a constant surface temperature or a constant heat flux (axially). It was noted in connection with the fully developed temperature profile solutions that the conductance for constant heat flux is considerably higher than the conductance for a constant surface temperature. Under fully developed temperature profile conditions with a constant heat flux, the surface temperature varies linearly along the length of the tube (as does the mean fluid temperature). It thus appears that the convection conductance is, at least to some degree, dependent on how the surface temperature varies along the tube. Engineers should understand fully under what conditions the conductance varies markedly with tube length and under what conditions it can be treated as at least approximately a constant, since the simpler methods of heat-exchanger analysis are based on a conductance constant with length, or at least a mean overall conductance with respect to length. It is the purpose of this section to investigate this effect.

Because of the linear, homogeneous nature of the energy equation (9-4), a sum of solutions is again a solution. It is thus possible to construct a solution for any kind of arbitrary surface temperature variation with length by merely breaking up the surface temperature into a number of constant-temperature steps and summing, or superposing, the constant-surface temperature, thermal-entry-length solutions for each step. This superposition technique can be employed for turbulent flow as

---

† x^+ as defined here is twice the $\bar{x}$ employed in Ref. 4.

well as laminar flow, and for flow over external surfaces as well as flow inside tubes. All that is needed as a starting point is a "step-function" solution, that is, a solution for the case where the fluid and surface are initially at the same temperature and then, at some point $x = 0$, the surface temperature steps to some different temperature and remains constant thereafter. Equation (9-36) is such a solution for a circular tube: it is based on a fully developed velocity profile, and $x = 0$ is taken as the point where heat transfer starts, not at the entrance to the tube.

The method of superposition can best be understood by reference to the diagram in Fig. 9-12. Let the fluid enter the tube at a uniform temperature t_e, and let the velocity profile be fully developed at the point $x^+ = 0$. Let the surface temperature vary in any arbitrary manner, starting at $x^+ = 0$, and let it be represented by a series of infinitesimal steps, or finite steps, or both. The fluid temperature at x^+ and any r^+ can then be determined by summing the contribution of each of the steps, either infinitesimal or finite. For the summation to be considered, x^+ is the point at which it is desired to determine the fluid temperature, and thus x^+ is treated as a constant. A dummy length variable ξ is used to designate the location of each step; ξ thus varies from 0 to x^+.

Consider now the solution for the temperature of the fluid at x^+ resulting from a single, finite step in wall temperature at $x^+ = 0$. This is Eq. (9-36), where

$$\theta(x^+, r^+) = \frac{t_0 - t}{t_0 - t_e}$$

Now if the surface temperature step occurs at ξ, as in Fig. 9-12b, the same solution applies, but the temperature at x^+ becomes

$$\theta(x^+ - \xi, r^+) = \frac{t_0 - t}{t_0 - t_e}$$

Suppose we consider next the temperature at x^+ resulting from an infinitesimal step in surface temperature at ξ (Fig. 9-12c). The same solution θ applies, and the infinitesimal increase in fluid temperature at x^+ can be calculated as follows:

$$\frac{dt_0 - dt}{dt_0} = \theta(x^+ - \xi, r^+)$$

from which

$$dt = [1 - \theta(x^+ - \xi, r^+)] \, dt_0$$

Similarly, the finite temperature rise at x^+ resulting from a finite step in surface temperature Δt_0 is

$$\Delta t = [1 - \theta(x^+ - \xi, r^+)] \, \Delta t_0$$

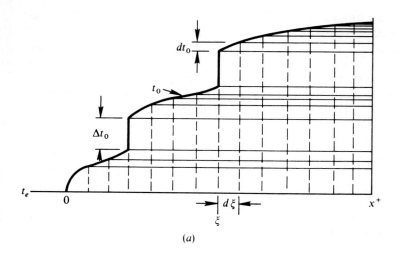

(a)

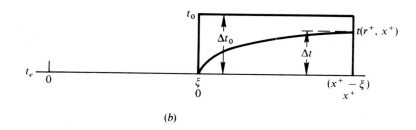

(b)

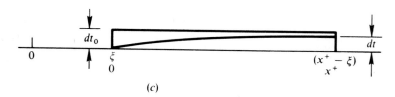

(c)

FIGURE 9-12

The complete solution for the fluid temperature at x^+ is simply a summation of the contributions of all the infinitesimal and finite surface-temperature steps from $\xi = 0$ to $\xi = x^+$. The linearity and homogeneity of the energy equation (9-4) ensure that such a sum of solutions still satisfies the differential equation. Examination of the result demonstrates that the boundary conditions are truly satisfied; that is, the fluid temperature is a uniform t_e at $x^+ = 0$, and at $r^+ = 1$, and follows the specified surface temperature t_0.

This summation can be represented by an ordinary Riemann integral wherever t_0 is continuous, plus a summation of the contribution of the discontinuities, if any. (The combination of the Riemann integral and the summation may be represented by what is known as a Stieltjes integral.) Thus the fluid temperature at any x^+ and r^+ is

$$t - t_e = \int_{\xi=0}^{\xi=x^+} [1 - \theta(x^+ - \xi, r^+)] \, dt_0 + \sum_{i=1}^{k} [1 - \theta(x^+ - \xi_i, r^+)] \, \Delta t_{0,i}$$

where $\Delta t_{0,i}$ is the temperature difference across the discontinuity at ξ_i. For example, with two steps, $\Delta t_{0,1} = t_1 - t_e$ and $\Delta t_{0,2} = t_2 - t_1$.

The integral may be evaluated by substituting $(dt_0/d\xi) \, d\xi$ for dt_0. The derivative is presumably a known function of ξ, and θ is a known step-function solution such as Eq. (9-36). Then[†]

$$t - t_e = \int_0^{x^+} [1 - \theta(x^+ - \xi, r^+)] \frac{dt_0}{d\xi} \, d\xi + \sum_{i=1}^{k} [1 - \theta(x^+ - \xi_i, r^+)] \, \Delta t_{0,i}$$

$$(9\text{-}43)$$

The final step is the evaluation of the local surface heat-transfer rate at x^+. A heat-transfer conductance and a Nusselt number can then be evaluated, or not, as desired. Often the heat-transfer rate alone is what is really desired. From Fourier's law,

$$\dot{q}_0''(x^+) = k \left(\frac{\partial t}{\partial r} \right)_{r=r_0} = \frac{k}{r_0} \left(\frac{\partial t}{\partial r^+} \right)_{r^+=1}$$

Differentiating Eq. (9-43) with respect to r^+ at $r^+ = 1$, and substituting,

$$\dot{q}_0''(x^+) = -\frac{k}{r_0} \left[\int_0^{x^+} \theta_{r^+}(x^+ - \xi, 1) \frac{dt_0}{d\xi} \, d\xi + \sum_{i=1}^{k} \theta_{r^+}(x^+ - \xi_i, 1) \, \Delta t_{0,i} \right]$$

$$(9\text{-}44)$$

The derivative θ_{r^+} may be obtained by differentiation of Eq. (9-36) with respect to r^+:

$$\theta_{r^+}(x^+, 1) = \sum_{n=0}^{\infty} C_n R_n'(1) \exp(-\lambda_n^2 x^+)$$

$$= -2 \sum_{n=0}^{\infty} G_n \exp(-\lambda_n^2 x^+)$$

[†] This problem can also be approached by direct application of Duhamel's theorem, developed for transient boundary conditions for solid conduction problems, an analogous situation: see Ref. 15.

Then

$$\theta_{r^+}(x^+ - \xi, 1) = -2 \sum_{n=0}^{\infty} G_n \exp\left[-\lambda_n^2(x^+ - \xi)\right] \tag{9-45}$$

For any given application the integral and summation of Eq. (9-44) must be evaluated. This can always be done numerically, but for certain elementary surface-temperature distributions an analytic expression for the heat flux can be readily obtained. As an example of an application of the method, a solution is developed for the case of a step in surface temperature at $x^+ = 0$ followed by a linear variation in surface temperature either increasing or decreasing. Thus the surface temperature is as shown in Fig. 9-13.

Substituting Eq. (9-45) into Eq. (9-44), with b for $dt_0/d\xi$, and including the one step at $x^+ = 0$ with $\Delta t_0 = a$, we obtain

$$\dot{q}_0''(x^+) = -\frac{k}{r_0} \left\{ \int_0^{x^+} \left[-2 \sum_{n=0}^{\infty} G_n \exp\left[-\lambda_n^2(x^+ - \xi)\right]\right] b\, d\xi \right.$$

$$\left. - 2a \sum_{n=0}^{\infty} G_n \exp\left(-\lambda_n^2 x^+\right) \right\}$$

By performing the integration, taking limits, and noting that $\sum_{n=0}^{\infty} (G_n/\lambda_n^2) = \frac{1}{8}$ (see Eq. (9-38)), the result reduces to

$$\dot{q}_0''(x^+) = \frac{2k}{r_0} \left[\frac{b}{8} - b \sum_{n=0}^{\infty} \frac{G_n}{\lambda_n^2} \exp\left(-\lambda_n^2 x^+\right) + a \sum_{n=0}^{\infty} G_n \exp\left(-\lambda_n^2 x^+\right)\right]$$

$$\tag{9-46}$$

A local Nusselt number may now be evaluated by first integrating Eq. (9-46) from 0 to x^+, to determine the total heat transfer to this point, and then applying an energy balance to evaluate the mixed mean fluid

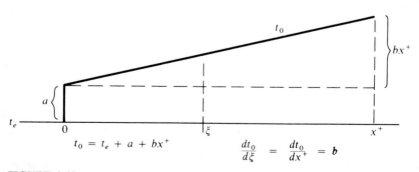

$$t_0 = t_e + a + bx^+ \qquad \frac{dt_0}{d\xi} = \frac{dt_0}{dx^+} = b$$

FIGURE 9-13
Example of a step–ramp surface temperature.

temperature at x^+:

$$\dot{q} = \frac{4\pi r_0^3 V \rho c}{k} \int_0^{x^+} \dot{q}_0'' \, dx^+$$

$$= \pi r_0^2 V \rho c (t_m - t_e)$$

By combining these equations and evaluating the integral, the mixed mean temperature becomes

$$t_m - t_e = bx^+ + a - 8b \sum_{n=0}^{\infty} \frac{G_n}{\lambda_n^4} + 8b \sum_{n=0}^{\infty} \frac{G_n}{\lambda_n^4} \exp\left(-\lambda_n^2 x^+\right)$$

$$- 8a \sum_{n=0}^{\infty} \frac{G_n}{\lambda_n^2} \exp\left(-\lambda_n^2 x^+\right) \qquad (9\text{-}47)$$

Next a local conductance h_x can be defined and a local Nusselt number evaluated:

$$\dot{q}_0''(x^+) = h_x(t_0 - t_m)$$

and

$$\text{Nu}_x = \frac{b - 8b \sum\limits_{n=0}^{\infty} \dfrac{G_n}{\lambda_n^2} \exp\left(-\lambda_n^2 x^+\right) + 8a \sum\limits_{n=0}^{\infty} G_n \exp\left(-\lambda_n^2 x^+\right)}{16b \sum\limits_{n=0}^{\infty} \dfrac{G_n}{\lambda_n^4} - 16b \sum\limits_{n=0}^{\infty} \dfrac{G_n}{\lambda_n^4} \exp\left(-\lambda_n^2 x^+\right) + 16a \sum\limits_{n=0}^{\infty} \dfrac{G_n}{\lambda_n^2} \exp\left(-\lambda_n^2 x^+\right)}$$

$$(9\text{-}48)$$

A number of interesting features of this result are worth noting. First note that for large x^+ all summations containing the exponential go to zero, and then the constant b cancels, which results in

$$\text{Nu}_x = \frac{1}{16 \sum\limits_{n=0}^{\infty} (G_n/\lambda_n^2)} \qquad (9\text{-}49)$$

If Table 9-3 is employed to evaluate the series, the sum is 0.01433 and the result is then

$$\text{Nu}_x = \frac{1}{16 \times 0.01433} = 4.364$$

This is immediately recognized as the previously derived Nusselt number for constant heat rate for a fully developed velocity and temperature profile; it is precisely the situation far downstream, when the surface temperature varies linearly.

Note also that for $b = 0$, Eq. (9-48) reduces again to the constant-surface-temperature case.

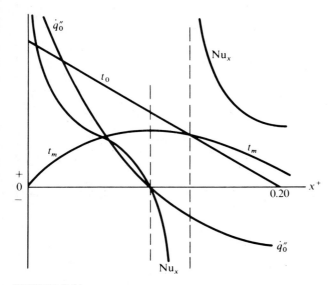

FIGURE 9-14
The effects of a positive value of a and a negative value of b in Fig. 9-12.

Equation (9-48) is useful to demonstrate the rather strange things that can happen to the heat-transfer coefficient, and thus the Nusselt number, under variable-surface-temperature conditions. Consider a tube in which the surface temperature starts out above the entering fluid temperature and then decreases linearly along the tube, finally reaching the fluid entering temperature at $x^+ = 0.20$. Thus

$$t_0 = t_e + (1 - 5x^+)(t_{0,e} - t_e)$$

Equations (9-46)–(9-48) are then directly applicable if b is set equal to $-5a$. Surface temperature, mixed mean fluid temperature, heat flux, and local Nusselt number are then readily evaluated as functions of x^+. The results are illustrated in Fig. 9-14. Quite obviously the convection conductance (in the Nusselt number) loses its useful significance when it behaves in such a manner as this.

THE EFFECT OF AXIAL VARIATION OF HEAT FLUX

Just as the constant-surface-temperature solution was used as the basic building block for any arbitrary axial variation in surface temperature, so also can be the constant-heat-flux solutions be used to build up solutions for any arbitrary axial variation in surface heat flux. We do not develop this in detail but indicate only the results and the calculating procedures (see, for example, Ref. 10).

Let $\dot{q}_0''(x^+) = \dot{q}_0''(\xi)$ be any arbitrary axial heat-flux function that may be specified. Then the surface temperature can be calculated from

$$t_0(x^+) - t_e = \frac{r_0}{k} \int_0^{x^+} g(x^+ - \xi)\dot{q}_0''(\xi)\, d\xi \qquad (9\text{-}50)$$

where

$$g(x^+) = 4 + \sum_m \frac{\exp(-\gamma_m^2 x^+)}{\gamma_m^2 A_m} \qquad (9\text{-}51)$$

The eigenvalues and constants for the circular tube in laminar flow are given in Table 9-5.

As before, we can easily evaluate the mixed mean fluid temperature by considering the total heat transferred up to x^+:

$$t_m(x^+) - t_e = \frac{4r_0}{k} \int_0^{x^+} \dot{q}_0''(\xi)\, d\xi \qquad (9\text{-}52)$$

and then, if desired, a conductance and Nusselt number can be readily defined and evaluated.

An interesting and important example of an application where the heat flux varies in a specified manner along the tube is the case of a nuclear reactor where the heat flux can vary sinusoidally along the cooling tubes.

$$\frac{\dot{q}_0''}{\dot{q}_{0_{max}}''} = \sin \frac{\pi x}{L} \qquad (9\text{-}53)$$

where L is the total tube length. By substituting (9-53) into (9-52), the mean fluid temperature becomes

$$t_m - t_e = \frac{4\dot{q}_{0_{max}}'' r_0}{k\beta}(1 - \cos \beta x^+) \qquad (9\text{-}54)$$

where $\beta = r_0 \operatorname{Re} \operatorname{Pr} \pi/L$. Substituting into Eq. (9-50) yields the surface temperature:

$$t_0 - t_e = \frac{\dot{q}_{0_{max}}'' r_0}{k}\left[\frac{4}{\beta}(1 - \cos \beta x^+) + \sin \beta x^+ \sum_m \frac{1}{A_m(\gamma_m^4 + \beta^2)}\right.$$

$$\left. - \beta \cos \beta x^+ \sum_m \frac{1}{\gamma_m^2 A_m(\gamma_m^4 + \beta^2)} + \beta \sum_m \frac{\exp(-\gamma_m^2 x^+)}{\gamma_m^2 A_m(\gamma_m^4 + \beta^2)}\right]$$

$$(9\text{-}55)$$

Figure 9-15 shows a typical plot of these results. Also shown by the dashed line is the surface temperature that would be predicted employing

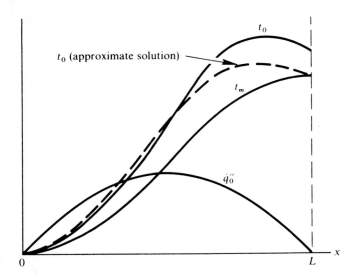

FIGURE 9-15
The effects of heat flux varying sinusoidally with tube length.

the given heat flux and fluid temperature but using a constant conductance based on the constant-heat-flux solutions. Note that the "exact" solution yields a higher peak surface temperature and that the peak occurs at a different place. How much these differences amount to depends, of course, on all the various system parameters. Although it is certainly simple, in applications of this type, to make use of a constant conductance, it is important to know the kind of error that is introduced by this simplification.

An analysis of these examples of surface temperature and heat flux varying in the axial direction yields the following conclusions:

1. An increasing $t_0 - t_m$ and/or $\dot{q}_0''$ in the flow direction tends to yield a high heat-transfer coefficient h.
2. A decreasing $t_0 - t_m$ and/or $\dot{q}_0''$ in the flow direction tends to yield a low heat-transfer coefficient h.

These conclusions apply only for continuous tubes, but the effects can be fairly important for laminar flow; the same conclusions apply for turbulent flow, but the effects are much less pronounced except for very low-Prandtl-number fluids for reasons that will be apparent later.

The constant-heat-rate annulus solutions given in Table 9-11 can also be used as a basis for calculating variable-heat-flux problems. The following equations for the inner and outer surface temperatures result directly from superposition:

$$t_i(x^+) - t_m = \frac{D_h}{k} \left[\int_0^{x^+} \frac{1}{\mathrm{Nu}_{ii}\,(x^+ - \xi)}\, d\dot{q}_i''(\xi) - \int_0^{x^+} \frac{\theta_i^*(x^+ - \xi)}{\mathrm{Nu}_{ii}\,(x^+ - \xi)}\, d\dot{q}_o''(\xi) \right]$$

(9-56)

$$t_o(x^+) - t_m = \frac{D_h}{k} \left[\int_0^{x^+} \frac{1}{\mathrm{Nu}_{oo}\,(x^+ - \xi)}\, d\dot{q}_o''(\xi) - \int_0^{x^+} \frac{\theta_o^*(x^+ - \xi)}{\mathrm{Nu}_{oo}\,(x^+ - \xi)}\, d\dot{q}_i''(\xi) \right]$$

(9-57)

The fluid mixed mean temperature t_m is determined by integrating the prescribed heat flux from $\xi = 0$ to $\xi = x^+$, as in Eq. (9-52).

These integrals are Stieltjes integrals, and for computational purposes can each be split into a Riemann integral plus a summation, as in Eq. (9-43). For example,

$$\int_0^{x^+} \frac{1}{\mathrm{Nu}_{ii}\,(x^+ - \xi)}\, d\dot{q}_i''(\xi) = \int_0^{x^+} \frac{1}{\mathrm{Nu}_{ii}\,(x^+ - \xi)} \frac{d\dot{q}_i''}{d\xi}\, d\xi$$
$$+ \sum_{j=1}^{k} \frac{1}{\mathrm{Nu}_{ii}\,(x^+ - \xi_j)} (\Delta \dot{q}_i'')_j$$

where $(\Delta \dot{q}_i'')_j$ is the heat-flux difference across the discontinuity at ξ_j. The integral is then used for continuously varying heat flux, and all discontinuities in heat flux are contained in the summation. The problem collapses back to the constant-heat-flux problem if $d\dot{q}_i''/d\xi = 0$, $\xi_j = 0$, and $k = 1$.

Note that the first term of Eq. (9-57) can also be used for the *circular-tube* variable-heat-flux problem, employing Table 9-6 with x^+ replaced by $x^+ - \xi_j$, and this is then an alternative to Eq. (9-50).

COMBINED HYDRODYNAMIC-AND THERMAL-ENTRY LENGTH

All the laminar-flow solutions considered so far have been based on the idealization that the velocity profile is fully developed, which, according to Eq. (7-20) for a circular tube, occurs when x/D is greater than approximately $\frac{1}{20}$ Re. On the other hand, it was shown in connection with Eq. 9-41 that for a Prandtl number greater than about 5 the velocity profile develops so much faster than the temperature profile that even if both temperature and velocity are uniform at the tube entrance, the fully developed velocity idealization introduces little error. At low Prandtl numbers, however, the temperature profile develops more rapidly than the velocity profile, and the fully developed velocity idealization is no longer valid for a thermal-entry-length solution. If x/D is greater than

$\frac{1}{20}$ Re, a fully developed velocity profile exists regardless of Prandtl number; but for applications where a *mean* conductance with respect to length is required, as in heat-exchanger analysis, the effects of the undeveloped velocity profile near the entrance can be substantial even for tubes much longer than $x/D = \frac{1}{20}$ Re.

The development of the velocity profile in a circular tube was considered in Chap. 7, where the solution by Langhaar was discussed. The applicable energy differential equation for calculation of the temperature field under conditions of a developing velocity field, axisymmetric heating, and negligible axial conduction must be Eq. (9-1), with the appropriate terms omitted:

$$\frac{1}{r}\frac{\partial}{\partial r}\left(r\frac{\partial t}{\partial r}\right) = \frac{u}{\alpha}\frac{\partial t}{\partial x} + \frac{v_r}{\alpha}\frac{\partial t}{\partial r} \tag{9-58}$$

The Prandtl number becomes a parameter in the solution for the combined hydrodynamic and thermal-entry-length problem because, while the velocity development is independent of Prandtl number, the temperature profile is Prandtl-number-dependent. Recall that when the velocity profile was assumed to be fully developed, the Prandtl number did not explicitly appear in the solution.

Several workers have solved Eq. (9-58) numerically, either using Langhaar's entry-region velocity profiles or generating the velocity solution numerically; and various degrees of approximation have been employed, resulting in considerable discrepancy among different solutions. Some results for a *constant surface temperature* for three different values of Prandtl number are given in Table 9-12. These come from Ref. 5 and are based on the numerical solutions of Hornbeck.[16,17] It is instructive to compare the results for Pr = 5 with the fully developed velocity-profile solution of Table 9-4. It is apparent that for Pr > 5 the eigenfunction solution for the thermal-entry length is a good approximation even if the velocity profile is developing simultaneously. For lower Prandtl numbers the Nusselt number for the combined entry length is always higher than for the thermal-entry length alone.

A rather different procedure was employed by Heaton, Reynolds, and Kays,[18] who linearized Eq. (9-58) by an approximation first introduced by Langhaar in connection with the velocity problem, and who were then able to obtain a generalized entry-region temperature profile which could be used in the energy integral equation. They obtained solutions for the entire family of circular-tube annuli for *constant heat rate* and various Prandtl numbers. An abstract of the results of Heaton, Reynolds, and Kays is presented in Table 9-13.

The annulus and parallel-planes results are applicable to Eqs. (9-28) and (9-29), and therefore different rates of heating from the two surfaces may be calculated. (See Table 9-13.) These results are *not* applicable to

TABLE 9-12
Nusselt numbers for the circular tube; constant surface temperature; combined thermal- and hydrodynamic-entry length

x^+	Nu$_x$			Nu$_m$		
	Pr = 0.7	Pr = 2	Pr = 5	Pr = 0.7	Pr = 2	Pr = 5
0.001	16.8	14.8	13.5	30.6	25.2	22.1
0.002	12.6	11.4	10.6	22.1	19.1	16.8
0.004	9.6	8.8	8.2	16.7	14.4	12.9
0.006	8.25	7.5	7.1	14.1	12.4	11.0
0.01	6.8	6.2	5.9	11.3	10.2	9.2
0.02	5.3	5.0	4.7	8.7	7.8	7.1
0.05	4.2	4.1	3.9	6.1	5.6	5.1
∞	3.66	3.66	3.66	3.66	3.66	3.66

TABLE 9-13
Nusselt numbers and influence coefficients for the circular-tube annulus family; constant heat rate; combined thermal- and hydrodynamic-entry length

Pr	x^+	Circular tube Nu	Parallel planes Nu$_{11}$	θ_1^*	Circular-tube annulus $r^* = 0.50$† Nu$_{ii}$	Nu$_{oo}$	θ_i^*	θ_i^*
0.01	0.002	24.2	24.2	0.048	—	24.2	—	0.0322
	0.010	12.0	11.7	0.117	—	11.8	—	0.0786
	0.020	9.10	8.80	0.176	9.43	8.90	0.252	0.118
	0.10	6.08	5.77	0.378	6.40	5.88	0.525	0.231
	0.20	5.73	5.53	0.376	6.22	5.60	0.532	0.238
	∞	4.36	5.39	0.346	6.18	5.04	0.528	0.216
0.70	0.002	17.8	18.5	0.037	19.22	18.30	0.0513	0.0243
	0.010	9.12	9.62	0.096	10.47	9.45	0.139	0.0630
	0.020	7.14	7.68	0.154	8.52	7.50	0.228	0.0998
	0.10	4.72	5.55	0.327	6.35	5.27	0.498	0.207
	0.20	4.41	5.40	0.345	6.19	5.06	0.527	0.215
	∞	4.36	5.39	0.346	6.18	5.04	0.528	0.216
10.0	0.002	14.3	15.6	0.0311	16.86	15.14	0.045	0.0201
	0.010	7.87	9.20	0.092	10.20	8.75	0.136	0.0583
	0.020	6.32	7.49	0.149	8.43	7.09	0.224	0.0943
	0.10	4.51	5.55	0.327	6.35	5.20	0.498	0.204
	0.20	4.38	5.40	0.345	6.19	5.05	0.527	0.215
	∞	4.36	5.39	0.346	6.18	5.04	0.528	0.216

† For other radius ratios see Ref. 18.

problems where the heat flux varies axially. Such problems can be solved by superposition only if a family of step-function solutions is available for surface heat-flux steps at all points in the hydrodynamic-entry length.

It should finally be added that any of these variable-surface-temperature and variable-heat-flux problems can also be readily solved by direct numerical solution of the appropriate differential equations. This is today a very practicable alternative, even in engineering applications, especially for the simpler geometries such as circular tubes, circular-tube annuli, and flat ducts. It is the preferred procedure when the specified surface temperature and/or heat flux is a complex function. There is no particular advantage to be gained using superposition techniques if numerical calculations must be used to evaluate the functions. Direct numerical procedures also have the virtue that such further complications as variable fluid properties can be readily introduced.

PROBLEMS

9-1. Starting from the appropriate momentum and energy differential equations, evaluate the Nusselt number for both surfaces of a parallel-planes duct in which there is fully developed laminar flow (both velocity and temperature developed) and in which there is heating from both surfaces but the heat flux from one surface is twice the flux from the other surface: and again when the heat flux from one surface is five times the flux from the other surface. The heat-transfer rate per unit of duct length is constant.

Check your results with those given in the text (Table 9-1).

9-2. With a low-Prandtl-number fluid, the temperature profile in a tube develops more rapidly than the velocity profile. Thus, as the Prandtl number approaches zero, the temperature profile can approach a fully developed form before the velocity profile has even started to develop (although this is a situation of purely academic interest). Convection solutions based on a uniform velocity over the cross section, as described, are called *slug-flow* solutions.

Develop an expression for the slug-flow, fully developed temperature-profile Nusselt number for constant heat rate per unit of tube length for a concentric circular-tube annulus with a radius ratio of 0.60 for the case where the inner tube is heated and the outer tube is insulated.

Compare with the results in Table 9-1 and discuss.

9-3. Consider a 0.6 cm inside-diameter. 1.2 m long circular tube, wound by an electric resistance heating element. Let the function of the tube be to heat an organic fuel from 10 to 65°C. Let the mass flow rate of the fuel be 1.26×10^{-3} kg/s. The following average properties may be treated as constant:

$$Pr = 10$$
$$\rho = 753 \text{ kg/m}^3$$
$$c = 2.092 \text{ kJ/(kg} \cdot \text{K)}$$
$$k = 0.137 \text{ W/(m} \cdot \text{K)}$$
$$\mu = 0.00065 \text{ Pa} \cdot \text{s}$$

Calculate and plot both tube surface temperature and fluid mean temperature as functions of tube length. What is the highest temperature experienced by any of the fluid?

9-4. Consider fully developed, constant-property laminar flow between parallel planes with constant heat rate per unit of length and a fully developed temperature profile. Suppose heat is transferred to the fluid on one side and *out* of the fluid on the other at the *same rate*. What is the Nusselt number on each side of the passage? Sketch the temperature profile. Suppose the fluid is an oil for which the viscosity varies greatly with temperature, but all the other properties are relatively unaffected by temperature. Is the velocity profile affected? Is the temperature profile affected? Is the Nusselt number affected? Explain.

9-5. Consider a concentric circular-tube annulus, with outer diameter 2.5 cm and inner diameter 1.25 cm, in which air is flowing under fully developed, constant-heat-rate conditions. Heat is supplied to the inner tube, and the outer tube is externally insulated. The radiation emissivity of both tube surfaces is 0.8. The mixed mean temperature of the air at a particular point in question is 260°C. The inner-tube surface temperature at this point is 300°C. What is *total* heat flux from the inner-tube surface at this point? What is the outer-tube surface temperature at this point? What percentage of the heat supplied to the inner tube is transferred directly to the air, and what percentage indirectly from the outer surface?

The Reynolds number is sufficiently low that the flow is laminar. Assume that the air is transparent to the thermal radiation. Make use of any of the material in the text as needed.

9-6. The heat flux along a cooling tube in a typical nuclear power reactor may often be approximated by

$$\dot{q}_0'' = a + b \sin \frac{\pi x}{L}$$

where L is the length of the tube and x is the distance along the tube.

A particular air-cooled reactor is to be constructed of a stack of fuel plates with a 3 mm air space between them. The length of the flow passage will be 1.22 m, and the heat flux at the plate surfaces will vary according to the above equation with $a = 900 \text{ W/m}^2$ and $b = 2500 \text{ W/m}^2$. The air mass velocity is to be 7.5 kg/(s $\cdot$ m^2). The air enters the reactor at 700 kPa and 100°C.

Prepare a scale plot of heat flux, air mean temperature, and plate surface temperature as a function of distance along the flow passage. Although the heat flux is not constant along the passage, the passage length-to-gap ratio is sufficiently large that the constant-heat-rate heat-transfer solution for the conductance h is not a bad approximation. Therefore assume h is a constant. We are most interested here in the peak surface temperature; and if this occurs in a region where the heat flux is varying only slowly, the approximation is still better. This is a point for discussion.

The properties of air at 250°C may be used and treated as constant.

9-7. A lubricating oil flows through a long 0.6 cm inner-diameter tube at a mean velocity of 6 m/s. If the tube is effectively insulated, calculate and plot the temperature distribution, resulting from *frictional* heating, in terms of the pertinent parameters. Let the fluid properties be those of a typical engine oil at 100°C. Start with Eq. (4-28).

9-8. Consider a journal bearing using the oil of Prob. 9-7. Let the journal diameter be 7·6 cm, the clearance be 0.025 cm, and the rpm be 3600. Neglecting end effects, and assuming no flow of oil into or out of the system, calculate the temperature distribution in the oil film on the assumption that there is no heat transfer into the journal but that the bearing surface is maintained at 80°C. Calculate the rate of heat transfer per square meter of bearing surface. Assume no eccentricity, that is, no load on the bearing. How much power is needed to rotate the journal if the bearing is 10 cm long?

9-9. Consider uniform-temperature laminar flow in a circular tube with a fully developed velocity profile. At some point $x^+ = 0$ the surface temperature is raised above the fluid temperature by an amount a. It remains constant at this value until a point $x^+ = x_1^+$ is reached, where the surface temperature is again raised an amount b, remaining constant thereafter. Develop a general expression for the surface heat flux, and for the mean fluid temperature θ_m, in the part of the tube following the second step in surface temperature. Use variable-surface-temperature theory.

9-10. Consider laminar flow in a circular tube with a fully developed velocity profile. Let heat be added at a constant rate along the tube from $x^+ = 0$ to $x^+ = 0.10$. Thereafter let the tube surface be adiabatic. Calculate and plot the tube surface temperature as a function of x^+.

9-11. Consider laminar flow in a circular tube with a fully developed velocity profile. Let the tube surface be alternately heated at a constant rate per unit of length and adiabatic, with each change taking place after intervals of x^+ of 0.020. How large must x^+ be for the effects of the original entry length to damp out? How does the Nusselt number vary along the heated segments after the effects of the original entry length have damped out?

9-12. Evaluate and plot both local and mean Nusselt numbers for fully developed laminar flow in a square tube. At what value of x^+ does the local Nusselt number come within two percent of the asymptotic value?

9-13. Consider the problem posed by Eq. (9-53). Let $\beta = 6$ and let the tube be circular. Evaluate and plot the local Nusselt number as a function of x/L. Explain physically the reasons for the behavior noted.

9-14. Consider fully developed laminar flow in a circular tube in which the heat flux varies axially according to the relation

$$\dot{q}_0'' = a + b \sin \frac{\pi x}{L}$$

where L is the total length of the tube, and a and b are constants (this is an approximation for a nuclear reactor cooling tube). Derive a general expression for the mean fluid temperature t_m as a function of x, and the tube surface temperature t_0 as a function of x, using variable-heat-flux theory [that is, an expression corresponding to Eq. (9-55) for the simple sinusoidal variation of heat flux].

9-15. Consider fully developed laminar flow in a circular-tube annulus with $r^* = 0.50$. Let there be heat transfer from the inner tube only (outer tube insulated), and let the heat flux on the inner tube vary as in Prob. 9-14. Describe in detail a computing procedure for evaluating both the inner- and outer-tube surface temperatures as functions of length along the tube.

9-16. Helium flows through a thin-walled 1.25 cm diameter circular tube at a mean velocity of 6 m/s under the following conditions at a particular point along the tube:

$$P = 345 \text{ kPa}, \qquad t_m = 200°C$$

The tube is exposed on one side to an infinite *plane* that emits black-body radiation at 1100°C, while the remainder of the surrounding space is effectively nonradiating. Assuming that the tube wall is sufficiently thin that peripheral conduction in the wall is negligible, assuming that the outer-wall surface is a black body, and evaluating radiation *from* the tube as if the entire tube were at a uniform temperature of 300°C (re-radiation will be relatively small and an exact solution would require iteration), calculate the net heat flux to the tube and estimate the temperature distribution in the wall around the tube. Assume that the heat-transfer resistance of the wall is negligible in the radial direction and that fluid properties are constant.

9-17. Consider steady flow in a tube with a fully developed velocity profile at the tube entrance. Let the fluid temperature at the tube entrance be uniform at t_e. Then let the tube surface temperature vary axially according to the relation

$$t_0 - t_e = \frac{a}{b} [\exp (bx^+) - 1]$$

where a and b are arbitrary constants.

Derive an expression for the local Nusselt number as a function of x^+.

Show that all members of this family of solutions lead to Nusselt numbers that are independent of x^+ at sufficiently large values of x^+. What are the implications of this result? Discuss how the constant b affects the asymptotic Nusselt number.

9-18. Consider fully developed laminar flow with constant properties in a circular tube. Let there be heat transfer to or from the fluid at a constant rate per unit of tube length. Determine the Nusselt number if the effect of frictional heating (viscous mechanical energy dissipation) is included in the analysis. How does frictional heating affect the Nusselt number? What are the significant new parameters? Consider some numerical examples and discuss the results.

9-19. Consider fully developed laminar flow with constant properties in a circular tube. Let there be heat transfer to or from the fluid at a constant rate per unit of tube length. Additionally, let there be heat generation within the fluid (perhaps by nuclear reaction) at a rate S, W/m^3, that is everywhere the same. Determine an expression for the Nusselt number as a function of the pertinent parameters. (What are they?) Evaluate the convection conductance in the usual manner, based on heat flux through the surface, surface temperature, and fluid mixed mean temperature.

9-20. Consider fully developed laminar flow with constant properties in a circular tube. Let the surface be insulated, but let there be heat generation within the fluid at a rate S, W/m^3, that is everywhere the same. From an examination of the applicable energy differential equation alone, deduce the approximate shape of the temperature profile within the fluid, and determine whether the highest temperature of the fluid at any axial position occurs at the tube wall or at the tube centerline. Explain the reasons for the result.

9-21. *Computer analysis of laminar thermal entry flow in a circular pipe with* $t_0 = const$ *and constant fluid properties:* Calculate the flow and construct plots to compare developement of the Nusselt number with $x^+ = 2(x/D_h)/Re\,Pr$ over the range $x^+ = 0$–0.3. Compare the results with Table 9-4.

Using the fully developed temperature-profile criterion associated with Eq. (9-7), evaluate the temperature profiles at various x^+ locations to demonstrate the concept of thermally fully developed flow. Carry out this problem for Prandtl number values of 0.01, 1.0, and 100 to show the independence of Pr as the thermally fully developed flow condition is met. Feel free to investigate any other attribute of the entry region or thermally fully developed region of the flow. You can choose how to set up the problem in terms of a suitable choice of Reynolds number, dimensions, and fluid properties (constant). For initial conditions let the velocity profile be hydrodynamically fully developed and the temperature profile be flat at some value t_e.

9.22. *Computer analysis of laminar thermal entry flow in a circular pipe with* $\dot{q}_0'' = const$ *and constant fluid properties:* Calculate the flow and construct plots to compare development of the Nusselt number with $x^+ = 2(x/D_h)/Re\,Pr$ over the range $x^+ = 0$–0.3. Compare the results with Table 9-6.

Follow the second paragraph of Prob. 9-21 for further specification of this problem.

9-23. *Computer analysis of laminar thermal entry flow between parallel plates with* $t_0 = const$ *and with* $\dot{q}_0'' = const$, *along with constant fluid properties:* Calculate the flow and construct plots to compare development of the Nusselt number with $x^+ = 2(x/D_h)/\mathrm{Re}\,\mathrm{Pr}$ over the range $x^+ = 0\text{--}0.3$. Compare the results with the entries for $b/a = \infty$ in Tables 9-9 and 9-10, respectively.

Follow the second paragraph of Prob. 9-21 for further specification of this problem. Let the centerline of the parallel planes be a symmetry line.

9-24. *Computer analysis of Prob. 9-3 with* $\dot{q}_0'' = const$ *and constant fluid properties:* Calculate the flow, starting with initial conditions of flat velocity and temperature profiles at the inlet. Compare the results.

9-25. *Computer analysis of Prob. 9-6 with* $\dot{q}_0'' = f(x)$ *and constant fluid properties:* Calculate the flow, starting with inital conditions of flat velocity and temperature profiles at the inlet. Compare the results. Discuss the assumption of $h = const$, required for the analysis in Prob. 9-6.

9-26. *Computer analysis of laminar combined hydrodynamic- and thermal-entry flow in a circular pipe with* $t_0 = const$ *over the distance* $0 \leqslant x^+ < 0.2$, *followed by analysis of the effect of axial variation of surface temperature,* $t_0 = f(x)$ *from Fig. 9-14 over the distance* $0.2 \leqslant x^+ \leqslant 0.4$, *with constant fluid properties:* Calculate the flow for $\mathrm{Pr} = 0.7$ and compare the results with Table 9-12 and Fig. 9-14. Compare the combined entry Nusselt number distribution with the therml-entry-length distribution. Discuss the behavior of the various variables shown in Fig. 9-14 in terms of the temperature profiles obtained as a part of the computer analysis. You can choose how to set up the problem in terms of a suitable choice of Reynolds number, dimensions, and fluid properties (constant). For initial conditons let the velocity and temperature profiles at the inlet both be flat.

REFERENCES

1. Sparrow, E. M., and S. V. Patankar: *J. Heat Transfer*, vol. 99, pp. 483–485, 1977.
2. Reynolds, W. C.: *Trans. ASME,* ser. C, vol. 82, p. 108, 1960.
3. Reynolds, W. C.: *Int. J. Heat Mass Transfer*, vol. 6, pp. 445–454, 1963.
4. Lundberg, R. E., W. C. Reynolds, and W. M. Kays: NASA TN D-1972, Washington, August 1963.
5. Shah, R. K., and A. L. London: "Laminar Flow Forced Convention in Ducts," *Advances in Heat Transfer*, Academic Press, New York, 1978.
6. Shah, R. K., and A. L. London: *J. Heat Transfer*, vol. 96, pp. 159–165, 1974.
7. Bhatti, M. S.: "Fully Developed Temperature Distribution in a Circular Tube With Uniform Wall Temperature," Unpublished paper, Owens–Corning Fiberglass Corporation, Granville, Ohio, 1985 (cited in Ref. 8).
8. Kakac, S., R. K. Shah, and W. Aung: *Handbook of Single-Phase Convective Heat Transfer*, John Wiley, New York, 1987.
9. Michelsen, M. L., and J. Villadsen, *Int. J. Heat Mass Transfer*, vol. 17, pp. 1391–1402, 1974.
10. Sellars, J. R., M. Tribus, and J. S. Klein: *Trans. ASME*, vol. 78, pp. 441–448, 1956.
11. Seigel, R., E. M. Sparrow, and T. M. Hallman: *Appl. Sci. Res.,* ser. A, vol. 7, pp. 386–392, 1958.

12. Cess, R. D., and E. C. Shaffer: *J. Aero-Space Sci.,* p. 538, 1959.
13. Dennis, S. C. R., A. McD. Mercer, and G. Poots: *Q. Appl. Math.,* vol. 17, pp. 285–297, 1959.
14. Wibulswas, P.: Ph.D. thesis, London University, 1966.
15. Carslaw, H. S., and J. C. Jaeger: *Conduction of Heat in Solids,* 2d ed., Oxford University Press, London, 1959.
16. Hornbeck, R. W.: ASME Paper no. 65-WA HT-36, 1965.
17. Hornbeck, R. W.: NASA SP-297, Washington, 1973.
18. Heaton, H. S., W. C. Reynolds, and W. M. Kays: *Int. J. Heat Mass Transfer,* vol. 7, pp. 763–781, 1964.

CHAPTER
10

HEAT TRANSFER: THE LAMINAR EXTERNAL BOUNDARY LAYER

Heat transfer between a body and a fluid flowing with steady laminar motion over that body is considered in this chapter. It is assumed that the thermal and momentum boundary layers that develop along the surface of the body are not influenced by the development of boundary layers on any adjacent surfaces; this is the distinguishing feature of the problems considered in this chapter as compared with the internal flow problems of Chap. 9. This chapter covers a very broad range of applications, including flow over airfoils, turbine blades, flow inside nozzles, and flow at the stagnation points of cylinders and spheres.

We assume that all body forces are negligible, so that the fluid is forced over the body by some external means unrelated to the temperature field in the fluid. All the laminar flow solutions considered in this chapter are based on an idealization of constant fluid properties, unaffected by temperature. The influence of temperature-dependent properties is discussed later in Chap. 15. Constant properties also imply no concentration gradients, so mass diffusion is neglected here but taken up later in Chap. 20. Consideration is also restricted to velocity sufficiently low that the viscous dissipation term in the energy equation can be neglected; the high-velocity boundary layer is considered in Chap. 16.

159

We consider first constant free-stream velocity flow along a flat, constant-temperature surface, then the same problem with $u_\infty = Cx^m$ and with injection or suction of fluid at the surface. For all these problems the applicable energy differential equation of the boundary layer is Eq. (4-39). These problems have in common the fact that similarity solutions are obtainable in much the same fashion as for the laminar momentum boundary layer in Chap. 8. In fact, the velocity solutions from Chap. 8 are employed in solving the energy equation.

Next a solution is developed for the thermal boundary layer on a flat plate with constant free-stream velocity but with an unheated starting length. In this case the energy integral equation (6-20) is employed to obtain an approximate solution. Then, by employing superposition, the unheated starting-length solution is used to develop solutions for any arbitrarily specified surface temperature or heat flux. These solutions are restricted to the case of constant free-stream velocity.

Finally, an approximate solution is developed for variable free-stream velocity flow over a body of arbitrary shape.

CONSTANT FREE-STREAM VELOCITY FLOW ALONG A CONSTANT-TEMPERATURE SEMI-INFINITE PLATE

This is perhaps the most elementary of the laminar boundary-layer heat-transfer solutions. Specification of a constant-temperature surface means that a thermal boundary layer will originate together with the momentum boundary layer at the leading edge of the plate. The system to be analyzed, the coordinate system, the development of the thermal boundary layer, and the expected temperature profiles are indicated in Fig. 10-1.

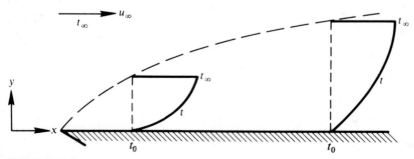

FIGURE 10-1
Temperature profiles in the thermal boundary layer.

The applicable energy differential equation of the boundary layer is Eq. (4-39). Let us first define a nondimensional temperature τ:

$$\tau = \frac{t_0 - t}{t_0 - t_\infty} \tag{10-1}$$

Then Eq. (4-39) becomes

$$\alpha \frac{\partial^2 \tau}{\partial y^2} = u \frac{\partial \tau}{\partial x} + v \frac{\partial \tau}{\partial y} \tag{10-2}$$

We seek a solution to Eq. (10-2) subject to the boundary conditions

$$\tau = 0 \quad \text{at} \quad y = 0$$
$$\tau = 1 \quad \text{at} \quad y \to \infty$$
$$\tau = 1 \quad \text{at} \quad x = 0$$

We already have a solution for the hydrodynamic counterpart of this problem, Eqs. (8-1) and (8-2), and associated boundary conditions; this solution is given in Eq. (8-15) and Table 8-1.

Before discussing a general solution to Eq. (10-2) for the specific boundary conditions, it is worth noting one particular solution that can be readily obtained, simply by analogy. Note that Eq. (10-2) is very similar to the corresponding momentum equation of the boundary layer, Eq. (8-1). In fact, the same function will satisfy both equations and boundary conditions if $\alpha = v$. But $\alpha = v$ corresponds to $\Pr = 1$. Thus for a fluid with $\Pr = 1$ the nondimensional velocity and temperature profiles are similar and grow along the plate at the same rate.

Similar velocity and temperature profiles can be easily shown to lead to a simple relation between the Stanton number and the friction coefficient:

$$\text{St} = c_f/2 \tag{10-3}$$

We already know c_f for this case from Eq. (8-15). Thus, purely by inspection, we already have one heat-transfer solution to the laminar boundary layer. Let us now proceed to a more general solution to Eq. (10-2).

The shape of the nondimensional temperature profiles and the fact that at $\Pr = 1$ the temperature profiles are identical to the velocity profiles suggest that similarity solutions may be obtainable for this problem. We could go through the identical procedure that we used for the velocity part of this problem, but it seems rather likely that we will obtain the same similarity parameter as before. Let us assume that this is so; the validity of this assumption will depend on whether we are successful in obtaining a solution.

Let us assume that

$$\tau = \tau(\eta)$$

where

$$\eta = \frac{y}{\sqrt{vx/u_\infty}}$$

To introduce the velocities u and v, which have already been evaluated in Chap. 8, let us use the stream function ψ and the relation between ψ and ζ as given in Eqs. (8-9) and (8-10):

$$u = \frac{\partial \psi}{\partial y}, \qquad v = -\frac{\partial \psi}{\partial x}, \qquad \psi = \sqrt{vxu_\infty}\,\zeta$$

Performing the indicated differentiation and substituting in Eq. (10-2), we obtain

$$\tau'' + \frac{\text{Pr}}{2}\zeta\tau' = 0 \tag{10-4}$$

where

$$\tau = \tau(\eta), \qquad \zeta = \zeta(\eta)$$

The ζ function was found in Chap. 8 from solutions to the Blasius equation.

We thus seek solutions to this ordinary differential equation subject to the boundary conditions

$$\tau(0) = 0, \qquad \tau(\infty) = 1$$

The fact that we were successful in reducing the partial differential equation (10-2) to an ordinary differential equation (10-4) indicates that similarity solutions do indeed exist for the constant-wall-temperature boundary condition.

Equation (10-4) may be directly integrated as follows:

$$\frac{d\tau'}{d\eta} + \frac{\text{Pr}}{2}\zeta\tau' = 0, \qquad \frac{d\tau'}{\tau'} + \frac{\text{Pr}}{2}\zeta\,d\eta = 0$$

$$\tau' = C_1 \exp\left(-\frac{\text{Pr}}{2}\int_0^\eta \zeta\,d\eta\right) \tag{10-5}$$

$$\tau = C_1 \int_0^\eta \left[\exp\left(-\frac{\text{Pr}}{2}\int_0^\eta \zeta\,d\eta\right)\right]d\eta + C_2 \tag{10-6}$$

In applying the boundary condition $\tau = 0$ at $\eta = 0$ it is apparent that

$$C_2 = 0$$

For the boundary condition $\tau = 1$ at $\eta = \infty$

$$C_1 = \frac{1}{\int_0^\infty \left[\exp \left(-\frac{\text{Pr}}{2} \int_0^\eta \zeta \, d\eta \right) \right] d\eta} \tag{10-7}$$

Therefore

$$\tau(\eta) = \frac{\int_0^\eta \left[\exp \left(-\frac{\text{Pr}}{2} \int_0^\eta \zeta \, d\eta \right) \right] d\eta}{\int_0^\infty \left[\exp \left(-\frac{\text{Pr}}{2} \int_0^\eta \zeta \, d\eta \right) \right] d\eta} \tag{10-8}$$

Equation (10-8) can now be readily solved for any particular Prandtl number, since we already have a tabulation of $\zeta(\eta)$, Table 8-1. We can also define a local conductance, or heat-transfer coefficient, h_x as follows:

$$h_x = \frac{\dot{q}_0''}{t_0 - t_\infty} \qquad \text{(positive } \dot{q}'' \text{ in the positive } y \text{ direction)}$$

$$\dot{q}_0'' = -k \left(\frac{\partial t}{\partial y} \right)_0 = -k(t_\infty - t_0) \left(\frac{\partial \tau}{\partial y} \right)_0 = k(t_0 - t_\infty) \left(\frac{d\tau}{d\eta} \frac{\partial \eta}{\partial y} \right)_0$$

$$= \frac{k(t_0 - t_\infty)}{\sqrt{\nu x / u_\infty}} \tau'(0)$$

To put the conductance in nondimensional form, a local Nusselt number Nu_x can then be defined:

$$\text{Nu}_x = \frac{h_x x}{k} = \frac{x}{\sqrt{\nu x / u_\infty}} \tau'(0) = \frac{\tau'(0)}{\sqrt{\nu / x u_\infty}} = \text{Re}_x^{1/2} \tau'(0)$$

From Eq. (10-5) it is apparent that

$$\tau'(0) = C_1$$

and

$$\text{Nu}_x = \frac{\text{Re}_x^{1/2}}{\int_0^\infty \left[\exp \left(-\frac{\text{Pr}}{2} \int_0^\eta \zeta \, d\eta \right) \right] d\eta} \tag{10-9}$$

Again, since we already know $\zeta(\eta)$, we can readily evaluate the integrals for any particular Prandtl number.

The results of such calculations, based in part on the calculations of Elzy and Sisson,[23] are given in Table 10-1.

Over the Prandtl number range 0.5–15 these results are very well approximated by the equation

$$\text{Nu}_x = 0.332 \, \text{Pr}^{1/3} \, \text{Re}_x^{1/2} \tag{10-10}$$

We could have anticipated the result for $\text{Pr} = 1$ from Eq. (10-3).

TABLE 10-1
Values of $Nu_x Re_x^{-1/2}$ for various Prandtl numbers; heat transfer to the laminar constant-property boundary layer ($u_\infty, t_\infty, t_0 = $ const)

Pr	0.001	0.01	0.1	0.5	0.7	1.0	7.0	10.0	15.0	50	100	1000
$Nu_x Re_x^{-1/2}$	0.0173	0.0516	0.140	0.259	0.292	0.332	0.645	0.730	0.835	1.247	1.572	3.387

Temperature profiles for this solution are shown on Fig. 10-2. The curve for Pr = 1.00 is identical to the velocity profile. Note in particular the strong dependence of the thermal boundary-layer thickness on Prandtl number. Low values of the Prandtl number result in a thermal boundary thickness greater than the momentum boundary-layer thickness, while high values of the Prandtl number result in thermal boundary-layer thickness smaller than the momentum boundary-layer thickness.

Simple closed-form solutions can easily be obtained for fluids whose Prandtl number is either very high or very low.

For very low Prandtl numbers the thermal boundary layer will develop very much faster than the velocity boundary layer, and little error will be introduced if the velocity everywhere in the thermal boundary layer is assumed to be the free-stream velocity u_∞. If we

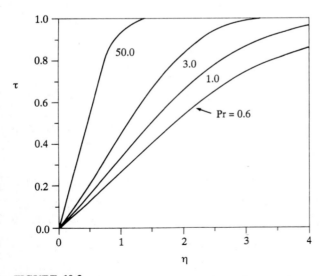

FIGURE 10-2
Temperature profiles for the laminar constant-property boundary layer, with constant free-stream velocity and constant surface temperature.

transpose Eq. (10-4) and differentiate with respect to η, we obtain

$$\frac{d(\tau''/\tau')}{d\eta} + \frac{\text{Pr}}{2}\zeta' = 0$$

But $\zeta' = u/u_\infty$. If u is everywhere equal to u_∞ then $\zeta' = 1$. Thus

$$\frac{d(\tau''/\tau')}{d\eta} + \frac{\text{Pr}}{2} = 0$$

This equation can now be integrated three times with respect to η. An additional necessary boundary condition is that $\tau''(0) = 0$, which can be obtained from the original partial differential equation (10-2), noting that at the wall surface $v = 0$ and $\partial \tau / \partial x = 0$. Carrying through as before, we obtain the following for the local Nusselt number:

$$\text{Nu}_x = 0.565\,\text{Pr}^{1/2}\,\text{Re}_x^{1/2} \quad \text{for } \text{Pr} \ll 1 \tag{10-11}$$

When the Prandtl number is very high, the velocity boundary layer will be very much thicker than the thermal boundary layer. If we assume that the thermal boundary layer is entirely within the part of the velocity boundary layer in which the velocity profile is approximately linear, we can get a simple solution to Eq. (10-4). From Table 8-1 we note that

$$\zeta''(0) = 0.3321$$

Then letting $\zeta'' = \zeta''(0)$ in the region of interest and integrating twice with respect to η, we get

$$\zeta = \frac{0.3321}{2}\eta^2$$

Substituting in Eq. (10-4) and completing the integration, we obtain

$$\text{Nu}_x = 0.339\,\text{Pr}^{1/3}\,\text{Re}_x^{1/2} \quad \text{for } \text{Pr} \gg 1 \tag{10-12}$$

Equation (10-12) corresponds closely to the exact solution for $\text{Pr} \geqslant 10$.

Note that these solutions can also be expressed in terms of a Stanton number, since $\text{Nu} = \text{St}\,\text{Pr}\,\text{Re}$. Thus Eq. (10-10) becomes

$$\text{St}\,\text{Pr}^{2/3} = 0.332\,\text{Re}_x^{-1/2} \tag{10-13}$$

The heat-transfer coefficient h decreases in the flow direction as the boundary layer becomes thicker. As $x \to 0$, on the other hand, h becomes indefinitely large, since the temperature gradient at the surface is infinite at $x = 0$.

Equation (10-10), or Eq. (10-13), can be used to evaluate the local heat-transfer rate at any point along the plate, but often we are interested in the total heat transfer. From each of the above solutions we note that

$$h = Cx^{-1/2}$$

To calculate total heat-transfer rates, we would like a mean conductance with respect to x. Thus

$$h_m = \frac{1}{x} \int_0^x h \, dx = \frac{C}{x} \int_0^x x^{-1/2} \, dx = 2h \qquad (10\text{-}14)$$

Then the complete result for the mean conductance is

$$\mathrm{Nu}_m = 0.664 \, \mathrm{Pr}^{1/3} \, \mathrm{Re}_x^{1/2} \qquad (10\text{-}15)$$

or

$$\mathrm{St}_m \, \mathrm{Pr}^{2/3} = 0.664 \, \mathrm{Re}_x^{-1/2}$$

Equations (10-13) and (10-15) are frequently referred to as the *Pohlhausen solutions*.

It is sometimes convenient to be able to express the Stanton number as a function of a Reynolds number based on *enthalpy thickness* rather than distance x along the surface. The *energy integral equation* (6-24) is applicable to this particular case. If Eq. (10-13) is substituted into Eq. (6-24), and the equation is then integrated, a relation between enthalpy thickness and distance is obtained. Substituting this result back into Eq. (10-13) yields

$$\mathrm{St} \, \mathrm{Pr}^{4/3} = 0.2205/\mathrm{Re}_{\Delta_2} \qquad (10\text{-}16)$$

FLOW WITH $u_\infty = Cx^m$ ALONG A CONSTANT-TEMPERATURE SEMI-INFINITE PLATE

For this problem Eq. (4-39) is again applicable; and since the temperature boundary conditions are the same, Eq. (10-2) with the indicated boundary conditions is again applicable. Since similarity solutions were obtained for the momentum equation, it might be expected that similarity solutions would be obtained for the energy equation. Using the same substitution of variables as above, the energy equation is transformed to an ordinary differential equation similar to Eq. (10-4), but with Pr replaced by Pr $(m + 1)$. The solution becomes Eq. (10-8) with Pr replaced by Pr $(m + 1)$. The solution for ζ is obtained from the appropriate wedge solution of Chap. 8, and a few of the results of this calculation by Eckert[1] are given in Table 10-2.

It is of interest to evaluate the mean conductance along the plate and compare it with the local conductance as tabulated above. For any value of m and Pr the results for the local Nusselt number are given in Table 10-2 in the form

$$\mathrm{Nu}_x \, \mathrm{Re}_x^{-1/2} = \mathrm{const}$$

TABLE 10-2
Values of $\mathrm{Nu}_x\,\mathrm{Re}_x^{-1/2}$ for various Prandtl numbers; heat transfer to the laminar constant-property boundary layer (t_∞, $t_0 = \text{const}$; $u_\infty = Cx^m$)

	Pr				
m	0.7	0.8	1.0	5.0	10.0
−0.0753	0.242	0.253	0.272	0.457	0.570
0	0.292	0.307	0.332	0.585	0.730
0.111	0.331	0.348	0.378	0.669	0.851
0.333	0.384	0.403	0.440	0.792	1.013
1.0	0.496	0.523	0.570	1.043	1.344
4.0	0.813	0.858	0.938	1.736	2.236

from which

$$\frac{hx}{k}\left(\frac{xu_\infty}{v}\right)^{-1/2} = \text{const}, \qquad h = \frac{(\text{const})kx^{-1/2}u_\infty^{1/2}}{v^{1/2}}$$

But for the wedge problem $u_\infty = Cx^m$. Then

$$h = \frac{(\text{const})k}{v^{1/2}}x^{(m-1)/2}$$

Note that for $m = 1$, the two-dimensional stagnation point, h is a constant and does not vary with x. Since all temperature profiles are similar, a constant h can only mean that the thermal boundary layer is of constant thickness for $m = 1$. Note that for $m < 1$ (including negative m and $m = 0$, the flat-plate case), h starts indefinitely large at $x = 0$ and decreases along the plate. For $m > 1$, h starts at zero and increases with x.

The mean conductance h_m from $x = 0$ to x is defined as

$$h_m = \frac{1}{x}\int_0^x h\,dx$$

Then

$$h_m = \frac{1}{x}\int_0^x \frac{Ck}{v^{1/2}}x^{(m-1)/2}\,dx$$

$$= \frac{2}{m+1}\frac{Ck}{v^{1/2}}x^{(m-1)/2}$$

Thus

$$\frac{h_m}{h} = \frac{2}{m+1} \tag{10-17}$$

For $m = 1$ and a Prandtl number not too far removed from 1 the Nusselt number varies as about the 0.4 power of the Prandtl number, so that to a good approximation the solution for $m = 1$ may be expressed as

$$\mathrm{Nu}_x = 0.57 \, \mathrm{Re}_x^{1/2} \, \mathrm{Pr}^{0.4} \tag{10-18}$$

By a transformation† of the coordinate system it is possible to obtain a similar solution for a three-dimensional axisymmetric stagnation point such as occurs as the nose of a blunt-nosed axisymmetric body:[3]

$$\mathrm{Nu}_x = 0.76 \, \mathrm{Re}_x^{1/2} \, \mathrm{Pr}^{0.4} \tag{10-19}$$

In order to make use of either Eq. (10-18) or Eq. (10-19), it is necessary to know the variation of free-stream velocity in the vicinity of the stagnation point. In the general case this must be evaluated from the applicable potential flow solution for the region outside the boundary layer. However, a good approximation for the stagnation point for any rounded-nosed two-dimensional or axisymmetric body may be obtained from the potential flow solutions for flow normal to a cylinder and a sphere, respectively. For a cylinder the velocity, referring to Fig. 10-3, is

$$u_\infty = \frac{2Vx}{R} \quad \text{for small } \frac{x}{R} \tag{10-20}$$

Similarly, for the sphere

$$u_\infty = \frac{3Vx}{2R} \quad \text{for small } \frac{x}{R} \tag{10-21}$$

Substituting Eq. (10-20) into Eq. (10-18), and Eq. (10-21) into Eq. (10-19), it is found that in both cases x cancels (indicating, as we have already seen, that the conductance is constant in the stagnation region), and we can re-form the Nusselt numbers and Reynolds numbers with R

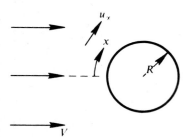

FIGURE 10-3
Nomenclature for flow normal to a cylinder or sphere.

---

† Mangler transformation; see Ref. 2, p. 235.

as the length dimension and V the velocity, yielding

$$\mathrm{Nu}_R = 0.81 \, \mathrm{Re}_R^{1/2} \, \mathrm{Pr}^{0.4} \tag{10-22}$$

for the two-dimensional stagnation point and

$$\mathrm{Nu}_R = 0.93 \, \mathrm{Re}_R^{1/2} \, \mathrm{Pr}^{0.4} \tag{10-23}$$

for the axisymmetric stagnation point.

Equations (10-22) and (10-23) can now be used as an approximation for the stagnation-point heat transfer for any rounded-nosed blunt body by simply using the radius of curvature at the nose for R and the oncoming velocity for V.

Making use of the energy integral equation, Eq. (10-19) for the axisymmetric stagnation point can also be expressed in terms of the local *enthalpy thickness* Reynolds number:

$$\mathrm{St} = 0.45/\mathrm{Re}_{\Delta_2} \quad (\text{for } \mathrm{Pr} = 0.7) \tag{10-24}$$

FLOW ALONG A CONSTANT-TEMPERATURE SEMI-INFINITE PLATE WITH INJECTION OR SUCTION

It was seen in Chap. 8 that similar solutions to the momentum equation of the boundary layer are obtainable for v_0 nonzero (normal velocity at the surface) for $u_\infty = Cx^m$ provided that v_0 varies in the following manner along the surface:

$$v_0 \propto x^{(m-1)/2}$$

The differential equation is unchanged; only the boundary conditions are different. The same is true of the energy differential equation, except that the boundary conditions are also unchanged; that is, Eq. (10-4) is again applicable, as well as its symbolic solution, Eqs. (10-8) and (10-9). However, for $m = 0$, Pr must be replaced by $\mathrm{Pr}\,(m+1)$, as discussed previously in connection with the wedge solutions. It is only necessary to employ the appropriate values of $\zeta(\eta)$ from the solutions to the momentum equation for various values of the blowing parameter $v_0 \, \mathrm{Re}_x^{1/2}/u_\infty$.

Some of the available results are given in Tables 10-3 and 10-4.[4-8,23]

Solutions for the axisymmetric stagnation point can be deduced again by employing the Mangler transformation. Some results are given in Table 10-5.[9]

Examination of the Nusselt numbers tabulated in Tables 10-3 to 10-5 shows that injection causes a decrease in the heat-transfer rate while suction increases heat transfer. Thus, injection can be used as a means

TABLE 10-3
Values of $Nu_x Re_x^{-1/2}$ for various Prandtl numbers and rates of blowing or suction; heat transfer to the laminar constant-property boundary layer ($t_\infty, t_0, u_\infty = $ const)

$\frac{v_0}{u_\infty}\sqrt{\frac{\rho u_\infty x}{\mu}}$	Pr = 0.7	Pr = 0.8	Pr = 1.0
−2.50	1.850	2.097	2.59
−0.750	0.772	0.797	0.945
−0.250	0.429	0.461	0.523
0	0.292	0.307	0.332
0.250	0.166	0.166	0.165
0.375	0.107	0.103	0.0937
0.500	0.0517	0.0458	0.0356
0.619			0

for cooling surfaces exposed to extremely hot gases such as rocket nozzles or rocket vehicle nose cones. Note that these solutions contain the condition that the fluid being injected is the same as the free-stream fluid, and that the temperature of the injected fluid is at the surface temperature. For suction the same is also true, because a boundary condition is that the fluid immediately adjacent to the surface is at surface

TABLE 10-4
Values of $Nu_x Re_x^{-1/2}$ for various rates of blowing or suction and various values of m; heat transfer to the laminar constant-property boundary layer ($t_\infty, t_0 = $ const; Pr = 0.7)

$\frac{v_0}{u_\infty} Re_x^{1/2} \frac{2}{m+1}$	m							
	−0.0476	−0.0244	0	0.0256	0.0526	0.111	0.333	1
−5	2.489	2.489	2.551	2.583	2.617	2.689	2.946	3.61
−2	1.108	1.108	1.136	1.151	1.167	1.2	1.319	1.622
−1	0.674	0.676	0.695	0.706	0.717	0.74	0.819	1.016
−0.5	0.468	0.472	0.487	0.497	0.506	0.526	0.59	0.741
0	0.267	0.278	0.293	0.303	0.313	0.331	0.384	0.496
0.1	0.226	0.24	0.256	0.267	0.277	0.296	0.347	0.452
0.2	0.182	0.202	0.22	0.231	0.243	0.261	0.311	0.409
0.3		0.164	0.185	0.198	0.209	0.228	0.277	0.368
0.4		0.124	0.15	0.166	0.178	0.197	0.245	0.33
0.5		0.0687	0.117	0.135	0.148	0.168	0.214	0.293
0.6			0.0849	0.106	0.12	0.141	0.186	0.259
0.8			0.024	0.0569	0.0727	0.0938	0.135	0.198
1			.	0.023	0.0376	0.057	0.0937	0.146

TABLE 10-5
Values of $\mathrm{Nu}_x \, \mathrm{Re}_x^{-1/2}$ for various rates of blowing for the axisymmetric stagnation point; heat transfer to the laminar constant-property boundary layer $(t_\infty, t_0 = \mathrm{const}; \mathrm{Pr} = 0.7)$

$\dfrac{v_0}{u_\infty} \sqrt{\dfrac{\rho u_\infty x}{\mu}}$	0	0.567	1.154
$\mathrm{Nu} \, \mathrm{Re}_x^{-1/2}$	0.664	0.419	0.227

temperature. As injection is increased, the Nusselt number ultimately become zero, which simply means that the boundary layer is completely blown off so that the fluid temperature gradient at the surface is zero. If the suction rate is increased indefinitely, the heat-transfer rate ultimately becomes equal to the change of enthalpy of the fluid sucked in, since it is changed to surface temperature right at the surface. The problem of injection of a fluid *different* from the free-stream fluid becomes one of mass diffusion in addition to heat transfer. The solution to the mass-diffusion problem is based on the same basic similarity solutions as presented here, but the whole problem must be cast in a somewhat different form.

In Chap. 21 this same problem is formulated in a different manner, with the results presented in Eq. (21-7) and Table 21-2 actually including all of Tables 10-3 to 10-5 as well as other interpolated values of the parameters.

NONSIMILAR THERMAL BOUNDARY LAYERS

The previous solutions have all been based on the energy and momentum differential equations of the boundary layer, but only those particular boundary conditions leading to similarity solutions have been considered. A comprehensive set of solutions to the laminar boundary-layer similarity equations for a perfect gas is given by Dewey and Gross.[10] As with the momentum boundary layer, several classes of solution methods exist for solving the nonsimilar thermal boundary-layer problem. The *first* class is a numerical solution whereby the momentum and energy differential equations are finite-differenced and then integrated. With this method, any number of other boundary conditions can be handled. The *second* class comprises the approximate boundary-layer solutions employing the energy integral equation. The *third* class of solutions employs the concept of local similarity and local nonsimilarity. This method is described by Sparrow and Yu.[11] The *fourth* class of solutions employs the concept of linear superposition similar to that developed for heat transfer in tube flow. In the following sections methods belonging to the second and fourth classes will be developed.

CONSTANT FREE-STREAM VELOCITY FLOW ALONG A SEMI-INFINITE PLATE WITH UNHEATED STARTING LENGTH

As an example of the use of the energy integral equation of the boundary layer, the laminar incompressible boundary layer on a flat plate with no pressure gradient is solved for the case where heating starts at a point $x = \xi$ on the plate, rather than $x = 0$ where the hydrodynamic boundary layer starts. A quite satisfactory approximate solution may be obtained using integral methods.[12] Later the result is used to solve the laminar boundary layer on a flat plate with no pressure gradient for the case where the surface temperature varies in an arbitrary manner along the surface.

The hydrodynamic and thermal boundary layers will build up on the plate as shown in Fig. 10-4. It is assumed that the thermal boundary layer is always thinner than the hydrodynamic boundary layer (the implications of this assumption will become apparent later).

In Chap. 8 a reasonably good solution to the hydrodynamic boundary layer under these conditions was obtained using a cubic parabola to approximate the velocity profile:

$$\frac{u}{u_\infty} = \frac{3}{2}\frac{y}{\delta} - \frac{1}{2}\left(\frac{y}{\delta}\right)^3$$

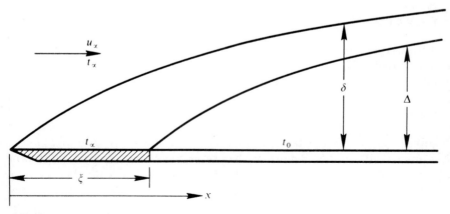

FIGURE 10-4
Boundary-layer development on a plate with an unheated starting length.

A reason for choosing the cubic parabola was that it yields a zero second derivative at the wall, a condition demanded by the boundary-layer differential equation. The same condition obtains for the energy differential equation (10-2), suggesting that a cubic parabola might also be satisfactory for the thermal boundary layer. On this basis, let

$$\theta = t_0 - t, \qquad \theta_0 = t_0 - t_\infty$$

and let

$$\frac{\theta}{\theta_0} = \frac{3}{2}\frac{y}{\Delta} - \frac{1}{2}\left(\frac{y}{\Delta}\right)^3 \qquad (10\text{-}25)$$

As is characteristic of the integral methods, the thermal boundary layer is treated as if it had a finite thickness Δ, whereas in reality it is recognized that this is not the case. The assumption of a finite thickness is one of the idealizations of the method.

Equations (8-32) and (10-25) can now be used to develop an expression for the enthalpy thickness of the thermal boundary layer Δ_2 by simply substituting into Eq. (6-15) and integrating. The integration need be carried through only the thermal boundary layer, because the integrand is zero for $y > \Delta$. Thus

$$\Delta_2 = \frac{3}{20}\frac{\Delta^2}{\delta} - \frac{3}{280}\frac{\Delta^4}{\delta^3}$$

The ratio of the thermal and hydrodynamic boundary-layer thicknesses at any point x is now defined as

$$r = \frac{\Delta}{\delta}, \qquad \Delta = r\delta$$

Making this substitution for Δ and simplifying,

$$\Delta_2 = 3\delta(\tfrac{1}{20}r^2 - \tfrac{1}{280}r^4)$$

from which we obtain the following for $d\Delta_2/dx$:

$$\frac{d\Delta_2}{dx} = 3\delta(\tfrac{1}{10}r - \tfrac{1}{70}r^3)\frac{dr}{dx} + 3(\tfrac{1}{20}r^2 - \tfrac{1}{280}r^4)\frac{d\delta}{dx}$$

If $r < 1$, the second term in each parenthetical expression is small, and we obtain, approximately,

$$\frac{d\Delta_2}{dx} = \frac{3\delta r}{10}\frac{dr}{dx} + \frac{3r^2}{20}\frac{d\delta}{dx}$$

The applicable form of the energy integral equation is Eq. (6-21), and the heat-flux term $\dot{q}_0''$ can be expressed in terms of the temperature gradient at the surface. Thus

$$\frac{d\Delta_2}{dx} = \frac{\dot{q}_0''}{\rho u_\infty c(t_0 - t_\infty)} = \frac{-k(\partial t/\partial y)_0}{\rho u_\infty c(t_0 - t_\infty)}$$

Substituting these expressions for $d\Delta_2/dx$ and evaluating the gradient at the surface from Eq. (10-25), we finally obtain the following ordinary differential equation:

$$2r^2\delta^2 \frac{dr}{dx} + r^3\delta \frac{d\delta}{dx} = \frac{10\alpha}{u_\infty}, \qquad \alpha = \frac{k}{\rho c}$$

Note that δ and $\delta\, d\delta/dx$ are already available from the earlier solution of the hydrodynamic boundary layer in Chap. 8, Eq. (8-33). Substituting for these terms, the following ordinary differential equation for r is obtained, noting that $\nu/\alpha = \mathrm{Pr}$:

$$r^3 + 4r^2x \frac{dr}{dx} = \frac{13}{14} \frac{1}{\mathrm{Pr}}$$

This differential equation can readily be solved to yield

$$r^3 = Cx^{-3/4} + \frac{13}{14} \frac{1}{\mathrm{Pr}}$$

The constant is evaluated from the boundary conditions, $r = 0$ at $x = \xi$. The final result for r is

$$r = \frac{\mathrm{Pr}^{-1/3}}{1.026} \left[1 - \left(\frac{\xi}{x}\right)^{3/4} \right]^{1/3} \tag{10-26}$$

Note that for no unheated starting length ($\xi = 0$), the ratio of boundary-layer thicknesses becomes

$$r = \frac{1}{1.026\, \mathrm{Pr}^{1/3}} \tag{10-27}$$

At $\mathrm{Pr} = 1$ the two boundary layers have close to the same thickness; at high Prandtl numbers the hydrodynamic boundary layer is thicker than the thermal boundary layer; at low Prandtl numbers the thermal boundary layer is evidently the thicker, but the solution is not applicable for this situation because Δ was assumed to be always less than δ. Thus care must be exercised in using this solution for Prandtl numbers below 1, although in the gas Prandtl number range (0.5–1.0) it is applicable over a reasonable range. For example, for $\mathrm{Pr} = 0.7$, $r \leq 1$ for $0.2 < \xi/x < 1$. As is seen below, even for $\xi/x = 0$ the final solution for the Nusselt number is very close to the known exact solution in the gas range, so that it would appear that the entire solution is a very good approximation in this Prandtl number range.

The heat-transfer rate at the surface and a heat-transfer coefficient can now be evaluated from the assumed temperature profile, since δ and r are now known:

$$\dot{q}_0'' = -k\left(\frac{\partial t}{\partial y}\right)_0 = h\theta_0$$

Substituting Eq. (10-25) for θ, Eq. (8-33) for δ, and Eq. (10-26) for r, and solving for h, we get

$$h = \frac{0.332k \, \mathrm{Pr}^{1/3}}{\sqrt{vx/u_\infty}[1 - (\xi/x)^{3/4}]^{1/3}}$$

A local Nusselt number can then be formed, and the final result is

$$\mathrm{Nu}_x = \frac{0.332 \, \mathrm{Pr}^{1/2} \, \mathrm{Re}_x^{1/2}}{[1 - (\xi/x)^{3/4}]^{1/3}} \tag{10-28}$$

For the case of no unheated starting length ($\xi = 0$), Eq. (10-28) becomes

$$\mathrm{Nu}_x = 0.332 \, \mathrm{Pr}^{1/3} \, \mathrm{Re}_x^{1/2} \tag{10-29}$$

The result is seen to be identical to the exact solution, Eq. (10-10).

This development provides a striking example of the power of the integral methods to yield quite adequate solutions for a problem where an exact solution might be considerably more difficult. Equation (10-28) is used later to build up flat-plate laminar boundary-layer solutions for any type of wall-temperature variation rather than just the constant surface temperature that had to be assumed in the above analysis.

CONSTANT FREE-STREAM VELOCITY FLOW ALONG A SEMI-INFINITE PLATE WITH ARBITRARILY SPECIFIED SURFACE TEMPERATURE

All the boundary-layer solutions considered so far have been for either a constant wall temperature or an unheated starting length followed by a step in wall temperature to a constant wall temperature. The method of superposition can be used in the same manner as for tube flow to develop heat-transfer solutions for the boundary layer for an arbitrary wall-temperature variation because of the linearity of the energy differential equation of the boundary layer, Eq. (4-39) or Eq. (10-2).

Let $\theta(\xi, x, y)$ be a solution to Eq. (4-39) for constant-property, constant free-stream velocity flow along a flat plate for the step-function boundary condition $t_0 = t_\infty$ for $x < \xi$, $t_0 = $ constant (different from t_∞) for $x > \xi$. Then θ will be defined such that

$$\frac{t_0 - t}{t_0 - t_\infty} = \theta(\xi, x, y)$$

Then, by following the same line of reasoning as in the tube flow problem, a solution to Eq. (4-39) for any arbitrary variation in surface temperature t_0, but with free-stream temperature constant, is

$$t - t_\infty = \int_0^x [1 - \theta(\xi, x, y)] \frac{dt_0}{d\xi} d\xi + \sum_{i=1}^k [1 - \theta(\xi_i, x, y)] \Delta t_{0,i} \quad (10\text{-}30)$$

The heat flux from the wall surface is determined from

$$\dot{q}_0'' = -k \left(\frac{\partial t}{\partial y} \right)_0$$

Differentiating Eq. (10-30) with respect to y at $y = 0$, and substituting,

$$\dot{q}_0'' = k \left[\int_0^x \theta_y(\xi, x, 0) \frac{dt_0}{d\xi} d\xi + \sum_{i=1}^k \theta_y(\xi_i, x, 0) \Delta t_{0,i} \right]$$

For the single-step-function solution,

$$\dot{q}_0'' = -k \left(\frac{\partial t}{\partial y} \right)_{y=0} = k\theta_y(\xi, x, 0)(t_0 - t_\infty)$$

$$= h(t_0 - t_\infty)$$

$$h(t_0 - t_\infty) = k\theta_y(\xi, x, 0)(t_0 - t_\infty)$$

$$\theta_y(\xi, x, 0) = \frac{h(\xi, x)}{k}$$

Thus for arbitrary wall-temperature variation,

$$\dot{q}_0'' = \int_0^x h(\xi, x) \frac{dt_0}{d\xi} d\xi + \sum_{i=1}^k h(\xi_i, x) \Delta t_{0,i} \quad (10\text{-}31)$$

where $h(\xi, x)$ is the local unit conductance from the single-step-function solution.

The previously derived step-function solution for the laminar boundary layer, Eq. (10-29), can be rewritten as

$$h(\xi, x) = \frac{0.332k}{x} \Pr^{1/3} \operatorname{Re}_x^{1/2} \left[1 - \left(\frac{\xi}{x} \right)^{3/4} \right]^{-1/3} \quad (10\text{-}32)$$

Equation (10-32) substituted into Eq. (10-31) then provides the method for the calculation of heat-transfer rates from a flat plate with a laminar boundary layer and any axial wall-temperature distribution. It is neces-

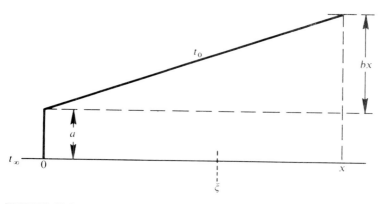

FIGURE 10-5
Example of a step–ramp surface temperature.

sary to insert only the desired $dt_0/d\xi$ as a function of ξ in the integral and any abrupt changes in wall temperature in the summation.

As an example of the method, consider a plate with a step in wall temperature at the leading edge, followed by a linear wall-temperature variation (Fig. 10-5). Then

$$t_0 = t_\infty + a + bx$$

$$\frac{dt_0}{dx} = b = \frac{dt_0}{d\xi}$$

Equation (10-31) can now be evaluated. There is only one step in wall temperature, at $\xi = 0$, so there is only one term in the summation. Thus

$$\dot{q}_0'' = \frac{0.332k}{x} \Pr^{1/3} \operatorname{Re}_x^{1/2} \left\{ \int_0^x \left[1 - \left(\frac{\xi}{x}\right)^{3/4} \right]^{-1/3} b \, d\xi + a \right\} \qquad (10\text{-}33)$$

Problems of this sort lead typically to integrals of the type

$$\int_0^1 Z^{m-1}(1 - Z)^{n-1} \, dZ = \beta(m, n) \qquad (10\text{-}34)$$

where the beta function

$$\beta(m, n) = \frac{\Gamma(m)\Gamma(n)}{\Gamma(m + n)} \qquad (10\text{-}35)$$

for $m > 0$ and $n < \infty$.

The following change of variable in the integral of Eq. (10-33) transforms the integral into the form of the beta function, Eq. (10-34):

$$u = 1 - \left(\frac{\xi}{x}\right)^{3/4}$$

The integral is then readily evaluated by use of beta function tables (see App. C), giving

$$\dot{q}_0'' = 0.332 \frac{k}{x} \Pr^{1/3} \operatorname{Re}_x^{1/2}(1.612bx + a)$$

A local heat-transfer coefficient and a local Nusselt number are then defined, leading to the final result

$$\mathrm{Nu} = \frac{0.332 \Pr^{1/3} \operatorname{Re}_x^{1/2}(1.61bx + a)}{a + bx} \tag{10-36}$$

For $b = 0$ this equation reduces to the constant-wall-temperature solution. Note that for $a = 0$ the solution reduces to a similar expression, but the Nusselt number is always 61 percent greater than for a constant wall temperature.† The importance of taking into consideration any variation of wall temperature is then apparent.

A solution for the case where the wall temperature can be expressed as a power series is also readily obtainable by the same methods. If the wall temperature is expressed by

$$t_0 = t_\infty + A + \sum_{n=1}^{\infty} B_n x^n$$

the heat flux at any point on the plate may be expressed by

$$\dot{q}_0'' = 0.332 \frac{k}{x} \Pr^{1/3} \operatorname{Re}_x^{1/2}\left(\sum_{n=1}^{\infty} nB_n \tfrac{4}{3}x^n \beta_n + A\right) \tag{10-37}$$

where

$$\beta_n = \frac{\Gamma(\tfrac{4}{3}n)\Gamma(\tfrac{2}{3})}{\Gamma(\tfrac{4}{3}n + \tfrac{2}{3})}$$

A heat-transfer coefficient and a Nusselt number could be easily evaluated, but the Nusselt number begins to lose its usefulness when the wall temperature varies markedly.

---

† For $\Pr = 0.7$ this case leads to $\mathrm{Nu}_x = 0.475 \operatorname{Re}_x^{1/2}$. An exact similarity solution can be obtained for this case, leading to $\mathrm{Nu}_x = 0.4803 \operatorname{Re}_x^{1/2}$, only differing by 1.1 percent. This provides some indication of the accuracy of this approximate procedure.

CONSTANT FREE-STREAM VELOCITY FLOW ALONG A SEMI-INFINITE PLATE WITH ARBITRARILY SPECIFIED SURFACE HEAT FLUX

In many cases of practical interest the heat-flux distribution is given and the wall-temperature distribution is to be found. For certain special cases the heat-transfer coefficient resulting from a step-function solution can be put in the form

$$h(\xi, x) = f(x)(x^\gamma - \xi^\gamma)^{-\alpha}$$

For these cases the solution for heat transfer can be modified to yield the wall temperature as a function of heat input.[13] The result is in the form

$$t_0(x) - t_\infty = \int_0^x \dot{q}_0''(\xi) g(\xi, x) \, d\xi \tag{10-38}$$

where $\dot{q}_0''(\xi) = \dot{q}_0''(x)$ is the arbitrarily prescribed surface heat flux and $g(\xi, x)$ is a modification of the function $h(\xi, x)$ found by the relation

$$g(\xi, x) = \frac{(x^\gamma - \xi^\gamma)^{\alpha-1} \gamma \xi^{\gamma-1}}{f(\xi)(-\alpha)! \, (\alpha - 1)!} \tag{10-39}$$

For a laminar boundary layer on a flat plate the step-function solution previously obtained is of the requisite form, and thus $g(\xi, x)$ can be readily evaluated from Eq. (10-28). The result is

$$g(\xi, x) = \frac{\text{Pr}^{-1/3} \text{Re}_x^{-1/2}}{6\Gamma(\frac{4}{3})\Gamma(\frac{5}{3})(0.332)k} \left[1 - \left(\frac{\xi}{x}\right)^{3/4}\right]^{-2/3} \tag{10-40}$$

Substituting Eq. (10-40) into Eq. (10-38) yields an expression that describes the surface temperature of a plate with any arbitrary flux variation:

$$t_0(x) - t_\infty = \frac{0.623}{k} \text{Pr}^{-1/3} \text{Re}_x^{-1/2} \int_0^x \left[1 - \left(\frac{\xi}{x}\right)^{3/4}\right]^{-2/3} \dot{q}_0''(\xi) \, d\xi \tag{10-41}$$

An obvious illustrative example is the case of constant heat rate per unit area. If $\dot{q}_0'' =$ constant in Eq. (10-41), the integral is easily converted to the form of the beta function and can then be evaluated. The wall temperature is then evaluated, and a convection conductance and a Nusselt number can be defined. The result is

$$\text{Nu}_x = 0.453 \, \text{Pr}^{1/3} \, \text{Re}_x^{1/2} \tag{10-42}$$

This result indicates a heat-transfer coefficient about 36 percent higher than at the comparable point on a plate with constant wall temperature, Eq. (10-10). Recall that the constant-wall-temperature solution was obtained using similarity analysis. For the constant-heat-flux boundary condition, no similarity solutions exist.

It is now of interest to employ the *energy integral equation* to determine the enthalpy thickness Δ_2 for this case, and then express the Stanton number as a function of *enthalpy thickness* Reynolds number. The following result is obtained:

$$\text{St Pr}^{4/3} = 0.205/\text{Re}_{\Delta_2} \tag{10-43}$$

Note that there is very little difference from the comparable result obtained for the case of a *constant surface temperature,* Eq. (10-16). In other words, when a *local* length parameter Δ_2 is used, the past history of the thermal boundary layer is of less importance. This is going to provide the key to an approximate procedure, employing the energy integral equation, that will be used later.

FLOW OVER A CONSTANT-TEMPERATURE BODY OF ARBITRARY SHAPE

A rigorous calculation of heat transfer through a laminar boundary layer on a body of arbitrary shape involves first solving the momentum equation to establish the velocity field and then solving the energy equation. This can be done to any degree of precision desired by using numerical finite-difference procedures,[14] or approximate solutions can be obtained based on the momentum and energy integral equations developed in Chap. 6.

A still simpler approach, but one introducing a greater degree of approximation, involves a solution to *either* the momentum or the energy integral equation, but not both. The most successful of the procedures, for a constant surface temperature but with free-stream velocity arbitrarily variable, is based on the energy integral equation.

Eckert[1] proposed a scheme, later modified by others, wherein it is assumed that the rate of growth of any of the thermal boundary-layer thicknesses Δ is a function of local parameters alone, that is,

$$\frac{d\Delta}{dx} = f\left(\Delta, u_\infty, \frac{du_\infty}{dx}, v, \text{Pr}\right)$$

Actually, any of the thermal boundary-layer thicknesses might be used, but for convenience let us use the conduction thickness: $\Delta_4 = k/h_x$. By dimensional analysis, the variation of Δ_4 can be expressed in nondimensional form as

$$\frac{u_\infty}{v}\frac{d\Delta_4^2}{dx} = f\left(\frac{\Delta_4^2}{v}\frac{du_\infty}{dx}, \text{Pr}\right) \tag{10-44}$$

The essence of the method lies in the assumption that the function f for an arbitrary variation of free-stream velocity is the same as for the family of wedge flows, since the latter can be readily evaluated via the similarity solutions.

In Table 10-2 there is tabulated, for each particular value of m and Prandtl number,

$$\mathrm{Nu}_x \, \mathrm{Re}_x^{-1/2} = c_1$$

from which

$$h \propto u_\infty^{1/2} x^{-1/2}$$

Employing

$$u_\infty = C x^m \quad \text{(the wedge condition)}$$

and

$$\Delta_4 = k/h$$

the following expressions can be easily developed for the wedge solutions:

$$\frac{u_\infty}{\nu} \frac{d\Delta_4^2}{dx} = \frac{1-m}{c_1^2} \tag{10-45}$$

$$\frac{\Delta_4^2}{\nu} \frac{du_\infty}{dx} = \frac{m}{c_1^2} \tag{10-46}$$

For each Prandtl number a plot of Eq. (10-44) can then be prepared from the data in Table 10-2. Such a plot is shown in Fig. 10-6 for $\mathrm{Pr} = 0.7$.

Eckert proposes to solve Eq. (10-44) numerically for any arbitrary free-stream velocity variation employing such a plot of the wedge solutions for the function f; and, of course, this can be readily done. Smith and Spalding[15] propose a further simplification, albeit a grosser approximation, by replacing the curve with a straight line, which leads to a simpler integration procedure. For $\mathrm{Pr} = 0.7$ the following linear relation fits the wedge data exactly for the stagnation point and for the flat plate; it is a fair fit in-between, but begins to depart markedly for strongly decelerating flows, although boundary-layer separation limits the curve in this direction:

$$\frac{u_\infty}{\nu} \frac{d\Delta_4^2}{dx} = 11.68 - 2.87 \frac{\Delta_4^2}{\nu} \frac{du_\infty}{dx} \tag{10-47}$$

Equation (10-47) then integrates and reduces to

$$\Delta_4^2 = \frac{11.68\nu \displaystyle\int_0^x u_\infty^{1.87} \, dx}{u_\infty^{2.87}} \tag{10-48}$$

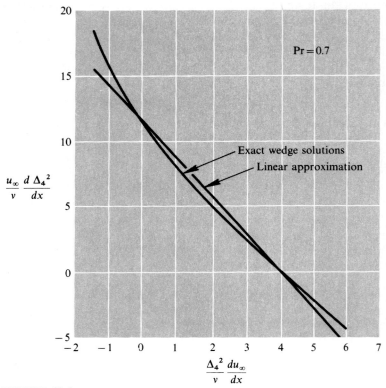

$$\frac{u_\infty}{v}\frac{d\,\Delta_4{}^2}{dx}$$

FIGURE 10-6

Equation (10-48) can now easily be evaluated for any desired variation of u_∞ with x, so the conduction thickness, and thus the local heat-transfer rate, can be readily calculated. However, before converting Eq. (10-48) to an explicit solution for the local conductance, we can transform the problem from the two-dimensional case to the *axisymmetric* case and thus have a much more useful solution. This is essentially the Mangler transformation.

The integral energy equation (6-18), with $t_0 - t_\infty$ and density constant, and with Δ_4 introduced, may be rewritten as

$$\frac{k}{(R\Delta_4)\rho u_\infty c} = \frac{1}{R^2}\frac{d(R\Delta_2)}{dx} - \frac{R\Delta_2}{u_\infty}\frac{1}{R^2}\frac{du_\infty}{dx} \qquad (10\text{-}49)$$

Now if R is constant, the problem reduces to the two-dimensional problem, and Eq. (10-49) becomes

$$\frac{k}{\Delta_4\rho u_\infty c} = \frac{d\Delta_2}{dx} + \frac{\Delta_2}{u_\infty}\frac{du_\infty}{dx} \qquad (10\text{-}50)$$

However, suppose that in Eq. (10-49) we make the following change of variables, where L is some arbitrary reference length:

$$\Delta_4' = \frac{R}{L}\Delta_4, \quad \Delta_2' = \frac{R}{L}\Delta_2, \quad dx' = \frac{R^2}{L^2}dx$$

Substituting, and noting that L cancels, we obtain

$$\frac{k}{\Delta_4'\rho u_\infty c} = \frac{d\Delta_2'}{dx'} + \frac{\Delta_2' du_\infty}{u_\infty dx'} \tag{10-51}$$

With this transformation, Eq. (10-51) is identical to the two-dimensional equation (10-50). A solution of the two-dimensional problem can be transformed into a solution to the axisymmetric problem by the same change of variables. In the present case Eq. (10-48) can be considered a solution to Eq. (10-51), and with the indicated change of variables becomes (note that L cancels)

$$R^2\Delta_4^2 = \frac{11.68\nu \int_0^x u_\infty^{1.87} R^2\, dx}{u_\infty^{2.87}}$$

Noting that $\Delta_4 = k/h$, transposing to form a Stanton number, and inserting Pr = 0.7, we obtain

$$St = 0.418\, \frac{\nu^{1/2}Ru_\infty^{0.435}}{\left(\int_0^x u_\infty^{1.87}R^2\, dx\right)^{0.5}} \tag{10-52}$$

Equation (10-52) has been derived for constant fluid properties. In many applications the fluid density varies markedly in the direction of flow. Examination of the manner in which u_∞ and ρ appear together in the integral energy equation suggests that the following form of Eq. (10-52) should be reasonably accurate for variable-density flow:

$$St = 0.418\, \frac{\mu^{1/2}RG_\infty^{0.435}}{\left(\int_0^x G_\infty^{1.87}R^2\, dx\right)^{0.5}} \tag{10-53}$$

where $G_\infty = u_\infty\rho$.

Equation (10-53) provides a very powerful and simple method for calculating the convection conductance for a laminar boundary layer on a constant-surface-temperature body of revolution with any variation of u_∞ along its surface. For the two-dimensional case R simply cancels. It is found that Eq. (10-53) reduces to the flat-plate solution, Eq. (10-13), and likewise reduces to both the two-dimensional and the axisymmetric

TABLE 10-6
Constants in Eq. (10-54) for various Prandtl numbers; heat transfer to the laminar constant-property boundary layer (t_∞, t_0 = const)

Pr	C_1	C_2	C_3
0.7	0.418	0.435	1.87
0.8	0.384	0.450	1.90
1.0	0.332	0.475	1.95
5.0	0.117	0.595	2.19
10.0	0.073	0.685	2.37

stagnation-point solutions, Eqs. (10-18) and (10-19). Thus, as indicated in Fig. 10-6, the method is exact for the flat plate and the stagnation points and is a reasonable approximation for any other case.

This development has been entirely confined to a fluid with a Prandtl number of 0.7, but it should be quite apparent how a similar solution can be obtained for any Prandtl number for which the wedge solutions are available (Table 10-2). The results for the other Prandtl numbers can be expressed in the form

$$\mathrm{St}_x = \frac{C_1 \mu^{1/2} R G_\infty^{C_2}}{\left(\int_0^x G_\infty^{C_3} R^2 \, dx \right)^{1/2}} \tag{10-54}$$

where the constants for various Prandtl numbers are given in Table 10-6.

In Chap. 21 this scheme is extended to the blowing and suction problem.

FLOW OVER A BODY OF ARBITRARY SHAPE AND ARBITRARILY SPECIFIED SURFACE TEMPERATURE

The variable-surface-temperature problem for a body over which the free-stream velocity varies must be handled by solving first the momentum equation for the specified free-stream velocity and then the energy equation for the varying surface temperature; thus, it is more difficult than the constant-surface-temperature problem where the free-stream velocity varies. A number of methods are presented in the literature, of which perhaps the most complete is that given by Spalding.[16] But in most applications of this type direct numerical solution of the differential equations of the boundary layer proves to be the more practicable approach; and if fluid properties are not constant, this is even more true.

A point is ultimately reached where approximate methods become so complex that it is simpler to seek close-to-exact solutions by numerical integration.

FLOW OVER BODIES WITH BOUNDARY-LAYER SEPARATION

In the presence of a positive pressure gradient in the flow direction (decelerating free-stream velocity) the velocity in the boundary layer decreases in such a way that the velocity gradient at the wall surface can go to zero. In such cases the entire boundary layer separates from the surface, leaving a region of reversed flow or stall near the wall. Separation can and does frequently occur on smooth, continuous surfaces, and virtually always occurs where there is an abrupt step in the surface.

Detailed flow-field and heat-transfer studies of separated flows have followed the advent of large-memory computers and boundary-layer diagnostic tools such as the laser–Doppler anemometer. Some of the flows of interest are laminar separation bubbles on airfoils, shock-induced separation in supersonic flow, and flow over bluff bodies. Reviews of laminar separated flows are given by Brown and Stewartson[17] and Williams.[18] Two simple bluff-body flows are considered here.

For flow normal to a circular cylinder or flow normal to a sphere, a laminar boundary layer forms on the upstream surface (at least for Reynolds numbers sufficiently small that transition to a turbulent boundary layer does not occur). Calculation of local heat-transfer rates at the stagnation point and over the entire upstream surface can be accomplished by methods already discussed. However, at a point just before the maximum diameter, boundary-layer separation occurs (it occurs somewhat later for a turbulent boundary layer), and over the remainder of the cylinder or sphere the local heat-transfer rate fluctuates as vortices are shed into the wake.

The average conductance over the entire surface (front and rear) has been measured by numerous experimenters. An extensive summary of available data for cylinders in cross flow is given by Morgan.[19] These results can be presented in terms of a Nusselt number based on the cylinder diameter and a Reynolds number based on cylinder diameter and the upstream normal velocity. If the following form of the relation is used, the constants C_1 and C_2 recommended by Morgan for air are as given in Table 10-7:

$$\text{Nu}_d = C_1 \, \text{Re}_d^{C_2} \tag{10-55}$$

Heat transfer with flow past spheres in the Reynolds number range 3600–52,000 has been studied by Raithby and Eckert.[20] They recommend

TABLE 10-7
Constants in Eq. (10-55) for various Reynolds numbers; flow normal to a circular cylinder; air

Re_d	C_1	C_2
$10^{-4}-4 \times 10^{-3}$	0.437	0.0895
$4 \times 10^{-3}-9 \times 10^{-2}$	0.565	0.136
$9 \times 10^{-2}-1$	0.800	0.280
$1-35$	0.795	0.384
$35-5 \times 10^3$	0.583	0.471
$5 \times 10^3-5 \times 10^4$	0.148	0.633
$5 \times 10^4-2 \times 10^5$	0.0208	0.814

the use of Eq. (10-55) with $C_1 = 0.257$ and $C_2 = 0.588$ if the sphere is supported from behind. For cross-supported spheres the C_1 coefficient increases by 13 percent.

It has been previously noted that for a laminar boundary layer the Nusselt number (for a fixed Reynolds number) varies approximately as $Pr^{1/3}$, except at very low Pr. Since the boundary layers on the cylinders and spheres under consideration are essentially laminar, Eq. (10-55) for air can be extended to high-Prandtl-number fluids on the basis of this approximation.

There is evidence that the cylinder and sphere behavior is rather sensitive to the turbulence intensity in the upstream flow, and this effect apparently is due to a disturbance of the boundary layer in the vicinity of the upstream stagnation point.[19,20]

Flow normal to *banks* of circular tubes is a configuration extensively used in the construction of heat exchangers. The effect of adjacent and preceding tubes on the behavior of any one tube in the bank is substantial and varies greatly with tube spacing. Extensive data are available, and the reader is referred to Kays and London[21] and to an extensive review by Zukauskas.[22]

The flow passages in high-performance heat exchangers are frequently constructed with extended surface, or fins, and the flow is interrupted in such a way that a laminar boundary layer forms on each fin segment, only to separate and begin again on the next fin. The flow pattern and the heat-transfer mechanism around each fin are very complex and do not lend themselves readily to boundary-layer analysis. Thus the heat-exchanger designer relies on overall experimental data. Data on a few such heat-exchanger surfaces may be found in Chap. 19, and for a large variety in Ref. 21.

PROBLEMS

10-1. Derive Eqs. (10-11), and (10-12) in the text. (See App. C for tables of error and gamma functions.)

10-2. For flow along a plate with constant free-stream velocity and constant fluid properties develop a similarity solution for a constant-temperature plate for a fluid with $Pr = 0.01$. Compare with the approximate results of Prob. 10-2. Note that numerical integration is required.

10-3. Develop an approximate solution of the energy equation for flow at a two-dimensional stagnation point for a fluid with very low Prandtl number, using the assumption that the thermal boundary layer is very much thicker than the momentum boundary layer. From this, develop an equation for heat transfer at the stagnation point of a circular cylinder in cross flow, in terms of the oncoming velocity and the diameter of the tube.

10-4. Repeat Prob. 10-2 for a two-dimensional stagnation point and compare with the results of Prob. 10-3. The following results from the momentum equation solution for the stagnation point are needed:

η	$\zeta(\eta)$
0	0
0.5	0.12
1.0	0.46
1.5	0.87
2.0	1.36
3.0	2.35
$\eta > 3.0$	$\zeta = \eta - 0.65$

10-5. Consider constant-property flow along a surface with constant free-stream velocity. Let the temperature difference between the wall surface and the fluid, $t_0 - t_\infty$, vary as x^m, where m is a constant. Show that a similarity solution to the energy equation is obtainable under these conditions. Carry out the necessary calculations to obtain the Nusselt number as a function of Reynolds number for $Pr = 0.7$ and $m = 1$.

10-6. Using the approximate solution developed in the text for a laminar boundary layer with constant free-stream velocity and a simple step in surface temperature at some arbitrary point, develop a solution for $Pr = 0.7$ and $t_0 - t_\infty$ varying directly with x, using superposition theory. Compare with the exact result from Prob. 10-5.

10-7. Consider liquid sodium at 200°C flowing normal to a 2.5 cm diameter tube at a velocity of 0.6 m/s. Using the results of Prob. 10-3, calculate the "conduction" thickness of the thermal boundary layer at the stagnation point. Calculate the corresponding "shear" thickness of the momentum boundary layer at the stagnation point and discuss the significance of the results.

10-8. Let air at 540°C and 1 atm pressure flow at a velocity of 6 m/s normal to a 2.5 cm diameter cylinder. Let the cylinder be of a thin-walled porous material so that air can be pumped inside the cylinder and out through the pores in order to cool the walls. Let the cooling air be at 40°C where it actually enters the porous material.

   The objective of the problem is to calculate the cylinder surface temperature in the region of the stagnation point for various cooling-air flow rates, expressed as the mass rate of cooling air per square meter of cylinder surface. The problem is to be worked first for no radiation and then assuming that the cylinder surface is a black body radiating to a large surrounding (say a large duct) at 540°C.

   The same cooling air could be used to cool the surface internally by convection without passing through the surface out into the main stream. Assuming that the cooling air is again available at 40°C and is ducted away from the surface at surface temperature, calculate the surface temperature as a function of cooling-air rate per square meter of cylinder surface area for this case and compare with the results above.

10-9. Let air at a constant velocity of 7.6 m/s, a temperature of 90°C, and 1 atm pressure flow along a smooth, flat surface. Let the plate be divided into three sections, each 10 cm in flow length. The first 10 cm section is maintained at 40°C, the second at 80°C, and the third at 40°C. Evaluate and plot the heat flux at all points along the 30 cm of plate length, and find the local heat-transfer coefficient.

10-10. Repeat Prob. 10-9 but let the surface temperature vary sinusoidally from 40°C at the leading and trailing edges to 80°C at the centerline.

10-11. The potential flow solution for the velocity along the surface of a cylinder with flow normal at a velocity V is

$$u_\infty = 2V \sin \theta$$

where θ is measured from the stagnation point. Assuming that this is a reasonable approximation for a real flow on the upstream side of the cylinder, calculate the local Nusselt number as a function of θ for $0 < \theta < \frac{1}{2}\pi$ for a fluid with $Pr = 0.7$ and prepare a plot. Compare these results with the experimental data for the average Nusselt number around a cylinder. What can you conclude about the heat-transfer behavior in the wake region on the rear surface of the cylinder?

10-12. The potential flow solution for the velocity along the surface of a sphere with flow normal at a velocity V is

$$u_\infty = \frac{3}{2}V \sin \theta$$

Do the same problem as the preceding one for the sphere.

10-13. Let air at a constant velocity of 7.6 m/s, a temperature of −7°C, and 1 atm pressure flow along a smooth, flat surface. The plate is 15 cm long (in the flow direction). The entire surface of the plate is adiabatic except for a 2.5 cm wide strip, located between 5 and 7.5 cm from the leading edge, which is electrically heated so that the heat-transfer rate per unit of area on this strip is uniform. What must be the heat flux from this strip such that the temperature of the surface at the trailing edge of the plate is above 0°C? Plot the temperature distribution along the entire plate surface. Discuss the significance of this problem with respect to wing de-icing. (A tabulation of incomplete beta functions, necessary for this solution, is found in App. C)

10-14. Let air at a constant velocity of 15 m/s, a temperature of 300°C, and 1 atm pressure flow along a smooth, flat surface. Let the first 15 cm of the surface be cooled by some internal means to a uniform temperature of 90°C. How does the surface temperature vary for the next 45 cm?

   Hint: Note that the first 15 cm must be treated as a surface-temperature-specified problem, while the last 45 cm must be treated as a surface-heat-flux-specified problem.

10-15. Water at 10°C flows from a reservoir through the convergent nozzle shown in Fig. 10-7 into a circular tube. The mass flow rate of the water is 9 g/s. Making a plausible assumption as to the origin of the boundary layer in the nozzle and assuming that the nozzle surface is at a uniform temperature different from 10°C, calculate the heat-transfer coefficient at the throat of the nozzle. How does this compare with the coefficient far downstream in the tube if the tube surface is at a uniform temperature?

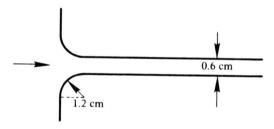

FIGURE 10-7

10-16. *Computer analysis of the laminar thermal boundary layer over a flat plate with zero pressure gradient, t_0 const, and constant properties:* Calculate the boundary-layer flow for Pr = 0.7, 1.0, and 5.0, and evaluate the concepts of boundary-layer thermal and velocity profile similarity, paying particular attention to the Pr = 1.0 case. Compare the Nusselt number results based on x-Reynolds number with the results in the text. Examine several consecutive thermal and velocity profiles to evaluate the enthalpy thick-ness distribution to assess the validity of the energy integral Eq. (6-24). You can choose how to set up the problem in terms of a suitable choice of initial x-Reynolds number, dimensions, and fluid properties (constant). For the initial conditions let the velocity profile be constructed from Table 8-1 and obtain the initial temperature profile from Eq. (10-6). Alterna-tively, you could make the temperature profile be proportional to the velocity profile, although this is strictly correct only for Pr = 1.0.

10-17. *Computer analysis of the laminar thermal boundary layer over a flat plate with a pressure gradient based on $u_\infty = Cx^m$, $t_0 = const$, and constant* *properties:* Calculate the boundary-layer flow for $m = 0.111$ and 1.0, and $\text{Pr} = 0.7$, 1.0, and 5.0, and evaluate the concepts of boundary-layer thermal and velocity profile similarity, paying particular attention to the $\text{Pr} = 1.0$ case. Compare the Nusselt number results based on x-Reynolds number with the results in the text. For the $m = 1.0$ case examine the development of the momentum and thermal boundary layers and discuss the validity of $h = \text{const}$ for this case. Examine several consecutive thermal and velocity profiles to evaluate the enthalpy thickness distribution to assess the validity of the energy integral Eq. (6-18). You can choose how to set up the problem in terms of a suitable choice of initial x-Reynolds number, dimensions, and fluid properties (constant). For the initial conditions let the velocity profile be constructed as described in Prob. 8-8, and obtain the initial temperature profile either from Eq. (10-6) if you solve the momentum problem for $\zeta(\eta)$, or by assuming the temperature profile is proportional to the velocity profile. This latter option is quite approximate.

10-18. *Computer analysis of Prob.* 10-9 *with* $t_0 = f(x)$ *as specified in that problem statement and constant properties:* Calculate the boundary-layer flow and compare with the superposition theory. For the initial conditions follow the instructions in Prob. 10-16.

10-19. *Computer analysis of Prob.* 10-10 *with* $t_0 = f(x)$ *as specified in that problem statement and constant properties:* Calculate the boundary-layer flow and compare with the superposition theory. For the initial conditions follow the instructions in Prob. 10-16.

10-20. *Computer analysis of Prob.* 10-13 *with* $\dot{q}_0'' = f(x)$ *as specified in the problem statement and constant properties:* Calculate the boundary-layer flow and compare with the superposition theory. For the initial conditions, follow the instructions in Prob. 10-16 with regard to the velocity profile, and make the temperature profile proportional to the velocity profile.

REFERENCES

1. Eckert, E. R. G.: *VDI-Forschungsh.,* vol. 416, pp. 1–24, 1942
2. Schlichting, H.: *Boundary Layer Theory,* 6th ed., McGraw-Hill, New York, 1968.
3. Reshotko, E., and C. B. Cohen: NACA (now NASA) TN 3513, Washington, July 1955.
4. Mickley, H. S., R. C. Ross, A. L. Squyers, and W. E. Stewart: NACA TN 3208, Washington, 1954.
5. Emmons, H. W., and D. Leigh: Combustion Aerodynamics Laboratory Report 9, Harvard University, Cambridge, Mass., 1953.
6. Brown, W. B., and P. L. Donoughe: NACA TN 2489, Washington, 1951.
7. Donoughe, P. L., and J. N. B. Livingood: NACA TN 3151, Washington, 1954.
8. Livingood, J. N. B., and P. L. Donoughe: NACA TN 3588, Washington, 1955.
9. Howe, J. T., and W. A. Mersman: NASA TN-D-12, Washington, D.C., 1959.
10. Dewey, C. F., Jr, and J. F. Gross: *Advances in Heat Transfer,* vol. 4, pp. 317–446, Academic Press, New York, 1967.
11. Sparrow, E. M., and H. S. Yu: *J. Heat Transfer,* vol. 93, pp. 328–334, 1971.

12. Eckert, E. R. G., and R. M. Drake, Jr: *Heat and Mass Transfer*, pp. 173–176, McGraw-Hill, New York, 1959.
13. Klein, J., and M. Tribus: ASME Paper 53-SA-46, ASME Semiannual Meeting, 1953; also *Symposium on Heat Transfer*, University of Michigan Press, 1953, and *Trans. ASME*, February 1956 and May 1957.
14. Blottner, F. G.: AGARD-LS-73 (NTIS AD-A013 269), von Kármán Institute, Brussels, February 1975.
15. Smith, A. G., and D. B. Spalding: *J. Roy. Aero. Soc.*, vol. 62, p. 60, 1958.
16. Spalding, D. B.: *J. Fluid Mech.*, vol. 4, p. 22, 1958.
17. Brown, S. N., and K. Stewartson: *Annual Review of Fluid Mechanics*, vol. 1, pp. 45–72, 1969.
18. Williams, J. C., III: *Annual Review of Fluid Mechanics*, vol. 9, pp. 113–144, 1977.
19. Morgan, V. T.: *Advances in Heat Transfer*, vol. 11, pp. 199–264, Academic Press, New York, 1975.
20. Raithby, G. D., and E. R. G. Eckert: *Int. J. Heat Mass Transfer*, vol. 11, pp. 1133–1152, 1968.
21. Kays, W. M., and A. L. London: *Compact Heat Exchangers*, 2d ed., McGraw-Hill, New York, 1964.
22. Zukauskas, A.: *Advances in Heat Transfer*, vol. 8, pp. 93–160, 1972.
23. Elzy, E., and R. M. Sisson: "Tables of Similar Solutions to the Equations of Momentum, Heat and Mass Transfer in Laminar Boundary Later Flow," Bulletin no. 40, Engineering Experiment Station, Oregon State University, Corvalis, Oregon, February 1967.

CHAPTER
11

MOMENTUM TRANSFER: THE TURBULENT BOUNDARY LAYER

In this chapter the stability of the laminar boundary layer is discussed, and a transition to a turbulent type of boundary layer is described. The characteristics of the turbulent boundary layer are first discussed in general terms. Following this, the Prandtl mixing-length theory is introduced as one method for solving the boundary-layer equations. This leads to the concept of a viscous sublayer and the development of the *law of the wall* for the near-wall region using a two-layer mixing-length model.

At this point an *approximate* solution to the momentum equation of the boundary layer is developed, leading to algebraic equations for the momentum thickness and friction coefficient. Next the Van Driest model for the sublayer is introduced, leading to a continuous *law of the wall*. The mixing-length theory is then extended over the entire boundary layer, and a method for solving the complete boundary-layer equations using finite-difference methods is described.

A turbulence model employing the *turbulence kinetic energy* equation is next described, and this leads to a so-called two-equation model based on the turbulence energy equation and another equation based on the turbulence dissipation rate, the $k-\varepsilon$ model of turbulence.

There then follows a discussion of the concept of "equilibrium" turbulent boundary layers, the effects of transpiration, surface roughness, axial surface curvature, and free-stream turbulence.

TRANSITION OF A LAMINAR BOUNDARY LAYER TO A TURBULENT BOUNDARY LAYER

An important characteristic of a laminar boundary layer, and of a laminar flow in a pipe or duct, is that under certain conditions the flow is unstable in the presence of small disturbances, and a transition to a fundamentally different kind of flow, a turbulent flow, can occur. The viscous forces, largely responsible for the characteristics of a laminar flow, have the effect of restoring the flow to its previous state when it is subjected to an external disturbance; on the other hand, inertia forces associated with the velocity changes caused by a disturbance have quite the opposite effect. Inertia forces tend to be destablizing and thus to amplify local disturbances. The Reynolds number is a ratio of inertia to viscous forces, and thus one might well expect that the stability of a laminar flow is in considerable part associated with the value of the Reynolds number, stable laminar flows being associated with low Reynolds numbers.

For a simple laminar boundary layer on a flat, smooth surface with no pressure gradient, it is found in typical laboratory wind tunnel experiments that a transition to a turbulent type of boundary layer tends to occur when the length Reynolds number Re_x reaches the range 300,000–500,000. But this particular range of critical Reynolds numbers seems to be associated with the surface roughness and free-stream disturbances characteristic of typical laboratory wind tunnels for low-speed flows. Laminar boundary layers have been observed to persist to Reynolds numbers several orders or magnitude higher if the free stream is devoid of disturbances and the surface is very smooth. Such conditions frequently occur during the reentry of space vehicles into the Earth's atmosphere.

At low Reynolds numbers it is found experimentally that a laminar boundary layer under a zero pressure gradient is stable to even very large disturbances if the length Reynolds number is less than about 60,000. Similarly, it is found that fully developed laminar flow in a pipe or duct is stable to large disturbances if the Reynolds number based on hydraulic diameter is less than about 2300.

There exists a theory of *viscous stability*—for that is what we are discussing here—and for small disturbances (which permit a linearized theory) it is found that the laminar external boundary layer for zero pressure gradient is stable for Re_x less than about 60,000. It can also be shown that a "favorable" pressure gradient (accelerating flow) leads to higher critical Reynolds numbers, and an "unfavorable" pressure gradient (decelerating flow) leads to lower critical Reynolds numbers. Suction or blowing at the surface might be expected to have similar effects, with suction being stabilizing and blowing destabilizing.

The length Reynolds number Re_x is not a particularly convenient parameter to use as a transition criterion, since it has useful significance only for flows of constant free-stream velocity (no pressure gradient). If it is assumed that the phenomenon is primarily a local one then a local Reynolds number criterion might be more generally useful. Either the displacement or momentum thickness Reynolds numbers might be used, but since the momentum thickness Reynolds number must generally be evaluated in any case in most applications, its use is suggested here.

Let us evaluate the momentum thickness Reynolds number Re_{δ_2} for the laminar, zero-pressure-gradient boundary layer as it reaches its lower critical Reynolds number, $Re_x = 60,000$. From Eq. (8-20)

$$\delta_2 = 0.664\sqrt{vx/u_\infty}$$

Then

$$Re_{x,\text{crit}} = 60,000 = \frac{u_\infty x}{v}$$

Combining, and eliminating x,

$$60,000 = \frac{(u_\infty \delta_2/v)^2}{0.442}$$

But

$$u_\infty \delta_2/v = Re_{\delta_2}$$

Thus

$$Re_{\delta_2,\text{crit}} = 162 \qquad (11\text{-}1)$$

This value of Re_{δ_2} is now proposed as a general criterion for transition whenever substantial free-stream disturbances are present, regardless of pressure gradient or mass transfer at the surface, and this seems to correspond reasonably well with the experimental facts. The laminar boundary layer can, of course, persist to higher values of Re_{δ_2} under undisturbed conditions. A strongly accelerated laminar boundary layer may have a somewhat higher critical Reynolds number, but such an acceleration inhibits the growth of Re_{δ_2} (it can actually decrease), so the point is usually moot.

It is interesting to note that if the equivalent of the momentum thickness Reynolds number is evaluated for fully developed laminar flow in a *flat duct* at a value of the hydraulic diameter Reynolds number equal to 2300, almost the same value of Re_{δ_2} is found as for the simple external boundary layer. This further suggests the general applicability of Eq. (11-1).

THE QUALITATIVE STRUCTURE OF THE TURBULENT BOUNDARY LAYER

The breakdown of the laminar boundary layer discussed in the preceding section does not occur everywhere across the flow at the same distance x. It occurs at certain favorable "spots" and then spreads laterally and downstream until the entire boundary layer is engulfed. There is thus a three-dimensional aspect to transition, and there is a definable "transition region," which can be as long or longer than the preceding laminar region. Ultimately however, a fully developed turbulent boundary layer is established, and this new boundary layer is again, on the average, two-dimensional to the same order as the preceding laminar boundary layer.

We are not going to attempt here to develop a theory for the *transition region,* other than to suggest that in engineering design the extent of this region can usually be assumed to be about the same length as the laminar region, and that the friction, heat-transfer, and mass-transfer coefficients can usually be assumed to (on average) vary continuously from the laminar to the fully developed turbulent region. With this said, we are going to skip forward to the fully developed turbulent region. Let us first see what it looks like and what seems to be happening.

The important characteristics of a turbulent flow adjacent to a solid surface differ little whether we are speaking of a turbulent boundary layer on an external surface or fully developed turbulent flow in a pipe or duct. Experimentally, at least two regions are observed: (1) a predominantly viscous region immediately adjacent to the wall surface where the momentum and heat transfer can be largely accounted for by the simple mechanisms of viscous shear and molecular conduction; and (2) a fully turbulent region, comprising most of the boundary layer, where velocity is nowhere independent of time, where "eddy" motion is observed, and where momentum and heat are transported normally to the flow direction at rates that are generally very much greater than can be accounted for by viscous shear and molecular conduction alone.

Examination of the fully turbulent region reveals that the velocity at any point seems to consist of a relatively large time-averaged velocity, upon which is superimposed a smaller fluctuating velocity with instantaneous components in all directions. The existence of velocity components normal to the mean velocity vector means that fluid is moving, at least momentarily, in the normal direction; this fluid carries momentum with it, and it may carry thermal energy, or any other property for which there is a mean gradient in the normal direction; and, in fact, this is the primary mechanism whereby such effects are transported in the normal direction.

Further examination of the fully turbulent region reveals that the velocity fluctuations are the result of *vorticity* in the fluid. In fact, virtually every particle is a part of fluid eddies that are turning over in various directions, and this is what we mean by a turbulent flow.

The thin, viscous region close to the wall, which we call the *viscous sublayer*, is in some ways of more importance than the fully turbulent region. It is in this sublayer that events leading to the production of turbulence occur. A simple description of the sublayer is that is essentially like a continually developing laminar boundary layer, which grows until its local Reynolds number becomes supercritical, at which point it becomes locally unstable and suffers a local breakdown. This breakdown is identified with a "burst" of turbulence from the wall region. The bursts are sufficiently frequent in both space and time to cause the sublayer to maintain, on the average, a constant *thickness* Reynolds number.

A "burst" resulting from local instability is a three-dimensional event that results in the ejection from the wall region of a relatively large element of slow-moving fluid, which moves out into the fully turbulent region, and its replacement by a similar element of higher-velocity fluid. Although there is some momentum (and heat) transfer within the sublayer associated with these events, viscous shear (and molecular conduction) remains the predominant transport mechanism in the inner part of the sublayer, especially immediately adjacent to the wall, simply because of the relative infrequency of the event.

These elements colliding with the higher-velocity fluid in the fully turbulent region outside of the sublayer provide the primary source of the turbulence kinetic energy in that region, i.e., the kinetic energy of the eddies. As we move out through the turbulent region, more turbulence kinetic energy is then generated through the interaction of the turbulent eddies with the mean velocity gradient. While this production of turbulence kinetic energy is taking place, the viscosity of the fluid is acting to decay the eddies and transform the turbulence energy into thermal energy. A *new state of stability* is established in which the production of turbulence energy is equal to the dissipation of that energy, plus a certain part that is convected and diffused outwards.

In a sense, then, the turbulent boundary layer is a stable region possessing the property of diffusing momentum, and other properties, very much more rapidly than by simple molecular processes. Although an instability in the sublayer is an essential feature, the overall structure is completely stable, self-adjusting, and experimentally repeatable.

As we move along the surface in the downstream direction, the turbulent boundary layer grows, much like the laminar boundary layer but more rapidly, and its momentum thickness Reynolds number grows. But the sublayer maintains essentially the same *thickness* Reynolds number (obviously a *critical* Reynolds number having to do with its viscous stability), and thus becomes a smaller and smaller fraction of the overall boundary layer.

THE CONCEPTS OF EDDY DIFFUSIVITY AND EDDY VISCOSITY

A useful, though not totally necessary, concept is now introduced. To perform turbulent boundary-layer calculations, we need either information or a theory to evaluate the term $\overline{u'v'}$ in Eq. (5-28). This is frequently referred to in the literature as the turbulence "closure" problem, and it is usually solved by solving an algebraic equation, or a differential equation, for $\overline{u'v'}$. In Chap. 5, $\overline{u'v'}$ is defined as an apparent turbulent shear stress, and it seems plausible that it should go to zero if there is no gradient in the mean velocity profile, just as the viscous shear stress goes to zero. In actual fact, this is not precisely so, but it is close enough for many practical applications. Thus it seems reasonable to state that

$$\overline{u'v'} \propto \frac{\partial \bar{u}}{\partial y}$$

As a proportionality factor, we define the eddy diffusivity for momentum ε_M:

$$\overline{u'v'} = -\varepsilon_M \frac{\partial \bar{u}}{\partial y} \qquad (11\text{-}2)$$

This is nothing more than a definition; we still have the problem of evaluating ε_M.

Let us now examine the implications of this definition. When Eq. (11-2) is substituted into Eq. (5-28), the momentum differential equation for the constant-property turbulent boundary layer becomes

$$\bar{u}\frac{\partial \bar{u}}{\partial x} + \bar{v}\frac{\partial \bar{u}}{\partial y} - \frac{\partial}{\partial y}\left[(\nu + \varepsilon_M)\frac{\partial \bar{u}}{\partial y}\right] + \frac{1}{\rho}\frac{d\bar{P}}{dx} = 0 \qquad (11\text{-}3)$$

ε_M is dimensionally the same as ν, and the turbulent contribution to diffusion in the y direction can be readily compared with the molecular contribution. In most cases it is found that $\varepsilon_M \gg \nu$ in the fully turbulent region, while $\nu \gg \varepsilon_M$ in the viscous sublayer close to the wall.

It is sometimes convenient to define an eddy viscosity μ_t. By analogy with the corresponding molecular quantities,

$$\mu_t = \rho \varepsilon_M \qquad (11\text{-}4)$$

THE PRANDTL MIXING-LENGTH THEORY

The Prandtl mixing-length theory is by all odds the simplest, as well as the oldest, of the turbulence closure models. Despite its simplicity, it provides a remarkably adequate basis for many engineering applications. We will carry the mixing-length theory as far as seems practicable, and then develop a more sophisticated model based on the turbulence kinetic energy equation and an equation for the turbulence dissipation rate (the $k-\varepsilon$ model). The differences between these two models will then be demonstrated by some actual comparisons with experimental data.

The defining equation for the mixing length, by itself, provides an adequate starting hypothesis for development of the mixing-length theory. However, the following development is given not as a *proof* of anything, but rather as a plausible argument for the *form* of the mixing-length equation.

Let us idealize the turbulent fluctuating-velocity components u' and v' as simple harmonic functions having the same period θ_1 but differing amplitudes and phase angles. That is,

$$u' = u'_{max} \sin \frac{2\pi\theta}{\theta_1}$$

$$v' = v'_{max} \sin \frac{2\pi(\theta + \alpha_1)}{\theta_1}$$

Next, form the time average of the product of u' and v':

$$\overline{u'v'} = \frac{1}{\theta_1} \int_0^{\theta_1} u'v' \, d\theta = \frac{u'_{max}v'_{max}}{2} \cos \frac{2\pi\alpha_1}{\theta_1}$$

From this result we see that if u' and v' are in phase, or nearly in phase, the "turbulent shear stress" $\overline{u'v'}$ is a finite quantity, although it is also possible for $\overline{u'v'}$ to be zero (if $\alpha_1/\theta_1 = \frac{1}{4}$). The first condition evidently does obtain when there is a *gradient* in the mean velocity profile, because we do observe substantial turbulent shear stresses in such situations.

Now let us define a "mixing length" l in the manner illustrated in Fig. 11-1. If l is small relative to the other dimensions of the system then, approximately,

$$u'_{max} = l \left| \frac{\partial \bar{u}}{\partial y} \right|$$

Next let us assume that

$$v'_{max} = ku'_{max} = kl \left| \frac{\partial \bar{u}}{\partial y} \right|$$

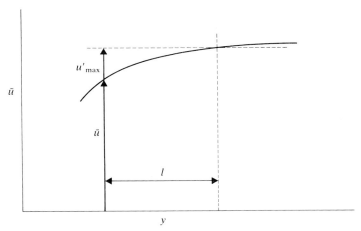

FIGURE 11-1
Graphical definition of the mixing length.

where k is simply a local constant. Then, for the case $\alpha_1 = 0$, we obtain

$$\overline{u'v'} = \tfrac{1}{2}u'_{\max}v'_{\max} = \tfrac{1}{2}kl^2\left(\frac{\partial \bar{u}}{\partial y}\right)^2$$

Next, we absorb k and $-\tfrac{1}{2}$ into l, which has yet to be determined anyway, and we obtain the defining equation for the mixing length:

$$\overline{u'v'} = -l^2\left(\frac{\partial \bar{u}}{\partial y}\right)^2 \qquad (11\text{-}5)$$

If we prefer to express the mixing length in terms of ε_M and/or μ_t, we can then use Eqs. (11-2) and (11-4) to obtain

$$\varepsilon_M = l^2\left|\frac{\partial \bar{u}}{\partial y}\right|, \qquad \mu_t = \rho l^2\left|\frac{\partial \bar{u}}{\partial y}\right| \qquad (11\text{-}6)$$

We are still faced with the proposition of assigning a value to l, but here simple dimensional reasoning can be useful. One can ask the question, on what system dimension would the mixing length be expected to scale? In the region not too distant from the wall the only significant length dimension is the *distance* from the wall, and thus it is reasonable to assume that the mixing length might scale on that distance. If we let the proportionality factor be κ, we can then obtain

$$l = \kappa y \qquad (11\text{-}7)$$

The validity of all these assumptions rests on whether κ does, in fact, turn out to be a constant when measured at various points in the

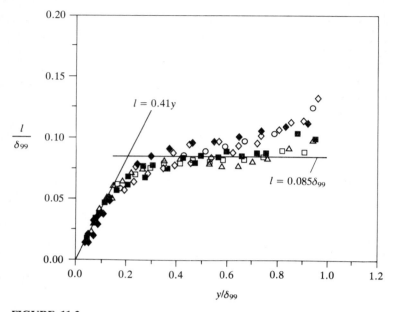

FIGURE 11-2

Mixing-length measurements of Andersen[1] for no pressure gradient, adverse pressure gradient, blowing, and suction.

boundary layer. Figure 11-2 shows some typical experimental measurements of l at five different stations along a surface (δ_{99} is the boundary-layer thickness at the point where $\bar{u}/u_\infty = 0.99$). In the region near the wall the data do indeed appear to be well represented by Eq. (11-7), with κ taking a value of about 0.41 (κ is usually called the *von Kármán constant*).

It should be added that the mixing length is not a quantity that can be "measured" directly. It is actually determined from measurements of the turbulent shear stress and the velocity gradient, employing Eq. (11-5).

There remains a region still closer to the wall, the *viscous sublayer,* where Eq. (11-7) is no longer valid, but the measurements shown on Fig. 11-2 do not go this close to the wall. We will discuss this region later.

In the outer part of the boundary layer it is apparent from Fig. 11-2 that the mixing length is proportional to total boundary-layer thickness rather than the distance from the wall. For y/δ_{99} greater than about 0.7 the value of the mixing length becomes relatively unimportant, so the outer region can be adequately approximated by

$$l = \lambda \delta_{99} \tag{11-8}$$

where λ is typically about 0.085.

Similar behavior is noted for adverse pressure gradients, favorable pressure gradients, blowing and suction, and the observed values for κ and λ seem little affected by these differing boundary conditions. Strictly speaking, these observations are valid for so-called *equilibrium* boundary layers, a subject that is discussed later, but the turbulent boundary layer tends to approach equilibrium, at least in the inner region, remarkably quickly.

WALL COORDINATES

Before discussing the viscous sublayer, we will find it convenient to introduce a new coordinate system that is meaningful in the region close to the wall. But first we will develop an expression for the distribution of the *total shear stress* near the wall.

From the definition of ε_M and the viscosity coefficient, the total apparent shear stress, molecular plus turbulent, can be expressed as

$$\frac{\tau}{\rho} = (\nu + \varepsilon_M)\frac{\partial \bar{u}}{\partial y} \tag{11-9}$$

Substituting this into Eq. (11-3), we obtain

$$\rho\bar{u}\frac{\partial \bar{u}}{\partial x} + \rho\bar{v}\frac{\partial \bar{u}}{\partial y} - \frac{\partial \tau}{\partial y} + \frac{d\bar{P}}{dx} = 0 \tag{11-10}$$

[Note that Eq. (11-10) does not include any assumption about the shear mechanism; in fact, it is also valid for a steady laminar boundary layer.]

Let us now confine our attention to a region relatively close to the wall such that $\rho\bar{u}\,\partial\bar{u}/\partial x$ is sufficiently smaller than the other terms that it can be neglected. We are talking about a region that generally extends outside the viscous sublayer, and in some cases can include as much as one-third of the entire boundary layer. We call this a *Couette flow* assumption and speak of a *Couette flow* region.

Under this assumption, $\rho\bar{u}\,\partial\bar{u}/\partial x = 0$, $\bar{u} = \bar{u}(y)$ alone, Eq. (11-10) becomes an ordinary differential equation, and $\bar{v} = v_0$, the value of the normal velocity component at the wall surface (which would be 0 unless there is blowing or suction at the surface):

$$\rho v_0\frac{d\bar{u}}{dy} + \frac{d\tau}{dy} + \frac{d\bar{P}}{dx} = 0 \tag{11-11}$$

This equation can now be integrated with respect to y between the limits $\tau = \tau_0$ and $\bar{u} = 0$ at $y = 0$, and corresponding values at any distance y:

$$\frac{\tau}{\tau_0} = 1 + \frac{\rho v_0\bar{u}}{\tau_0} + \frac{d\bar{P}}{dx}\frac{y}{\tau_0} \tag{11-12}$$

Next we will introduce a set of nondimensional variables that are based on quantities that are significant in the near-wall region; these are what are referred to as *wall coordinates*. First, from the definition of the friction coefficient we define a *shear velocity*, or *friction velocity*, u_τ for the constant-property boundary layer:

$$\frac{\tau_0}{\rho} = \frac{c_f u_\infty^2}{2} = u_\tau^2$$

Thus

$$u_\tau = \sqrt{\tau_0/\rho}$$

Note that u_τ has the dimensions of velocity. Employing u_τ as a characteristic velocity, the following nondimensional variables can be formed (although these nondimensional variables can also be expressed in terms of the free-stream velocity and the friction coefficient, they are fundamentally independent of the former):

$$u^+ = \frac{\bar{u}}{u_\tau} = \frac{\bar{u}/u_\infty}{\sqrt{c_f/2}}$$

$$y^+ = \frac{y u_\tau}{\nu} = \frac{y\sqrt{\tau_0/\rho}}{\nu} = \frac{y u_\infty \sqrt{c_f/2}}{\nu}$$

$$v_0^+ = \frac{v_0}{u_\tau} = \frac{v_0}{\sqrt{\tau_0/\rho}} = \frac{v_0/u_\infty}{\sqrt{c_f/2}}$$

$$p^+ = \frac{\mu\, d\bar{P}/dx}{\rho^{1/2}\tau_0^{3/2}}$$

When these are substituted into Eq. (11-12), we obtain

$$\frac{\tau}{\tau_0} = 1 + v_0^+ u^+ + p^+ y^+ \tag{11-13}$$

Note that Eq. (11-13) is valid only in the region where the Couette flow assumption is valid; at the outer edge of the boundary layer τ/τ_0 must go to 0.

THE LAW OF THE WALL FOR THE CASE OF $p^+ = 0.0$ AND $v_0^+ = 0.0$

This is the simplest possible boundary layer, with constant free-stream velocity, no transpiration, and an aerodynamically smooth surface. If we consider only the Couette flow region, i.e., that near the wall, Eq. (11-13) reduces to

$$\tau/\tau_0 = 1.0$$

To obtain the law of the wall, we integrate Eq. (11-9) under this condition. Thus,

$$\frac{\tau_0}{\rho} = (v + \varepsilon_M)\frac{d\bar{u}}{dy} \tag{11-14}$$

Now we will propose a model of the turbulent boundary layer consisting of two distinct regions, a *viscous sublayer* for which we will assume $v \gg \varepsilon_M$, and a *fully turbulent* region for which $\varepsilon_M \gg v$. For the latter we will employ the Prandtl mixing-length model of turbulence. Later we will examine a considerably better model of the near-wall region, but we will still employ the mixing-length theory.

For the *viscous sublayer*, Eq. (11-14) reduces to

$$d\bar{u} = \frac{\tau_0}{\rho v}\,dy = \frac{\tau_0}{\mu}\,dy$$

Integrating from the wall,

$$\int_0^{\bar{u}} d\bar{u} = \frac{\tau_0}{\mu}\int_0^y dy$$

$$\bar{u} = \frac{\tau_0}{\mu}y$$

We now introduce the definitions of u^+ and y^+, and obtain

$$u^+ = y^+ \tag{11-15}$$

Next we must face the question of the *effective thickness* of the sublayer. Experimentally we find that the sublayer thickness can be expressed in terms of a critical value of y^+, and this value remains unchanged regardless of the total thickness of the boundary layer. The reason is that y^+ is simply a local thickness Reynolds number, and the stability of the viscous sublayer is such that this Reynolds number always remains a constant at the outer edge of the sublayer. It is a completely self-adjusting process related to the local collapses and "bursting" phenomena discussed earlier. If the effective value of y^+ at the outer edge of the sublayer exceeds this critical value, the sublayer becomes unstable and a "burst" ejects low-momentum fluid, replacing it with higher-momentum fluid from outside the sublayer, thereby momentarily decreasing the sublayer thickness.

For the fundamental case of $p^+ = 0$ and $v_0^+ = 0$ it is found that the critical value of y^+ for this *two-layer model* is 10.8. But note that this is an *effective* thickness that is obtained when using the rather artificial two-layer model. In actual fact, the viscous-dominated region can be much thicker, although locally and momentarily it can be thinner. The artificiality of the thickness will be seen shortly when we compare the results of this model with experimental data.

If we introduce a pressure gradient in the direction of flow, or introduce blowing or suction, or consider a rough surface, then the critical value of y^+ changes. We will discuss these effects later in connection with a better sublayer model.

For the *fully turbulent* region, where we intend to neglect v relative to ε_M, Eq. (11-14) becomes

$$\frac{\tau_0}{\rho} = \varepsilon_M \frac{d\bar{u}}{dy}$$

Now let us introduce the Prandtl mixing-length theory, Eqs. (11-6) and (11-7). Note that we are still confining attention to the region for which Eq. (11-7) is valid:

$$\frac{\tau_0}{\rho} = l^2 \left(\frac{d\bar{u}}{dy}\right)^2 = \kappa^2 y^2 \left(\frac{d\bar{u}}{dy}\right)^2$$

Next, we introduce the definitions of u^+ and y^+; after taking the root of the resulting equation, we obtain

$$\frac{du^+}{dy^+} = \frac{1}{\kappa y^+}$$

We then integrate from the outer edge of the viscous sublayer:

$$\int_{10.8}^{u^+} du^+ = \frac{1}{\kappa} \int_{10.8}^{y^+} \frac{dy^+}{y^+}$$

$$u^+ - 10.8 = \frac{1}{\kappa} \ln \frac{y^+}{10.8}$$

Setting $\kappa = 0.41$ and rearranging, we obtain the logarithmic equation that is generally called the *law of the wall*:

$$u^+ = 2.44 \ln y^+ + 5.0 \tag{11-16}$$

A comparison with experiment is shown on Fig. 11-3. Data from two different investigators, one using water and the other air, at widely differing momentum thickness Reynolds numbers are plotted, along with Eqs. (11-15) and (11-16), plotted on semi-logarithmic coordinates. Note that the data for these two velocity profiles depart in the outer part of the boundary layer, but they collapse together in the inner region and are well represented by Eq. (11-16) down to about $y^+ = 40$, where the effects of the sublayer begin to be seen (although the data of Wieghardt do not extend into the sublayer region). Numerous other experiments at widely differing values of velocity, fluid properties, and Reynolds number yield essentially the same results.

Note that very close to the wall, i.e., for $y^+ < 5$, the data do approach Eq. (11-15). However, the artificiality represented in the

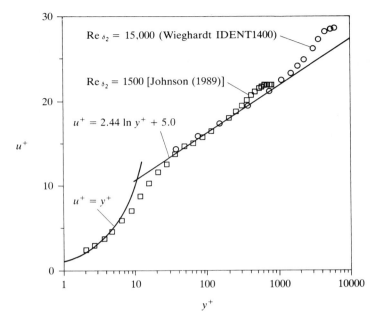

FIGURE 11-3
The law of the wall (constant free-stream velocity): turbulent boundary-layer profiles in wall coordinates.†

two-layer model is apparent, and what is meant by an *effective sublayer thickness* should be evident.

Note the use of semi-logarithmic coordinates to represent these data. This coordinate system results in a straight line for the region where the law of the wall is valid, but it also has the effect of expanding the region near the wall and compressing the outer region. Since most of the velocity change takes place in the inner region, the logarithmic scale provides a convenient way to critically examine the most important region, and for that reason it is extensively employed in presenting experimental data.

In the outer part of the boundary layer the departure of the two profiles from one another, and from Eq. (11-16), results from both Eqs. (11-7) and (11-13) no longer being valid—facts that were of course anticipated. As we go to higher and higher Reynolds numbers, the outer

---

† The term IDENT refers to the data-sets used at the 1968 Stanford University AFOSR Conference, as well as to data-sets tabulated in App. E.

edge of the boundary layer goes to higher and higher values of y^+, and the range of y^+ for which Eq. (11-16) is applicable increases. At the outer edge there is yet another interesting feature. For the case of no pressure gradient and no blowing or suction, the value of u^+ at the outer edge of the boundary layer is always approximately 2.3 above Eq. (11-16), a fact that is observed experimentally, and is also predictable if l in the outer region is evaluated from Eq. (11-8) and the full partial differential equation (11-3) is solved. The outer region in which there is a departure from the logarithmic law of the wall is frequently called the *wake region.*

Figure 11-4 shows the effects of streamwise pressure gradients, both positive and negative, on the velocity profile. Note that the *wake* is very different. For an accelerating flow (a favorable pressure gradient) the wake is diminished and can even be negative, i.e., u^+ can fall *below* the law of the wall, while a decelerating flow (adverse pressure) leads to a much larger wake. But notice that Eq. (11-16) is again approached in the inner region of the boundary layer. This is the reason why Eq. (11-16) is usually called the *law of the wall,* because it appears to be somewhat more general than the simple case of no pressure gradient for which it was derived.

Although a logarithmic form of the law of the wall generally fits the

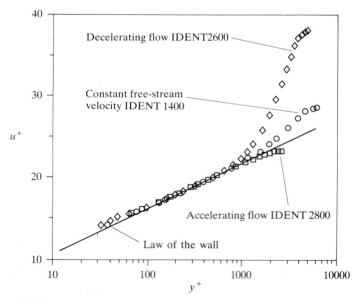

FIGURE 11-4
The effect of pressure gradient on the turbulent velocity profile.

data better, a *power law* is sometimes more convenient. A *power law* that fits the logarithmic form fairly well out to at least $y^+ = 1500$ is

$$u^+ = 8.75y^{+1/7}$$ (11-17)

We will use this equation to develop an approximate solution for the friction coefficient.

AN APPROXIMATE SOLUTION FOR THE TURBULENT MOMENTUM BOUNDARY LAYER

Equation (11-17), together with the momentum integral equation, can be used to obtain a relatively simple algebraic equation for the friction coefficient for a simple turbulent boundary layer. Since Eq. (11-17) was developed for the case of constant free-stream velocity, i.e., no pressure gradient, this result is likewise so restricted. However, it turns out to be a quite good approximation for rather strongly accelerated flows and a fair approximation for very mild decelerations. But it is not at all useful when there is blowing or suction.

First we make the assumption that Eq. (11-17) is valid over the entire boundary layer, i.e., we ignore the wake. Let the total boundary layer thickness be designated as δ, the location at which the velocity is u_∞. We are thus assuming that the boundary layer has a finite thickness, but since the thickness cancels out later, this is not a critical assumption. Then

$$\frac{u_\infty}{\sqrt{\tau_0/\rho}} = 8.75\left(\frac{\delta\sqrt{\tau_0/\rho}}{\nu}\right)^{1/7}$$

Solving for τ_0,

$$\tau_0 = 0.0225\rho u_\infty^2 \left(\frac{\delta u_\infty}{\nu}\right)^{-1/4}$$ (11-18)

Next, δ_1 and δ_2, the displacement and momentum thicknesses, respectively, can be evaluated substituting Eq. (11-17) into Eqs. (6-5) and (6-6), with the results

$$\frac{\delta_1}{\delta} = 0.125, \qquad \frac{\delta_2}{\delta} = 0.097$$

From these we get the following ratio, the *shape factor*, which is used later:

$$\delta_1/\delta_2 = 1.29 = H$$

Next, Eq. (11-18) can be expressed in terms of the *momentum thickness*:

$$\tau_0 = 0.0125 \rho u_\infty^2 \left(\frac{\delta_2 u_\infty}{\nu} \right)^{-1/4} \tag{11-19}$$

Now introducing the definition of the friction coefficient, we obtain an expression for the friction coefficient as a function of a Reynolds number based on the local momentum thickness, i.e.,

$$c_f/2 = 0.0125 \, \mathrm{Re}_{\delta_2}^{-1/4} \tag{11-20}$$

Next we would like to develop a way to evaluate δ_2 as a function of x along the surface. Let us assume that Eq. (11-19) is a reasonable approximation for the surface shear stress even when the free-stream velocity is increasing or decreasing in the direction of flow. This is a reasonable assumption for an accelerating flow, but for a decelerating flow (adverse pressure gradient) it is only adequate for very mild decelerations. The momentum integral equation, Eq. (6-7), for the case of $v_0 = 0$ and constant density is

$$\frac{\tau_0}{\rho u_\infty^2} = \frac{d\delta_2}{dx} + \delta_2 \left[\left(2 + \frac{\delta_1}{\delta_2} \right) \frac{1}{u_\infty} \frac{du_\infty}{dx} + \frac{1}{R} \frac{dR}{dx} \right]$$

Substituting Eq. (11-19) for the shear stress and taking a value of 1.29 for the shape factor H yields

$$0.0125 \left(\frac{u_\infty \delta_2}{\nu} \right)^{-1/4} = \frac{d\delta_2}{dx} + 3.29 \frac{\delta_2}{u_\infty} \frac{du_\infty}{dx} + \frac{\delta_2}{R} \frac{dR}{dx}$$

This can be rearranged to

$$d(\delta_2^{5/4} R^{5/4} u_\infty^{4.11}) = 0.0156 R^{5/4} u_\infty^{3.86} \nu^{1/4} \, dx$$

Integrating with the initial condition that at $x = 0$ at least one of δ_2, R, or u_∞ will be zero, and solving for δ_2, we obtain

$$\delta_2 = \frac{0.036 \nu^{0.2}}{R u_\infty^{3.29}} \left(\int_0^x R^{5/4} u_\infty^{3.86} \, dx \right)^{0.8} \tag{11-21}$$

Like its laminar flow counterpart, Eq. (8-42), this equation then involves a simple procedure that depends upon a given variation of u_∞ and R with x. With δ_2 established as a function of x, the shear stress, or the friction coefficient, is then determined from Eq. (11-19). However, it should again be emphasized that Eq. (11-21) should not be used for any but very mild adverse pressure gradients, although it is quite reasonable for favorable pressure gradients.

This equation implies that the turbulent boundary layer originates at $x = 0$, without a preceding laminar boundary layer and transition region. In such a case, and in Eqs. (11-22)–(11-24) to follow, the point

$x = 0$ is a fictitious *virtual origin* of the turbulent boundary layer—the point where the turbulent boundary layer would have originated were it not preceded by a laminar boundary and a transition region, provided that the same turbulent transport mechanisms were applicable down to zero Reynolds number.

Note that if a laminar boundary layer precedes the turbulent boundary layer, and if sufficient information is available to solve the laminar boundary layer and to determine the "point" of transition, then the value of δ_2 at transition as determined from the laminar boundary-layer solution would provide the necessary lower limit on δ_2 at transition in the equation preceding Eq. (11-21). Without this information, we will simply use Eq. (11-21) with the understanding that x is the distance from the *virtual origin* of the turbulent boundary layer. Practically speaking, if x is sufficiently large, the difference between the real and virtual origins is often sufficiently small that little error is introduced if the preceding laminar boundary layer is ignored, and that is the course we will follow for the moment. If such an assumption is not warranted, it is a relatively simple matter to start back at the equation preceding Eq. (11-21).

Let us now reduce Eq. (11-21) to the elementary case of constant u_∞ and constant (or very large) R:

$$\delta_2 = \frac{0.036\nu^{0.2}}{u_\infty^{3.29}}(u_\infty^{3.86}x)^{0.8}$$

or

$$\frac{\delta_2}{x} = \frac{0.036\nu^{0.2}}{u_\infty^{0.2}x^{0.2}} = 0.036\,\mathrm{Re}_x^{-0.2} \tag{11-22}$$

This result can now be substituted into Eq. (11-19), and, after introduction of the definition of the local friction coefficient, we obtain

$$c_f/2 = 0.0287\,\mathrm{Re}_x^{-0.2} \tag{11-23}$$

Equation (11-23) is in quite good agreement with experiments for Reynolds numbers up to several million, but then becomes increasingly lower than the experimental data for higher Reynolds numbers. The reason is that Eq. (11-17) is no longer a good approximation to the law of the wall at higher Reynolds numbers. Equation (11-17) could be modified for higher Reynolds numbers using a smaller exponent, but this hardly seems worthwhile, since empirical equations based on experimental data are available, and also since a solution to the full partial differential momentum equation (11-3) can be readily obtained using numerical methods and the same mixing-length theory. Probably the most definitive experimental correlation is that of Schultz-Grunow,[3] which is valid from $\mathrm{Re}_x = 5 \times 10^5$ to very high Reynolds numbers:

$$c_f/2 = 0.185(\log_{10}\mathrm{Re}_x)^{-2.584} \tag{11-24}$$

A CONTINUOUS LAW OF THE WALL:
THE VAN DRIEST MODEL

It is quite possible to improve upon the two-layer model of the turbulent boundary layer in a number of ways, but any model that involves a totally viscous sublayer ($\varepsilon_M = 0$), while perhaps satisfactory for solution of the momentum equation, can lead to substantial underprediction of heat transfer, especially at high Prandtl numbers. The reason is that a very small eddy diffusivity in the region $y^+ < 5.0$ can contribute greatly to the heat-transfer rate while having a negligible effect on momentum transfer.

The concept of a sublayer in which relatively large elements of fluid lift off the surface, to be replaced immediately by other fluid from the fully turbulent region—the "bursting" phenomenon that was discussed earlier—demands a sublayer model for which the eddy diffusivity retains a finite magnitude throughout the sublayer and goes to zero only at the wall itself. The idea that ε_M is nonzero, though very small, very near the wall does not mean that we are talking about small turbulent eddies. ε_M is a statistical quantity averaged over time and space, and very small ε_M in the sublayer merely implies a relatively infrequent event.

The Van Driest hypothesis[4] is a sublayer scheme that provides for an eddy diffusivity that is only 0 at $y = 0$, and that also has the virtue that it allows a continuous calculation through the sublayer and into the fully turbulent region with no discontinuities. It does a reasonable job of predicting the various flow parameters throughout the entire sublayer region. There are other schemes that will do much the same thing, so the particular function used in the Van Driest hypothesis should not be regarded as having any theoretical basis.

With this scheme, we use the Prandtl mixing length all the way to the wall instead of truncating it to zero at an assumed effective outer edge of the sublayer, but we simulate the sublayer by introducing a damping function into the mixing-length equation (11-7). An exponential function has the desired characteristics, so Van Driest proposed that

$$l = \kappa y (1 - e^{-y/A})$$

The constant A is an empirically determined effective sublayer thickness, but since it is a dimensional quantity, and we have previously argued that the sublayer thickness should be expressed as a local thickness Reynolds number, we should multiply the numerator and denominator of the exponent by u_τ / v so as to obtain y^+ / A^+. Then

$$l = \kappa y (1 - e^{-y^+/A^+}) \tag{11-25}$$

We now propose to evaluate the total shear stress using Eqs. (11-6) and (11-9) throughout the entire boundary layer, without neglecting either ε_M or v anywhere. If we confine attention to the region relatively near the wall, we can still make the Couette flow assumption if desired, but we are forced to make computer calculations using finite-difference methods, and simple algebraic solutions are no longer feasible.

There remains the problem of determining A^+. This is done by assuming various values of A^+ and carrying out calculations until the calculated values of u^+ outside of the sublayer correspond to the law of the wall, Eq. (11-16). The results of such calculations for $A^+ = 25.0$ are shown on Fig. 11-5.

$A^+ = 25.0$ is determined for a boundary layer with no axial pressure gradient and no blowing or suction; it is found that both pressure gradient and transpiration (as well as surface roughness and perhaps other effects) have a pronounced effect upon A^+. Since the latter is an effective sublayer thickness, these other parameters apparently influence the sublayer thickness. A favorable pressure gradient (accelerating flow) induces a thicker sublayer; an adverse pressure gradient has the opposite effect. Blowing decreases A^+, while suction increases it. Roughness, discussed later, decreases A^+. These results are not surprising when one contemplates the probable influence of these various effects on the stability of the viscous sublayer.

An empirical equation for A^+ as a function of the pressure gradient

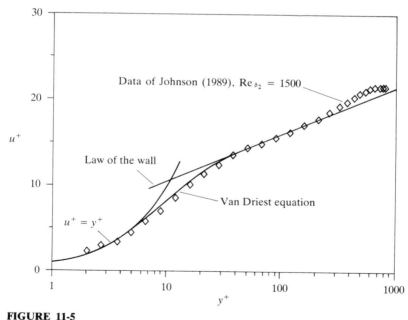

FIGURE 11-5
The Van Driest sublayer (constant free-stream velocity): the sublayer as calculated using the Van Driest equation with $A^+ = 25.0$.

parameter p^+ and the transpiration parameter v_0^+ (both in wall coordinates) is presented by Kays and Moffat.[5]

$$A^+ = \frac{25.0}{a\{v_0^+ + b[p^+/(1 + cv_0^+)]\} + 1} \tag{11-26}$$

where

$$a = 7.1, \quad b = 4.25, \quad c = 10.0$$

If $p^+ > 0$ then $b = 2.9$ and $c = 0.0$.

SUMMARY OF A COMPLETE MIXING-LENGTH THEORY

In the preceding sections we have discussed the elements of a complete mixing-length theory, and then approximate procedures that lead to simple algebraic solutions. For the sake of clarity it is now worthwhile summarizing the complete mixing-length theory, which does, however, require a finite-difference procedure and a digital computer for implementation.

The momentum boundary-layer equation to be solved, together with the appropriate version of the continuity equation and any desired initial and boundary conditions, is Eq. (11-3).

The eddy diffusivity ε_M is then replaced by Eq. (11-6). In the outer part of the boundary layer, the mixing length l is evaluated from Eq. (11-8), while in the inner part it is evaluated from Eq. (11-25). The intersection of the outer and inner parts is determined by equating Eqs. (11-7) and (11-8).

The effective sublayer thickness A^+ can be evaluated from Eq. (11-26).

One difficulty that arises when using Eq. (11-26) is that the sublayer thickness does not instantaneously change when p^+ and/or v_0^+ are abruptly changed. A new state of sublayer equilibrium may not obtain for some distance along the surface. The following rate equation provides a convenient and reasonably practicable way to handle this problem:

$$\frac{dA^+}{dx^+} = \frac{A_{eq}^+ - A^+}{C} \tag{11-27}$$

where A_{eq}^+ is the value determined from Eq. (11-26), $x^+ = x\sqrt{\tau_0/\rho}/v$, and C is an empirical constant. A value of $C = 4000$ appears to give reasonable results.

Figure 11-6 shows an example of a complete velocity profile calculated by the mixing-length procedure just described. (Also included is a velocity profile calculated using the k–ε model, which will be discussed later.) This is a case of constant free-stream velocity, which is

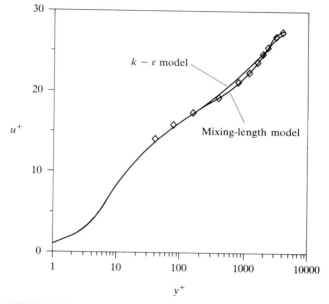

FIGURE 11-6
Prediction of a constant free-stream velocity profile (data of Wieghardt IDENT 1400, $Re_{\delta_2} = 10,600$).

easy for most models to handle, but it is worth noting that the "wake" is quite well reproduced by this very simple version of a mixing-length model.

Figure 11-7 shows the friction coefficient plotted as a function of Re_{δ_2} for this model, together with experimental data, and also Eq. (11-23). Note that the mixing-length model (and also the $k-\varepsilon$ model) predicts this case very well, but that Eq. (11-23) tends to underpredict $c_f/2$ at the higher Reynolds numbers. The reasons for this were discussed earlier.

Figure 11-8 shows a prediction of $c_f/2$ for an accelerating flow. Here again the mixing-length model fares rather well, and it is likewise quite adequate for blowing and suction, and strong acceleration. Where it fares less well is for adverse pressure gradients (as shown in Fig. 11-9) and for strongly nonequilibrium boundary layers, a topic that will be discussed later (as shown in Fig. 11-10).

A MODEL BASED ON THE TURBULENCE KINETIC ENERGY EQUATION

Some of the limitations of the mixing-length model of turbulence were noted in the last section. Although the mixing-length model seems quite

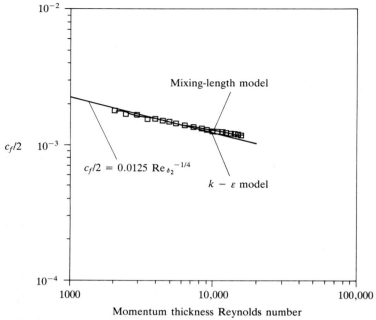

FIGURE 11-7
Friction coefficient for constant free-stream velocity (data of Wieghardt IDENT 1400).

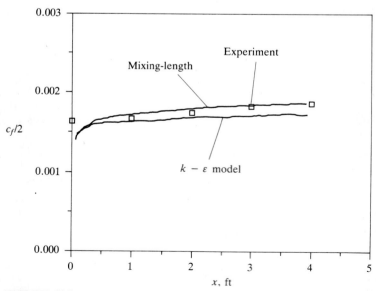

FIGURE 11-8
Friction coefficient for an accelerating flow. Accelerating flow IDENT 2800.

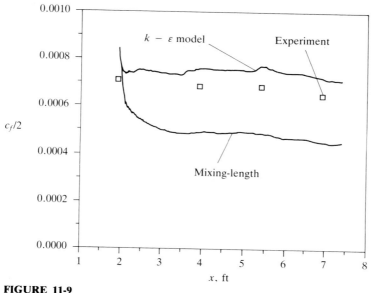

FIGURE 11-9
Friction coefficient for an adverse pressure gradient. Equilibrium decelerating flow IDENT 2600.

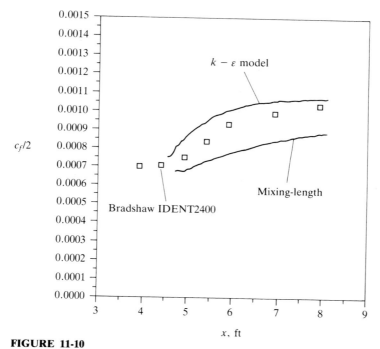

FIGURE 11-10
Friction coefficient for a strongly nonequilibrium boundary layer (constant free-stream velocity following an adverse pressure gradient). Bradshaw relaxing flow IDENT 2400.

adequate for accelerating flows, and it also seems adequate for blowing and suction (although this has not yet been examined), its limitations begin to show up when decelerating flows (adverse pressure gradients) are considered, and it becomes quite evident that it has serious limitations in nonequilibrium situations. An example of a nonequilibrium flow was shown in Fig. 11-10. In that case an adverse-pressure-gradient flow, where the velocity profile has a large wake, is followed by constant free-stream velocity. The level of turbulence in the outer part of the boundary layer is very much greater than it would be for a constant free-stream velocity, and the boundary layer in the constant free-stream section is rather slow to respond to this change of boundary conditions; the mixing-length model provides no provision for this possibility. The result is a considerable under-prediction of the friction coefficient. The mixing-length model also contains no simple way to account for the effects of free-stream turbulence.

When considering heat and mass transfer, situations may be encountered where there is a substantial level of turbulence but no gradient in the mean velocity profile. The mixing-length model would predict zero eddy diffusivity, which is obviously incorrect. An example of this possibility would be the region near the centerline of a duct where the velocity gradient is zero or near zero. The eddy diffusivity in this region can be quite substantial, which would have little effect on velocity-profile calculations, but could have very substantial effects on heat-transfer calculations if the duct were heated asymmetrically.

The difficulty is that it is only for a few relatively simple cases that the mixing length is a function of distance from the wall and/or boundary-layer thickness.

However, a model based on the *kinetic energy of turbulence* offers a possibility for predicting the eddy diffusivity under conditions such as those described, because it seems plausible that there is a relationship between eddy diffusivity and turbulence kinetic energy. Furthermore, Eq. (5-40), developed in Chap. 5, shows that the turbulence kinetic energy is a quantity that is produced and decays in the boundary layer, but that also diffuses across the boundary layer in much the same manner as heat. And, in fact, it seems to have some of the very characteristics that would be necessary to account for the apparent behavior of the eddy diffusivity in the examples cited above.

If the fluid density is treated as constant, the turbulence kinetic energy equation for the boundary layer, Eq. (5-40) becomes

$$\bar{u}\frac{\partial \bar{k}}{\partial x} + v\frac{\partial \bar{k}}{\partial y} - \frac{\partial}{\partial y}\left[v\frac{\partial \bar{k}}{\partial y} - \overline{(k' + P'/\rho)v'} \right] = -\overline{u'v'}\frac{\partial \bar{u}}{\partial y} - \varepsilon \quad (11\text{-}28)$$

The terms on the left-hand side are analogous to similar terms found in the momentum equation and the energy equation—they represent convection and diffusion of turbulence kinetic energy. The first term on the right-hand side accounts for the *production* of turbulence kinetic energy, while the second represents the *dissipation* of that energy into thermal energy through the action of viscosity.

Note that the equation tells us that it requires turbulence to produce turbulence, and that turbulence is only produced when there is a gradient in the mean velocity $\bar{u}$. The equation further implies that viscosity continually acts to decay the turbulent motion regardless of whether turbulence is being produced. The left-hand side implies that turbulence can be convected and diffused into a region where it is not being produced.

Equation (11-28) is valid only in the region where turbulence is fully developed. It is not applicable in the sublayer region, where viscous forces predominate.

We will now introduce some simplifications into Eq. (11-28). First the definition of eddy diffusivity, Eq. (11-2), provides a way of evaluating $\overline{u'v'}$. Next the similarity of the term $\overline{(k' + P'/\rho)v'}$ suggests that an eddy diffusivity for the turbulent transport of turbulent kinetic energy be defined:

$$\overline{\left(k' + \frac{P'}{\rho}\right)v'} = -\varepsilon_k \frac{\partial \bar{k}}{\partial y} \tag{11-29}$$

With these substitutions, Eq. (11-28) becomes

$$\bar{u}\frac{\partial \bar{k}}{\partial x} + \bar{v}\frac{\partial \bar{k}}{\partial y} - \frac{\partial}{\partial y}\left[(\nu + \varepsilon_k)\frac{\partial \bar{k}}{\partial y}\right] = \varepsilon_M\left(\frac{\partial \bar{u}}{\partial y}\right)^2 - \varepsilon \tag{11-30}$$

What kind of relationship could exist between the eddy diffusivity and the turbulence kinetic energy? The simplest possible relationship is suggested by dimensional analysis:

$$\varepsilon_M = al_t\bar{k}^{1/2} \tag{11-31}$$

A length scale l_t must be included to render this equation dimensionally homogeneous. This will be interpreted as a length associated with the size of the turbulent eddies, but of course information on this length will have to be supplied. The coefficient a is a quantity that will have to be established from experiments. Note that this equation can be rendered nondimensional by dividing both sides by the kinematic viscosity ν. In that case the terms on the right-hand side form a Reynolds number, which will be called the *Reynolds number of turbulence*, i.e.,

$$\mathrm{Re}_t = \frac{l_t\bar{k}^{1/2}}{\nu} \tag{11-32}$$

Now let us examine the region near the wall, but outside of the sublayer, where the law of the wall, Eq. (11-16), is applicable. In this region $\tau = \tau_0$, v can be neglected, and $\bar{u}$ is a function of y alone. Thus Eq. (11-9) becomes

$$\tau_0 = \varepsilon_M \frac{d\bar{u}}{dy} \tag{11-33}$$

Differentiation of Eq. (11-16) yields

$$\frac{du^+}{dy^+} = \frac{1}{\kappa y^+}$$

Then

$$\frac{d\bar{u}}{dy} = \frac{\tau_0^{1/2}}{\kappa y} \tag{11-34}$$

so that

$$\tau_0 = \varepsilon_M \frac{\tau_0^{1/2}}{\kappa y}$$

from which

$$\varepsilon_M = \kappa y \tau_0^{1/2} = Cy \tag{11-35}$$

Now compare this result with Eq. (11-31). In this region near the wall the only length dimension on which l_t might be expected to scale would be distance from the wall, y, just was was the case for the mixing length. If that is the case, Eq. (11-31) requires that $\bar{k}$ be a constant in this region. If $\bar{k}$ is a constant then the left-hand side of Eq. (11-30) goes to zero, and the right-hand side reduces to the statement that *production* of turbulence energy is equal to *dissipation* of turbulence energy. This would seem to be the basic characteristic of the law-of-the-wall region.

The idea that in the inner region the rate of production of turbulence energy is equal to the rate of dissipation of turbulence energy provides a clue as to how the dissipation rate ε might be evaluated. Equation (11-30) reduces to

$$\varepsilon_M \left(\frac{d\bar{u}}{dy} \right)^2 = \varepsilon \tag{11-36}$$

From Eq. (11-34),

$$\left(\frac{d\bar{u}}{dy} \right)^2 = \frac{\tau_0}{\kappa^2 y^2}$$

Using Eq. (11-33),

$$\left(\frac{d\bar{u}}{dy} \right)^2 = \frac{\varepsilon_M}{\kappa^2 y^2} \frac{d\bar{u}}{dy}$$

Then, squaring,

$$\left(\frac{d\bar{u}}{dy}\right)^2 = \frac{\varepsilon_M^2}{\kappa^4 y^4}$$

Substituting this result into Eq. (11-36) gives

$$\varepsilon = \frac{\varepsilon_M^3}{\kappa^4 y^4}$$

We now substitute Eq. (11-31), and set $y = l_t$, which we will now presume to be valid in the logarithmic region:

$$\varepsilon = \frac{a^3 l_t^3 \bar{k}^{3/2}}{\kappa^4 l_t^4} = \frac{a^3 \bar{k}^{3/2}}{\kappa^4 l_t} \tag{11-37}$$

It is now proposed that Eq. (11-37), as well as Eq. (11-31), be used *outside* of the logarithmic region, where $\bar{k}$ is not necessarily a constant and where l_t is not necessarily proportional to y. Obviously, we still need information on l_t, as well as the constant a.

To complete the model, we need to be able to evaluate ε_k, the eddy diffusivity for turbulence kinetic energy transport. If we assume that a direct relationship probably exists between ε_k and ε_M, we can define a turbulence Schmidt number Sc_k, and treat it as an experimental constant, i.e.,

$$Sc_k = \varepsilon_M / \varepsilon_k \tag{11-38}$$

Reasonable results for a two-dimensional boundary layer are obtainable from this model using the following information for l_t, a, and Sc_k:

$$l_t = \begin{cases} y & \text{for } y \leq \lambda \delta_{99}/\kappa \\ \lambda \delta_{99}/\kappa & \text{for } y > \lambda \delta_{99}/\kappa \end{cases}$$

$$\kappa = 0.41$$
$$\lambda = 0.085$$
$$a = 0.22$$

Of course it will be observed that the function for l_t is virtually identical to that for the mixing length, so one might question what has been accomplished. The point is that ε_M is now being evaluated from Eq. (11-31) rather than from Eq. (11-6), and this does provide the possibility for the boundary layer to respond to some of the problems discussed at the beginning of this section.

To use this model, it is necessary to solve Eq. (11-30) simultaneously with the momentum equation, so there is an additional partial differential equation to deal with; however, this is not difficult to do on a computer when performing numerical calculations, because the two equations have similar structures.

It is also necessary to provide some kind of independent solution for the sublayer region, where this model is not valid. The simplest procedure is to assume that the sublayer is not affected by variations in $\bar{k}$ in the outer region, i.e., the sublayer comes to equilibrium with the imposed boundary conditions more quickly than does the outer region. The mixing-length model can then be used to solve the momentum equation through the sublayer and for some predetermined distance into the logarithmic region. At this point, which can be arbitrarily chosen, ε_M is established from the mixing-length solution, and then Eq. (11-31) can be used to determine $\bar{k}$ at this point. Equation (11-30) is then solved from this point outwards using the calculated value of $\bar{k}$ as its inner boundary condition. The boundary condition on $\bar{k}$ at the outer edge of the boundary layer is either 0 or whatever is specified as the free-stream value of $\bar{k}$.

The unsatisfactory aspect of this model is the necessity to provide an empirical algebraic equation for the length scale l_t, since one would expect the length scale to be affected by the very same influences that cause $\bar{k}$ to depart from its equilibrium values. It is feasible to develop a differential equation for the length scale (actually the product $\bar{k}l_t$) similar in form to the equation for $\bar{k}$ (see, for example, Ref. 6), and the solution of such an equation simultaneously with Eq. (11-30) would in principle provide a more satisfactory model. However, a differential equation for the dissipation rate ε can also be developed (see Chap. 5), and the dissipation rate equation, together with Eq. (11-30), has been found to provide a more satisfactory model. This two-equation model, which is usually referred to as the $k-\varepsilon$ model, was apparently first proposed by Harlow and Nakayama,[7] and first popularized by Jones and Launder.[8] This model is the subject of the next section.

THE $k-\varepsilon$ MODEL

A transport equation for the dissipation rate ε was introduced in Chap. 5. Although this equation is significantly more complex than the k equation, its modeling proceeds in an almost identical manner. Two modeling issues must be addressed, namely the diffusion terms and the source and sink terms, i.e., the production and dissipation of ε. In the set of diffusion terms $T_\varepsilon + \Pi_\varepsilon + D_\varepsilon$ the pair $T_\varepsilon + \Pi_\varepsilon$ are modeled using the same eddy diffusivity concept as that leading to Eq. (11-29), and D_ε is modeled as a molecular-gradient diffusion term. The set $P_\varepsilon - \varepsilon_\varepsilon$ is modeled using dimensional analysis and the requirements that it reflect production of ε proportional to the production of k, and dissipation of ε that follows the trend of the decay of grid-generated turbulence as observed experimentally. With these models, the boundary-layer form of the dissipation equation (5-42) becomes, for the case of constant fluid properties,

$$\bar{u}\frac{\partial \varepsilon}{\partial x} + \bar{v}\frac{\partial \varepsilon}{\partial y} - \frac{\partial}{\partial y}\left[(v + \varepsilon_\varepsilon)\frac{\partial \varepsilon}{\partial y}\right] = C_1 \frac{\varepsilon}{k}\left[\varepsilon_m\left(\frac{\partial u}{\partial y}\right)^2\right] - C_2 \frac{\varepsilon^2}{k} \quad (11\text{-}39)$$

The eddy diffusivity for the turbulent transport of ε, ε_ε, can be assumed to be related to the eddy diffusivity for momentum, ε_M, so we will define a turbulent Schmidt number for the transport of dissipation rate:

$$Sc_\varepsilon = \varepsilon_M/\varepsilon_\varepsilon \quad (11\text{-}40)$$

The hope here, of course, is that Sc_ε can be treated as an experimentally determined constant.

To use this model, let us return to Eqs. (11-31) and (11-37). If these are each solved for l_t and then combined to eliminate l_t, we obtain an equation for ε_M in terms of $\bar{k}$ and ε (the constants have been combined into a new constant C_μ):

$$\varepsilon_M = C_\mu \bar{k}^2/\varepsilon \quad (11\text{-}41)$$

Thus if we have a solution to both the $\bar{k}$ equation (11-30) and the ε equation (11-39), we can determine ε_M. A set of constants recommended by Jones and Launder[8] after examination of a considerable body of experimental data is

$$C_\mu = 0.09, \quad C_1 = 1.44, \quad C_2 = 1.92$$
$$Sc_k = 1.0, \quad Sc_\varepsilon = 1.3$$

Once more, this equation is valid only *outside* of the sublayer. The sublayer region can be calculated using mixing-length theory, and then the value of ε at the point at which it is desired to patch the inner and outer regions together can be determined from Eq. (11-41). In other words, the wall region is handled in the same manner as described for the $\bar{k}$ equation. The boundary condition on ε at the outer edge of the boundary layer is determined by either using a value specified from what is known about the free stream, or, if the free-stream turbulence is specified as zero, an adequate procedure is to assume that the gradient of ε with respect to y is zero. Note that if the free-stream turbulence kinetic energy and the length scale for the free-stream turbulence are known, a boundary value for ε can be evaluated from Eq. (11-37).

The $k-\varepsilon$ model thus involves simultaneous solution of the momentum equation, continuity equation, $\bar{k}$ equation, and ε equation. Since these equations all have the same general structure, their simultaneous solution by numerical methods does not add greatly to the complexity and time required for solution.

Numerous proposals have been made to extend the $\bar{k}$ and ε equations all the way to the wall. This has generally involved adding additional terms to the right-hand sides of Eqs. (11-30) and (11-39), together with additional experimental constants. However, it is not at present clear which proposal is definitive. In the meantime the use of mixing-length theory for the sublayer, or simply using an algebraic equation such as the law of the wall, is adequate for many applications. The use of an algebraic equation is often referred to as "using a wall function."

Some examples using the $k-\varepsilon$ model are shown in Figs. 11-6 to 11-10. Figure 11-6 shows a comparison of a velocity profile, as predicted by the $k-\varepsilon$ model, with experiment for the case of constant free-stream velocity. Note that the model does predict a wake, although it does not fit the experimental data quite as well as the mixing-length model. The prediction for the friction coefficient, Fig. 11-7, for the same case is virtually identical to the results from the mixing-length model. Figure 11-8 shows friction-coefficient results for an accelerating flow. The mixing-length model appears to give slightly better results than the $k-\varepsilon$ model, but the differences are small. Where the $k-\varepsilon$ model appears to be an improvement is shown in Figs. 11-9 and 11-10. Figure 11-9 shows the case of an adverse pressure gradient (decelerating flow), while Fig. 11-9 shows a nonequilibrium case produced by following an adverse pressure gradient with a constant free-stream velocity section. In both cases the mixing-length model underpredicts the friction coefficient while the $k-\varepsilon$ model comes considerably closer to the data.

EQUILIBRIUM TURBULENT BOUNDARY LAYERS

In the preceding sections reference has been made to *equilibrium* boundary layers without precisely defining what the term implies. In the study of laminar boundary layers it was found that significant simplifications occurred when attention was confined to certain classes of flows, in particular to those flows that led to *similarity* solutions and geometrically similar velocity profiles.

In the study of *turbulent* boundary layers it would be convenient to be able to define some kind of similarity that would lead to a classifiable group of flows. The problem is not quite so straightforward as for laminar boundary layers. In our discussion of the law of the wall we observed a type of similarity in the *inner region* in (u^+, y^+) coordinates, but in that coordinate system the outer region is not self-similar, as can be seen when the two velocity profiles shown in Fig. 11-3 are compared. We are now going to seek a family of flows having *outer-region similarity*, even if we have to sacrifice inner-region similarity in order to define such flows.

Clauser[9] proposed that turbulent boundary layers having outer-region similarity be called *equilibrium boundary layers*, and that an equilibrium boundary layer be one for which the velocity profiles plotted in *velocity defect coordinates* be universal, i.e.

$$\frac{\bar{u} - u_\infty}{u_\tau} = f\left(\frac{y}{\delta_3}\right)$$

where this function is independent of position along the surface, and where

$$\delta_3 = -\int_0^\infty \frac{\bar{u} - u_\infty}{u_\tau} \, dy \tag{11-42}$$

Figure 11-11 shows a plot of three velocity profiles for a typical equilibrium boundary layer. (Actually, this one is an adverse-pressure-gradient boundary layer with blowing.)

Clauser also proposed a *shape factor* that would be a constant, independent of x, under these conditions:

$$G = \int_0^\infty \left(\frac{\bar{u} - u_\infty}{u_\tau}\right)^2 d\left(\frac{y}{\delta_3}\right) \tag{11-43}$$

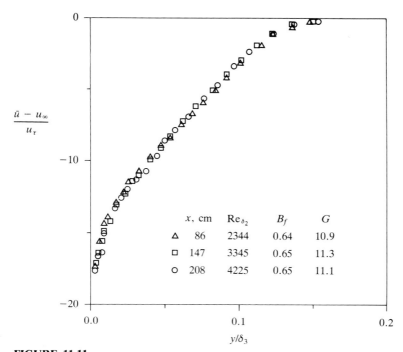

	x, cm	Re_{δ_2}	B_f	G
△	86	2344	0.64	10.9
□	147	3345	0.65	11.3
○	208	4225	0.65	11.1

FIGURE 11-11
Defect profiles for an adverse pressure gradient, transpired turbulent boundary layer (data of Andersen *et al.*[1]).

It is now instructive to examine the *momentum integral equation,* which can be put in the following form (for ρ constant):

$$\frac{d(u_\infty^2 \delta_2)}{dx} = \frac{\tau_0}{\rho}(1 + B_f + \beta) \qquad (11\text{-}44)$$

where

$$B_f = \frac{v_0/u_\infty}{c_f/2} = \frac{\rho v_0 u_\infty}{\tau_0}, \qquad \text{a blowing or transpiration parameter}$$

$$\beta = \frac{\delta_1}{\tau_0}\frac{dP}{dx}, \qquad \text{a pressure-gradient parameter}$$

B_f is really the same "blowing parameter" that we encountered in connection with the laminar similarity solutions, i.e., the parameter that had to be maintained constant (recall that $c_f/2$ varies as $\text{Re}_x^{-1/2}$ for laminar boundary layers). But its *physical* interpretation can be seen in the second term. It is the ratio of the *transpired momentum flux* to the *wall shear force,* and thus it is this ratio that is being maintained constant when the laminar similarity solutions for blowing or suction are obtained.

β is another force ratio, of *axial pressure force* to *wall shear force.* Note that the overall equation expresses the rate of growth of the momentum deficit of the boundary layer.

β is related to some of the other parameters that we have discussed:

$$\beta = p^+ H\, \text{Re}_{\delta_2}(c_f/2)^{1/2}$$

We find experimentally that if for a turbulent boundary layer B_f or β, or both, are constant, we also observe outer-region similarity and G is constant. As a matter of fact, we find that G is a unique function of $B_f + \beta$, as can be seen from the experimental data plotted on Fig. 11-12—data that include a variety of combinations of B_f and β.

Finally, the question remains as to how free-stream velocity or axial pressure-gradient must vary in order to maintain constant β. For a *laminar* boundary layer it was found that similarity solutions were obtained if

$$u_\infty = Cx^m$$

Not too surprisingly, the same free-stream velocity variation for a turbulent boundary layer leads to constant G, that is, to equilibrium boundary layers having outer region similarity.

Another rather special case of equilibrium boundary layers occurs in accelerating flows if an acceleration parameter K is maintained

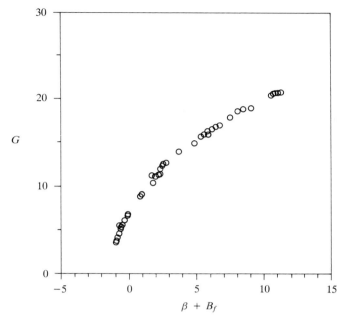

FIGURE 11-12
Clauser shape factor for a variety of equilibrium turbulent boundary layers, no pressure gradient with and without transpiration, and various adverse pressure gradients with and without transpiration (data of Andersen et al.[1]).

constant along the surface:

$$K = \frac{v}{u_\infty^2} \frac{du_\infty}{dx} \tag{11-45}$$

Note that K is related to the inner-region pressure-gradient parameter p^+ by

$$p^+ = \frac{-K}{(c_f/2)^{3/2}}$$

Using this parameter K, the momentum integral equation (for plane flows and constant density) may be written as

$$\frac{1}{u_\infty/v} \frac{d\,\mathrm{Re}_{\delta_2}}{dx} = \frac{c_f}{2} + \frac{v_0}{u_\infty} - K(1 + H)\,\mathrm{Re}_{\delta_2} \tag{11-46}$$

If K and v_0/u_∞ are maintained constant with respect to x, if K is positive, and if the last two terms sum to a negative number, then the boundary layer inevitably approaches an equilibrium condition for which

Re_{δ_2} is constant. We speak of this as an *asymptotic accelerating flow* (it is also called a *sink flow*). Actually, it is a boundary layer that has both *inner* and *outer* similarity; not only is Re_{δ_2} constant, but so also are $c_f/2$, H, G, and β.

Note that for $K = 0$ an asymptotic layer will be reached for *negative* values of v_0/u_∞, that is, constant suction. This is spoken of as an *asymptotic suction layer*.

If $v_0/u_\infty = 0$, the value of Re_{δ_2} at the point of equilibrium depends on K, large values of K leading to low Re_{δ_2}. Note that Re_{δ_2} can decrease in the flow direction if K is large. If K is sufficiently large, the equilibrium value of Re_{δ_2} can be below the critical Reynolds number for transition from a laminar to a turbulent boundary layer, in which case turbulence production ceases, the turbulence decays, and a laminar boundary layer reemerges. Experimentally, this phenomenon, which is often called *laminarization*, occurs when K exceeds about 3×10^{-6}, although "*laminar-like*" heat-transfer behavior is observed at considerably lower values of K. This phenomenon is frequently seen in highly accelerated flows in nozzles.

As can be seen from the relation following Eq. (11-45), a large value of K is accompanied by a large negative value of p^+. Referring now to Eq. (11-26) for the effective sublayer thickness A^+, a large negative value of p^+ causes A^+ to increase, i.e., the viscous sublayer becomes thicker. At approximately $K = 3 \times 10^{-6}$, A^+ becomes indefinitely large, which means that the viscous sublayer overwhelms the entire boundary layer, all turbulence production ceases, and the boundary layer becomes laminar. So this is another way to explain laminarization.

THE TRANSPIRED TURBULENT BOUNDARY LAYER

So far we have discussed only turbulent boundary layers for which the normal component of velocity v_0 is zero, i.e., the case of the impermeable wall. Nonzero values of v_0 can occur if the wall is porous and fluid is "blown" or injected into the boundary layer, or is withdrawn or "sucked." Evaporation or condensation or mass transfer in general lead to nonzero values of v_0. We use the term *transpiration*, or the *transpired boundary layer*, as a general description interchangeably with blowing, suction, mass transfer at the surface, etc.

For laminar boundary layers it was found that *similarity solutions* exist for nonzero values of v_0 provided that a blowing parameter, essentially B_f, is maintained constant. We have already noted that constant B_f results in an *equilibrium* turbulent boundary layer.

Transpiration alters the structure of the turbulent boundary layer rather considerably, affecting the shear-stress distribution [see Eq.

(11-13)], and also strongly affecting the sublayer thickness A^+ [see Eq. (11-26)]. The transpired boundary layer can be calculated quite adequately by finite-difference techniques using the mixing-length model discussed earlier. It is found that both κ and λ are essentially unchanged, so one only need account for the effect of transpiration on A^+. Alternatively, the $k-\varepsilon$ model can be used, but since the primary effect of transpiration is in the sublayer, and the $k-\varepsilon$ model described here is valid only in the outer region, it is still A^+ that provides the critical information. It is also quite possible to develop a transpiration *law of the wall* comparable to Eq. (11-16) but with an additional log-squared term.

Figure 11-13 shows two typical velocity profiles, in wall coordinates, for a turbulent boundary layer with strong blowing, and strong suction. Two things are worthy of note. First, in both cases there is definitely a "logarithmic" region close to the wall (a region that yields a straight line on semi-logarithmic coordinates), but it departs significantly from the law of the wall, Eq. (11-16). Second, the *wake* is very large for the blowing case and resembles the wake for an adverse pressure gradient; see Fig. 11-4. Suction has the opposite effects, and the wake completely disap-

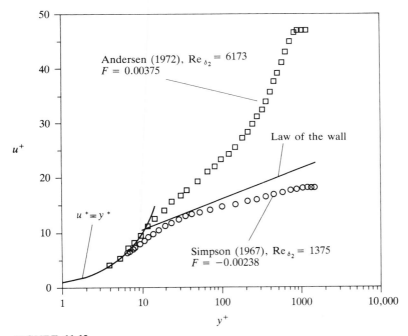

FIGURE 11-13
The effect of transpiration on velocity profiles, constant free-stream velocity (data of Andersen[10] and Simpson[11]).

pears under strong suction conditions, again similar to the effect of a favorable pressure gradient.

As an alternative to complete solution to the momentum partial differential equation, approximate closed-form algebraic solutions can be developed that are extremely useful and provide quick and easy ways to obtain engineering answers that are frequently quite adequate. The following Couette flow solution is rather crude theoretically, but has the virtue of fitting the experimental data surprisingly well.

Consider Eq. (11-3) under the conditions of no pressure gradient, and apply the Couette flow approximation:

$$v_0 \frac{d\bar{u}}{dy} - \frac{d}{dy}\left[(v + \varepsilon_M)\frac{d\bar{u}}{dy}\right] = 0$$

Let the boundary conditions be

$$\bar{u} = 0, \quad \frac{\tau_0}{\rho} = v\left(\frac{d\bar{u}}{dy}\right)_0 \quad \text{at } y = 0$$

$$\bar{u} = u_\infty \quad \text{at } y = \delta$$

Integrating once and applying the first boundary condition,

$$v_0\bar{u} - (v + \varepsilon_M)\frac{d\bar{u}}{dy} = -\frac{\tau_0}{\rho}$$

Rearranging, and integrating across the boundary layer,

$$\int_0^{u_\infty} \frac{d\bar{u}}{v_0\bar{u} + \tau_0/\rho} = \int_0^\delta \frac{dy}{v + \varepsilon_M}$$

that is,

$$\frac{1}{v_0} \ln\left(1 + \frac{v_0 u_\infty \rho}{\tau_0}\right) = \int_0^\delta \frac{dy}{v + \varepsilon_M}$$

But note that

$$\frac{v_0 u_\infty \rho}{\tau_0} = B_f = \frac{v_0/u_\infty}{c_f/2}$$

Then

$$\ln(1 + B_f) = \frac{c_f}{2} B_f u_\infty \int_0^\delta \frac{dy}{v + \varepsilon_M}$$

$$\frac{c_f}{2} = \frac{\ln(1 + B_f)}{B_f} \frac{1}{u_\infty} \left(\int_0^\delta \frac{dy}{v + \varepsilon_M}\right)^{-1}$$

This equation is indeterminate for $B_f = 0$, but

$$\lim_{B_f \to 0} \frac{\ln(1 + B_f)}{B_f} = 1.0$$

So,

$$\left(\frac{c_f}{2}\right)_0 = \frac{1}{u_\infty} \left(\int_0^\delta \frac{dy}{\nu + \varepsilon_M}\right)^{-1}$$

where subscript 0 refers to the $B_f = 0$ case.

Then if we make the assumption (obviously not proved) that the two integrals are independent of B_f, dividing the two equations yields

$$\frac{c_f/2}{(c_f/2)_0} = \frac{\ln(1 + B_f)}{B_f} \tag{11-47}$$

We find that this is an excellent representation of the available experimental data for the no-pressure-gradient case if $(c_f/2)_0$ is evaluated at the *same x Reynolds number*. Thus we can use Eq. (11-23) to obtain

$$\frac{c_f}{2} = 0.0287 \frac{\ln(1 + B_f)}{B_f} \mathrm{Re}_x^{-0.2} \tag{11-48}$$

By use of the momentum integral equation, this result can also be expressed as a function of the *momentum thickness Reynolds number*:

$$\frac{c_f}{2} = 0.0125 \left[\frac{\ln(1 + B_f)}{B_f}\right]^{1.25} (1 + B_f)^{0.25} \mathrm{Re}_{\delta_2}^{-0.25} \tag{11-49}$$

Experimental data indicate that the function $(c_f/2)/(c_f/2)_0$ implied by Eq. (11-49) applies equally well in pressure gradients if the comparison is made at the same value of Re_{δ_2}. (Experimental data supporting this point will be presented later in connection with the corresponding heat-transfer problem.) Furthermore, although the experiments for which these relations have been established were generally carried out at constant B_f (i.e., equilibrium boundary layers), it is found that they hold quite well if B_f varies along the surface, as, for example, when v_0/u_∞ is held constant rather than B_f. This is simply because the viscous sublayer and the inner region of the turbulent boundary layer come into "equilibrium" very quickly.

Because we have specified a constant-density fluid throughout this discussion, density canceled out of the defining equation for B_f, leaving a velocity ratio v_0/u_∞. If the numerator and denominator are multiplied by ρ, this becomes a mass-flux ratio, and it is really mass flux rather than velocity that is responsible for the phenomena observed. Since many applications, especially to gases, involve significant density differences across the boundary layer, a preferred definition for B_f is

$$B_f = \frac{\dot{m}''/G_\infty}{c_f/2}$$

where $\dot{m}''$ is the mass flux at the wall surface and G_∞ is the free-stream mass flux, that is $G_\infty = u_\infty \rho_\infty$. (Note that $\dot{m}''/G_\infty$ is also denoted by F.)

Another "blowing parameter" frequently used is

$$b_f = \frac{\dot{m}''/G_\infty}{(c_f/2)_0}$$

From Eq. (11-47) we then find that $b_f = \ln(1 + B_f)$, and then

$$\frac{(c_f/2)}{(c_f/2)_0} = \frac{b_f}{e^{b_f} - 1} \tag{11-50}$$

It is frequently more convenient to use b_f than B_f in the evaluation of c_f because of the implicit nature of such an equation as Eq. (11-48). However, Eq. (11-50) is only valid when u_∞ is a constant.

Two limits to the transpired boundary layer should be pointed out. One, the *asymptotic suction layer*, was described in the discussion of equilibrium boundary layers. From Eq. (11-46), for $K = 0$, a constant negative value of v_0/u_∞ (or $\dot{m}''/G_\infty$) leads to constant Re_{δ_2}, and then

$$c_f/2 = -\dot{m}''/G_\infty$$

The second limit occurs at large values of blowing, where the friction coefficient tends to *zero* and the boundary layer is literally blown off the wall surface, an occurrence similar to the separation of a boundary layer in an adverse pressure gradient. Two commonly used rules of thumb for "blow-off" are $\dot{m}''/G_\infty = 0.01$ and $b_f = 4.0$.

THE EFFECTS OF SURFACE ROUGHNESS

In all of the preceding discussion of the turbulent boundary layer, it has been assumed that the surface is aerodynamically smooth, without defining what that term implies. At this point it is appropriate to discuss a rational definition of roughness, so as to establish the limits of the theory already presented, and then to describe some of the effects of roughness.

Surface roughness can take on many shapes. The name of Nikuradse[12] is indelibly associated with the rational analysis of rough surfaces as a result of his experiments with turbulent flow in pipes that were artificially roughened with uniform grains of sand. Schlichting[13] introduced the concept of equivalent sand-grain roughness as a means of characterizing other types of roughness elements by referring to the equivalent net effect produced by Nikuradse's experiments. We will use the symbol k_s, a length dimension, to describe roughness element size, k_s being actually the size of sieve used by Nikuradse to sift the sand.

The effect of roughness on a turbulent boundary layer is primarily right at the wall, and thus a nondimensional expression of a roughness size is logically based on the shear velocity u_τ. This leads to a roughness Reynolds number Re_k as a nondimensional measure of surface roughness:

$$Re_k = u_\tau k_s / v$$

Three regimes can be identified from the experimental data in terms of values of Re_k. For $Re_k < 5.0$ the surface behaves as perfectly *smooth* (aerodynamically smooth). For $5.0 < Re_k < 70.0$ there is an increasing effect of roughness, but some of the smooth-surface characteristics persist, and this is called the range of *transitional roughness*. The range $Re_k > 70.0$ is the regime of a *fully rough* surface. A basic characteristic of the latter is that the friction coefficient c_f becomes independent of Reynolds number, which essentially means that viscosity is no longer a significant variable.

A mixing-length model can be readily developed for the fully rough region. The first effect of roughness is to destabilize the viscous sublayer, which results in an effectively thinner sublayer. At about $Re_k = 70$ the sublayer disappears entirely, which accounts for the fact that viscosity is no longer a significant variable. But this also means that the shear stress must be transmitted to the wall by some mechanism other than viscous shear. This different mechanism is quite obviously pressure drag directly on the roughness elements, i.e., a result of impact or dynamic pressure on the upstream side of each element.

If the Prandtl mixing-length scheme is used, it is apparent that Eq. (11-7) is no longer valid at the wall surface, because l, and thus ε_M, cannot go to zero at the wall, for otherwise there is no mechanism for transferring the shear stress to the wall. The eddy diffusivity and mixing length must apparently be finite at $y = 0$ for a fully rough surface. This fact can be readily modeled by a near-wall mixing-length equation

$$l = \kappa(y + \delta y_0) \tag{11-51}$$

where δy_0 might be expected to be proportional to k_s (actually, δy_0 turns out to be considerably smaller than k_s).

If we use wall coordinates, the nondimensional form of δy_0 is then

$$(\delta y_0)^+ = \frac{\delta y_0 \, u_\tau}{v} \tag{11-52}$$

Experimentally it is found that, approximately,

$$(\delta y_0)^+ = 0.031 \, Re_k \tag{11-53}$$

To develop a law of the wall for the fully rough region, we go back to the development leading to Eq. (11-16) and simply replace $l = \kappa y$ with Eq. (11-51). This then leads to

$$\frac{du^+}{dy^+} = \frac{1}{\kappa[y^+ + (\delta y_0)^+]}$$

Since there is no sublayer in the fully rough region, the lower limit of integration for this differential equation is at $y^+ = 0$:

$$\int_0^{u^+} du^+ = \frac{1}{\kappa} \int_0^{y^+} \frac{dy^+}{y^+ + (\delta y_0)^+}$$

that is,

$$u^+ = \frac{1}{\kappa} \ln \left[\frac{y^+}{(\delta y_0)^+} + 1 \right] \tag{11-54}$$

We then substitute Eq. (11-53):

$$u^+ = \frac{1}{\kappa} \ln \left(\frac{32.6 y^+}{Re_k} + 1 \right) \tag{11-55}$$

For $y^+ \geqslant Re_k$ the last term in the parentheses can be neglected and a simpler form results:

$$u^+ = \frac{1}{\kappa} \ln y^+ + \frac{1}{\kappa} \ln \frac{32.6}{Re_k} \tag{11-56}$$

This equation fits the available experimental data in the fully rough region for zero pressure gradient and no transpiration quite well, using $\kappa = 0.41$.[14]

An approximate expression for the friction coefficient under the same conditions can be readily developed from Eq. (11-56). Recall that it was noted in connection with Eq. (11-16) that at the outer edge of the boundary layer, for the case of zero pressure gradient and no transpiration, u^+ is always greater than the law of the wall by an additive 2.3. Assuming that the same wake effect obtains for a rough-surface boundary layer, Eq. (11-56) can be modified to yield the value of u^+ at the outer edge of the boundary layer:

$$u_\infty^+ = \frac{1}{\kappa} \ln \frac{\delta u_\tau}{\nu} + \frac{1}{\kappa} \ln \frac{32.6}{Re_k} + 2.3$$

Then noting that

$$u_\infty^+ = \frac{u_\infty}{u_\tau} = \frac{1}{\sqrt{c_f / 2}}$$

we obtain

$$\frac{1}{\sqrt{c_f / 2}} = \frac{1}{\kappa} \ln \frac{84 \delta}{k_s}$$

Then if $\delta_2 / \delta = 0.097$ still applies, which is a reasonable assumption,

$$\frac{c_f}{2} = \frac{0.168}{[\ln (864 \delta_2 / k_s)]^2} \tag{11-57}$$

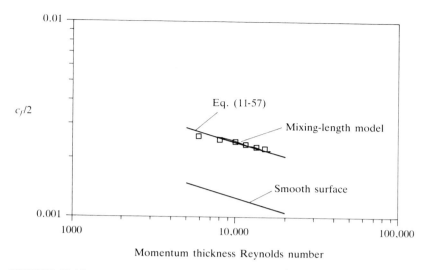

FIGURE 11-14

Comparison of Eq. (11-57), and the mixing-length model, with experimental data of Pimenta[14] for a rough surface composed of packed balls at constant free-stream velocity (IDENT 71374 fully rough surface, 1.27 mm spheres, $k_s = 0.787$ mm.)

Note that c_f depends only on boundary-layer thickness and the roughness size, and is independent of viscosity and velocity. Figure 11-14 shows a comparison of Eq. (11-57) with experiment. To express $c_f/2$ as a function of distance x along a surface, the momentum integral equation can be used.

Of course the comparison with theory in Fig. 11-14 only demonstrates that k_s is a constant, since the value of k_s used was in fact determined to fit the data. A more convincing demonstration is shown in Fig. 11-15 where a complete velocity profile is shown and compared with Eq. (11-56). Note that the effect of roughness is to move the $u^+ - y^+$ curve downwards, although it retains its original slope.

This model can be readily adapted to complete solution of the momentum differential equation of the boundary layer using finite-difference techniques. Referring to the section entitled "Summary of a complete mixing-length theory" (p. 212) it is only necessary to include Eqs. (11-51) and (11-53), and then modify the A^+ function, Eq. (11-26), to account for the reduction of A^+ caused by surface roughness.†

---

† For this type of calculation Eq. (11-53) is better approximated by $(\delta y_0)^+ = 0.031 \, (\text{Re}_k - 43)$.

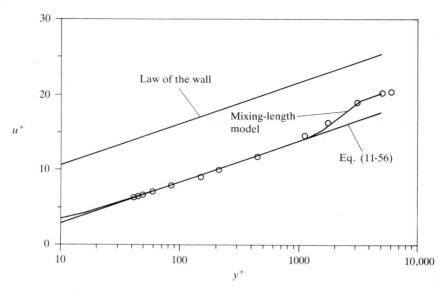

FIGURE 11-15
Comparison of theory with experiment for a velocity profile for a fully rough surface (data from IDENT 71374, packed spheres, 1.27 mm diameter, $k_s = 0.787$; air, constant $u = 39.82$ m/s, $Re_{\delta_2} = 11{,}509$, $Re_k = 101$).

Examples of such calculations are shown on both Figs. 11-14 and 11-15. The results are virtually identical to those provided by the more approximate theory, except that now the *wake* is reproduced, and of course other boundary conditions can be easily introduced.

A reasonably adequate procedure for the transitional roughness regime is to let A^+ vary from its smooth-surface value at $Re_k = 5$ to zero at $Re_k = 70$. In the absence of other information, a linear variation will usually suffice.

Blowing and suction also affect the way in which a surface responds to roughness. The effective value of Re_k is increased by blowing, and it is found that the following modification of the definition of Re_k provides a reasonable way to account for this fact:

$$Re_k = k_s(u_\tau \rho + 14\dot{m}'')/\mu$$

THE EFFECTS OF AXIAL CURVATURE

All of the boundary layers considered to this point have been on a flat surface. If, in the direction of flow, the surface is either convex or concave, substantial changes in the structure of the boundary layer can occur. Figure (11-16) shows some examples of the measured friction

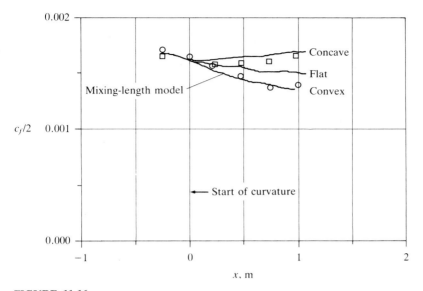

FIGURE 11-16
The effect of axial surface curvature on the friction coefficient (data of Gibson IDENT 8000/8100; curvature 0.41/m, $u = 23$ m/s).

coefficients for concave and convex surfaces having in both cases a constant radius of curvature of 2.44 m. Convex curvature causes the friction coefficient to decrease, while concave curvature has the opposite effect.

Convex curvature has a *stabilizing* effect on the boundary layer, while concave curvature is *destabilizing*. One would expect then that the sublayer would thicken with convex curvature and thin with concave curvature. While there is some evidence that this may be true, the effect is small. What happens to the outer part of the boundary layer is much more dramatic. Convex curvature can cause a substantial *reduction* in the eddy diffusivity over the entire outer region. Concave curvature causes just the reverse, i.e., an *increase* in eddy diffusivity over the entire outer region. A local nondimensional variable that can be used to describe this effect is the *curvature Richardson number* Ri, defined as follows:

$$\mathrm{Ri} = \frac{\bar{u}}{R} \bigg/ \frac{d\bar{u}}{dy}$$

where R is the radius of curvature in the flow direction, taken *positive* for convex curvature.

A fairly good mixing-length model for weak curvature ($\delta/R < 0.05$) can be constructed by assuming that the *effective* mixing length at any point in the boundary layer can be determined from the following linear function of Ri:

$$l_{\text{eff}} = l(1 - 10\,\text{Ri})$$

Thus convex curvature has the effect of reducing mixing length, and convex curvature of increasing it.

The rather small effect on sublayer thickness can be modeled by the following modification of the A^+ equation:

$$A_{\text{eff}}^+ = \left(1 + \frac{1000\nu}{Ru_\tau}\right)A^+$$

It is found that upon entering a curved region there is a substantial lag in the reduction or augmentation of the mixing length. The following rate equation has been found to provide a reasonable way to approximate this lag, introducing an effective value for the radius of curvature. Let $C = 1/R$, the curvature, and let C_{eff} be the *effective* value of the curvature. Then

$$\frac{dC_{\text{eff}}}{dx} = \frac{0.1(C - C_{\text{eff}})}{\delta_{99}}$$

where δ_{99} is the 99% boundary-layer thickness.

If a convex-curved region is followed by a region with no curvature, the boundary layer returns to equilibrium rather slowly. Again this effect can be modeled by a similar rate equation:

$$\frac{dC_{\text{eff}}}{dx} = \frac{800(C - C_{\text{eff}})}{\text{Re}_{\delta_2}\,\delta_{99}}$$

Figure 11-16 shows examples of friction-coefficient predictions for both convex and concave curvature using this mixing-length model.

THE EFFECTS OF FREE-STREAM TURBULENCE

It has been assumed in all of the preceding that turbulence in the free stream outside of the boundary layer is negligible. It is customary to describe the intensity of turbulence in terms of the root mean square of the velocity fluctuation in the flow direction. If the turbulence were *isotropic*, the velocity components in the other directions would be the same, but of course this is not always the case. The turbulence intensity can be expressed nondimensionally by normalizing with respect to other velocities. If the free-stream mean velocity is used for normalization, the turbulence intensity is then defined as

$$\text{Tu} = \frac{\sqrt{u_\infty'^2}}{u_\infty} \tag{11-58}$$

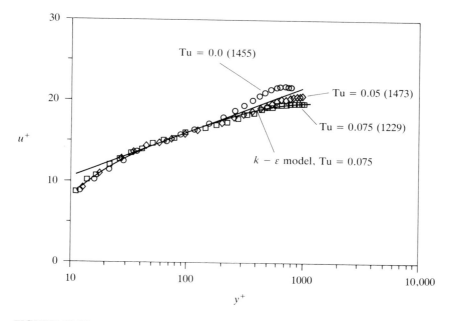

FIGURE 11-17
The effect of free-stream turbulence on velocity profiles; Re_{δ_2} in parentheses (data of Johnson and Johnston[2]).

The main effect of free-stream turbulence has been found to occur in the outer part of the boundary layer, the wake, where free-stream turbulence increases the eddy diffusivity. The sublayer and the inner part of the logarithmic region are generally unaffected, although it seems likely that sufficiently high free-stream turbulence would affect the entire boundary layer.

Figure 11-17 shows three examples of velocity profiles with free-stream turbulence 0.0, 0.05, and 0.075. There it can be seen rather clearly how the wake is depressed at increasingly higher values of Tu, and nearly all of the available experimental data show similar behavior. Note that the inner region seems virtually unaffected.

The value of u^+ at the outer edge of a turbulent boundary layer is always equal to $1/\sqrt{c_f/2}$, so it is apparent that free-stream turbulence has the effect of *increasing* the friction coefficient.

At the present time there seems to be no generally agreed correlation for the effect of Tu on c_f. Some researchers (see, for example, Hancock and Bradshaw[15]) feel that the enhancement of the friction coefficient is a function of both Tu and the *scale* of the free-stream turbulence. The problem is further complicated by the fact that free-stream turbulence decays in the flow direction at a rate that itself depends upon the scale. This decay can be calculated using the turbulence energy equation (11-30) and the dissipation equation (11-39). In both equations it need only be noted that there are no gradients in the y direction in the free stream, so from these we obtain a pair of ordinary differential equations for $\bar{k}$ and ε. Then Eq. (11-37) can be used to evaluate l_t, and in fact a similar procedure is often used to determine a length scale from the measurements of the decay of $\sqrt{\overline{u_\infty'^2}}$.

The idea that free-stream turbulence effects are a function of both the turbulence intensity and a length scale suggests that the $k-\varepsilon$ model might provide a useful tool for calculating such effects.

Figure 11-17 shows an example of a $k-\varepsilon$ calculation. It was assumed that the free-stream turbulence was isotropic, so that the turbulent kinetic energy could be evaluated from Tu. The value of free-stream ε used was based on the experimental value of dissipation length and Eq. (11-37). However, the results were relatively insensitive to free-stream length scale. If this were generally true, it would be fortunate, because estimating the free-stream dissipation rate is a vexing problem in design applications. On the other hand, there remains the apparent experimental evidence that this is indeed an important parameter. Although it appears that the $k-\varepsilon$ model yields reasonable results for Tu ≤ 0.075, it remains to be seen whether it is adequate at higher turbulence levels.

PROBLEMS

11-1. Using the Van Driest equation for the mixing length in the sublayer, determine u^+ as a function of y^+ for $p^+ = 0$ and $v_0^+ = 0$ by numerical integration of the momentum equation in the region where the Couette flow approximation is valid, for $A^+ = 22$, 25, and 27, and compare with the experimental data in Fig. 11-3. (It is presumed that a programmable computer is used for this problem.)

11-2. Develop a law of the wall for a transpired turbulent boundary layer (that is, $v_0 \neq 0$) based on the Prandtl mixing-length theory and a two-layer model of the Couette flow region near the wall. Note that you need to develop a new relation for both the viscous sublayer and the fully turbulent region, and the apparent thickness of the sublayer will be a consant to be determined from experiments.

The table at the top of the next page shows two sets of experimental points for turbulent velocity profiles for $p^+ = 0$ but $v_0^+ \neq 0$. Plot these profiles on semi-logarithmic paper and superimpose the equation you have derived for the fully turbulent region, determining the apparent sublayer thickness from the best fit to the data. Note that there is a "wake" or outer region for which your analysis does not apply. Finally, plot the apparent sublayer thickness y_{crit}^+ as a function of v_0^+ and discuss the significance of the results.

$v_0^+ = 0.1773$		$v_0^+ = -0.065$	
y^+	u^+	y^+	u^+
30.6	16.84	35.6	12.33
50.3	19.37	48.7	13.02
99.6	23.41	81.7	13.59
148.9	25,72	150.8	14.24
247.6	29.99	249.6	14.80
362.6	34.11	364.8	15.34
510.6	39.36	496.5	15.87
724.3	45.13	628.2	16.20
921.5	47.29	792.8	16.31
1053.0	47.32	990.4	16.31

11-3. Repeat Prob. 11-2 using the Van Driest equation for the sublayer mixing length and numerical integration of the momentum equation. Determine the values of A^+ that best fit the experimental data, and plot these as a function of v_0^+. (It is presumed that a programmable computer is used for this problem.)

11-4. Consider constant-property flow along a flat plate with constant u_∞. Let the boundary layer starting at the origin of the plate be laminar, but assume that a transition to a turbulent boundary layer takes place abruptly at some prescribed critical Reynolds number. Assuming that at the point of transition the momentum thickness of the turbulent boundary layer is the same as the laminar boundary layer (and this is a point for discussion), calculate the development of the turbulent boundary layer and the friction coefficient for the turbulent boundary layer. Plot the friction coefficient as a function of Reynolds number on log–log paper for transition Reynolds numbers (based on distance from the leading edge) of 300,000 and 1,000,000, and compare with the turbulent flow friction coefficient that would obtain were the boundary layer turbulent from the plate origin. On the basis of these results, determine a "virtual origin" of a turbulent boundary layer preceded by a laminar boundary layer; that is, the turbulent boundary layer will behave as if the boundary layer had been entirely a turbulent one starting at the virtual origin.

11-5. Redevelop Eq. (11–21) for the case where density and viscosity are functions of x.

11-6. A nuclear rocket nozzle of circular cross section has the geometry shown in Fig. 11-18. The working fluid is helium, and the stagnation pressure and temperature are 2100 kPa and 2475 K, respectively. Assuming one-dimensional isentropic flow, constant specific heats, and a specific heat ratio of 1.67, calculate the mass flow rate and the gas pressure, temperature, and density as functions of distance along the axis. Then, assuming that a laminar boundary layer originates at the corner where the convergence starts, calculate the momentum thickness of the boundary layer and the momentum thickness Reynolds number as functions of

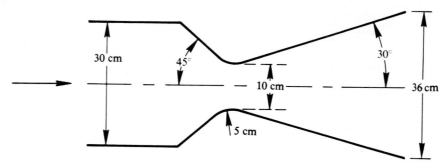

FIGURE 11-18

distance along the surface. Assume that a transition to a turbulent boundary layer takes place if and when the momentum thickness Reynolds number exceeds 162.

An approximate analysis may be carried out on the assumption of constant fluid properties, in which case let the properties be those obtaining at the throat. Alternatively, a better approximation can be based on the results of Prob. 11-5. In either case it may be assumed that the viscosity varies approximately linearly from $\mu = 5.9 \times 10^{-5}\,\text{Ns/m}^2$ at 1400 K to $\mu = 8.3 \times 10^{-5}\,\text{Ns/m}^2$ at 2500 K.

11-7. Work Prob. 11-6 but let the fluid be air and the stagnation pressure and temperature be 2100 kPa and 1275 K, respectively. Calculate the displacement thickness of the boundary layer at the throat of the nozzle. Is any correction to the mass flow rate warranted on the basis of this latter calculation?

11-8. A gas turbine blade, as illustrated in Fig. 11-19, has the following operating conditions:

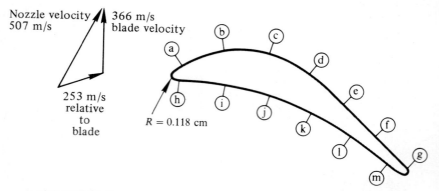

FIGURE 11-19

Fluid: air
Stagnation conditions:

$$P_s = 515.7 \text{ kPa}, \qquad T_s = 1139 \text{ K}$$

Conditions just upstream of the blade:

$$P = 473.7 \text{ kPa}, \qquad T = 1111 \text{ K}$$
$$V = 253 \text{ m/s}, \qquad G = 376 \text{ kg/(s} \cdot \text{m}^2)$$

Free-stream conditions along blade surface:

Point	Distance from leading edge, cm	G_∞, kg/(s · m²)	ρ, kg/m³	T, K
a	0.38	537	1.36	1072
b	1.40	586	1.31	1052
c	2.41	635	1.10	988
d	3.18	620	1.23	1031
e	4.19	576	1.31	1056
f	5.21	547	1.34	1068
g	6.22	537	1.36	1972
h	0.25	293	1.52	1122
i	1.27	317	1.51	1121
j	2.29	327	1.51	1118
k	3.30	352	1.50	1116
l	4.32	430	1.45	1100
m	5.33	537	1.36	1072

Calculate the momentum thickness and the momentum thickness Reynolds number along both surfaces of the blade. Assume that a transition to a turbulent boundary layer takes place when the momentum thickness Reynolds number exceeds 162.

Describe how the forces acting on the blade could be analyzed from the given data.

11-9. The following table is an actual velocity profile measured through a turbulent boundary layer on a rough surface made up of 1.27 mm balls packed in a dense, regular pattern. There is no pressure gradient or

y, cm	u, m/s	y, cm	u, m/s
0.020	12.94	0.660	27.38
0.030	14.08	1.10	31.10
0.051	15.67	1.61	34.15
0.081	17.31	2.12	37.21
0.127	19.24	2.82	39.37
0.191	20.87	3.58	39.68
0.279	22.68		
0.406	24.54		

transpiration. The fluid is air at 1 atm and 19°C; and $u_\infty = 39.7\,\mathrm{m/s}$, $\delta_2 = 0.376\,\mathrm{cm}$, $\mathrm{Re}_{\delta_2} = 9974$, $c_f/2 = 0.00243$. The distance y is measured from the plane of the tops of the balls. The objective of this problem is to analyze these data in the framework of the rough-surface theory developed in the text. What is the apparent value of k_s? Of Re_k? What is the roughness regime? What is the apparent value of κ? How does the wake compare with that of a smooth surface? Do the data support the theory?

The use of the plane of the tops of the balls as the origin for y is purely arbitrary. Feel free to move the origin if this will provide a more coherent theory.

11-10. *Computer analysis of the turbulent momentum boundary layer over a flat plate with zero pressure gradient and constant properties:* Use a mixing-length turbulence model with the Van Driest damping function to calculate a flat-plate boundary layer flow over the range of Re_{δ_2} from 2000 to 10,000. Compare the friction coefficient results based on x Reynolds number and momentum thickness Reynolds number with the results in the text. Evaluate the virtual origin concept as described in Prob. 11-4. Examine the development in the streamwise direction of the momentum thickness and assess the validity of the momentum integral, Eq. (6-7). Feel free to investigate any other attribute of the boundary-layer flow. You can choose how to set up the problem in terms of suitable choice of dimensions and fluid properties (constant).

For the initial condition at the specified starting Re_{δ_2}, construct a composite velocity using a continuous *law-of-the-wall* model in the inner region and a *power-law* model in the outer region. The inner-region profile is developed by converting Eq. (11-9) to *wall coordinates* and equating it to Eq. (11-13) for the case of no transpiration or pressure gradient. Upon separating variables and introducing the continuous *law of the wall* for ε_M, Eq. (11-6), along with Eq. (11-25), the inner-region velocity profile is developed. At some point within the logarithmic region (say $y^+ \approx 200$) the *power-law* form of the velocity profile, Eq. (11-17), can be matched to the inner-region profile, and then used to compute the velocity to the edge of the boundary layer.

REFERENCES

1. Anderson, P. S., W. M. Kays, and R. J. Moffat: *J. Fluid Mech.*, vol. 69, pp. 353–375, 1975.
2. Johnson, P. L., and J. P. Johnston: *Seventh Symposium on Turbulent Shear Flows, Stanford University, August 21–23, 1989*, pp. 20.2.1–20.2.6, Department of Mechanical Engineering, Stanford University, 1989.
3. Schultz-Grunow, F.: NACA (now NASA) TM-986, Washington, 1941.
4. Van Driest, E. R.: *J. Aero. Sci.*, vol. 23, p. 1007, 1956.
5. Kays, W. M., and R. J. Moffat: *Studies in Convection*, vol. 1, pp. 213–319, Academic Press, London, 1975.
6. Rodi, W., and D. B. Spalding: *Wärme- und Stoffübetragung*, Bd. 3, S. 85–95, 1970.
7. Harlow, F. H., and P. Nakayama: "Transport of Turbulence Decay Rate," Los Alamos Scientific Laboratory, University of California, Report LA-3854, 1968.

8. Jones, W. P., and B. E. Launder: *Int. J. Heat Mass Transfer,* vol. 15, p. 301, 1972.
9. Clauser, F. H.: *J. Aero. Sci.,* vol. 21, pp. 91–108, 1954.
10. Anderson, P. S., W. M. Kays, and R. J. Moffat: Report HMT-15, Thermosciences Division, Department of Mechanical Engineering, Stanford University, May 1972.
11. Simpson, R. L., W. M. Kays, and R. J. Moffat: Report HMT-2, Thermosciences Division, Department of Mechanical Engineering, Stanford University, December 1967.
12. Nikuradse, J.: *Forsch. Arb. Ing.-Wes.,* Nr. 361, 1933.
13. Schlichting, H.: *Ing.-Arch.,* vol. 7, pp. 1–34, 1936.
14. Pimenta, M. M., R. J. Moffat, and W. M. Kays: Report HMT-21, Thermosciences Division, Department of Mechanical Engineering, Stanford University, May 1979.
15. Hancock, P. E., and P. Bradshaw: *J. Heat Transfer,* vol. 105, pp. 284–289, 1983.

CHAPTER
12

MOMENTUM TRANSFER: TURBULENT FLOW IN TUBES

In this chapter we are primarily concerned with fully developed flow in a circular tube or pipe, but we will examine briefly the problem of noncircular tubes and the effects of surface roughness. The turbulence phenomenon involved represents nothing substantially different from that discussed in Chap. 11 in connection with the turbulent external boundary layer, and therefore the material of that chapter is heavily drawn on. Similarly, the peculiarities of a fully developed flow in a constant-area tube or pipe are discussed in Chap. 7 in the context of laminar flow, and much of that material is also used here.

FULLY DEVELOPED FLOW IN A CIRCULAR TUBE

The geometry under consideration is shown in Fig. 7-1. The applicable momentum differential equation can be readily deduced from Eq. (4-11) simply by comparison with Eq. (11-3). For steady, constant-property flow we obtain

$$\bar{u}\frac{\partial \bar{u}}{\partial x} + \bar{v}_r\frac{\partial \bar{u}}{\partial r} - \frac{1}{r}\frac{\partial}{\partial r}\left[(\nu + \varepsilon_M)r\frac{\partial \bar{u}}{\partial r}\right] + \frac{1}{\rho}\frac{d\bar{P}}{dx} = 0 \qquad (12\text{-}1)$$

At and near the entrance to the tube a turbulent boundary layer develops in much the same manner as an external boundary layer, provided that the Reynolds number is sufficiently high. But what happens in the entry region depends heavily on the geometric character of the entrance itself. If the tube is preceded by a smoothly converging nozzle, the boundary layer is often initially *laminar,* and then a *transition* to a turbulent boundary layer takes place somewhere farther downstream, assuming that the Reynolds number based on tube diameter, Eq. (7-14), is greater than the critical Reynolds number, about 2300. But if the entrance involves an abrupt or sharp-edged contraction, as is so often the case in engineering applications, no laminar boundary layer forms at all so long as Re > 2300, but the eddy diffusivity in the turbulent developing region is very much higher than that given by the turbulence theories discussed in Chap. 11.

It is thus evident that the velocity developing region for turbulent flow in a tube or pipe will only lend itself to an unambiguous analytical treatment for a highly idealized situation. If the entrance is a smoothly converging nozzle, so that the velocity is uniform at the beginning of the constant-area tube, and if it is assumed that a turbulent boundary layer originates at this point, then the entry region can be calculated by finite-difference solution of Eq. (12-1), using the same turbulence models as for the external boundary layer. The only thing new is the handling of the pressure gradient, which is now a dependent variable related to the acceleration of the fluid near the centerline of the tube caused by the displacement of fluid by the growing boundary layer.

Alternatively, the idealized entry region can be calculated by *integral methods* using an assumed velocity profile such as discussed in Chap. 11. A classic solution of this type is that developed by Latzko.[1] Of particular interest is the length of the developing region. Latzko's solution yields the following for the *entry length*:

$$(x/D)_{\text{entry}} = 0.623 \, \text{Re}^{0.25} \qquad (12\text{-}2)$$

where $\text{Re} = DG/\mu$ for a circular tube.

As an example, if $\text{Re}_D = 50,000$ then $x/D = 9.3$. This is obviously very much shorter than for a laminar flow, where up to 100 diameters is typical. Of course this is an idealized case, but in general the turbulent entry length is found to be of this order of magnitude. In most engineering applications, heat exchangers for example, the tubes are very much longer than this. For this reason the entry-length solution is usually not very important. The remainder of the chapter will be devoted to what will be termed *fully developed flow.*

The *unique feature* of flow in a constant-area tube is the fact that the growing boundary layers in the entry region must ultimately meet at the tube centerline (hence the *length* of the entry region). Thereafter no further growth is possible, the flow must adjust to this fact, and a *fully developed flow* results. Referring to Eq. (12-1), the radial velocity component $\bar{v}_r$ must be zero, and $\bar{u}$ must be a function of r alone. Then Eq. (12-1) becomes an ordinary differential equation:

$$\frac{1}{r}\frac{d}{dr}\left[(v + \varepsilon_M)r\frac{d\bar{u}}{dr}\right] = \frac{1}{\rho}\frac{d\bar{P}}{dx} \tag{12-3}$$

The problem is now much like the Couette flow problem considered for the region near the wall in Chap. 11. But, because of the pressure gradient, the shear stress cannot be independent of r; and, in fact, it is given by Eq. (7-12), i.e., it is the same as for a *laminar* flow.

Close to the wall, where r is near r_0 and τ is near τ_0, the velocity profile outside the viscous sublayer should differ little from the law of the wall, Eq. (11-16). The experimental data over a substantial portion of the flow area, again excepting the sublayer, are quite well represented by a slight modification of Eq. (11-16):

$$u^+ = 2.5 \ln y^+ + 5.5 \tag{12-4}$$

where $y = r_0 - r$. (This equation is frequently called the *Nikuradse equation*.)

In the context of the mixing-length theory, Eq. (12-4) implies a value of $\kappa = 0.40$, and a slightly thicker sublayer than for the simple external boundary layer, perhaps attributable to the negative pressure gradient that is always present for flow in a tube.

Equation (12-4) cannot be precisely valid at or near the tube centerline because the velocity gradient must there be zero due to symmetry. This is similar to the *wake* problem of the external boundary layer, although in this case the "wake" is smaller because of the negative pressure gradient. Furthermore, Eq. (12-4) implies $\varepsilon_M/v = 0$ at the centerline, which seems implausible. An empirical equation developed by Reichardt[2] for ε_M/v for the entire region outside of the sublayer is

$$\frac{\varepsilon_M}{v} = \frac{\kappa y^+}{6}\left(1 + \frac{r}{r_0}\right)\left[1 + 2\left(\frac{r}{r_0}\right)^2\right] \tag{12-5}$$

where $r = r_0 - y$. (Note that as $r \to r_0$, $\varepsilon_M/v \to \kappa y^+$, which is the same result as that obtained from simple mixing-length theory where there is no pressure gradient.) Note also that Eq. (12-5) provides a finite eddy diffusivity at the tube centerline.

Equation (12-5) is plotted in Fig. 12-1 for one particular value of Reynolds number. Note that the eddy diffusivity is finite at the tube centerline, although somewhat lower than its maximum. This is quite realistic; there is no *production* of turbulence at the centerline because there is no mean velocity gradient, but turbulence is continuously

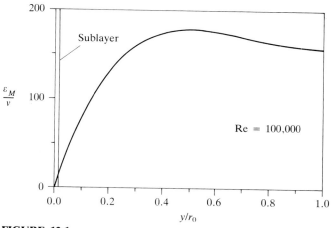

FIGURE 12-1
The distribution of eddy diffusivity for fully developed flow in a circular tube.

diffused towards the center of the tube from the region where it is produced, and it is dissipated near the centerline at the same rate. The $k-\varepsilon$ model of turbulence can be used to calculate the velocity profile, and the turbulence kinetic energy will show the same characteristics as seen for the eddy diffusivity in Fig. 12-1.

If we now solve for the velocity using Eq. (11-9), but neglecting ν; using the linear shear-stress variation, Eq. (7-12), together with Eq. (12-5); and then using the same sublayer thickness as is implied by Eq. (12-4), with $\kappa = 0.4$; we obtain

$$u^+ = 2.5 \ln \left[y^+ \frac{1.5(1 + r/r_0)}{1 + 2(r/r_0)^2} \right] + 5.5 \tag{12-6}$$

This equation is in slightly better agreement with experiment all the way to the centerline of the tube, and has zero slope there, which is at least more esthetically pleasing than Eq. (12-4). However the actual difference in the calculated values of u^+ is small and Eq. (12-4) is often preferred because of its simplicity.

Equation (12-3) can also be solved by finite-difference methods using essentially the same turbulence theories as discussed in Chap. 11. If the mixing-length theory is used, the sublayer can be calculated using the Van Driest function, although $A^+ = 26$ fits the data better than the value $A^+ = 25$ recommended for the simple external boundary layer. If Eq. (11-26) for A^+ is used, the pressure-gradient correction in that equation will in fact yield $A^+ \approx 26$ for Reynolds numbers (based on diameter) in the range 30,000–50,000, so the slightly higher value observed for A^+ for fully developed flow in a pipe or tube is probably due to the negative pressure gradient. For fully developed flow in a circular tube the following relationship for p^+ can be readily derived from the definition given in Chap. 11:

$$p^+ = \frac{14}{\text{Re}\,\sqrt{c_f/2}}$$

If this equation is substituted into the equation for A^+, Eq. (11-26), it can be seen that A^+ increases at low Reynolds numbers. Although Eq. (11-26) is an empirical equation, a particularly interesting feature is that if the friction coefficient for fully developed *laminar* flow, $c_f/2 = 8/\text{Re}$, is substituted, it will be found that A^+ becomes indefinitely large at $\text{Re} = 2000$.

In the region near the center of the tube the assumption of a constant mixing length as recommended for external boundary layers is not a good one. A better alternative is Eq. (12-5), or a simpler and adequate scheme is to use the value given by Eq. (12-5) for $r = 0$ for the entire inner region, starting at the point where the same value of ε_M/ν is calculated using $l = \kappa y[1 - \exp(-y^+/A^+)]$.

A relatively simple algebraic equation for the friction coefficient for fully developed flow can be developed by approximating Eq. (12-6) with a power law. It will be recalled that this procedure is used to obtain a closed-form equation for the friction coefficient for the external boundary layer, Eq. (11-20). The following is quite a good fit for moderate Reynolds numbers:

$$u^+ = 8.6 y^{+1/7} \tag{12-7}$$

Then at the centerline

$$u_c^+ = 8.6 y_c^{+1/7} \tag{12-8}$$

and

$$\frac{u}{u_c} = \left(\frac{y}{r_0}\right)^{1/7} \tag{12-9}$$

The mean velocity can then be determined by substitution of Eq. (12-9) into Eq. (7-6) to yield the ratio of mean velocity to centerline velocity:

$$\frac{V}{u_c} = 0.817 \tag{12-10}$$

Then, from Eq. (12-8),

$$\frac{u_c}{u_\tau} = 8.6 \left(\frac{r_0 u_\tau}{\nu}\right)^{1/7}$$

But $u_\tau = \sqrt{\tau_0/\rho} = V\sqrt{c_f/2}$. Then

$$\frac{u_c}{V\sqrt{c\rho/2}} = 8.6\left(\frac{r_0 V\sqrt{c_f/2}}{\nu}\right)^{1/7}$$

Introducing Eq. (12-10), and the definition of a Reynolds number based on tube diameter and mean velocity, $Re = DV/\nu$, we obtain

$$c_f/2 = 0.039\, Re^{-0.25} \tag{12-11}$$

(Note that henceforth Re will denote Reynolds number based on tube diameter.) Equation (12-11) fits the experimental data very well for $10^4 < Re < 5 \times 10^4$.

For higher Reynolds numbers the logarithmic form of velocity profile is more precise. If we assume that Eq. (12-4) is a reasonable approximation for the *entire* velocity profile, we can go through precisely the same procedure as was used to derive Eq. (12-11) and obtain a logarithmic equation for the friction coefficient. With a small modification of the coefficients to provide a better fit to the experimental data, we then obtain the "classical" Kármán–Nikuradse equation:

$$\frac{1}{\sqrt{c_f/2}} = 2.46 \ln (Re \sqrt{c_f/2}) + 0.30 \tag{12-12}$$

This is a little awkward to use, and a commonly employed empirical equation that closely fits the Kármán–Nikuradse equation over the range $3 \times 10^4 < Re < 10^6$ is

$$c_f/2 = 0.023\, Re^{-0.2} \tag{12-13}$$

Another equation that fits the data (and the Kármán–Nikuradse equation) over the range $10^4 < Re < 5 \times 10^6$ is that proposed by Petukhov:[3]

$$c_f/2 = (2.236 \ln Re - 4.639)^{-2} \tag{12-14}$$

The origin of this equation can be seen if for Eq. (12-12) a nominal value is given to $c_f/2$ in the logarithmic term, and then the equation is squared and inverted.

OTHER FLOW CROSS-SECTIONAL SHAPES

Most of the velocity change in a turbulent flow takes place very close to the wall surface and is relatively independent of the proximity of any other walls. For this reason, the shape of the flow tube cross-section has little effect on the shear stress at the wall, except where sharp corners are involved. Thus it would not be surprising if Eq. (12-12) were a good approximation for noncircular tubes, provided that some kind of equivalent tube diameter could be defined for evaluating the Reynolds number. A dimension that is independent of tube shape is the *hydraulic radius*, the ratio of flow area to wall surface perimeter:

$$r_h = \frac{\text{flow area}}{\text{perimeter}} = \frac{A_c L}{A} \qquad (12\text{-}15)$$

The hydraulic radius has the dimensions of length, and thus scales on system size, and for a circular tube is identically $\frac{1}{4}D$. Thus a *hydraulic diameter* can be defined as

$$D_h = 4r_h = \frac{4A_c L}{A} \qquad (12\text{-}16)$$

If D_h is substituted for D in Eqs. (12-11), (12-12), (12-13), or (12-14), these equations hold quite well for noncircular cylindrical tubes.

This simplification obviously breaks down for passages with sharp corners, such as triangular passages with small angles at one or two corners, because the sublayer thickness becomes large relative to the distance to an opposite surface. The limits of applicability of the hydraulic diameter concept for triangular tubes are discussed by Carlson and Irvine.[4] Hartnett, Joh, and McCornas[5] find that for rectangular passages the circular-tube equations using the hydraulic diameter concept work very well.

Experimentally determined friction coefficients for a large variety of tube shapes, including passages with various types of boundary-layer interruptions, are given by Kays and London.[6] Another very extensive source of data is the *Handbook of Single-Phase Convective Heat Transfer*.[7]

EFFECTS OF SURFACE ROUGHNESS

The effects of surface roughness on fully developed flow in a tube are essentially the same as discussed in Chap. 11 in connection with the external boundary layer. The three roughness regimes—smooth, transitional, and fully rough—are observed as before, and the only differences are associated with the definitions of mean velocity V, the friction coefficient $c_f/2$, and the Reynolds number DV/ν.

For tube flow the *roughness Reynolds number* can be more conveniently expressed in terms of the tube-diameter Reynolds number and friction coefficient as follows:

$$\begin{aligned} \mathrm{Re}_k &= \frac{k_s u_\tau}{\nu} = \frac{k_s V \sqrt{c_f/2}}{\nu} = \frac{k_s 2r_0 V \sqrt{c_f/2}}{2r_0 \nu} \\ &= \frac{\mathrm{Re}\sqrt{c_f/2}}{D/k_s} \end{aligned}$$

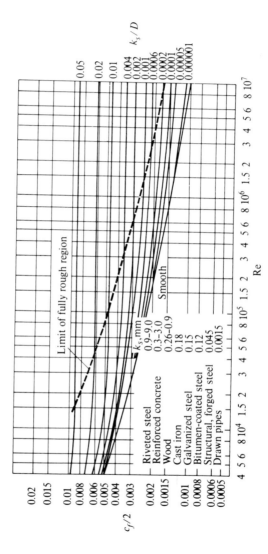

FIGURE 12-2
Friction coefficients for fully developed turbulent flow in smooth- and rough-walled circular tubes (Moody[8]).

Thus the criterion for a *smooth* surface, $Re_k < 5$, becomes

$$\frac{Re \sqrt{c_f/2}}{D/k_s} < 5$$

and similarly the *fully rough* surface corresponds to

$$\frac{Re \sqrt{c_f/2}}{D/k_s} > 70$$

An algebraic equation for friction coefficient for a *fully rough* surface can be developed from the fully rough-surface law of the wall, say, Eq. (11-56), by essentially the same procedure as was used to derive the Kármán–Nikuradse equation (12-12), yielding

$$\frac{1}{\sqrt{c_f/2}} = 2.46 \ln \frac{D}{k_s} + 3.22$$

from which

$$\frac{c_f}{2} = \left[2.46 \ln \left(\frac{D}{k_s} \right) + 3.22 \right]^{-2} \qquad (12\text{-}17)$$

Note that, as before, the friction coefficient for the fully rough case is independent of flow velocity and viscosity.

The friction coefficient for the entire range of roughness Reynolds numbers, together with data on values for k_s for typical types of pipe surface, are treated extensively by Moody.[8] An abstract of the Moody data is given in Fig. 12-2.

PROBLEMS

12-1. Develop an equation for the friction coefficient for fully developed turbulent flow between parallel planes, assuming that Eq. (12-7) is a reasonable approximation for the velocity profile.

12-2. Employing numerical integration to determine the ratio of mean velocity to centerline velocity, and Eq. (12-6) for the velocity profile, evaluate the friction coefficient for fully developed turbulent flow in a circular tube for two different Reynolds numbers: 30,000 and 150,000. Compare results with other relations for the friction coefficient given in the text. (It is presumed that a programmable computer is used for this problem.)

12-3. Develop a solution for the friction coefficient and velocity profile in the entry region of a flat duct, assuming that the velocity is uniform over the flow cross section at the entrance, that the boundary layer that develops on the two surfaces is turbulent from its very beginning, and that Eq. (12-7) is an adequate approximation for the velocity profile both in the entry region and in the fully developed region. Employ the momentum integral equation, and assume that the velocity profile in the entry region can be divided into two parts: a uniform-velocity core region and a boundary layer that ultimately engulfs the entire core. Note that the uniform velocity in the core region accelerates because of the displacement of the boundary layer, and that part of the pressure drop in the entry region is due to this acceleration. Determine the length of the entry region.

12-4. Develop Eq. (12-17) by the indicated procedures.

12-5. Employing Eq. (12-6), numerical integration, and a programmable computer, determine the ratio of mean velocity to centerline velocity for fully developed flow in a circular tube for a number of different Reynolds numbers, and compare with Eq. (12–10), which was developed assuming a $\frac{1}{7}$-power profile. Determine whether it is valid to neglect the contribution of the sublayer in these calculations.

12-6. The viscous sublayer behaves as if it were almost completely laminar out to a value of y^+ of about 5.0. With this idea in mind, calculate the ratio of such a sublayer thickness to pipe diameter for fully developed turbulent flow in a smooth-walled pipe. Using these results, discuss the significance of the data in Fig. 12-2.

12-7. Two water tanks, open to the atmosphere, are connected by a pair of parallel pipes each having an inside diameter of 2.5 cm. The pipes are 20 m long. One is a "smooth" tube, but the other is a galvanized iron pipe. What must be the elevation difference for the two tanks in order for the total flow rate for the two pipes to be 1.00 kg/s? (Neglect entrance and exit pressure-drop effects.)

12-8. *Computer analysis of the turbulent momentum flow in a circular tube with constant properties:* Use a mixing-length turbulence model with the Van Driest damping function to calculate the flow in a tube for Reynolds numbers 30,000 and 150,000, from the tube entrance to a location where the friction coefficient becomes constant. Examine a velocity profile at the hydrodynamically fully developed state to evaluate the *law of the wall* and the ratio of mean velocity to centerline velocity. Feel free to investigate any other attribute of the tube flow. You can choose how to set up the problem in terms of a suitable choice of dimensions and fluid properties (constant). For the initial condition at the tube entrance use a flat velocity profile.

12-9. *Computer analysis of the turbulent momentum flow between parallel plates with constant properties:* Use a mixing-length turbulence model with the Van Driest damping function to calculate the flow between parallel plates for hydraulic diameter Reynolds numbers 30,000 and 150,000, from the tube entrance to a location where the friction coefficient becomes constant. Examine a velocity profile at the hydrodynamically fully developed state to evaluate the *law of the wall* and the ratio of mean velocity to centerline velocity. Compare the profile shape to that described as part of Prob. 12-1. Feel free to investigate any other attribute of the tube flow. You can choose how to set up the problem in terms of a suitable choice of dimensions and fluid properties (constant). For the initial condition at the entrance use a flat velocity profile.

REFERENCES

1. Latzko, H.: *Z. Angew. Math. Mech.,* vol. 1, pp. 268–290, 1921.
2. Reichardt, H.: *Arch. Ges. Warmetechnik,* vol. 6/7, pp. 129–143, 1951.
3. Petukhov, B. S.: *Advances in Heat Transfer,* vol. 6, pp. 503–504, Academic Press, New York, 1970.
4. Carlson, L. W., and T. F. Irvine, Jr: *J. Heat Transfer,* vol. 83, pp. 441–444, 1961.
5. Hartnett, J. P., J. C. Y. Koh, and S. T. McCornas: *J. Heat Transfer,* vol. 86, pp. 82–88, 1962.
6. Kays, W. M., and A. L. London: *Compact Heat Exchangers,* 3d ed., McGraw-Hill, New York, 1984.
7. Kakac, S., R. K. Shah, and W. Aung: *Handbook of Single-Phase Convective Heat Transfer,* Wiley, New York, 1987.
8. Moody, L. F.: *Trans ASME,* vol. 66, p. 671, 1944.

CHAPTER
13

HEAT TRANSFER: THE TURBULENT BOUNDARY LAYER

The turbulent thermal boundary layer bears many similarities to the corresponding momentum boundary layer, the differences being attributable primarily to the variety of possible boundary conditions, and to the influence of the Prandtl number and the turbulent Prandtl number. However, the only new concept and new information required is that of the turbulent Prandtl number. This chapter will start by introducing the concept of eddy diffusivity for heat transfer, and then the turbulent Prandtl number. The Reynolds analogy will be developed as a very much simplified model for the latter. After examining the available experimental evidence for the turbulent Prandtl number, and following the development of a somewhat better model, a thermal law of the wall will be developed for a simple thermal boundary layer for air, and for water. This then leads to some useful though approximate algebraic solutions. Complete solutions to the energy differential equation using either the mixing-length or $k-\varepsilon$ models of turbulence developed in Chap. 11, along with finite-difference calculations, are then discussed together with a comparison with experimental data for a variety of boundary conditions, roughly paralleling the turbulent momentum boundary layers considered in Chap. 11.

Throughout this chapter all fluid properties are treated as constant, and velocities are considered sufficiently low so that viscous energy dissipation is negligible.

THE CONCEPTS OF EDDY DIFFUSIVITY FOR HEAT TRANSFER, EDDY CONDUCTIVITY, AND TURBULENT PRANDTL NUMBER

In order to perform turbulent boundary-layer heat-transfer calculations, we need some method for evaluating $\overline{v't'}$ in Eq. (5-34). A theory can be sought for direct evaluation of $\overline{v't'}$, or alternatively an eddy diffusivity can be introduced, analogous to the eddy diffusivity for momentum. This is the course that will be followed here. $\overline{v't'}$ is an apparent turbulent heat flux in the direction normal to the main flow. It seems plausible that it goes to zero if there is no mean temperature gradient in the normal direction, for otherwise a violation of the second law of thermodynamics becomes possible. Thus it seems reasonable to state that

$$\overline{v't'} \propto \frac{\partial \bar{t}}{\partial y}$$

As a proportionality factor we define the eddy diffusivity for heat transfer, ε_H:

$$\overline{v't'} = \varepsilon_H \frac{\partial \bar{t}}{\partial y} \tag{13-1}$$

Substituting into Eq. (5-34), we obtain

$$\bar{u}\frac{\partial \bar{t}}{\partial x} + \bar{v}\frac{\partial \bar{t}}{\partial y} - \frac{\partial}{\partial y}\left[(\alpha + \varepsilon_H)\frac{\partial \bar{t}}{\partial y}\right] = 0 \tag{13-2}$$

As was the case for the momentum boundary layer, we often find that $\varepsilon_H \gg \alpha$ in the fully turbulent region, while $\alpha \gg \varepsilon_H$ in the viscous sublayer very close to the wall. However, this is not always the case, and the problem is not totally analogous to that of the momentum boundary layer.

It is sometimes convenient to speak of an eddy conductivity k_t, which can be defined as follows:

$$k_t = \rho c \varepsilon_H \tag{13-3}$$

Thus the turbulent heat flux becomes

$$\dot{q}'' = -\rho c \varepsilon_H \frac{\partial \bar{t}}{\partial y} = -k_t \frac{\partial \bar{t}}{\partial y} \tag{13-4}$$

Since we have now defined an eddy viscosity and an eddy conductivity, we can define a turbulent Prandtl number analogous to the molecular Prandtl number, which is of course a fluid property that we have previously found to have a profound influence on the laminar thermal boundary layer:

$$\mathrm{Pr}_t = \frac{\mu_t c}{k_t} = \frac{\varepsilon_M}{\varepsilon_H} \tag{13-5}$$

If we now substitute for ε_H in Eq. (13-2), we obtain the following form for the energy equation:

$$\bar{u}\frac{\partial \bar{t}}{\partial x} + \bar{v}\frac{\partial \bar{t}}{\partial y} - \frac{\partial}{\partial y}\left[\left(\alpha + \frac{\varepsilon_M}{\mathrm{Pr}_t}\right)\right] = 0 \tag{13-6}$$

From Eq. (13-6) it is now seen that if one has a solution to the momentum equation, so that ε_M is known at every point, the only new information needed to solve the thermal boundary layer is information on Pr_t. Of course, one may choose not to introduce the concept of Pr_t, in which case it is necessary to either attempt to evaluate $\overline{v't'}$ directly, or else evaluate ε_H directly, and these are courses that some choose to follow. However, since the turbulent motions that give rise to $\overline{u'v'}$ also generate $\overline{v't'}$, it might be hoped that a simple relationship exists between ε_M and ε_H, i.e., that Pr_t might be a reasonably simple function. If this is the case then the problem of calculating the thermal boundary layer is greatly simplified, because the momentum boundary layer must be solved in any case.

In the next section we will develop a very simple model for Pr_t, and following this we will examine some of the experimental evidence for Pr_t.

THE REYNOLDS ANALOGY

The oldest and simplest model for Pr_t is the Reynolds analogy. Let us propose and examine a physical model of the turbulent momentum and energy exchange process. Refer to Fig. 13-1 and imagine an element of fluid of mass δm that moves in the y direction a distance l (which is a "mixing length") as a consequence of turbulent motion. Let the element be originally in velocity and temperature equilibrium with the surrounding fluid at $\bar{u}$ and $\bar{t}$, and after the "event" let it come into equilibrium with the surrounding fluid at $\bar{u} + \delta\bar{u}$ and $\bar{t} + \delta\bar{t}$, assuming that gradients in $\bar{u}$ and $\bar{t}$ with respect to y exists. The net result of this "event" is the transfer of momentum and thermal energy in the y direction completely independent of that transferred by purely molecular processes.

A comment is in order about the mechanisms whereby the element comes into equilibrium with the surrounding fluid. The momentum

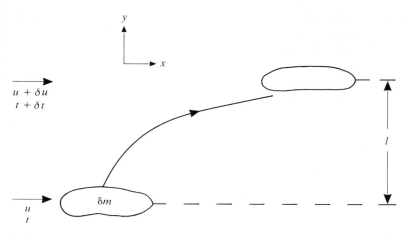

FIGURE 13-1
Turbulent exchange model on which the Reynolds analogy is based.

exchange is evidently primarily through pressure forces acting on the element as a result of contact with the surrounding fluid. Viscous forces are obviously present, but apparently play a negligible role because it has already been observed in the development of Eq. (11-16) that the turbulent shear stress is independent of viscosity. On the other hand, the corresponding heat-exchange process must be by molecular conduction, since there is no mechanism for heat transfer analogous to the pressure mechanism for momentum transfer. This difference in mechanisms, which is ignored in the Reynolds analogy, is considered later when we attempt to improve upon the latter.

Let us now compute the rate of x-momentum transfer, and the rate of heat transfer, implied by Fig. 13-1. Assume that the process at any one point in the flow is taking place continuously, i.e., one element following another, such that the effective continuous velocity of fluid in the y direction is $C\sqrt{\overline{v'^2}}$. Then consider the transfer rate across some area A parallel to the direction of the main flow. The rate of x-momentum transfer across A must be $(AC\sqrt{\overline{v'^2}}\rho)\,\delta\bar{u}$. Similarly, the rate of thermal energy transfer transfer across A must be $(AC\sqrt{\overline{v'^2}}\rho)c\,\delta\bar{t}$.

According to the momentum theorem, the effective shear force F acting across a mean streamline is equal to the rate of momentum transfer. Then the effective shear stress must be

$$\tau_t = \frac{F}{A} = C\sqrt{\overline{v'^2}}\,\rho\,\delta\bar{u}$$

Similarly, the effective heat flux is

$$\dot{q}_t'' = C\sqrt{\overline{v'^2}}\,\rho c\,\delta\bar{t}$$

Now if l is small relative to the other dimensions of the system,

$$\delta\bar{u} \approx l\frac{d\bar{u}}{dy}, \qquad \delta\bar{t} \approx l\frac{d\bar{t}}{dy}$$

Substituting,

$$\tau_t = C\sqrt{\overline{v'^2}}\,\rho l\frac{d\bar{u}}{dy}, \qquad \dot{q}_t'' = C\sqrt{\overline{v'^2}}\,\rho c l\frac{d\bar{t}}{dy}$$

Now rearrange to the *diffusivity* form

$$\frac{\tau_t}{\rho} = C\sqrt{\overline{v'^2}}\,l\frac{d\bar{u}}{dy}, \qquad \frac{\dot{q}_t''}{\rho c} = C\sqrt{\overline{v'^2}}\,l\frac{d\bar{t}}{dy}$$

But, by definition of ε_M and ε_H,

$$\frac{\tau_t}{\rho} = \varepsilon_M\frac{d\bar{u}}{dy}, \qquad \frac{\dot{q}_t''}{\rho c} = \varepsilon_H\frac{d\bar{t}}{dy}$$

From comparison of terms in the two sets of expressions, it is apparent that

$$\varepsilon_H = \varepsilon_M$$

This equality of diffusivities is the Reynolds analogy. But, going back to Eq. (13-5), it is seen that the Reynolds analogy also implies that

$$\mathrm{Pr}_t = 1$$

Such a simple result, if true, would indeed be fortunate. As we shall see when we examine the experimental evidence, things are not quite so simple, but, on the other hand, the Reynolds analogy is not far wide of the mark.

The Reynolds analogy has some consequences that are in themselves sometimes called the Reynolds analogy. For the case of molecular Prandtl number Pr equal to unity, $u_\infty = $ constant, and $t_\infty - t_0 = $ constant, it can be shown easily, just from examination of the differential equations, that $\mathrm{Pr}_t = 1$ leads to similar velocity and temperature profiles. Under these conditions, it can then be shown that

$$\mathrm{St} = c_f/2$$

This of course is exactly what obtains under the same conditions for the laminar boundary layer, and for basically the same reasons.

TURBULENT PRANDTL NUMBER: THE EXPERIMENTAL EVIDENCE

From the definition of Pr_t, Eq. (13-5), and the definitions of the eddy diffusivities, Pr_t can be expressed in terms of quantities that can in principle be measured at each point in the boundary layer:

$$\mathrm{Pr}_t = \overline{u'v'}\frac{d\bar{t}}{dy} \bigg/ \overline{v't'}\frac{d\bar{u}}{dy}$$

Four quantities are required: the mean velocity and temperature gradients, the turbulent shear stress, and the turbulent heat flux. It is extremely difficult to measure all four of these quantities accurately, and for this reason reliable data on local values of Pr_t are relatively rare. We will examine some of the available data shortly, but before doing so we should note that there are indirect methods for obtaining at least average values for Pr_t. It will be seen later that when temperature profiles are plotted in semi-logarithmic *wall coordinates* for the simple case of constant free-stream velocity and constant surface temperature, there is a *straight-line* region similar to that observed for the velocity profiles. From the *slope* of this line, as compared with the slope of the corresponding line from the velocity profiles, Pr_t can be deduced, and it further follows that Pr_t is evidently a constant in the *logarithmic region* of the boundary layer. Table 13-1 shows some results using this technique for several cases where the fluid is air, and one case where the fluid is water.

It is evident that, at least for these fluids with $\mathrm{Pr} = 0.7$ and 5.9, the turbulent Prandtl number in the logarithmic region, while not equal to 1

TABLE 13-1
Turbulent Prandtl number from logarithmic-region profile slopes; zero pressure gradient

Experiment	Re_{δ_2}	Re_{Δ_2}	Pr_t
	Fluid: air, $\mathrm{Pr} = 0.7$		
Reynolds[1]	Various to 6,400		0.73
Moffat[2]	4,419	4,756	0.85
Thielbahr[3]	1,572	1,684	0.85
Blackwell[4]	2,481	2,648	0.85
	2,609	3,008	0.78
	2,971	3,180	0.79
Gibson[5]	2,750	3,129	0.92
Simon[6]	4,445	1,756	0.77
	6,365	3,746	0.78
	4,957	4,261	0.82
Pimenta[7]	15,142	15,412	0.84
(fully rough surface)			
	Fluid: water, $\mathrm{Pr} = 5.9$		
Hollingsworth[8]	1,552	287	0.85

as implied by the Reynolds analogy, is certainly close to 1. The data range from about 0.7 to 0.9, with a preponderance of data around 0.85. Note that one case is for a "fully rough" surface, and roughness evidently has no effect. It is also worth noting that water with $Pr = 5.9$ yields about the same results as air with $Pr = 0.7$, so there is little if any effect of molecular Prandtl number, at least over this range.

Experiments with *liquid metals* flowing in pipes with Prandtl numbers in the range $0.001–0.02$ suggest that Pr_t is considerably greater than 1.00 at low values of the Prandtl number. The evidence for this is very indirect—it is simply that it is necessary to assume a high value for Pr_t in order to predict heat-transfer rates adequately by solution of the energy differential equation. Therefore there does appear to be a considerable Prandtl number effect, but it is confined to very low-Prandtl-number fluids.

Figure 13-2 shows some data for the logarithmic region presented by Zukaukas and Slanciauskas[9] for a fairly wide range of Prandtl number, all at constant free-stream velocity. These results are quite consistent with those in Table 13-1, but most importantly show little or no effect of Prandtl number up to $Pr = 65$. From this it can be inferred that Pr_t is probably not affected by Prandtl number at high values of the latter.

Local point values of Pr_t, obtained using the preceding equation, are shown on Fig. 13-3 for air and for water, from constant free-stream velocity experiments. The data for water extends only to $y^+ = 30$ because

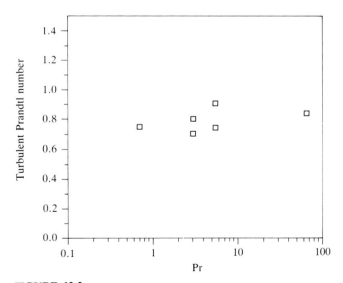

FIGURE 13-2
Turbulent Prandtl number measurements of Zukauskas and Stanciauskas[9] (data from logarithmic range of y^+).

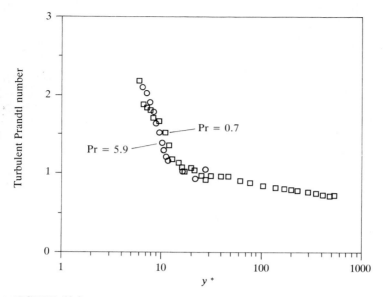

FIGURE 13-3
Point values for turbulent Prandtl number for air and water; constant u_∞ [data of Blackwell[4] (air) and Hollingsworth[8] (water)].

the thermal boundary layer is very much thinner than the momentum boundary layer at this Prandtl number, and the measurements become very inaccurate at higher values of y^+. However, the log-slope in the outer region (see Fig. 13-10) suggests $Pr_t = 0.85$. Several observations can be made. Although Pr_t is reasonably constant in the "logarithmic" or "law-of-the-wall" region, it apparently goes to much higher values in the sublayer, and air and water yield almost identical results.

In the "wake" region (and this refers only to the air data), Pr_t tends to decrease, and this and other experiments suggest that it approaches about 0.7 at the outer edge of the boundary layer. However, in calculating the thermal boundary layer, the behavior in the outer region is of minor significance.

Of much more significance is the sharp rise of Pr_t in the sublayer, indicating that the mechanisms of heat and momentum transfer differ greatly there, a conclusion that is probably not surprising. Successful calculation of the thermal boundary layer for moderate and high Prandtl numbers depends critically on including these high values of Pr_t. The data presented here extend only down to $y^+ = 6$. For $Pr < 10$ the value of Pr_t for $y^+ < 6$ is not particularly important, but for higher-Prandtl-number fluids this region can become of critical importance, and at present there are simply no data.

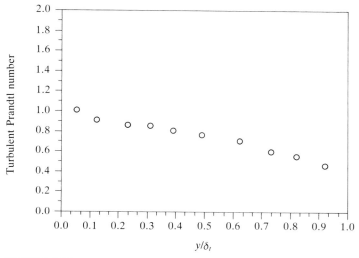

FIGURE 13-4
Point values for turbulent Prandtl number for air; constant u_∞ (data of Roganov *et al.*;[10] procedure and Reynolds number unknown).

    Additional data on point values of Pr_t for air at constant free-stream velocity are given by Roganov *et al.*,[10] and are shown on Fig. 13-4. These are quite consistent with the data on Fig. 13-3. However, these data do not extend significantly into the sublayer, although a beginning of a rise in Pr_t as the wall is approached is seen.

    Figure 13-5 shows the results of three experiments with positive transpiration (blowing). These are all very similar to the results on Fig. 13-3 and indicate that transpiration has little or no effect on Pr_t.

    Figure 13-6 shows the effect on an adverse pressure gradient on Pr_t. It does appear that an adverse pressure gradient *decreases* Pr_t, and one would then infer that a favorable pressure gradient (negative) would *increase* Pr_t. This is not surprising, given the mechanisms that must be involved, but the evidence is scanty.

TURBULENT PRANDTL NUMBER: ANALYTIC RESULTS

Using supercomputers, it has become feasible to calculate turbulent flows over a limited domain through a full solution of the time-dependent Navier–Stokes and energy differential equations, i.e., to solve for the fluctuating velocities and temperatures as functions of time. From such solutions it is then possible to evaluate all of the statistical properties of the flow, including the turbulent Prandtl number. Rogers *et al.*[11] carried

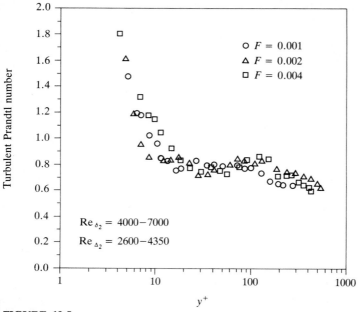

FIGURE 13-5
Effect of blowing on turbulent Prandtl number for air, with a mild pressure gradient (data of Blackwell[4]).

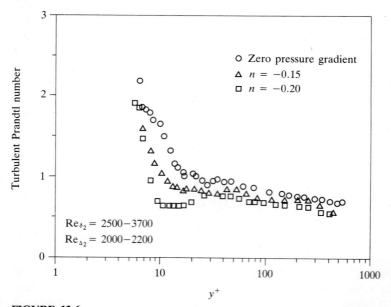

FIGURE 13-6
Effect of an adverse pressure gradient on turbulent Prandtl number for air (data of Blackwell[4]).

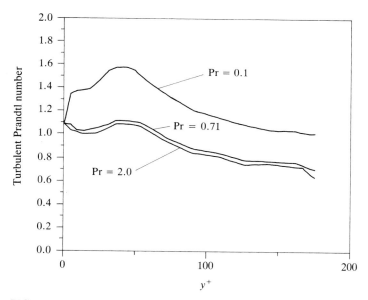

FIGURE 13-7
Turbulent Prandtl number for fully developed flow in a flat duct, calculated by direct numerical simulation.[12]

out such calculations for a uniform turbulent shear layer, which would correspond to the logarithmic region of the boundary layer, and found Pr_t to be about 0.85. Kim and Moin[12] carried out calculations for fully developed flow in a flat duct, and the results are shown on Fig. 13-7. Three different Prandtl numbers were considered: 0.1, 0.71, and 2.0.

These results show many of the characteristics seen in the experimental data—with the important exception that the steep rise in Pr_t in the sublayer does not appear. However, the calculations were made for a very low Reynolds number, which results in a relatively large *favorable* pressure gradient. It may be that a favorable pressure gradient has an influence on Pr_t in the sublayer. These results do show a substantial increase in Pr_t at low molecular Prandtl numbers, and confirm the idea that Pr_t can be considerably higher for liquid metals, which generally have molecular Prandtl numbers well below 0.1. This would seem to suggest that it is the *thermal conductivity* of the low-Pr fluids that causes a high Pr_t. (Note that high Pr_t implies a *low eddy conductivity*.) Of particular interest is the fact that at the wall Pr_t apparently approaches a value (about 1.09) that is independent of Prandtl number. It is almost impossible to make this kind of a measurement in a real boundary layer. Kim and Moin[12] call attention to the fact that Deissler[13] presents a turbulence analysis that predicts that Pr_t should approach 1, independently of Prandtl number, under these conditions.

More recent calculations[14] using the same techniques, but considering an actual external turbulent boundary layer with no pressure gradient, yield results very similar to those of Kim and Moin. However, once more the Reynolds numbers are very low: 653 for the momentum thickness and 773 for the enthalpy thickness. At these low Reynolds numbers there is only a very short "logarithmic" region in the velocity and temperature profiles, and a comparison with the law of the wall suggests that the sublayer is somewhat thicker than is the case at the higher Reynolds numbers where the experimental data on turbulent Prandtl number have been obtained. Thus it is not certain whether the relatively low values for turbulent Prandtl number in the sublayer observed in the calculated results do not result from these very low Reynolds numbers. Until full simulation calculations can be carried out at Reynolds numbers in the range 2000–3000, it may not be possible to explain this discrepancy between the calculations and experiments. However, as of this writing, the costs of full simulation calculations at such Reynolds numbers appear to be prohibitive.

Yakhot *et al.*[15] present a calculation of Pr_t, based on fundamental turbulence theory, that is entirely devoid of empirical input. At Prandtl numbers of 0.7 and higher, and/or at high Reynolds numbers, their result yields a constant Pr_t of about 0.85 (although a later consideration by the same authors suggests about 0.72). At very low Prandtl numbers (the liquid metal range) their results yield very much higher Pr_t, but values that decrease and approach 0.85 at high Reynolds number (or high y^+), all of which is consistent with the experimental data. However, the Yakhot equation does not yield the high values of Pr_t in the sublayer that one sees in the experimental data, suggesting that it is only valid in the logarithmic region of the boundary layer.

A CONDUCTION MODEL FOR TURBULENT PRANDTL NUMBER

At this point our concern is with the development of a simple computing equation for Pr_t that includes the sublayer effect, but that is also based on a physical model that is easy to comprehend. The Reynolds analogy was a first attempt to develop a model for the calculation of turbulent Prandtl number. The need for something better became evident when it was realized that Pr_t can become considerably higher than 1 for very low-Prandtl-number fluids.

A number of attempts have been made to predict turbulent Prandtl number through highly idealized analyses (see, for example, Ref. 16). Most prediction models for low Prandtl number fluids are modifications of an idea, first suggested by Jenkins,[17] that a turbulent eddy while moving transverse to the mean direction of flow might lose heat at a different rate than it loses momentum. Jenkins' original analysis assumed that the eddy lost heat by simple molecular conduction and lost momentum by the action of viscous shear. Thus the ratio of v to α, that is, the molecular Prandtl number, became the primary parameter, and for Pr = 1 the results came back to the simple Reynolds analogy. Further reflection on this problem suggests that turbulent eddies transfer momentum by the action of pressure forces, and that viscous forces are not involved. The success of the mixing-length theory, in which viscosity is not a variable, would suggest this to be the case. On the other hand, there is no mechanism other than molecular conduction whereby heat can be transferred to or from an eddy. Since the transfer mechanisms must be different, it is not surprising that Pr_t different from 1 is observed even for a fluid with Pr = 1. It should be perfectly possible for an eddy to lose all its x momentum while it still has a velocity in the y direction, and thereby to carry heat a greater distance than momentum. This would account for observed values of Pr_t less than 1. On the other hand, if the fluid has a relatively *high thermal conductivity*, the eddy could lose a substantial amount of heat by conduction before it has traversed the distance of the momentum mixing length, and Pr_t would be greater than 1.

These ideas can be incorporated into a relatively simple *conduction model* for Pr_t. The algebraic details are somewhat lengthy, and the theoretical basis certainly rests on insecure footing, so only the final result is presented here. Suffice it to say that the model illustrated in Fig. 13-1 is modified to include *heat conduction* to or from the fluid element δm as it moves. The important point is that the following result contains the variables that untuitively seem correct, and when the two free constants are specified by comparison with experiment, the final equation fits the available experimental data reasonably well.

$$Pr_t = \cfrac{1}{\cfrac{1}{2\,Pr_{t_\infty}} + C\,Pe_t\sqrt{\cfrac{1}{Pr_{t_\infty}}} - (C\,Pe_t)^2\left[1 - \exp\left(-\cfrac{1}{C\,Pe_t\sqrt{Pr_{t_\infty}}}\right)\right]}$$

(13-7)

where $Pe_t = (\varepsilon_M/v)\,Pr$, Pr_{t_∞} is the value of Pr_t far from the wall (an experimental constant), and C is an experimental constant.

Pe_t is a turbulent *Péclet number*, and although it is conveniently expressed as the product of the Prandtl number and the eddy-viscosity ratio (which is effectively a turbulent Reynolds number), note that the viscosity v cancels.

Equation (13-7), with $C = 0.3$ and $Pr_{t_\infty} = 0.85$, is plotted on Fig. 13-8 for three Prandtl numbers: 0.7, 0.1, and 0.01. Also included is the experimental data for air from Fig. 13-3. Note that it fits the experimental data for air quite well, *including the rise in Pr_t in the sublayer*. The latter

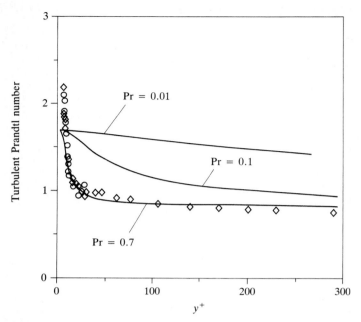

FIGURE 13-8
A plot of Eq. (13-7) for a boundary layer with no pressure gradient.

fact can only be fortuitous, because we have already seen that Pr_t in the sublayer is apparently independent of Prandtl number, and thus can have nothing to do with heat conduction. Furthermore, Eq. (13-7) with these constants yields $Pr_t = 1.7$ at the wall, which is probably much too high. However, for fluids with $Pr < 1$ the value of Pr_t at $y^+ < 7$ or 8 has a negligible effect on calculations. For very low-Prandtl-number fluids Eq. (13-7) with these constants yields values of Pr_t over the entire boundary layer that are quite consistent with the available data. As a consequence, this equation provides a reasonable way to evaluate Pr_t for any fluid for $Pr < 1$. For gases it can be thought of as merely an empirical fit to the data; for low-Prandtl-number fluids it contains at least some theoretical basis. It can be used for fluids with Pr up to at least 6 (water), *provided that 0.7 is substituted for Pr in the equation,* since we have seen that the turbulent Prandtl number for water is virtually identical to that for air. For higher Prandtl numbers the values of Pr_t very close to the wall (i.e., $y^+ < 6$) become increasingly important, and there are simply no data at present available. The best procedure would probably be to patch on a function that goes to 1 at the wall.

Although there may be a pressure-gradient effect on Pr_t, no attempt has been made to include such effects in Eq. (13-7).

COMPLETE SOLUTION OF THE ENERGY EQUATION

We now have all the ingredients needed to solve the energy differential equation of the boundary layer for a very large variety of boundary conditions, including all of those considered for the momentum equation in Chap. 11, and also virtually any kind of temperature boundary condition. It will always be necessary to solve the momentum equation, and its solution will provide values for ε_M. Then Pr_t is the only additional information required. However, before examining particular solutions, which involve finite-difference procedures and computer implementation, it will prove worthwhile to consider some simple cases where approximate algebraic closed-form solutions are obtainable, and at the same time some of the available experimental data will be examined.

A LAW OF THE WALL FOR THE THERMAL BOUNDARY LAYER

Equation (13-2) can be written in the following form if $\dot{q}''$ is used to represent the total heat flux, laminar plus turbulent:

$$\bar{u}\frac{\partial \bar{t}}{\partial x} + \bar{v}\frac{\partial \bar{t}}{\partial y} + \frac{1}{\rho c}\frac{\partial \dot{q}''}{\partial y} = 0 \qquad (13\text{-}8)$$

where

$$\dot{q}'' = -\rho c(\alpha + \varepsilon_H)\frac{\partial \bar{t}}{\partial y} \qquad (13\text{-}9)$$

Let us now restrict consideration to the region near the wall, and then let the wall temperature t_0 be a constant. The axial gradient $\partial \bar{t}/\partial x$ must then be small, $\bar{t} = \bar{t}(y)$, and we are again introducing the Couette flow approximation. Additionally, $v = v_0$, and thus Eq. (13-8) becomes

$$v_0\frac{d\bar{t}}{dy} + \frac{1}{\rho c}\frac{d\dot{q}''}{dy} = 0$$

Integrating between the limits $\bar{t} = t_0$, $\dot{q}'' = \dot{q}_0''$ at $y = 0$, and $\bar{t}$, $\dot{q}''$ at $y = y$, we obtain

$$v_0(\bar{t} - t_0) + \frac{1}{\rho c}(\dot{q}'' - \dot{q}_0'') = 0$$

from which

$$\frac{\dot{q}''}{\dot{q}_0''} = 1 + \frac{\rho c v_0(t_0 - \bar{t})}{\dot{q}_0''}$$

We then introduce the definition of v_0^+,

$$v_0^+ = \frac{v_0}{\sqrt{\tau_0/\rho}}$$

and let the remaining terms on the right-hand side, which must be nondimensional, be represented by a nondimensional temperature t^+:

$$t^+ = \frac{(t_0 - \bar{t})\sqrt{\tau_0/\rho}}{\dot{q}_0''/\rho c} \tag{13-10}$$

Then

$$\frac{\dot{q}''}{\dot{q}_0''} = 1 + v_0^+ t^+ \tag{13-11}$$

Note the similarity to the comparable equation for τ/τ_0, Eq. (11-13), except that there is no p^+ term. In fact, the whole concept of analogous behavior between heat and momentum transfer breaks down in a pressure gradient. Remember, however, that Eq. (13-11) is only applicable relatively near the wall. At the outer edge of the boundary layer $\dot{q}''/\dot{q}_0''$ must go to zero.

Now let us consider a particular case, $v_0^+ = 0$. Then

$$\dot{q}'' = \dot{q}_0''$$

so that

$$\frac{\dot{q}_0''}{\rho c} = -(\alpha + \varepsilon_H)\frac{d\bar{t}}{dy}$$

and thus

$$\int_{t_0}^{\bar{t}} d\bar{t} = -\frac{\dot{q}_0''}{\rho c}\int_0^y \frac{dy}{\alpha + \varepsilon_H}$$

Let us now introduce y^+ as a variable. After rearrangement,

$$\frac{(t_0 - \bar{t})\sqrt{\tau_0/\rho}}{\dot{q}_0''/(\rho c)} = \int_0^{y^+} \frac{dy^+}{1/\mathrm{Pr} + \varepsilon_H/\nu}$$

But the left-hand side is t^+, Eq. (13-10). Thus we obtain

$$t^+ = \int_0^{y^+} \frac{dy^+}{1/\mathrm{Pr} + \varepsilon_H/\nu} \tag{13-12}$$

Following the same procedures as were used for the momentum boundary layer in Chap. 11, let us first consider a *two-layer model* with a fully viscous sublayer and a fully turbulent region. For computer solution of the complete energy differential equation we will abandon the two-layer model in favor of a continuous model, but the two-layer model provides some easy insights into the structure of the thermal boundary layer, and also leads to some useful closed-form solutions.

For the simple momentum boundary layer (no pressure gradient, no transpiration) we found that the effective thickness of the sublayer was $y^+ = 10.8$. For a thermal boundary layer the effective thickness will vary with the Prandtl number of the fluid because the relative contribution of the eddy diffusivity to the heat-transfer conductance in the actual sublayer will vary with Prandtl number, as will be seen below. Only for a fluid with $Pr = 1$ might one expect to find 10.8, but, because of the high and varying *turbulent Prandtl number* in the sublayer, even this would not be the case. Sufficient data are available to determine the effective sublayer thickness for the thermal boundary layer for air ($Pr = 0.7$), and water at $Pr = 5.9$ (Table 13-2).

Let us now split the integral of Eq. (13-12) into two parts corresponding to a viscous sublayer and a fully turbulent region. Then for the region $y^+ > y^+_{crit}$,

$$t^+ = \int_0^{y^+_{crit}} \frac{dy^+}{1/Pr + \varepsilon_H/\nu} + \int_{y^+_{crit}}^{y^+} \frac{dy^+}{1/Pr + \varepsilon_H/\nu} \qquad (13\text{-}13)$$

If we were to follow precisely the procedure used in Chap. 11, we would now neglect ε_H/ν in the first integral, and $1/Pr$ in the second. But the term $1/Pr$ can vary tremendously depending on the particular fluid involved. Prandtl numbers for viscous fluids like oils can easily exceed 100, while those for liquid metals can be as low as 0.001. The magnitude of this term in each of these integrals can then have a large influence on the relative importance of all the terms in the denominators.

In Chap. 11 we noted that in reality the eddy diffusivity in the sublayer is not zero, it is merely small. But if Pr is large, a small diffusivity in the sublayer can be significant; and in fact if Pr is much greater than 1, significant error is introduced if ε_H/ν is neglected in the sublayer. For the present we would like to neglect this term. The use of a critical value of y^+ that is a function of Pr is a way of compensating for this effect, but this means that we are restricted to just two Prandtl numbers, 0.7 and 5.9. However, at very low Pr the $1/Pr$ term so dominates the first integral that the ε_H/ν term can be neglected in any case.

TABLE 13-2
No pressure gradient, no transpiration

Pr	y^+_{crit}
0.7	13.2
5.9	7.55

In the second integral, the fully turbulent region, the situation is reversed. The 1/Pr term can become large for low-Pr fluids, and can become significantly larger than ε_H/ν. Physically, this means that *molecular conduction* is a significant and even the dominant mechanism in the fully turbulent region for a liquid metal, and thus cannot be neglected. The relative magnitude of these terms can be easily investigated by assuming that the Reynolds analogy is applicable and then evaluating the eddy diffusivity for momentum from the mixing-length theory developed in Chap. 11:

$$\varepsilon_M/\nu = \kappa y^+ \tag{13-14}$$

where $\kappa = 0.41$. For Pr greater than about 0.5 the molecular-conduction term can reasonably be neglected, but not for lower Prandtl numbers.

Let us now restrict consideration to air, Pr = 0.7. In this case we can neglect ε_H/ν in the first integral, and the 1/Pr term in the second. Substituting Eq. (13-14), the definition of Pr_t, and a value of 13.2 for y_{crit}^+, Eq. (13-13) becomes

$$t^+ = \int_0^{13.2} Pr \, dy^+ + \int_{13.2}^{y^+} \frac{Pr_t \, dy^+}{\kappa y^+}$$

If we assume that Pr_t can be treated as a constant in the fully turbulent region, performing the integration in this equation gives

$$t^+ = 13.2 \, Pr + \frac{Pr_t}{\kappa} \ln \frac{y^+}{13.2}$$

Assuming that a reasonable average value for Pr_t in the Couette flow region is 0.85, and that $\kappa = 0.41$, we obtain a thermal "law of the wall" for gases:

$$t^+ = 2.075 \ln y^+ + 13.2 \, Pr - 5.34 \tag{13-15}$$

Figure 13-9 shows a comparison of Eq. (13-15) with two temperature profiles for air—each from different experiments and for different Reynolds numbers. Note that, like the corresponding u^+-y^+ plot, there is a *wake*, and that the actual viscous sublayer extends to about $y^+ = 30$. As the Reynolds number increases, the wake extends to higher values of y^+, but in the *inner* region both profiles collapse onto the same function. Outside of the sublayer this function is very well represented by Eq. (13-15).

A comment about the Reynolds numbers should be made here. The Reynolds numbers indicated on Fig. 13-9 are based on *enthalpy thickness*. The data on the figure are from experiments for which the virtual origins of both the momentum boundary layers and the thermal boundary layers are at the same point. We will speak of such boundary layers as *equilibrium thermal boundary layers*. If the virtual origins differ, the

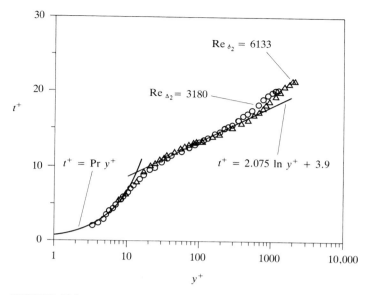

FIGURE 13-9
A thermal law of the wall for air: equilibrium thermal boundary layers; enthalpy thickness Reynolds numbers are indicated.

temperature profiles will be affected, although most of the difference will generally be noted in the wake, or outer region. For equilibrium thermal boundary layers with *gases* the ratio of enthalpy thickness to momentum thickness tends to be approximately constant, about 1.10–1.15.

If we now repeat the same analysis for water with $Pr = 5.9$ and $Pr_t = 0.85$, we obtain

$$t^+ = 2.075 \ln y^+ + 7.55\, Pr - 3.95 \qquad (13\text{-}16)$$

Figure 13-10 shows a comparison of Eq. (13-16) with experimental data for water at $Pr = 5.9$. These data were obtained using very small temperature differences so that the effect of temperature on the Prandtl number and the viscosity is believed to have been small. This is a problem that besets heat-transfer experiments with liquids, and is discussed in more detail in Chap. 15.

Another potential difficulty with these data is that the *virtual origins* of the momentum and thermal boundary layers are not the same, and it is not clear what effect this might have. It will be noted that there is a *negative wake*, but this is to be expected because the thermal boundary layer is very much thinner than the momentum boundary layer at this Prandtl number, and does not effectively extend into the wake region of the momentum boundary layer. Note the very small enthalpy thickness

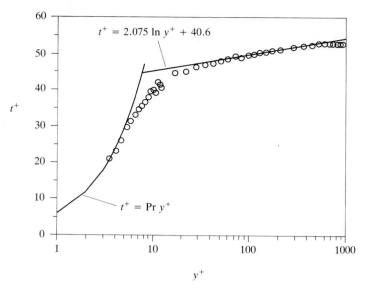

FIGURE 13-10
A thermal law of the wall for a fluid (water) with $Pr = 5.9$; $Re_{\delta_2} = 1552$, $Re_{\Delta_2} = 287$.

Reynolds number as compared with the momentum thickness Reynolds number, and also the fact that a far greater part of the heat-transfer resistance is in the sublayer, as compared with the air boundary layer. Since the Prandtl number here is only 5.9, it can be appreciated that for very high-Prandtl-number fluids (say $Pr = 100$) the heat-transfer resistance will be almost entirely in the sublayer, and the temperature profile in the fully turbulent part of the boundary layer will be virtually flat. In that case the heat-transfer mechanisms in the fully turbulent region become irrelevant. Another conclusion is that at very high Prandtl numbers the eddy diffusivity in the sublayer, though small, can play a critical role, and it will be necessary to know accurately both the eddy diffusivity for momentum and the turbulent Prandtl number for very small values of y^+.

EFFECT OF PRESSURE GRADIENT ON TEMPERATURE PROFILE

Figure 13-11 shows the effect of axial pressure gradient on the nondimensional temperature profiles. Note that each of the temperature profiles has what could be called a "logarithmic" region, but—unlike the corresponding velocity profiles (Fig. 11-4)—the profiles *do not collapse* together in the logarithmic region. Thus the term thermal "law of the

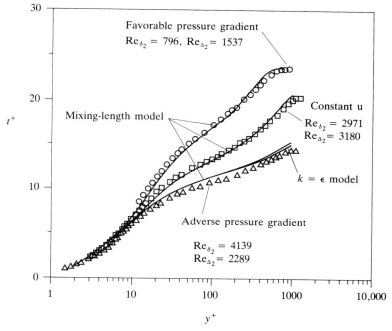

FIGURE 13-11
Effect of pressure gradient on temperature profiles (data of Thielbahr[3] and Blackwell[4]).

wall" does not have the same connotation as it does for the momentum equation. Equations (13-15) and (13-16) are applicable only for a constant free-stream velocity (no pressure gradient).

Also shown on Fig. 13-11 are temperature profiles predicted by solution of the energy differential equation using the mixing-length model previously discussed, and, in the case of the adverse pressure gradient, the k–ε model. These predictions all employ Eq. (11-26) for evaluation of A^+, and Eq. (13-7) for the turbulent Prandtl number. Note that there is only a minor improvement in the profile predicted by the k–ε model, but the use of Eq. (11-26) is critical, especially in the favorable-pressure-gradient case.

A HEAT-TRANSFER SOLUTION FOR CONSTANT FREE-STREAM VELOCITY AND SURFACE TEMPERATURE

With an algebraic "wall law" available for both the momentum boundary layer and the thermal boundary layer (at least for constant free-stream velocity), it becomes possible to develop an algebraic equation for the heat-transfer coefficient (i.e., the Stanton number) for the constant free-stream velocity case, and this will lead to algebraic solutions for even more complex cases. However, it should be kept in mind that complete solution of the momentum and energy differential equations, using finite-difference techniques and the turbulence models already developed, can be used for any of these problems, and does not involve any of the approximations that characterize the algebraic solutions. Nevertheless, the algebraic solutions are extremely useful, both to provide quick answers to engineering problems and to provide insight into how the variables affect the answers.

The first assumption to be made here is that the momentum and thermal boundary layers have the same thickness. For a *laminar* boundary layer this is not the case, the relative thickness depending primarily on the Prandtl number. But since for the turbulent boundary layer the momentum layer provides the primary transport mechanism (the eddy diffusivity) in the outer region, it is not possible for the thermal layer to have a thickness that is significantly different, except for a very low-Prandtl-number fluid or for a case where the *virtual origins* of the two boundary layers are very different. We will *not* be considering a very low-Prandtl-number fluid, and we *will* be considering an equilibrium thermal boundary layer, i.e., a thermal boundary layer having the same virtual origin as the momentum boundary layer.

A second assumption will be that the "strength" of the wake for the thermal boundary layer for air is the same as that for the momentum boundary layer. For the latter, for constant free-stream velocity, it is 2.3, i.e., the departure of u^+ from the law of the wall at the outer edge of the boundary layer. Thus we will assume that the value of t^+ at the outer edge is 2.3 above the thermal wall law. The data on Fig. 13-9 indicate that this is a good assumption.

With these assumptions, u^+ at the outer edge, from Eq. (11-16), becomes

$$u_\infty^+ = 2.44 \ln y_\infty^+ + 5.00 + 2.3$$

Similarly, Eq. (13-15) becomes

$$t_\infty^+ = 2.075 \ln y_\infty^+ + 13.2 \, \text{Pr} - 5.34 + 2.3$$

We now solve each equation for $\ln y_\infty^+$ and equate. From the definition of u^+, it can be seen that

$$u_\infty^+ = \frac{1}{\sqrt{c_f/2}}$$

Similarly, from Eq. (13-10) it can be shown that

$$t_\infty^+ = \frac{\sqrt{c_f/2}}{\text{St}}$$

We substitute and solve for St:

$$\text{St} = \frac{c_f/2}{\sqrt{c_f/2}(13.2\,\text{Pr} - 8.66) + 0.85}$$

We now substitute for $c_f/2$:

$$\text{St} = \frac{0.0287\,\text{Re}_x^{-0.2}}{0.169\,\text{Re}_x^{-0.1}(13.2\,\text{Pr} - 8.66) + 0.85} \tag{13-17}$$

For $0.5 < \text{Pr} < 1.0$ and $5 \times 10^5 < \text{Re}_x < 5 \times 10^6$ the denominator of the equation is quite well approximated by $\text{Pr}^{0.4}$. Thus a much simpler expression for *gases* is

$$\text{St}\,\text{Pr}^{0.4} = 0.0287\,\text{Re}_x^{-0.2} \tag{13-18}$$

We can now use the energy integral equation (6-24) to obtain an equation for St as a function of Re_{Δ_2}:

$$\text{St} = \frac{d\Delta_2}{dx}$$

We substitute for St from Eq. (13-18), solve for Δ_2 as a function of x, and then substitute for x in Eq. (13-18). The final result is

$$\text{St}\,\text{Pr}^{0.5} = 0.0125\,\text{Re}_{\Delta_2}^{-0.25} \tag{13-19}$$

A comparison of Eq. (13-19) for $\text{Pr} = 0.7$ with experimental data for air is shown on Fig. 13-12. For Re_{Δ_2} from 2000 to 7000 it fits the data very well, although it tends to overpredict a bit at lower Reynolds numbers. Also shown on Fig. 13-12 are the results of a full solution to the energy differential equation using the mixing-length model and the turbulent Prandtl equation developed earlier. The k–ε model yields virtually the same result. As can be seen, the full solution fits the data somewhat better over a wider range of Reynolds number.

Since we also have available a thermal "law of the wall" for *water* with $\text{Pr} = 5.9$, the identical procedures can be used to develop equations for St for this value of Prandtl number. However, Pr for water varies greatly with temperature, from about 7 for cold water to near 1 for very hot water. Thus an equation for $\text{Pr} = 5.9$ is not very useful in and of itself. Hollingsworth[8] has developed a reasonable interpolation for Prandtl numbers from 0.7 to 5.9 using the two "wall laws" developed above as a basis and assuming that the critical thickness of the sublayer is

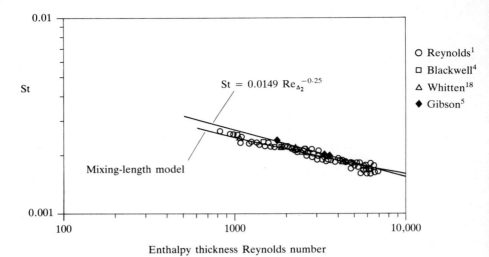

FIGURE 13-12
Experimental equilibrium boundary-layer data for air; constant free-stream velocity and constant surface temperature (data of Reynolds,[1] Blackwell,[4] Gibson,[5] and Whitten[18]).

a simple function of Pr. These equations are

$$St = 0.02426 \, Pr^{-0.895} \, Re_x^{-n}, \qquad n = 0.1879 \, Pr^{-0.18} \qquad (13\text{-}20)$$

$$St = C\left(\frac{1-n}{C}\right)^{n/(n-1)} Re_{\Delta_2}^{n/(n-1)}, \qquad C = 0.02426 \, Pr^{-0.895} \qquad (13\text{-}21)$$

From Eq. (13-21), St is plotted as a function of Prandtl number on Fig. 13-13 for various values of the enthalpy thickness Reynolds number. One point to note here is that the slopes of the curves for the higher Reynolds numbers vary from the approximately -0.5 indicated by Eq. (13-19) for gases to a slope near -0.75 at Pr $= 6$. This larger slope is consistent with the larger slopes that have consistently been observed in pipe-flow experiments with high-Pr liquids. At Pr $= 0.7$ these equations reduce to the previously developed equations for air.

CONSTANT FREE-STREAM VELOCITY FLOW ALONG A SEMI-INFINITE PLATE WITH UNHEATED STARTING LENGTH

A number of approximate procedures have been proposed for determining heat transfer through a turbulent boundary layer with an unheated starting length, that is, on a plate on which the surface temperature is the same as that of the fluid up to some distance ξ, after which it is maintained at a different temperature, t_0. The problem is shown on Fig. 13-14.

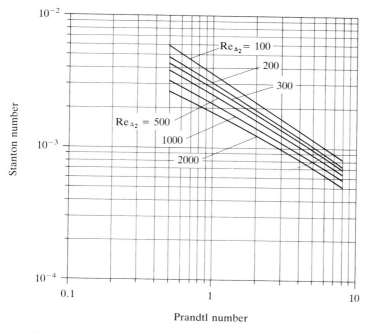

FIGURE 13-13
Stanton number for equilibrium turbulent thermal boundary layers at higher Prandtl numbers; enthalpy thickness Reynolds numbers are indicated.

This is the same problem that was solved for the laminar boundary layer in Chap. 10. The primary significance of this solution is that it provides a basic building block for constructing solutions for nonisothermal surfaces. Note, however, that this problem as well as the nonisothermal-surface problem can also be solved by finite-difference solution of the energy differential equation, and in that case no

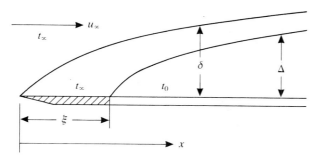

FIGURE 13-14
Boundary-layer development on a plate with an unheated starting length.

approximations need be introduced. But the objective here is to develop algebraic solutions, accepting some degree of approximation in return for a simply usable result.

To solve the laminar counterpart of this problem, the energy *integral* equation was used, and the same general procedure can be used here. One has to be careful in employing the energy integral equation for a turbulent boundary layer, however, because a simple assumed equation for the temperature profile may be quite adequate over most of the boundary layer but may give totally erroneous results in the sublayer and particularly at the wall. This same difficulty was encountered in obtaining a solution for the turbulent *momentum* boundary layer in Chap. 11. In that case a $\frac{1}{7}$-power equation was used to evaluate the integral; but since such a profile gives an infinite velocity gradient at the wall, it is obviously inadequate to describe the sublayer region, and a different procedure had to be used to evaluate the shear stress. The procedure used here, developed by Reynolds,[1] is one of several that have been proposed. It is developed for a fluid with $\Pr = 1$, but the effect of unheated starting length is probably not highly dependent on Prandtl number, at least for Pr near 1, and the results are in good agreement with experiments with air.

A basic assumption is that the velocity and temperature profiles can be expressed by

$$\frac{\bar{u}}{u_\infty} = \left(\frac{y}{\delta}\right)^{1/7} \tag{13-22}$$

$$\frac{t_0 - \bar{t}}{t_0 - t_\infty} = \left(\frac{y}{\Delta}\right)^{1/7} \tag{13-23}$$

It is assumed that the local shear stress and heat flux can be expressed by Eqs. (11-9) and (13-9) and that the molecular Prandtl number is 1, so that $\alpha = \nu$. It is also assumed that the turbulent Prandtl number is 1, so that $\varepsilon_H = \varepsilon_M$. The total diffusivities, $\alpha + \varepsilon_H$ and $\nu + \varepsilon_M$, are then the same for both momentum and heat transfer, and this is true in both the fully developed turbulent region and in the sublayer.

Since the algebraic details of this development are somewhat lengthy, only the method and the more important intermediate results will be discussed. As a first step, the momentum integral equation is developed in a form slightly different from that in Chap. 6. The momentum theorem is applied to a control volume extending from a point y in the boundary layer out to the free stream. Thus, if y is set equal to zero, the original momentum integral equation is obtained. Then substitution of Eq. (13-22) leads to an expression for the variation of *total* shear stress through the boundary layer:

$$\frac{\tau}{\tau_0} = 1 - \left(\frac{y}{\delta}\right)^{9/7} \tag{13-24}$$

The next step is to combine Eqs. (11-9), (13-22), and (13-24), and solve for the total diffusivity:

$$\nu + \varepsilon_M = 7\nu \frac{c_f}{2} \frac{\delta}{x} \operatorname{Re}_x \left[1 - \left(\frac{y}{\delta}\right)^{9/7}\right]\left(\frac{y}{\delta}\right)^{6/7} \tag{13-25}$$

In the sublayer this equation yields a *total* diffusivity that goes to zero at the wall, which is obviously incorrect. But when it is used together with Eq. (13-22), which is equally incorrect *at the wall*, a correct result can be obtained there.

For the next step Eq. (13-9) is solved for the heat flux using Eq. (13-25) to supply the diffusivity and Eq. (13-23) to give the temperature gradient (thus the artificial diffusivity is again used to compensate for the equally artificial $\frac{1}{7}$-power profile in the sublayer):

$$\frac{\dot{q}''}{\rho c u_\infty(t_0 - t_\infty)} = \frac{c_f}{2}\left[1 - \left(\frac{y}{\delta}\right)^{9/7}\right]\left(\frac{\Delta}{\delta}\right)^{-1/7} \tag{13-26}$$

We now set $y = 0$, and have an expression for the surface heat flux, and then the local Stanton number:

$$\operatorname{St} = \frac{\dot{q}_0''}{\rho c u_\infty(t_0 - t_\infty)} = \frac{c_f}{2}\left(\frac{\Delta}{\delta}\right)^{-1/7} \tag{13-27}$$

For the final step the energy integral equation (6-24) is solved for Δ_2 as a function of x, using Eq. (13-27) for the Stanton number, and using Eqs. (13-22) and (13-23) to evaluate Δ_2 from Eq. (6-15). An ordinary differential equation results that is readily solved. Introducing the fact that the momentum boundary-layer thickness varies as the $\frac{4}{5}$ power of x, we obtain

$$\operatorname{St} = \frac{c_f}{2}\left[1 - \left(\frac{\xi}{x}\right)^{9/10}\right]^{-1/9} \tag{13-28}$$

The bracketed term contains the unheated starting-length effect; for $\xi = 0$ we obtain $\operatorname{St} = c_f/2$, which is precisely what Eqs. (11-23) and (13-18) reduce to for $\operatorname{Pr} = 1$. If we then assume that the starting-length correction for a step-wall temperature is independent of Prandtl number, we can simply multiply the right-hand side of Eq. (13-18) by the correction term, and we have a simple equation applicable to *gases*:

$$\operatorname{St} \operatorname{Pr}^{0.4} = 0.0287 \operatorname{Re}_x^{-0.2}\left[1 - \left(\frac{\xi}{x}\right)^{9/10}\right]^{-1/9} \tag{13-29}$$

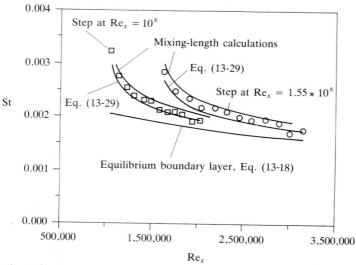

FIGURE 13-15
Effects of a step change in surface temperature (data of Reynolds[1]).

So many simplifying assumptions were introduced into this development that its validity must rest heavily on experimental verification. Fortunately, extensive experimental data are available.[1] A comparison of Eq. (13-29) with the data is shown on Fig. 13-15 for two cases. Also included on the figure are calculations based on the energy differential equation and the mixing-length model. The experimental uncertainty in these data is estimated at about plus or minus five percent; both methods of calculation fall within this band.

CONSTANT FREE-STREAM VELOCITY FLOW ALONG A SEMI-INFINITE PLATE WITH ARBITRARILY SPECIFIED SURFACE TEMPERATURE

The preceding step-function solution (unheated-starting-length solution) can now be used to solve the case of an *arbitrary* wall-temperature variation in exactly the same manner as was done for the laminar boundary layer. Again the method of superposition can be used to add step-function solutions to fit a prescribed wall temperature variation. Equation (10-31), previously derived, is directly applicable. Since the method of solution is identical to that used for the laminar boundary layer, only the final results for one particular type of problem are given.

Consider the case where the wall temperature can be expressed as a power series of the form

$$t_0 = t_\infty + A + \sum_{n=1}^{\infty} B_n x^n \qquad (13\text{-}30)$$

The heat flux at any point along the plate may then be expressed [using Eq. (13-29) for the step solution] as

$$\dot{q}_0''(x) = 0.0287 \frac{k}{x} \mathrm{Pr}^{0.6} \, \mathrm{Re}_x^{0.8} \left(\sum_{n=1}^{\infty} n B_n x^n \tfrac{10}{9} \beta_n + A \right) \qquad (13\text{-}31)$$

where β_n can be evaluated in terms of gamma functions:

$$\beta_n = \frac{\Gamma(\tfrac{8}{9})\Gamma(\tfrac{10}{9}n)}{\Gamma(\tfrac{8}{9} + \tfrac{10}{9}n)}$$

Of course these results are restricted to the Prandtl-number range of *gases*, since Eq. (13-29) is so restricted.

CONSTANT FREE-STREAM VELOCITY FLOW ALONG A SEMI-INFINITE PLATE WITH ARBITRARILY SPECIFIED HEAT FLUX

This problem again can be handled in exactly the same manner as for the laminar boundary layer. The step-function solution, Eq. (13-29), has the requisite form, and the function $g(\xi, x)$ evaluated from Eq. (10-39) becomes

$$g(\xi, x) = \frac{\tfrac{9}{10}\, \mathrm{Pr}^{-0.6}\, \mathrm{Re}_x^{-0.8}}{\Gamma(\tfrac{8}{9})\Gamma(\tfrac{1}{9})(0.0287k)} \left[1 - \left(\frac{\xi}{x}\right)^{9/10} \right]^{-8/9} \qquad (13\text{-}32)$$

Substituting into Eq. (10-38), the expression for the wall temperature resulting from an arbitrarily specified heat-flux variation $\dot{q}_0''(\xi)$ is

$$t_0(x) - t_\infty = \frac{3.42}{k} \mathrm{Pr}^{-0.6}\, \mathrm{Re}_x^{-0.8} \int_0^x \left[1 - \left(\frac{\xi}{x}\right)^{9/10} \right]^{-8/9} \dot{q}_0''(\xi)\, d\xi \qquad (13\text{-}33)$$

The simplest example for which the heat flux is specified is that where it is taken as *constant*. With $\dot{q}_0''(\xi)$ a constant in Eq. (13-33), the integral can be readily evaluated by use of the beta function previously mentioned in connection with the laminar flow problem. The final result is

$$\mathrm{St}\, \mathrm{Pr}^{0.4} = 0.030\, \mathrm{Re}_x^{-0.2} \qquad (13\text{-}34)$$

This result is approximately 4 percent higher than the comparable equation for constant surface temperature, Eq. (13-18). Recall that the difference was 36 percent for the laminar boundary layer. The turbulent boundary layer is thus much less sensitive to variations of wall temperature than the laminar boundary layer. For high-Prandtl-number fluids this is even more the case, while for very low-Prandtl-number fluids the large laminar-like effects begin to reappear.

AN APPROXIMATE SOLUTION FOR VARYING FREE-STREAM VELOCITY

The turbulent boundary-layer solutions considered up to this point have been concerned with a flat plate over which the free-stream velocity is constant. Obviously, more interesting problems occur when the velocity varies along the surface, such as flow over an airfoil, a turbine blade, or inside a variable-area passage such as a rocket nozzle.

For a laminar boundary layer a nearly exact solution was developed for this problem for a constant surface temperature, building from the wedge-flow solutions. For the turbulent boundary layer no such convenient building blocks are available. On the other hand, a characteristic of the turbulent boundary layer is that it adjusts rather quickly to changes in boundary conditions, and this suggests that the integral boundary layer equations might be used to obtain at least approximate solutions for situations where the boundary conditions are not constant.

To be precise, it should be necessary to solve both the momentum and energy boundary-layer equations. A number of approximate procedures to do this have been proposed, but it turns out that for calculation of heat transfer the energy equation is by far the more important. Furthermore, a procedure based on the energy equation alone turns out to be remarkably adequate over a wide range of operating conditions. This procedure, suggested by Ambrok,[19] is described now.

For the case of constant free-stream velocity and constant surface temperature we developed Eq. (13-19) for gases, wherein the Stanton number is expressed as a function of the enthalpy thickness Reynolds number [Eq. (13-21) extends this idea to $Pr = 6$]. In this equation the heat-transfer coefficient is expressed entirely in terms of *local* variables. If the thermal boundary layer responds very quickly to changes in boundary conditions, and if the local pressure gradient and local surface temperature gradient are not significant variables, then it is possible that Eq. (13-19) might be valid even if boundary conditions change in the flow direction. To test this hypothesis, let us examine some experimental data for a positive velocity gradient (favorable pressure gradient) and for a negative velocity gradient (adverse pressure gradient). Figure 13-16 shows the free-stream velocities for two such experiments. Both are for essentially *equilibrium* momentum and *equilibrium* thermal boundary layers. IDENT 51368 is a moderately strong acceleration, for which the

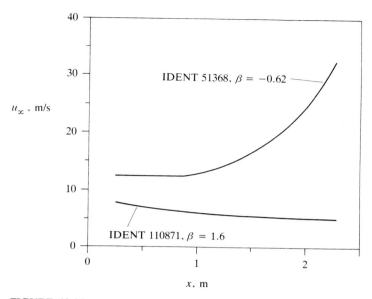

FIGURE 13-16
Free-stream velocity variations for two equilibrium boundary layers: IDENT 51368 (Thielbahr[3]) and IDENT 110871 (Blackwell[4]).

acceleration parameter $K = 0.57 \times 10^{-6}$ and $\beta = -0.62$. IDENT 110871 is a moderately strong deceleration for which $\beta = 1.6$. Figure 13-17 shows a plot of Stanton number versus enthalpy thickness Reynolds numbers for these two experiments, together with Eq. (13-19) (for air). In both cases Eq. (13-19) represents the data remarkably well. Furthermore, if Eq. (13-29) for a step change in surface temperature is converted to enthalpy thickness Reynolds number (using the energy integral equation), it too will correlate quite well with the data on Fig. 13-17. All of this suggests that the range of approximate validity of Eq. (13-19) is far greater than for the simple case of constant free-stream velocity and constant surface temperature.

If this indeed is the case then the energy integral equation in its more general form, Eq. (6-20), can be used together with Eq. (13-19) to solve the thermal boundary layer for arbitrarily specified variations in free-stream velocity and surface temperature, and also for specified variations in the radius of a body of revolution such as a nozzle or flow over a body of revolution. If we substitute Eq. (13-19) into Eq. (6-20), we obtain, after some algebraic manipulation,

$$d[G_\infty \Delta_2 R(t_0 - t_\infty)]^{1.25} = 0.0156 \, \text{Pr}^{0.5} R^{1.25} \mu^{0.25} G_\infty (t_0 - t_\infty)^{1.25} \, dx \quad (13\text{-}35)$$

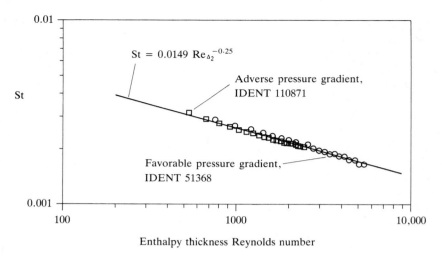

FIGURE 13-17
Demonstration of the generality of Eq. (13-19): effect of pressure gradient on St for air.

Let $x = 0$ designate the *virtual origin* of the thermal boundary layer, which is essentially a point where any one of the variables within the exact differential is zero, and then integrate between the limits 0 and x. Introducing the definition of local Stanton number, we obtain

$$ St = 0.0287 \, Pr^{-0.4} \frac{R^{0.25}(t_0 - t_\infty)^{0.25}\mu^{0.2}}{\left[\int_0^x R^{1.25}(t_0 - t_\infty)^{1.25}G_\infty \, dx \right]^{0.2}} \qquad (13\text{-}36) $$

Alternatively, if a laminar boundary layer precedes the turbulent boundary layer, the enthalpy thickness of the laminar boundary layer at the presumed "point" of transition can be used as a lower limit of integration. (Note that the viscosity μ could have been left in the integrand, but it appears to such a small power that the error introduced by treating it as a constant is certainly less than the overall uncertainty of the analysis.)

The form of this result is similar to the comparable result for a laminar boundary layer, Eq. (10-53). Of particular interest is the fact that use of the energy equation has made it possible to account for a varying surface temperature, approximate though the result may be. Since we already have a reasonably accurate variable-surface-temperature solution for a constant free-stream velocity, we can at least check Eq. (13-36) for this case. It will be found that Eq. (13-36) provides a quite reasonable prediction for variable surface temperature.

It is now worth noting that the energy integral equation, upon which this result is based, is nothing more than a way of assuring that the first law of thermodynamics is satisfied in the boundary layer. The success of the method suggests that satisfying the first law in the boundary layer is of paramount importance.

Equation (13-36), despite the assumptions that went into its development, is in quite good agreement with experimental data for a wide range of applications, including rocket nozzles and missile bodies. It fails badly only for *strongly accelerated* flows, which can be characterized by large magnitudes of the acceleration parameter K defined by Eq. (11-45). For $K > 10^{-6}$ the favorable pressure gradient causes the viscous sublayer to thicken, which results in an increase in heat-transfer resistance in the sublayer and a decrease in Stanton number. This effect can be seen in Eq. (11-26) for A^+, the effective sublayer thickness. It was noted in Chap. 11 that for $K > 3.5 \times 10^{-6}$ the boundary layer begins to *relaminarize* as the viscous sublayer engulfs the entire boundary layer.

Note, of course, that Eq. (13-36) is based on *constant fluid properties* and *the absence of viscous dissipation.* We will see in Chaps. 15 and 16 how to take these effects into consideration quite simply. However, we have already taken account of one of the major compressible-flow effects, the variation of density ρ_∞ in the flow direction. We have done this by employing the free-stream mass velocity G_∞ as an independent variable. Since $G_\infty = u_\infty \rho_\infty$, the separate variations of u_∞ and ρ_∞ are immaterial.

STRONGLY ACCELERATED BOUNDARY LAYERS

Figure 13-18 shows some experimental data for a strongly accelerated boundary layer. For enthalpy thickness Reynolds numbers between 1000 and 2000 the free-stream is accelerated at $K = 2.0 \times 10^{-6}$. Note that the Stanton number decreases markedly as a result of the acceleration, although it recovers quickly when the acceleration is taken off. Also shown on the figure are the results of a calculation using the mixing-length model, including Eq. (11-26) for A^+, and the model obviously predicts the data quite well. It is worth noting that the same acceleration causes the friction coefficient to *increase* rather than decrease. The boundary layer shown on Fig. 13-18 is still a turbulent boundary layer, and it is not until K reaches about 3.5×10^{-6} that it will become entirely laminar, in which case the Stanton number simply continues to decrease. This boundary layer is an unusual turbulent boundary layer in that it has an abnormally thick sublayer.

The reduction in Stanton number caused by strong acceleration is frequently observed in the convergent section of small supersonic nozzles. In large nozzles it is seldom that K reaches these high values.

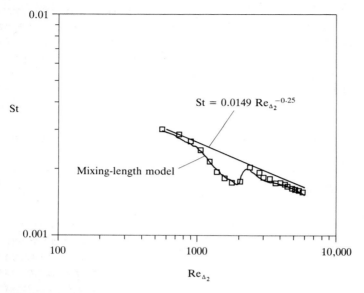

FIGURE 13-18
Heat transfer through a strongly accelerated turbulent boundary layer (data of Kearney,[20] IDENT 91069).

THE TRANSPIRED TURBULENT BOUNDARY LAYER

The effect of transpiration on the turbulent boundary layer is very similar to the effect on the momentum boundary layer described in Chap. 11. The *Couette flow* analysis described in Chap. 11 can be developed in precisely the same manner for the Stanton number as for the friction coefficient. The only difference is that a heat-transfer blowing parameter B_h is introduced rather than the friction blowing parameter B_f:

$$B_h = \frac{v_0/u_\infty}{\mathrm{St}} = \frac{\dot{m}''/G_\infty}{\mathrm{St}}$$

Then

$$\frac{\mathrm{St}}{\mathrm{St}_0} = \frac{\ln{(1 + B_h)}}{B_h} \tag{13-37}$$

If comparison between St and St_0 is made at the same *x Reynolds number* for the case of constant free-stream velocity, Eq. (13-37) fits the experimental data remarkably well. Thus, using Eq. (13-18), we obtain for a gas

$$\mathrm{St}\,\mathrm{Pr}^{0.4} = 0.0287\,\mathrm{Re}_x^{-0.2}\,\frac{\ln{(1 + B_h)}}{B_h} \tag{13-38}$$

If we now use the energy integral equation, we can express the Stanton number as a function of the enthalpy thickness Reynolds number rather than the x Reynolds number (holding B_h constant):

$$\text{St Pr}^{0.5} = 0.0125 \, \text{Re}_{\Delta_2}^{-0.25} \left[\frac{\ln(1 + B_h)}{B_h} \right]^{1.25} (1 + B_h)^{0.25} \qquad (13\text{-}39)$$

It then follows that if St_0 is defined as the Stanton number for no transpiration at the *same value* of Re_{Δ_2} then

$$\left. \frac{\text{St}}{\text{St}_0} \right|_{\Delta_2} = \left[\frac{\ln(1 + B_h)}{B_h} \right]^{1.25} (1 + B_h)^{0.25} \qquad (13\text{-}40)$$

The range of applicability of Eq. (13-40) is remarkable, and the same equation applies for c_f / c_{f_0} if B_h is replaced by B_f. Figure 13-19 shows a set of experimental results for both St and c_f that include blowing, suction, free-stream acceleration, and deceleration. All are very well approximated by Eq. (13-40).

These results are based on experiments carried out with B_h and B_f held constant. These are the conditions for an *equilibrium thermal*

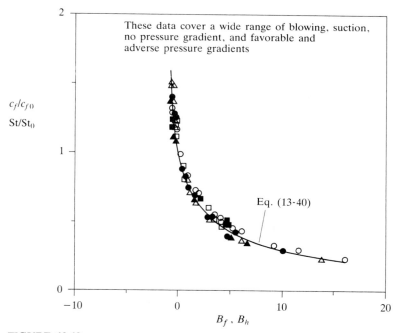

FIGURE 13-19
The effect of transpiration on Stanton number and friction coefficient for a wide range of flows (data of Kays and Moffat[21]).

boundary layer with blowing or suction, i.e., conditions that lead to outer-region similarity in both the velocity and temperature profiles. (Note that earlier we have used the term *equilibrium thermal boundary layer* for a thermal boundary layer at constant free-stream velocity having the same virtual origin as the momentum boundary layer.) However, the turbulent boundary layer comes to local equilibrium so quickly that any of Eqs. (13-38), (13-39), or (13-40) is quite applicable under conditions where B_h varies quite markedly along the surface. A rather extreme example is shown on Fig. 13-20, where a step in blowing is suddenly applied. The data are seen to closely approach the new equilibrium condition very soon after the step.

Also shown on Fig. 13-20 are the results of a full solution of the energy differential equation using the mixing-length model, which in-cludes Eq. (11-26) for A^+. The mixing-length calculations do fit the data slightly better.

Experiments in which $\dot{m}''/G_\infty$ is held constant along the surface, rather than B_f and B_h, yield results for c_f and St that are virtually indistinguishable from the constant-B_f and constant-B_h experiments. This is in marked contrast to the behavior of the laminar boundary layer, where a substantial difference is seen.

By following the same procedures as for the friction coefficient in

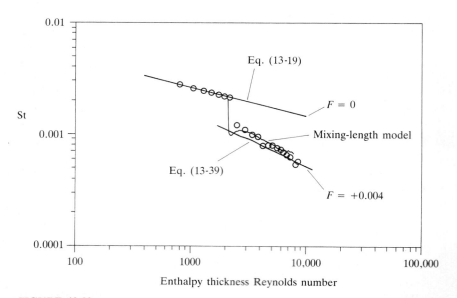

FIGURE 13-20
The response of Stanton number to a step increase in blowing, with $F = \dot{m}''/G_\infty$ going from 0 to +0.004; the fluid is air (data of Whitten, IDENT 50167).

Chap. 11, an alternative blowing parameter b_h can be defined such that

$$\left|\frac{St}{St_0}\right|_x = \frac{b_h}{e^{b_h} - 1} \tag{13-41}$$

where

$$b_h = \frac{v_0/u_\infty}{St_0} = \frac{\dot{m}''/G_\infty}{St_0}$$

St_0 is here defined as the Stanton number for no transpiration at the same *x Reynolds number,* and thus Eq. (13-41) is valid only for the case of constant free-stream velocity. But it does provide an explicit relation for St rather than the implicit equation (13-37).

For the more general case where free-stream velocity, temperature difference, and body radius, as well as transpiration rate, all vary in any arbitrary manner along a surface, a reasonable approximate procedure is to use Eq. (13-36) and simply multiply it by Eq. (13-37) to account for the effects of transpiration. A slightly more rational procedure would be to use Eq. (13-39) in place of Eq. (13-19) in the development of Eq. (13-36). However, if the pressure gradient is large, these approximate methods can lead to significant error.

Figure 13-21 shows an example of a somewhat unexpected effect of a combination of transpiration and varying free-stream velocity. From the beginning of the plate out to $Re_{\Delta_2} = 4500$ the free-stream velocity is held constant, but $\dot{m}''/G_\infty$ is maintained constant at $+0.0062$. This is a quite high rate of blowing, and the experimental uncertainty for these data is at least ± 10 percent. Nevertheless, Eq. (13-39) approximates the data reasonably well. At $Re_{\Delta_2} = 4500$ and thereafter the free-stream is accelerated at a constant rate $K = 0.77 \times 10^{-6}$. It was seen earlier that strong acceleration causes the Stanton number to *decrease.* Yet here it is seen that an acceleration superimposed on positive transpiration has the effect of *increasing* the Stanton number. Thus it is apparent that the various effects of changing boundary conditions are not necessarily additive.

Also shown in Fig. 13-21 are the results of a solution of the energy equation using the mixing-length model. This model does seem to show the increase in Stanton number accompanying acceleration.

There are two major areas of technical application of transpired turbulent boundary-layer theory: transpiration cooling and mass transfer. Positive transpiration, or blowing, through a porous surface provides a very effective way to cool a solid surface and protect it from a hot mainstream fluid. Not only does the transpired fluid absorb thermal energy from the porous surface as it passes through, but transpiration has the effect of substantially reducing the heat-transfer coefficient and therefore the heat-transfer rate.

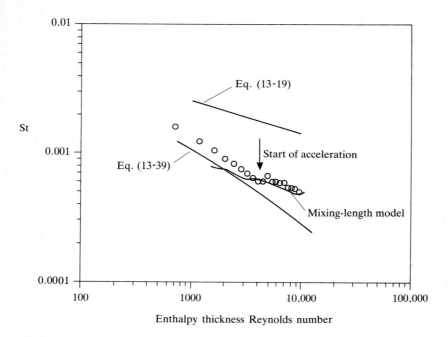

FIGURE 13-21
Combined effects of blowing and acceleration: blowing, $F = 0.0062$ (constant); constant u_∞, then $K = 0.77 \times 10^{-6}$ (data of Whitten, IDENT 30868).

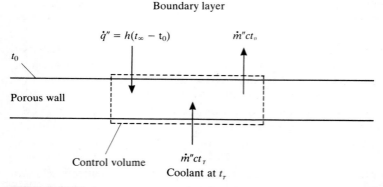

FIGURE 13-22
Control volume for analysis of transpiration cooling.

Figure 13-22 shows the control volume that must be analyzed in order to employ the transpired boundary-layer solutions for transpiration cooling. Assuming that the transpiration rate $\dot{m}''$ is known and the heat transfer coefficient h has been evaluated, an energy balance on the indicated control volume provides a relationship between the free-stream temperature t_∞, the surface temperature t_0, and the temperature of the coolant, which is denoted by t_T. (The enthalpy of the cooling fluid is represented as the product of the specific heat and the temperature.) But implicit in this analysis, as well as in all of the transpired boundary-layer solutions, including those for laminar boundary layers, is the assumption that the temperature of the transpired fluid is at t_0 when it leaves the porous surface. In other words, it is assumed that the porous surface is a 100 percent effective heat exchanger. If the wall is thick and the pores, or holes, are very small, this may well be the case—but this is something that should be analyzed if the wall is very thin. Another critical assumption implicit in the transpired boundary-layer solutions is that the pores are infinitely small and infinitely close together. In the case of the turbulent boundary layer this effectively means that both the pores and the pore spacing are small relative to the sublayer thickness. Practically speaking, this probably means that the pore size and spacing should be no greater than one y^+ unit, although this a problem that has not been investigated. It is related to a problem that is discussed in the next section.

In all of the previous development it has been assumed that the transpired fluid is chemically the same as the mainstream fluid. An important variation arises when the transpired fluid is chemically different, and this is sometimes a practicable cooling alternative. The associated momentum-transfer problem is basically unchanged, but the heat-transfer problem is altered, and a new transfer mechanism, *mass diffusion*, is introduced. In addition to solution of the momentum and energy equations of the boundary layer, the mass-diffusion equation of the boundary layer must be solved. But this then brings us to the general subject of mass transfer, which is the subject of Chaps. 20, 21, and 22. Since the transpired boundary layer does involve mass transfer across the fluid–solid interface, in reality it is part of the mass-transfer spectrum of problems, and the results developed in this section are used in the mass-transfer chapters even though the problem is formulated somewhat differently.

FULL-COVERAGE FILM COOLING

In cooling problems, of which the cooling of gas-turbine blades is a good example, it is often not practicable to construct the surface with a porous material of the type discussed above. An alternative is to drill an array of

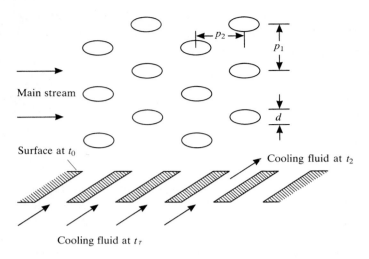

FIGURE 13-23
An example of a surface used for full-coverage film cooling.

small holes through the surface. The holes can be drilled normally to the surface, or a more effective arrangement is to drill them at an angle as in Fig. 13-23. But these holes are usually very large relative to the sublayer, or even the total boundary-layer thickness.

The term *full-coverage film cooling* is used here because a similar arrangement, but usually with one or two rows of holes only, is frequently used for cooling where the primary objective is to cool the surface *downstream* from the cooling holes by injecting a blanket of cool fluid into the boundary layer. The latter arrangement will be termed simply *film cooling,* and is discussed in the next section. The primary objective of full-coverage film cooling is to cool the surface *between* the cooling holes, and this may indeed involve the entire surface. Of course, if a full-coverage region is followed by an impermeable region, the latter will be cooled in the same manner as in simple film cooling, so the distinction is a bit academic. But full-coverage film cooling is obviously closely related to transpiration cooling, and, as the holes and hole spacing are made smaller, the behavior of the transpired boundary layer will be approached.

In Fig. 13-23 the temperature of the cooling fluid at the surface is denoted by t_2. Unlike the case of transpiration cooling, t_2 will usually be different from t_0 because the tubes will be too short to provide sufficient heat-transfer surface area to bring the cooling fluid to surface temperature. Thus a new variable enters the problem. Fortunately, the linearity of the energy differential equation of the boundary layer provides a simple way to handle this difficulty. If experiments are performed at two different values of t_2, the behavior at any value of t_2 can be determined. Let θ be defined by

$$\theta = \frac{t_2 - t_\infty}{t_0 - t_\infty} \tag{13-41}$$

Now suppose two experiments are performed: one for $\theta = 0$ and the other for $\theta = 1$. It can be readily shown by superposition that at a particular point x along the surface

$$St(\theta) = St|_{\theta=0} - \theta(St|_{\theta=0} - St|_{\theta=1}) \tag{13-42}$$

Note that $\theta = 1$ corresponds to the case studied for the transpired boundary layer. With one other experiment, in this case letting the coolant enter at free-stream temperature, a heat-transfer coefficient for *any* value of t_2 can be evaluated.

It should be added that the heat-transfer coefficient implied is a local *average* over the solid surface between the holes; locally, the heat-transfer coefficient may vary considerably; but if the surface material has a reasonable thermal conductivity, the effect of local variations will damp out.

The number of geometric variables involved in an arrangement such as that illustrated in Fig. 13-23 is so large that it is not practicable to present any kind of comprehensive survey of the available data here. However, in order to illustrate the kind of results that are obtainable, one particular case at one Reynolds number will be presented. For more information the reader is referred to the report by Crawford, Kays, and Moffat.[22]

Figure 13-24 shows some results for a full-coverage plate with $p_1 = p_2$, $p_1/d = 5$, and the tubes at a 30° angle from the surface. This particular experiment was at constant free-stream velocity, at $Re_x = 1,750,000$, and the thermal boundary layer had a delayed starting length with its origin at $Re_x = 1,250,000$. This was also the point where the rows of injection holes started. The Reynolds number Re_d based on hole diameter and free-stream velocity was 12,971. The holes had $d = 1.03$ cm, and the boundary-layer total thickness at the start of injection was about 2 cm. Thus the holes were large relative to the boundary layer.

The St_0 on Fig. 13-24 is the Stanton number for no injection as given by Eq. (13-29). For this particular experiment there is a relatively large thermal boundary-layer starting-length effect, so it is appropriate to use Eq. (13-29) as a base. The ratio St/St_0 is plotted as a function of a blowing parameter M defined as

$$M = \frac{\rho_2 u_2}{\rho_\infty u_\infty} \tag{13-43}$$

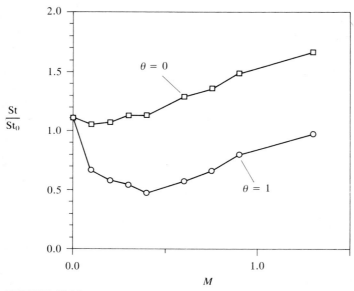

FIGURE 13-24
An example of experimental data for full-coverage film cooling (data of Crawford *et al.*,[22] taken at 11th row of holes).

M is thus similar to the transpiration mass-flux ratio $\dot{m}''/G_\infty$, but is much larger because it acts over a much smaller area.

It will be observed that the data for $\theta = 1$ initially *decrease* and behave very much like transpiration, i.e., the heat-transfer coefficient is decreased as blowing is increased. However, at about $M = 0.4$ it starts to rise, and continues to do so thereafter. The reason for this is that apparently the cooling fluid is begining to pass completely through the sublayer instead of providing a blanket of cool fluid along the surface, and additionally it is increasing the turbulence in the boundary layer.

The effect of t_2 being less than t_0 can be investigated through Eq. (13-42). A complete thermal analysis of the wall, including evaluation of heat transfer within the tubes and on its underside, is necessary before any conclusions can be drawn.

FILM COOLING

Another effective way to protect a surface from a hot fluid stream is to inject a cooler fluid under the boundary layer, thus forming a protective layer or film along the surface. The injected fluid can enter the boundary layer in a number of ways, and the number of possible geometrical arrangements is considerable. Figure 13-25 shows an example of the use

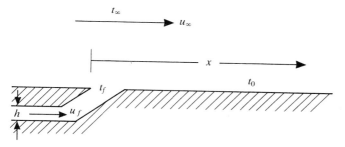

FIGURE 13-25
Example of a film-cooling scheme.

of a *slot* in the surface through which the cooling fluid may be injected. But the slot could be replaced by a row of holes, or two or three rows of holes, and the angle of injection could vary from normal to the surface to virtually parallel to it. Fluid could be injected through a section of *porous surface,* and also the region downstream of the *full-coverage* film-cooling region discussed in the preceding section has the same general characteristics as does the slot.

Most film-cooling experiments have been concerned with an *adiabatic* surface in the region downstream of the injection slot, and it is the temperature of this surface t_{aw} relative to the free-stream temperature t_∞ and the cooling-fluid temperature t_f that is of concern. A film-cooling effectiveness η is defined as

$$\eta = \frac{t_{aw} - t_\infty}{t_f - t_\infty} \tag{13-44}$$

It is found that η is primarily a function of a blowing-rate parameter $M = \rho_f u_f / \rho_\infty u_\infty$, the width or height of the injection slot h, and the distance x.

For essentially the scheme illustrated in Fig. 13-25, using air and a constant free-stream velocity, Weighardt[23] presents the correlation

$$\eta = 21.8(x/Mh)^{-0.8} \tag{13-45}$$

for $0.22 < M < 0.74$ and $x/h > 100$.

For applications where there is heat transfer to the surface downstream from the injection slot, the conventional procedure is to evaluate t_{aw} from η, and then calculate the heat-transfer rate using the heat-transfer coefficient that would obtain if there were no injection and the temperature difference were $t_{aw} - t_0$ rather than $t_\infty - t_0$. Metzger and Fletcher[24] have pointed out that this is not a good approximation immediately downstream of the point of injection, but high precision is generally not particularly important in that region.

Many other correlations for film-cooling are available in the literature, as well as many other geometric variations. A comprehensive review of the correlations is given by Goldstein.[25] It is worth noting that if the slot in Fig. 13-25 is arranged parallel to the surface, and M is allowed to go to infinity, the problem of the *wall jet* is approached (see Ref. 26).

Transpiration cooling, full-coverage film cooling, film cooling by slot or hole injection, and the wall jet all obviously involve closely related phenomena. If the geometry is simple, complete solution of the boundary-layer equations is sometimes practicable, and when this is the case it is certainly the preferred method. However, if boundary-layer separation is involved, as is frequently the case, or three-dimensional effects are present, turbulence models developed for the particular application must be used, and this is beyond the scope of the present discussion.

THE EFFECTS OF SURFACE ROUGHNESS

All of the preceding discussion has been based on the assumption of an *aerodynamically smooth* surface. The effect of a rough surface on the momentum boundary layer is discussed in Chap. 11, and a mixing-length model for a *fully rough* surface ($Re_k > 70$) is developed. The same procedure can now be applied to thermal boundary layer, but an importance difference is encountered. Whereas a turbulent shear stress can be transmitted directly to the roughness elements as a result of dynamic or pressure forces on the area of the elements normal to the mean direction of flow, there exists no comparable mechanism for heat transfer. Thus heat can be transferred by *eddy conductivity* down to the plane of the roughness elements, but the actual final heat transfer to the solid surface must be through the mechanism of molecular conduction. A consequence of this fact is that, even in the fully rough regime, molecular thermal conductivity remains a significant variable; and thus molecular Prandtl number remains a variable.

A mixing-length model incorporating this fact can be readily developed. Let us consider the fully rough regime for the elementary case of no pressure gradient and no transpiration. We develop a rough surface thermal law of the wall paralleling the development of Eq. (13-15). We start with Eq. (13-12), but, since there is no sublayer, we assume that $\varepsilon_H \gg 1/Pr$ all the way down to $y^+ = 0$, so we neglect $1/Pr$. (This is obviously not valid for very low-Pr fluids, but for Pr of the order of magnitude of unity or larger this is an assumption of the same order as was made for the corresponding momentum-transfer problem in Chap. 11.) Next we take the lower limit of t^+ as a finite number δt_0^+ instead of zero. δt_0^+ represents the temperature difference (nondimensional) across which heat is transferred by conduction through what may be a semistagnant fluid in the roughness cavities at the surface. Thus Eq. (13-12) becomes

$$t^+ - \delta t_0^+ = \int_0^{y^+} \frac{dy^+}{\varepsilon_H / \nu}$$

The turbulent Prandtl number is then introduced to relate ε_H to ε_M, and ε_M is obtained from the corresponding development in Chap. 11:

$$\frac{\varepsilon_H}{\nu} = \frac{1}{\mathrm{Pr}_t} \frac{\varepsilon_M}{\nu} = \frac{\kappa}{\mathrm{Pr}_t}(y^+ + \delta y_0^+)$$

Integrating, and introducing Eq. (11-53) for δy_0^+ (treating Pr_t as constant), we have

$$t^+ = \delta t_0^+ + \frac{\mathrm{Pr}_t}{\kappa} \ln \frac{32.6 y^+}{\mathrm{Re}_k}$$

If we now examine the definition of δt_0^+ [defined similarly to t^+ in Eq. (13-10)] and define a local heat-transfer coefficient based on δt_0, we see that δt_0^+ can be expressed in terms of a Stanton number based on u_τ:

$$\delta t_0^+ = \frac{\rho c u_\tau}{h_k} = \frac{1}{\mathrm{St}_k}$$

Then

$$t^+ = \frac{1}{\mathrm{St}_k} + \frac{\mathrm{Pr}_t}{\kappa} \ln \frac{32.6 y^+}{\mathrm{Re}_k} \tag{13-46}$$

St_k must be determined from experiment, and it is undoubtedly a function of the type of roughness present. Dipprey and Sabersky[27] find that the available data can be correlated by an equation of the form

$$\mathrm{St}_k = C \, \mathrm{Re}_k^{-0.2} \, \mathrm{Pr}^{-0.44} \tag{13-47}$$

The data of Pimenta, Moffat, and Kays[7] for a rough surface composed of closely packed *spheres* yield a value of $C = 0.8$ when $\mathrm{Pr}_t = 0.9$ is used in Eq. (13-46).

Since we now have a thermal law of the wall for a fully rough surface for the case of constant u_∞ and $v_0 = 0$, as well as the corresponding momentum law of the wall, Eq. (11-56), the conventional Stanton number under these conditions can be determined in precisely the same manner as was used to develop Eq. (13-17). The result is

$$\mathrm{St} = \frac{c_f / 2}{\mathrm{Pr}_t + \sqrt{c_f / 2} / \mathrm{St}_k} \tag{13-48}$$

Equation (11-57) can then be used to determine $c_f / 2$.

300 CONVECTIVE HEAT AND MASS TRANSFER

Equation (13-48) would suggest that for a fully rough surface St is independent of Prandtl number. However, note that St_k is a function of Prandtl number, and as the latter becomes very large, this term can become the dominant one. Physically this means that the heat-transfer resistance offered by the molecular-conduction process in the cavities between the roughness elements becomes the major source of heat-transfer resistance in the boundary layer. Large values of $c_f/2$ have the same effect.

An example of the use of Eq. (13-48) is shown on Fig. 13-26. However, the value for C in Eq. (13-47) was *determined* from these data, so the close correspondence between the data and Eq. (13-48) only demonstrates that C is indeed a constant. A better demonstration of the validity of the model is shown in the comparison of temperature profiles discussed below.

Note that a rough surface yields Stanton numbers significantly higher than a smooth surface—in this case about 60 percent higher. Thus roughening a surface is an effective way to increase heat-transfer rates.

The same mixing-length model can be used to solve the complete boundary-layer differential equations by finite-difference methods. The corresponding momentum boundary-layer problem is discussed in Chap. 11. The complete absence of a sublayer is modelled by setting $A^+ = 0$. Since there is no sublayer, the high values of Pr_t seen in the sublayer for a smooth surface are probably not present, and use of $Pr_t = 0.9$ throughout

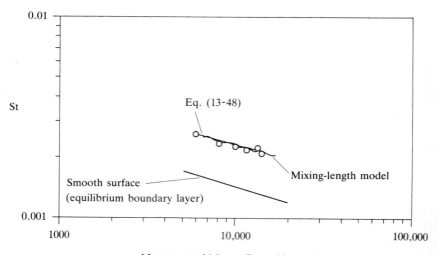

Momentum thickness Reynolds number

FIGURE 13-26
Comparison of fully rough-surface model with experiment (data from IDENT 71374; packed spheres, diameter 1.27 mm; $k_s = 0.787$ mm; air at constant u).

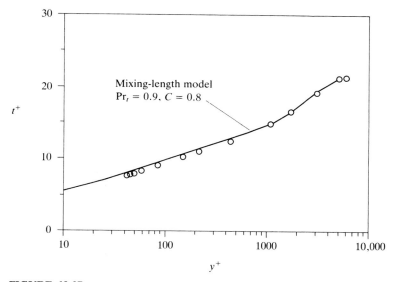

FIGURE 13-27

Comparison of a temperature profile calculated from the mixing-length model with experimental data (data from IDENT 71374; packed spheres, diameter 1.27 mm; $k_s = 0.787$; $Re_{\delta_2} = 11,509$; $Re_{\Delta_2} = 11,558$; $Re_k = 101$).

the boundary layer seems appropriate. The results of such a calculation are also shown on Fig. 13-26. They are virtually the same as Eq. (13-48) (the "wavyness" is the result of stability problems).

Figure 13-27 is of more significance. The model reproduces the nondimensional temperature profile extremely well.

THE EFFECTS OF AXIAL CURVATURE

The effects of axial curvature on the momentum boundary layer, and in particular on the friction coefficient, are discussed in Chap. 11. Curvature has a very similar effect upon the Stanton number. In fact, the data of Gibson[5] shown on Fig. 13-28 are based on the same experiments for which the friction coefficient is shown on Fig. 11-16, and the results are qualitatively the same. Convex curvature causes a *reduction* in St, while concave curvature has the opposite effect.

The same mixing-length model, in which the effective mixing length is reduced or increased by a linear function of the local Richardson number, and A^+ changes slightly, can be employed, i.e.,

$$l_{eff} = l(1 - 10\,\text{Ri})$$

$$A_{eff}^+ = \left(1 + \frac{1000\nu}{Ru_\tau}\right)A^+$$

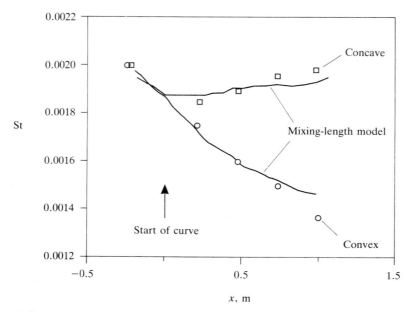

FIGURE 13-28
The effect of axial surface curvature on the Stanton number (data of Gibson,[5] IDENT 8000/8100; air at $u_\infty = 23.25$ m/s; $R = 0.41$ m; $Cr = 0.26 \times 10^{-6}$; at $x = 0$, $Re_{\delta_2} = 3540$, $Re_{\Delta_2} = 4180$).

Calculations employing this model, with no change in turbulent Prandtl number, are shown on Fig. 13-28, with quite reasonable results.

The data shown on Fig. 13-28 are for a relatively mild curvature. The strength of the curve can be expressed in terms of a curvature parameter Cr, which is simply the inverse of a Reynolds number based the radius of curvature of the surface, R, and the free-stream velocity:

$$Cr = v/u_\infty R$$

For both of the experiments shown on Fig. 13-28, $Cr = 0.26 \times 10^{-6}$.

Figure 13-29 shows experimental data for a much stronger case of convex curvature, $Cr = 1.21 \times 10^{-6}$. Precisely the same mixing-length model shown on this figure again yields quite satisfactory results.

Experiments by Hollingsworth[8] for concave curvature, using water, and with a considerably larger value of Cr, yield results for $c_f/2$ that can again be predicted quite well with the same model. However, Hollingsworth's heat-transfer results for strong concave curvature suggest that turbulent Prandtl number is affected by concave curvature, although the effect is not large in the early stages of a curved region. It should be recalled, however, that for water with $Pr = 5.9$ the thermal boundary

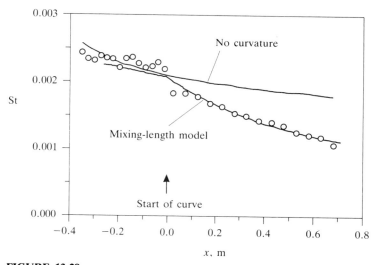

FIGURE 13-29
Effect of convex streamwise curvature (data of Simon;[6] air at $u_\infty = 28$ m/s; $R = 2.22$ m; $Cr = 1.21 \times 10^{-6}$; at $x = 0$, $Re_{\delta_2} = 4860$, $Re_{\Delta_2} = 2500$).

layer is largely within the sublayer, and this turbulent Prandtl-number effect may be unique to the sublayer.

An important question for technical applications is that of the range of validity of the flat-surface results, i.e., how much curvature can be tolerated before it is necessary to take it into consideration. This is similar to the question about the range of validity of the aerodynamically smooth surface correlations and models. Insufficient information is available to answer this question precisely, but the above results suggest that $Cr \ll 0.26 \times 10^{-6}$ would be a necessary condition.

THE EFFECTS OF FREE-STREAM TURBULENCE

The effects of free-stream turbulence on the Stanton number are very similar to the corresponding effects on the friction coefficient (see Fig. 11-17). The wake is diminished, or completely eliminated, and St *increases.*

Some experimental results by Blair[28] are shown on Fig. 13-30, extending up to a nominal value of turbulent intensity $Tu = 0.06$, indicating an increase in St of about 20 percent. Also shown in this figure are the results of calculations using the $k-\varepsilon$ model of turbulence, assuming isotropic free-stream turbulence. These results, together with the corresponding results for the friction coefficient (Fig. 11-17), suggest

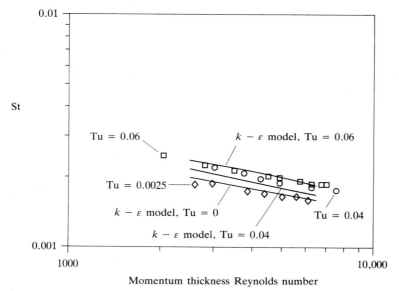

FIGURE 13-30
The effects of free-stream turbulence on heat transfer (data of Blair,[28] for air).

that the k–ε model is quite adequate up to least Tu = 0.075. An interesting result of such calculations is that the increase in St resulting from free-stream turbulence seems to be relatively independent of free-stream turbulence length-scale, despite experimental evidence that this may not be the case.[29]

The maximum value of Tu within the turbulent boundary layer with no free-stream turbulence is typically in the range 0.09–0.11. The experiments cited above are for values of the free-stream turbulence that are below the turbulence level generated within the boundary layer as a result of sublayer instability. Thus free-stream turbulence has its major effect on the outer part of the boundary layer, i.e., on the wake. But when free-stream turbulence is greater than that naturally generated by the boundary layer itself, it is probably not surprising that the situation is changed.

At high values of the free-stream turbulence, Maciejewski[30] has found, based on an examination of a large amount of data under widely varying conditions, that (for air) the heat-transfer coefficient is almost directly proportional to $\sqrt{\overline{u'^2}}$, and relatively independent of Reynolds number. If a Stanton number St' is defined based on $\sqrt{\overline{u'^2}}$ as the significant velocity then Maciejewski suggests the following correlation:

$$\text{St}' = 0.018 \quad \text{for Tu} > 0.20 \tag{13-49}$$

PROBLEMS

13-1. The development leading to Eq. (13-17) was not expected to be applicable to a high-Prandtl-number fluid. The objective of this problem is to develop a heat-transfer solution for the turbulent boundary layer for no pressure gradient or transpiration, applicable at $Pr = 100$. Use the Van Driest mixing-length equation and $Pr_t = 0.9$, and integrate Eq. (13-12) out to about $y^+ = 100$ numerically. (It is desirable to use a programmable computer or calculator.) Then for $y^+ > 100$ neglect the $1/Pr$ term and integrate as in the text. Compare the results with Eqs. (13-15) and (13-17).

13.2 Consider heat transfer to a turbulent boundary layer with no pressure gradient or transpiration but with a vanishingly small-Prandtl-number fluid. Why is this problem simpler than for the turbulent boundary layer at moderate and high Prandtl numbers? What closed-form solution already in hand should be a good approximation? Why?

13-3. The approximate solution in the text for the development of a thermal boundary layer under an already existing momentum boundary layer, Eq. (13-29), is not valid very close to the step in surface temperature. An alternative possibility in this region may be based on the fact that for a short distance from the step the thermal boundary layer is entirely within the almost completely laminar part of the sublayer, say out to $y^+ = 5$. A heat-transfer solution for this region can then be obtained in much the same manner as for a laminar boundary layer, but with the local turbulent boundary-layer wall shear stress used to establish the velocity profile. Develop a heat-transfer solution for this region, assuming that the shear stress is a constant throughout, and compare the results with Eq. (13-29). What is the range of validity of the result? What is the influence of the Prandtl number? This problem can be solved exactly using similarity methods or integral methods (see Chaps 8. and 10).

13-4. Starting with Eq. (13-34), determine the Stanton number as a function of the Prandtl number and *enthalpy thickness* Reynolds number for the case of *constant heat flux* along a surface. Compare with Eq. (13-19).

13-5. Using Eq. (13-29) and making any mathematical approximations that seem appropriate, determine the Stanton number as a function of the Prandtl number, *enthalpy thickness* Reynolds number, and *momentum thickness* Reynolds number. Note that Eq. (13-29) provides for the possibility of the ratio Δ_2/δ_2 varying from 0 to 1.

13-6. Consider the development of a turbulent boundary layer in a convergent axisymmetric nozzle. Let both the free-stream and surface temperatures be constant. As an approximation, treat the flow as one-dimensional, so that the mass velocity G may be calculated as the mass flow rate divided by the cross-sectional area of the duct, πR^2. Assume that the thermal boundary layer originates at the start of the convergence of the nozzle. Then take the case where the nozzle throat diameter is one-fifth the duct diameter at the start of convergence, x is a linear function of R, and the convergence angle is 45°. Derive an expression for the Stanton number at the nozzle throat as a function of the Prandtl number and a Reynolds number based on throat diameter and throat mass velocity. How sensitive is this expression to convergence angle? Would shapes other than the straight wall of this example yield significantly different results? Compare your results with the corresponding expression for fully developed turbulent flow in a circular pipe (see Chap. 14).

13-7. Air at a temperature of $-7°C$ and 1 atm pressure flows along a flat surface (an idealized airfoil) at a constant velocity of 46 m/s. For the first 0.6 m the surface is heated at a constant rate per unit of surface area; thereafter, the surface is adiabatic. If the total length of the plate is 1.8 m, what must be the heat flux on the heated section so that the surface temperature at the trailing edge is not below $0°C$? Plot the surface temperature along the entire plate. Discuss the significance of this problem with respect to wing de-icing. (A tabulation of incomplete beta functions, necessary for this problem, is found in App. C.)

13-8. Work Prob. 13-7 but divide the heater section into two 0.3 m strips, with one at the leading edge and the other 0.9 m from the leading edge. What heat flux is required such that the plate surface is nowhere less than $0°C$?

13-9. In film cooling the primary effect is believed to be due to the energy put into or taken out of the boundary layer, rather than the mass of fluid injected. If this is the case, it should be possible to approximate the effect of slot injection by simulating the slot with a strip heater in which the total heat rate is set equal to the product of the injection mass flow rate and the enthalpy of the injected fluid. Using the methods of the preceding two problems, carry out an investigation of the case represented by Eq. (13-45).

13-10. Consider a film-cooling application of the type described in Prob. 13-9 (Fig. 13-25). The objective of this problem is to investigate methods of calculating heat transfer to or from the surface downstream of the injection slot (that is, the heater or heat extractor, if the above analogy is employed) when the surface temperature is maintained at some temperature different from the "adiabatic wall temperature." Since heat transfer is zero when the surface temperature is equal to t_{aw}, it would seem reasonable to define a heat-transfer coefficient based on $t_{aw} - t_0$. Such a coefficient would be useful if it turned out that it did not differ substantially from that given by, for example, Eq. (13-18). The investigation can be carried out by simulating the injection slot as in Prob. 13-9 and then specifying a constant but substantially smaller heat flux along the remainder of the plate.

13-11. An aircraft oil cooler is to be constructed using the skin of the wing as the cooling surface. The wing may be idealized as a flat plate over which air at 71 kPa and $-4°C$ flows at 61 m/s. The leading edge of the cooler may be located 0.9 m from the leading edge of the wing. The oil temperature and oil-side heat-transfer resistance are such that the surface can be at approximately $54°C$, uniform over the surface. How much heat can be dissipated if the cooler surface measures 60 cm by 60 cm? Would there be any substantial advantage in changing the shape to a rectangle 1.2 m wide by 0.3 m in flow length?

13-12. Consider a constant free-stream velocity flow of air over a constant-surface-temperature plate. Let the boundary layer be initially a laminar one, but let a transition to a turbulent boundary layer take place in one case at $Re_x = 300,000$ and in another at $Re_x = 10^6$. Evaluate and plot (on log–log paper) the Stanton number as a function of Re_x out to $Re_x = 3 \times 10^6$. Assume that the transition is abrupt (which is not actually very realistic). Evaluate the Stanton number for the turbulent part using the energy integral equation and an analysis similar to that used to develop Eq. (13-36), matching the enthalpy thicknesses of the laminar and turbulent boundary layers at the transition point. Also plot the Stanton number for a turbulent boundary layer originating at the leading edge of the plate. Where is the "virtual origin" of the turbulent boundary layer when there is a preceding laminar boundary layer? What is the effect of changing the transition point? How high must the Reynolds number be in order for turbulent heat-transfer coefficients to be calculated with 2 percent accuracy without considering the influence of the initial laminar portion of the boundary layer?

13-13. A 12 m diameter balloon is rising vertically upward in otherwise still air at a velocity of 3 m/s. When it is at 1500 m elevation, calculate the heat-transfer coefficient over the entire upper hemispherical surface, making any assumptions that seem appropriate regarding the free-stream velocity distribution and the transition from a laminar to a turbulent boundary layer.

13-14. A round cylindrical body 1.2 m in diameter has a hemispherical cap over one end. Air flows axially along the body, with a stagnation point at the center of the end cap. The air has an upstream state of 1 atm pressure, 21°C, and 60 m/s. Under these conditions, evaluate the local heat-transfer coefficient along the cylindrical part of the surface to a point 4 m from the beginning of the cylindrical surface, assuming a constant-temperature surface. Make any assumptions that seem appropriate about an initial laminar boundary layer and about the free-stream velocity distribution around the nose. It may be assumed that the free-stream velocity along the cylindrical portion of the body is essentially constant at 60 m/s, although this is not strictly correct in the region near the nose.

    Then calculate the heat-transfer coefficient along the same surface by idealizing the entire system as a flat plate with constant free-stream velocity from the stagnation point. On the basis of the results, discuss the influence of the nose on the boundary layer at points along the cylindrical section and the general applicability of the constant free-stream velocity idealization.

13-15. Problem 11-9 is concerned with the momentum boundary layer on a rough surface, and it involves analyzing some experimental data. The corresponding measured temperature profile is given in the following table. The surface temperature is 35.22°C, constant along the surface, and the free-stream temperature is 19.16°C. $St = 0.00233$. How do the results compare with the theory developed in the text?

y, cm	t, °C	y, cm	t, °C
0.020	29.02	0.660	24.17
0.030	28.64	1.10	22.84
0.051	28.14	1.61	21.62
0.081	27.57	2.12	20.36
0.127	26.97	2.82	19.46
0.191	26.36	3.58	19.16
0.279	25.76		
0.406	25.13		

13-16. Consider a flat surface that is 30 cm square. Hot air at 800 K is flowing across this surface at a velocity of 50 m/s. The hot-air density is 0.435 kg/m³. The boundary layer on the plate is turbulent, and at the leading edge of the section of interest the momentum thickness Reynolds number is 1100. There is no thermal boundary layer in the region preceding the 30 cm section of interest; i.e., the surface on which the momentum boundary layer has developed is adiabatic and thus at 800 K.

Cooling air is available at 290 K, at a rate of 0.0037 kg/s. The density of the collant is 1.2 kg/m³. Investigate what can be done with three methods of cooling the 30 cm section: (1) convection from the rear surface of the plate; (2) transpiration; and (3) film cooling. In the first assume that the surface is sufficiently thin that the conduction resistance is negligible, that the effective heat-transfer coefficient on the rear surface is 25 W/(m² · K), and that the effective coolant temperature is 290 K over the whole surface. For the second case let $\dot{m}''$ be uniform everywhere.

Although the surface temperature varies in the direction of flow for each of the cooling methods used, ignore this effect on the heat-transfer coefficient, i.e., use constant-surface-temperature theory to determine h.

13-17. *Computer analysis of the turbulent thermal boundary layer over a flat plate with zero pressure gradient, $t_0 = const$, and constant properties:* Use a mixing-length turbulence model with the Van Driest damping function, and the variable turbulent Prandtl number model to calculate a flat-plate boundary-layer flow over the distance Re_{δ_2} from 2000 to 10,000 with initial $Re_{\Delta_2} \approx Re_{\delta_2}$. Compare the Stanton numbers based on x Reynolds number and enthalpy thickness Reynolds number with the results in the text. Examine a temperature profile to evaluate the thermal *law of the wall*. Examine the development in the streamwise direction of the enthalpy thickness and assess the validity of the energy integral, Eq. (6-24). Feel free to investigate any other attribute of the boundary-layer flow. You can choose how to set up the problem in terms of a suitable choice of dimensions and fluid properties (constant).

For the initial velocity profile at the specified starting Re_{δ_2} refer to Prob. 11-10. The initial thermal profile construction is similar in concept, using a continuous *law-of-the-wall* model in the inner region and a *power-law* model in the outer region. The inner-region profile is developed by converting Eq. (13-9) to *wall coordinates* and equating it to Eq. (11-11) for the case of no transpiration. Upon separating variables and introducing the turbulent Prandtl number definition, Eq. (13-5), along with the continuous *law of the wall* for ε_M, Eq. (11-6) and the Pr_t model, Eq. (13-7), the inner-region temperature profile is developed. At the point within the log-region where the *power-law* form of the velocity profile is applied, a similar power-law form of the thermal profile is used to compute the temperature to the edge of the boundary layer, assuming $\delta = \Delta$. Note that the resulting Re_{Δ_2} will only be approximately the same as the initial Re_{δ_2}.

13-18. *Computer analysis of Prob.* 13-7 *with* $\dot{q}_0'' = f(x)$ *as specified in that problem statement and constant properties:* Calculate the boundary-layer flow and compare with the superposition theory. For the initial conditions, follow the instructions in Prob. 13-17 with regard to the velocity and temperature profiles. Start the calculations at a suitable x location where the flow is turbulent, or start the calculations as a laminar boundary layer, along with a transition model such as abruptly switching to turbulent flow at an assumed Reynolds number.

13-19. *Computer analysis of Prob.* 13-11 *with* $t_0 = f(x)$ *as specified in the problem statement and constant properties:* Calculate the boundary-layer flow and compare with the superposition theory. For the initial conditions follow the instructions in Prob. 13-17 with regard to the velocity profile (note that there is no temperature profile in the initial region). Start the calculations at a suitable x location where the flow is turbulent, or start the calculations as a laminar boundary layer, along with a transition model such as abruptly switching to turbulent flow at an assumed Reynolds number.

REFERENCES

1. Reynolds, W. C., W. M. Kays, and S. J. Kline: NASA Memo 12-1-58W, Washington, December 1958.
2. Moffat, R. J., and W. M. Kays: Report HMT-1, Thermosciences Division, Department of Mechanical Engineering, Stanford University, August 1967.
3. Thielbahr, W. H., W. M. Kays, and R. J. Moffat: Report HMT-5, Thermosciences Division, Department of Mechanical Engineering, Stanford University, April 1969.
4. Blackwell, B. F., W. M. Kays, and R. J. Moffat: Report HMT-16, Thermosciences Division, Department of Mechanical Engineering, Stanford University, August 1972.
5. Gibson, M. M., and C. A. Verriopoulos: *Experiments in Fluids,* vol. 2, pp. 73–80, Springer-Verlag, New York 1984.
6. Simon, T. W., R. J. Moffat, J. P. Johnston, and W. M. Kays: Report HMT-32, Thermosciences Division, Department of Mechanical Engineering, Stanford University, November 1980.
7. Pimenta, M. M., R. J. Moffat, and W. M. Kays: Report HMT-21, Thermosciences Division, Department of Mechanical Engineering, Stanford University, May 1975.
8. Hollingsworth, D. K., W. M. Kays, and R. J. Moffat: Report HMT-41, Thermosciences Division, Department of Mechanical Engineering, Stanford University, September 1989.
9. Zukauskas, A., and A. Slanciauskas: *Heat Transfer in Turbulent Fluid Flows,* p. 131, Hemisphere, Washington, 1987.

10. Roganov, P. S., V. P. Zabolotsky, E. V. Shihov, and A. I. Leontiev: *Int. J. Heat Mass Transfer,* vol. 27, pp. 1251–1259, 1984.
11. Rogers, M. M., P. Moin, and W. C. Reynolds: Report TF-25, Thermosciences Division, Department of Mechanical Engineering, Stanford University, August 1986.
12. Kim, J., and P. Moin: *Proceedings of 6th Symposium on Turbulent Shear Flows, Toulouse, September 7–9, 1987,* pp. 5-2-1–5-2-6.
13. Deissler, R. G.: *Int. J. Heat Mass Transfer,* vol. 6, p. 257, 1963.
14. Bell, D.: NASA–Ames Laboratories, Sunnyvale, California, personal communication, 1992.
15. Yakhot, V., S. A. Orszag, and A. Yakhot: *Int. J. Heat Mass Transfer,* vol. 30, pp. 15–22, 1987.
16. Reynolds, A. J.: *Int. J. Heat Mass Transfer,* vol. 18, pp. 1055–1068, 1975.
17. Jenkins, R.: *Proceedings of Heat Transfer and Fluid Mechanics Institute, Stanford University,* p. 147, 1951.
18. Whitten, D. G., W. M. Kays, and R. J. Moffat: Report HMT-3, Thermosciences Division, Department of Mechanical Engineering, Stanford University, December 1967.
19. Ambrok, G. S.: *Sov. Phys. Tech. Phys.,* vol. 2, p. 1979, 1957.
20. Kearney, D. W., R. J. Moffat, and W. M. Kays: Report HMT-12, Thermosciences Division, Department of Mechanical Engineering, Stanford University, April 1970.
21. Kays, W. M., and R. J. Moffat: *Studies in Convection,* pp. 223–319, Academic Press, London.
22. Crawford, M. E., W. M. Kays, and R. J. Moffat: Report HMT-30, Thermosciences Division, Department of Mechanical Engineering, Stanford University, August 1979; also NASA Report CR-3219, January 1980.
23. Wieghardt, K.: AAF Translation F-TS-919-RE, August 1946.
24. Metzger, D. E., and D. D. Fletcher: *J. Aircraft,* vol. 8, pp. 33–38. 1971.
25. Goldstein, R. J.: *Advances in Heat Transfer,* vol. 7, pp. 321–379, Academic Press, New York, 1971.
26. Meyers, G. E., J. J. Schauer, and R. H. Eustis: *J. Heat Transfer,* vol. 85, pp. 209–214, 1963.
27. Dipprey, D. F., and D. H. Sabersky, *Int. J. Heat Mass Transfer,* vol. 6, pp. 329–353, 1963.
28. Blair, M. F.: *J. Heat Transfer,* vol. 105, pp. 33–40, 1983.
29. Hancock, P. E., and P. Bradshaw: *J. Heat Transfer,* vol. 105, pp. 284–289, 1983.
30. Maciejewski, P. K., and R. J. Moffat: Report HMT-42, Thermosciences Division, Department of Mechanical Engineering, Stanford University, 1989.

CHAPTER

14

HEAT TRANSFER: TURBULENT FLOW INSIDE TUBES

For a turbulent flow in smooth tubes we will consider, as nearly as possible, all the same boundary conditions as were covered in Chap. 9 for a laminar flow. By employing the same nomenclature and essentially the same differential equations, we can use many of the results of the laminar flow analyses. The turbulent transport mechanism and some techniques for calculating heat-transfer rates in a turbulent shear layer have been discussed at considerable length in Chaps. 11 and 13 in the framework of the turbulent external boundary layer. Some of the peculiarities of the tube-flow problem were discussed in Chap. 12. Thus there is little that is fundamentally new to be discussed in this chapter; we are concerned primarily with analytic and experimental results for particular boundary conditions of interest, using techniques previously developed. We will restrict consideration to constant fluid properties and negligible viscous dissipation; these latter effects will be considered in later chapters.

To find an appropriate energy differential equation for the case of a *circular tube*, let us go back to the equation for a steady laminar flow in a circular tube, Eq. (4-35). We will evaluate the enthalpy from $di = c\,dt$ and restrict the problem to axisymmetric heating $(\partial t/\partial \phi = 0)$ and

311

negligible axial conduction ($\partial^2 t / \partial x^2 \approx 0$). Thus

$$u\rho c \frac{\partial t}{\partial x} + v_r \rho c \frac{\partial t}{\partial r} - \frac{1}{r} \frac{\partial}{\partial r} \left(rk \frac{\partial t}{\partial r} \right) = 0$$

From this equation we simply *infer* the corresponding equation for a *turbulent* flow in a circular tube by comparison with Eq. (13-2), and at the same time we will restrict the problem to the case of constant fluid properties:

$$\bar{u} \frac{\partial \bar{t}}{\partial x} + \bar{v}_r \frac{\partial \bar{t}}{\partial r} - \frac{1}{r} \frac{\partial}{\partial r} \left[r(\alpha + \varepsilon_H) \frac{\partial \bar{t}}{\partial r} \right] = 0 \qquad (14\text{-}1)$$

We already have at our disposal all the necessary tools to solve this equation for any desired thermal boundary conditions and any velocity initial conditions. Either the Prandtl mixing-length or the k–ε methods of Chap. 11 offer a method for evaluating ε_M, or for the fully developed flow case Eq. (12-5) can be used directly. The sublayer region can be calculated using the Van Driest function, Eq. (11-25). The turbulent Prandtl number results of Chap. 13 should be equally applicable to flow in a tube and thus offer a method for evaluating ε_H from ε_M. Of course the same methods must be used to solve the corresponding momentum equation (12-1) in order to evaluate $\bar{u}$ and $\bar{v}_r$. All of this requires use of finite-difference methods and a digital computer. But, given these tools, the problem can be solved for the most complex set of boundary conditions.

However, solutions for certain particular boundary conditions can frequently be obtained by simpler methods or directly from experiments, and very often such solutions are both quite adequate and more convenient for engineering analysis. Furthermore, the study of such solutions can enhance understanding of the phenomena involved in a way that is generally not possible from numerical solutions.

It was noted in Chap. 12 that for a turbulent flow the velocity profile tends to become *fully developed* after an axial distance of 10–15 diameters, and that the behavior in the *entry region* is very dependent on the character of the tube entrance. Since the fully developed velocity region is often of much more importance than the entry region (and this is in sharp contrast to the behavior of a *laminar flow*), it is worthwhile taking advantage of the mathematical simplifications that obtain. For fully developed flow $\bar{v}_r = 0$, $\bar{u}$ is a function of r alone, and Eq. (14-1) reduces to

$$\frac{1}{r} \frac{\partial}{\partial r} \left[r(\alpha + \varepsilon_H) \frac{\partial \bar{t}}{\partial r} \right] = \bar{u} \frac{\partial \bar{t}}{\partial x} \qquad (14\text{-}2)$$

Equation (14-2), and the corresponding momentum equation, can now be solved, using the turbulence models discussed above, and finite-difference computations, for *any thermal boundary conditions*, so long as the velocity profile is assumed to be *fully developed* at the point where heat transfer starts.

One class of problem that can be solved using Eq.(14-2) is the *thermal-entry-length problem*, where the entry fluid temperature is uniform, but the velocity profile is already fully developed at the point where heat transfer starts. This problem can be solved by finite-difference methods, but Eq. (14-2) can also be solved by the classical method of separation of variables in much the same manner as for the corresponding laminar flow problem, the so-called *turbulent Graetz problem*. (This procedure involves a numerical solution, but it is a "one-time" solution). The Graetz problem is considered later in this chapter for two cases—where the surface temperature is constant, and where the surface heat flux is constant.

But first let us consider a further simplification of Eq. (14-2). Refer to the section in Chap. 9 where the *fully developed temperature profile* is discussed. For the two thermal boundary conditions, *constant surface temperature* and *constant heat rate* (constant surface heat flux), a fully developed temperature profile, in a nondimensional sense, obtains following an entry region; and again the axial distance required is usually not more than 10–15 diameters (exceptions to this rule are the very low-Prandtl-number fluids, the liquid metals). These two boundary conditions are commonly encountered in engineering applications, and the tubes are frequently very much longer than 10–15 diameters. Furthermore, it is shown in Chap. 9 that this situation leads to a heat-transfer coefficient h that is invariant with axial distance. *Heat-exchanger theory* is based on an assumption that h is independent of axial distance, since this is a commonly observed phenomenon. Thus the fully developed temperature-profile problem is worthy of study, and indeed more so for a turbulent flow than for a laminar flow, where entry lengths are often large.

CIRCULAR TUBE WITH FULLY DEVELOPED VELOCITY AND TEMPERATURE PROFILES, CONSTANT HEAT RATE, PRANDTL NUMBERS 0.6–6.0

This is the thermal boundary condition that occurs in a counterflow heat exchanger when the fluid capacity rates are the same, and it also occurs in a tube heated by electric resistance heating. It *can* occur as a result of nuclear heating or radiation from a high-temperature source. The axial temperature distribution is shown in Fig. 9-2.

When both the velocity and temperature profiles are fully developed, Eq. (9-8) is applicable and Eq. (14-2) becomes

$$\frac{1}{r}\frac{\partial}{\partial r}\left[r(\alpha + \varepsilon_H)\frac{\partial \bar{t}}{\partial r}\right] = \bar{u}\frac{dt_m}{dx} \tag{14-3}$$

Equation (14-3), and the associated momentum equation (12-3), are essentially ordinary differential equations, and are relatively easy to solve simultaneously by finite-difference methods using the mixing-length and eddy diffusivity methods of Chap. 12 and the turbulent Prandtl number, Eq. (13-7). Such calculations have been carried out for the Prandtl number range 0.7–5.9 using Eq. (13-7) but with the Prandtl number in that equation held at 0.7, because it was noted in Chap. 13 that the Pr_t distribution for water at $Pr = 5.9$ is virtually identical to that for air, $Pr = 0.7$. The resulting Nusselt numbers correspond closely to the generally accepted experimental data over this range of Prandtl number.

From these calculations it has then been possible to empirically fit a universal temperature profile in the same form as Eqs. (13-15) and (13-16), analogous to the Nikuradse equation (12-4) for velocity:

$$t^+ = 2.2 \ln y^+ + 13.39 \, Pr^{2/3} - 5.66 \tag{14-4}$$

Since this equation is based on turbulent Prandtl number data at just two points, $Pr = 0.7$ and 5.9, it may not be valid far outside of this Prandtl number range; the $\frac{2}{3}$ power of the Prandtl number is somewhat arbitrary, but this is a function that has long been found to correlate experimental data to very high Prandtl number. Thus Eq. (14-4) is proposed as a *thermal law of the wall* for flow in a circular tube over the Prandtl number range 0.6–6.0, but one that may be reasonably valid at much higher Prandtl numbers. Note that this range covers all of the gases, as well as many light liquids, including water. It is certainly not valid for the very low-Prandtl-number fluids, the liquid metals. Note also that Eq. (14-4) is not valid inside the viscous sublayer, but this is also true of Eq. (12-4). Furthermore, neither Eq. (12-4) nor Eq. (14-4) is precisely valid at the tube centerline (see the discussion of this point in Chap. 12), but the manner in which these equations are now going to be used minimizes this discrepancy.

The temperature profiles implied by Eq. (14-4) are plotted on Fig. 14-1 for a single value of Reynolds number and three values of the Prandtl number. Note that the effect of increasing Prandtl number is to give a more "square" temperature profile, while a low Prandtl number yields a more rounded profile that is similar to that for laminar flow. Still higher and lower values of Prandtl number continue these trends. Note the location of the major source of heat-transfer resistance. At high Prandtl numbers the resistance is increasingly in the sublayer, whereas at low Prandtl numbers it is distributed over the entire flow cross section.

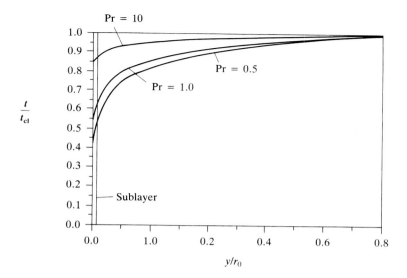

FIGURE 14-1
Effect of Prandtl number on turbulent flow temperature distribution; $Re = 100,000$.

The reasons for these differences can be seen in the "total conductance" term in Eq. (14-3), $\alpha + \varepsilon_H$. If this term is rendered nondimensional by dividing it by the kinematic viscosity v, it becomes

$$1/Pr + \varepsilon_H/v$$

The relative importance of turbulent eddy diffusion and molecular diffusion is seen to depend directly on the Prandtl number. Along any radius, the molecular-conduction term $1/Pr$ is constant, while the eddy-conduction term ε_H/v varies from a relatively large magnitude near the center of the tube to zero at the wall. The shape of the temperature profile and the whole character of turbulent flow heat transfer are determined by the relative magnitude of these two terms. At very low Prandtl numbers the conduction term dominates everywhere; and not only is the temperature profile similar to a laminar profile, but also the thermal-entry length and the response of the fluid to axial changes in surface temperature are both quite similar to laminar behavior. At higher Prandtl numbers we find very short thermal-entry lengths and quick response to axial surface-temperature changes, both attributable to the concentration of the thermal resistance close to the wall and the rapid diffusion of heat over the entire fluid once the sublayer has been penetrated.

The effect of increased Reynolds number is to decrease the thickness of the sublayer and increase the eddy diffusivity in the outer region, both of which lead toward more "square" temperature profiles with a still greater part of the resistance concentrated in the sublayer.

Let us now use Eq. (14-4) to develop an equation for the Nusselt number. Assuming that Eqs. (12-4) and (14-4) are valid at the tube centerline, we can write

$$u_\infty^+ = 2.5 \ln r_0^+ + 5.5$$

$$t_\infty^+ = 2.2 \ln r_0^+ + 13.39 \, \mathrm{Pr}^{2/3} - 5.66$$

We now combine these equations, eliminating the term $\ln r_0^+$, and substitute the definitions of u_∞^+ and t_∞^+. Then we assume, for simplicity, that

$$\frac{t_c - t_0}{t_m - t_0} = 1, \quad \frac{u_c}{V} = 1$$

where t_c and u_c are the centerline temperature and velocity, respectively, t_m is the mean temperature, and V is the mean velocity.

These ratios vary somewhat with both Re and Pr, but in any case are generally equal to or greater than 0.9. Then, if the definitions of c_f and St are introduced, we obtain

$$\frac{2.5\sqrt{c_f/2}}{\mathrm{St}} + 26.25 - 33.475 \, \mathrm{Pr}^{2/3} = \frac{2.2}{\sqrt{c_f/2}}$$

Solving for the Stanton number,

$$\mathrm{St} = \frac{c_f/2}{0.88 + 13.39(\mathrm{Pr}^{2/3} - 0.78)\sqrt{c_f/2}}$$

or, in Nusselt number form,

$$\mathrm{Nu} = \frac{\mathrm{Re} \, \mathrm{Pr} \, c_f/2}{0.88 + 13.39(\mathrm{Pr}^{2/3} - 0.78)\sqrt{c_f/2}} \tag{14-5}$$

(Note that in this chapter Re will always denote the Reynolds number based on the tube diameter.)

Equation (12-13) can be used for the friction coefficient, with the result

$$\mathrm{Nu} = \frac{0.023 \, \mathrm{Re}^{0.8} \, \mathrm{Pr}}{0.88 + 2.03(\mathrm{Pr}^{2/3} - 0.78) \, \mathrm{Re}^{-0.1}} \tag{14-6}$$

This expression is unnecessarily cumbersome, and in the Prandtl number range of *gases* (0.5–1.0) is approximated quite well by

$$\mathrm{Nu} = 0.022 \, \mathrm{Pr}^{0.5} \, \mathrm{Re}^{0.8} \tag{14-7}$$

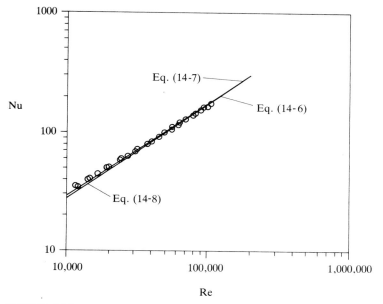

FIGURE 14-2
Comparison of analysis with experiment for fully developed velocity and temperature profiles in a circular tube, constant heat rate (data of Kays and Leung,[1] for air, $Pr = 0.702$).

A comparison of Eqs. (14-6) and (14-7) with some experimental data for air is shown in Fig. 14-2. The comparison is seen to be quite good, at least over the indicated range of Reynolds number.

CIRCULAR TUBE WITH FULLY DEVELOPED FLOW, HIGHER PRANDTL NUMBERS

A great many fluids of considerable technical interest have Prandtl numbers substantially higher than those considered in the preceding section. Oils, for example, frequently have Prandtl numbers in excess of 100, and for some very viscous fluids this figure may reach 1000. Essentially any fluid with $Pr > 10$ can be considered to have a "high" Prandtl number. The reason for treating these fluids separately lies in a fact pointed out in the preceding section: as the Prandtl number increases, the major part of the heat-transfer resistance appears closer and closer to the wall, i.e., the region where the temperature changes from its wall-surface value to its centerline value is very close to the wall. For Prandtl numbers greater than about 10 almost the entire temperature profile is inside the sublayer, and in fact most if it is inside $y^+ = 5.0$.

There is then a major difficulty in attempting to integrate the energy differential equation, because (1) it is not known how precisely the Van Driest equation (or any other turbulence theory) predicts eddy diffusivity close to the wall, and (2) little or nothing is known about the turbulent Prandtl number in this region. Although the eddy diffusivity is very small for $y^+ < 5$, it becomes increasingly important as the Prandtl number increases, as can be seen in the total conduction term $1/\text{Pr} + \varepsilon_M/\nu$. The Van Driest equation is perfectly adequate for the *momentum equation* because it is not necessary to know the eddy diffusivity close to the wall with high precision; however, such is not the case for high-Prandtl-number heat transfer. The uncertainties in turbulent Prandtl number near the wall were discussed in Chap. 13. It is suspected that it has a value near 1, but it has not yet been practicable to measure it.

One can carry out calculations using the turbulence models that we have discussed, and the results for the Nusselt number appear reasonable, but there is usually sufficient discrepancy from the experimental data that one would like to correct the model. But it is not possible to say whether the difficulty is in the model used for the eddy diffusivity or in that used for the turbulent Prandtl number.

For these reasons we will not attempt to calculate temperature profiles for high-Prandtl-number fluids, but will instead simply examime what appear to be the most reliable experimental correlations for the Nusselt number. Later we will introduce techniques for calculating thermal-entry lengths and dealing with problems where surface temperature and/or heat flux vary axially, and we will then be obliged to use functions calculated on the basis of assumed models, uncertain as they may be. But in that case any model is better than none, because there is simply no other way to handle that kind of problem.

One of the conclusions that can be reached using crude models is that for $\text{Pr} > 1$ the Nusselt number becomes virtually independent of any but very substantial axial variations of surface temperature. Thus there is virtually no difference between a Nusselt number determined from a constant-surface-temperature experiment and one determined from a constant-heat-rate experiment. As will be recalled, this was definitely not true for laminar flows, and it is likewise not true for very low-Prandtl-number fluids in turbulent flow. The experimental correlations that will now be described are thus applicable to either of the two fundamental thermal boundary conditions, and will usually be adequate for still other variations of surface temperature or heat flux.

One further point should be noted. Most high-Prandtl-number fluids are liquids with high viscosity. The viscosity is virtually always a strong function of temperature. Unless experiments are performed with very small temperature differences, it is always difficult to infer what the Nusselt number would be for the case of constant fluid properties. Since most experiments have been performed with substantial temperature differences, the experimental uncertainty associated with this problem probably accounts for the considerable scatter of data in the literature. (The problems of the influence of temperature-dependent fluid properties are discussed in greater detail in Chap. 15.)

Kakac, Shah, and Aung[2] have examined a great many correlations for fully developed turbulent flow in a circular tube, and conclude that the following equation attributable to Gnielinski[3] correlates the available data somewhat better than any other over the range of Prandtl number from 0.5 to 2000, and Reynolds number from 2300 to 5×10^6:

$$\mathrm{Nu} = \frac{(\mathrm{Re} - 1000)\, \mathrm{Pr}\, c_f/2}{1.0 + 12.7\sqrt{c_f/2}(\mathrm{Pr}^{2/3} - 1.0)} \tag{14-8}$$

Note that the *form* of this equation is virtually the same as Eq. (14-5). In fact, the only difference is what is apparently a low-Reynolds-number correction, so Eq. (14-8) has at least some theoretical foundation.

Sleicher and Rouse[4] suggest a somewhat more convenient empirical formulation that gives close to the same results:

$$\mathrm{Nu} = 5 + 0.015\, \mathrm{Re}^a\, \mathrm{Pr}^b \tag{14-9}$$

where

$$a = 0.88 - \frac{0.24}{4 + \mathrm{Pr}}, \quad b = 0.333 + 0.5e^{-0.6\,\mathrm{Pr}}$$

The range of validity of this equation is stated to be

$$0.1 < \mathrm{Pr} < 10^4, \quad 10^4 < \mathrm{Re} < 10^6$$

Equations (14-8) and (14-9) are stated to be valid for gases as well as high-Prandtl-number fluids. For air, $\mathrm{Pr} = 0.7$, Eq. (14-8) yields Nusselt numbers that are generally within 2 percent of Eqs. (14-6) and (14-7). A comparison of Eq. (14-8) with experiment and with Eqs. (14-6) and (14-7) is shown on Fig. 14-2. Equation (14-9) tends to be 3–4 percent low in this Prandtl number range.

Another equation that has long been popular is that of Dittus and Boelter[5] as interpreted by Kakac *et al.*:[2]

$$\mathrm{Nu} = \begin{cases} 0.024\, \mathrm{Re}^{0.8}\, \mathrm{Pr}^{0.4} & \text{for heating} \\ 0.026\, \mathrm{Re}^{0.8}\, \mathrm{Pr}^{0.4} & \text{for cooling} \end{cases}$$

Of course the Dittus–Boelter equation contains a condition regarding the direction of heat transfer (probably a variable-properties correction), but in any case it tends to overpredict the Nusselt number for gases by at least 20 percent, and to underpredict Nusselt number for the higher-Prandtl-number fluids by 7–10 percent. It is no longer recommended that this equation be used.

VERY LOW-PRANDTL-NUMBER HEAT TRANSFER, LIQUID METALS

The very low-Prandtl-number fluids are all liquid metals and have values of Prandtl number less than 0.1, ranging down to less than 0.001. The high thermal conductivity of these fluids accounts for the very low Prandtl number, since the viscosity and specific heat of liquid metals do not differ greatly from other common liquids. Also, the high thermal conductivity accounts for the unusual heat-transfer characteristics of liquid metals. These characteristics are related to the fact that often $\alpha \gg \varepsilon_H$, or nondimensionally $1/\text{Pr} \gg \varepsilon_H/\nu$; that is, molecular conduction is the dominant diffusion mechanism even in the fully turbulent part of the flow field. As $\text{Pr} \to 0$, turbulent eddy diffusion becomes unimportant. This means that a liquid metal has many of the heat-transfer characteristics of a laminar flow, while at the same time its momentum-transfer characteristics are no different than those of any other fluid. For example, fully developed temperature profiles tend to take on a parabolic shape, and this can be seen in Fig. 14-1, discussed earlier. The thermal-entry length tends to be large, similar to laminar flow, and axial variations in surface temperature and heat flux tend to have a substantial effect on the convective heat-transfer coefficient.

There is evidence that the turbulent Prandtl number Pr_t is considerably higher for a liquid metal than for other fluids, a fact discussed in Chap. 13 and that must be related to high molecular conductivity.

Although the Nusselt number tends to be low for a liquid metal, again like laminar flow, the heat-transfer coefficient tends to be very high because of the high molecular conductivity. In fact, this is one of the main reasons why liquid metals are of considerable interest as heat-transfer fluid media. High heat-transfer coefficients, however, mean that wall-surface contamination can have a large influence on the effective heat-transfer coefficient, and surface wetting problems can be similarly important. These facts have led to difficulties and large uncertainties in experiments. The results discussed below are *believed* to represent the behavior of a clean system with complete wetting of the surface, but some uncertainty does persist.

Finally it should be noted that *longitudinal conduction* can be an important factor in liquid-metal heat transfer. It will be recalled that the term in the differential energy equation that accounts for longitudinal conduction, $\partial^2 \bar{t}/\partial x^2$, has been omitted in all of the equations that we have used. As a simple rule of thumb, longitudinal conduction may be a factor resulting in the lowering of the effective heat-transfer coefficient if $\text{Re Pr} < 100$.

Any of Eqs. (14-1)–(14-3) can be solved by computer-aided finite-difference calculations using the turbulent transport models that we

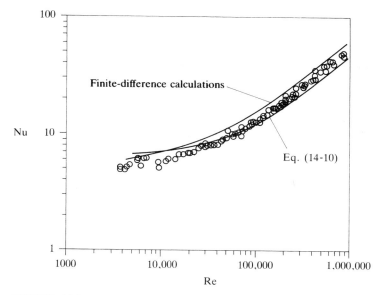

FIGURE 14-3

A comparison of the data of Skupinski *et al.*[6] with Eq. (14-10) and with the results of direct integration of the energy equation. These data were obtained using an NaK mixture having a Prandtl number of 0.0153.

have discussed. The only real uncertainty lies in the values of Pr_t used. Quite reasonable results for the Nusselt number are obtained using Eq. (13-7) for Pr_t, but, as discussed in Chap. 13, there is little direct information available on Pr_t.

An empirical equation that corresponds fairly well to the available experimental data is proposed by Sleicher and Rouse:[4]

$$\text{Nu} = 6.3 + 0.0167\,\text{Re}^{0.85}\,\text{Pr}^{0.93} \quad \text{(constant heat rate)} \quad (14\text{-}10)$$

In Fig. 14-3 the data of Skupinski, Tortel, and Vautrey[6] for a liquid metal with $Pr = 0.0153$ and constant heat rate per unit of tube length are compared with Eq. (14-10) and with finite-difference calculations using Eq. (13-7) for Pr_t. The fact that the finite-difference calculations lie consistently about 15 percent above the data over the entire range of Reynolds number suggests that the discrepancy lies in the data rather than in the turbulent Prandtl number used in the calculations, because at the low-Reynolds-number end of the data the turbulent Prandtl number has little influence.

We now have equations covering the entire Prandtl number spectrum for *fully developed constant heat rate in a circular tube.* For

TABLE 14-1
Nusselt number for fully developed turbulent flow in a circular tube with constant heat rate

Pr	Laminar	Re 10,000	30,000	100,000	300,000	1,000,000
0	4.364	6.30	6.30	6.30	6.30	6.30
0.001		6.37	6.47	6.78	7.53	9.71
0.003		6.49	6.78	7.64	9.70	15.77
0.01		6.88	7.77	10.40	16.73	35.32
0.03		7.91	10.39	17.69	35.27	86.92
0.5		22.91	56.90	146.99	345.90	883.02
0.7		27.42	69.51	182.99	436.91	1,130.64
1		32.81	84.86	227.70	552.05	1,449.75
3		53.82	146.10	411.93	1,041.71	2,855.87
10		86.45	242.16	707.11	1,846.82	5,246.01
30		128.74	366.12	1088.07	2,890.81	8,375.94
100		195.53	560.62	1682.54	4,514.18	13,233.20
1000		425.59	1226.81	3706.15	10,009.84	29,581.97

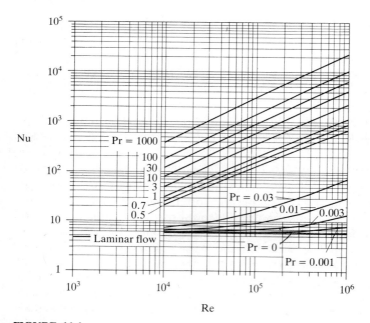

FIGURE 14-4
Nusselt number for fully dveloped velocity and temperature profiles; circular tube, constant heat rate.

convenience, results from Eqs. (14-8) and (14-10) are presented in Table 14-1 and Fig. 14-4.

Note that as the Prandtl number approaches very low values, the Nusselt number approaches a constant value. Lowering the Reynolds number has a similar effect. In either case the influence of the molecular-conduction mechanism is being observed; and, of course, for fully developed *laminar* flow the Nusselt number is indeed a constant.

CIRCULAR TUBE, FULLY DEVELOPED PROFILES, CONSTANT SURFACE TEMPERATURE

Following the pattern used in Chap. 9 for laminar flow in a tube, we have just considered the case of fully developed constant heat rate per unit of tube length. As was the case for laminar flow, another useful boundary condition for which a fully developed temperature profile is physically possible is that of a *constant surface temperature*. This condition is very frequently encountered in heat exchangers where one fluid capacity rate is very much greater than the other or when one fluid is boiling or condensing.

The applicable energy differential equation is derivable from Eq. (14-2) by comparison with Eq. (9-19):

$$\frac{1}{r}\frac{\partial}{\partial r}\left[r(\alpha + \varepsilon_H)\frac{\partial \bar{t}}{\partial r}\right] = \bar{u}\frac{t_0 - \bar{t}}{t_0 - t_m}\frac{dt_m}{dx} \qquad (14\text{-}11)$$

This equation can be integrated numerically using the same turbulence models that have been used previously, although an iterative technique will have to be employed.

Alternatively, the fully developed constant-surface-temperature solutions may be obtained as special cases of the thermal-entry-length solutions to be described later, i.e., solutions to Eq. (14-2) for a uniform surface temperature and a uniform fluid temperature at the tube entrance. Sleicher and Tribus[7] were the first to present such a solution, and although the values of the turbulent Prandtl number used were higher than what is now believed to be correct at very low Prandtl numbers, this does not have a large effect on the particular results of interest here. The fully developed Nusselt numbers for both *constant surface temperature* and *constant heat rate* can be deduced from the Sleicher and Tribus results, and thus the ratio of Nusselt numbers for the two cases can be determined. These are presented in Fig. 14-5, where subscript Ⓗ refers to constant heat rate and Ⓣ to constant surface temperature.

Note that only at very low Prandtl number is there a significant difference between the constant-heat-rate and constant-surface-

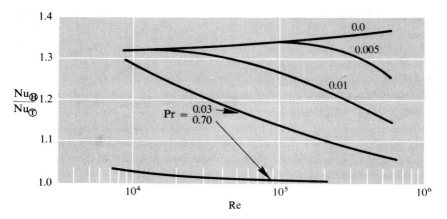

FIGURE 14-5
Ratio of Nusselt number for *constant heat rate* to Nusselt number for *constant surface temperature* for fully developed conditions in a circular tube (Sleicher and Tribus).[7]

temperature results. At $Pr = 0.7$ the difference is only a few percent, and at higher Prandtl numbers the difference is even less significant. In the liquid-metal region (very low Prandtl number), however, the difference is substantial and actually in some cases becomes larger than the laminar flow difference [see Eqs. (9-17) and (9-21)]. The reasons lie again in the effect of Prandtl number on the distribution of thermal resistance. At very low Prandtl numbers, where the heat-transfer mechanism is predominantly molecular diffusion, the resistance is distributed over the entire flow cross section, and the different boundary conditions yield differently shaped temperature profiles. At high Prandtl numbers the resistance is primarily very close to the wall, yielding a quite "square" temperature profile regardless of the heating boundary condition. This fact suggests also that the Nusselt number for a high-Prandtl-number fluid is relatively insensitive to a variation of surface and/or heat flux in the direction of flow, whereas the Nusselt number for a low-Prandtl-number fluid is greatly affected by axial variation of surface temperature and/or heat flux. Such is indeed the case, as is shown later.

Figure 14-5 can now be used in conjunction with the data in Table 14-1, or Fig. 14-4, to completely establish the Nusselt numbers for fully developed velocity and temperature profiles for *constant surface temperature* over a wide range of Reynolds and Prandtl numbers.

Algebraic empirical correlations can also be developed for fully developed *constant surface temperature*. For *gases* Eq. (14-7) can be modified by lowering the coefficient a few percent to yield

$$Nu = 0.021 \, Pr^{0.5} \, Re^{0.8} \qquad (14\text{-}12)$$

For the very low Prandtl-number fluids Sleicher and Rouse[4] suggest

$$\text{Nu} = 4.8 + 0.0156\,\text{Re}^{0.85}\,\text{Pr}^{0.93} \qquad (14\text{-}13)$$

EFFECT OF PERIPHERAL HEAT-FLUX VARIATION

In the examples previously considered the heat flux and surface temperature have both been assumed to be uniform around the tube periphery. In many applications the heat flux and surface temperature are not uniform around the tube, and thus it becomes important to know how the heat-transfer coefficient varies around the tube.

This problem is discussed in Chap. 9 for laminar flow. The corresponding turbulent flow problem has been considered and treated in a similar manner by Gartner, Johannsen, and Ramm.[8] The energy equation for the case of fully developed constant heat rate per unit of tube length, but with nonaxisymmetric heating, becomes

$$\frac{1}{r}\frac{\partial}{\partial r}\left[r(\varepsilon_{Hr} + \alpha)\frac{\partial \bar{t}}{\partial r}\right] + \frac{1}{r^2}\frac{\partial}{\partial \phi}\left[(\varepsilon_{H\phi} + \alpha)\frac{\partial \bar{t}}{\partial \phi}\right] = \bar{u}\frac{dt_m}{dx} \qquad (14\text{-}14)$$

where ε_{Hr} and $\varepsilon_{H\phi}$ are the eddy diffusivities for heat transfer in the radial and circumferential directions, respectively. There is experimental evidence that $\varepsilon_{H\phi}$ is greater than ε_{Hr}, especially near the wall, and this fact was taken into consideration in the development of a solution. A theoretical solution for the diffusivities was used, which compares reasonably with the experimental data, and also with the radial diffusivity and turbulent Prandtl number data used in this book.

In Ref. 8, Eq. (14-14) has been solved for any arbitrary heat-flux variation that can be expressed by a Fourier expansion. The complete solution for an arbitrarily prescribed heat flux around the periphery requires an extensive tabulation of functions. However, the simpler case of a *cosine* heat-flux variation may be presented very briefly, and since the thermal conductivity of the tube wall usually tends to smooth out peripheral temperature variations to a certain extent, most applications can at least be approximated by the simple cosine case.

Let the prescribed heat flux be

$$\dot{q}_0''(\phi) = \dot{q}_0''(1 + b\cos\phi) \qquad (14\text{-}15)$$

Then the Nusselt number around the periphery is

$$\text{Nu}(\phi) = \frac{1 + b\cos\phi}{1/\text{Nu} + \frac{1}{2}S_1 b\cos\phi} \qquad (14\text{-}16)$$

The functions S_1 are shown in Table 14-2 for a wide range of Reynolds and Prandtl numbers. The Nusselt number Nu in Eq. (14-16) is

TABLE 14-2
Circumferential heat-flux functions S_1 for fully developed turbulent flow in a circular tube with constant heat rate

	Re					
Pr	Laminar	10^4	3×10^4	10^5	3×10^5	10^6
0	1.000	1.000	1.000	1.000	1.000	1.000
0.001		0.9989	0.9937	0.9561	0.8059	0.4853
0.003		0.9929	0.9613	0.8005	0.5042	0.2300
0.01		0.9499	0.7915	0.4567	0.2185	0.0867
0.03		0.7794	0.4705	0.2055	0.0888	0.0336
0.7		0.1116	0.0442	0.0161	0.00644	0.00232
3		0.0440	0.0168	0.00594	0.00234	0.000824
10		0.0233	0.00879	0.00305	0.00120	0.000415
30		0.0145	0.00544	0.00184	0.000744	0.000250
100		0.00941	0.00354	0.00110	0.000484	0.000149

that for uniform peripheral temperature and heat flux as given in Table 14-1.

As an example, consider $\text{Pr} = 0.7$ and $\text{Re} = 100{,}000$, with $b = 0.20$. From Table 14-1, $\text{Nu} = 183$; and from Table 14-2, $S_1 = 0.0161$. Substituting in Eq. (14-16), we find that the local Nusselt number will vary from 170 to 208. Thus the effect is rather pronounced even for only a 20 percent variation in heat flux. As might be expected, the effects are greater than this at low Prandtl numbers and less at high Prandtl numbers.

FULLY DEVELOPED TURBULENT FLOW BETWEEN PARALLEL PLANES AND IN CONCENTRIC CIRCULAR-TUBE ANNULI

The annulus geometry and its limiting case, flow between parallel planes, provide a particularly interesting heat-transfer problem because of the possibility of asymmetric heating from the two surfaces. Most applications do in fact involve asymmetry. The laminar flow annulus problem is discussed in Chap. 9, where the method of solution is outlined and the use of superposition for asymmetric heating is also treated.

The turbulent flow annulus problem can in principle be solved by the same superposition method. However, finding a satisfactory turbulence model presents new difficulties. A mixing-length model would be satisfactory near each of the two wall surfaces, but one has no basis for specifying the mixing length near the center of the passage, and this was also true for the simple circular tube. Presumably the $k-\varepsilon$ model might be satisfactory, but most of the analytic work on turbulent flow in the circular-tube annulus was carried out before the $k-\varepsilon$ model was invented.

Kays and Leung[1] carried out extensive calculations based on an empirical eddy-diffusivity equation, which was in turn based on empirical information on the ratio of the shear stresses at the two wall surfaces (which are, of course, different). It can readily be shown by a simple force balance on a fluid control volume that the point of zero shear stress in the flow, which is also the point of maximum velocity, is directly related to the surface shear-stress ratio. Measurements of the point of maximum velocity resulted in the following relation for the radius of maximum velocity (zero shear stress), which was found to be independent of Reynolds number:

$$\frac{r_m}{r_o} = r^{*0.343}(1 + r^{*0.657} - r^*) \tag{14-17}$$

where $r^* = r_i/r_o$ and r_m is the radius of maximum velocity.

With this information, a complete velocity profile could be constructed assuming that Eq. (12-4) is valid when applied separately starting from each surface. The sublayer was modeled using an equation suggested by Deissler.[9] Finally an assumption was made that the eddy diffusivity in the region near the maximum-velocity point could be approximated by a linear connection between the maximum eddy diffusivities on either side of it, as determined from the velocity profile and the shear-stress distribution. It is necessary to make some kind of assumption about the central-region eddy diffusivity because asymmetric heating involves heat transfer across that region. Fortunately it is not a critical assumption.

Calculations were carried out over a wide range of Reynolds and Prandtl numbers for the case of one surface heated and the other insulated, and are presented in the same form as the analogous laminar annulus solutions. Thus Eqs. (9-28) and (9-29) are directly applicable, and any heat-flux ratio on the two surfaces can be solved. The results of these calculations are presented in Tables 14-3 to 14-5. They are in excellent agreement with experiments performed by Kays and Leung for air ($\text{Pr} = 0.7$), including asymmetric heating. At high and low Prandtl numbers the same uncertainty regarding Pr_t as was encountered for the circular tube still obtains. At very low Prandtl numbers the equation used for Pr_t yields values that are somewhat higher than are now believed to be correct, so the results for the Nusselt number may be somewhat low. At $\text{Pr} \geqslant 0.7$, $\text{Pr}_t = 0.9$ was used.

Note that the Nusselt and Reynolds numbers for the circular-tube annulus are based on the *hydraulic diameter*, which is defined as $4 \times \text{area}/(\text{perimeter})$. For flow between parallel planes the *hydraulic diameter* is then $2 \times \text{spacing}$.

TABLE 14-3
Nusselt numbers and influence coefficients for $r^* = 0.20$; fully developed turbulent flow in a circular-tube annulus; constant heat rate

	Laminar		Re									
			10^4		3×10^4		10^5		3×10^5		10^6	
Pr	Nu_{ii}	θ_i^*	Nu_{ii}	θ_i^*	Nu_{ii}	θ_i^*	Nu_{ii}	θ_i^*	Nu_{ii}	θ_i^*	Nu_{ii}	θ_i^*
					Inner wall heated							
0.0	8.50	0.905	8.40	1.01	8.30	1.03	8.30	1.02	8.30	1.04	8.30	1.02
0.001			8.40	1.01	8.40	1.04	8.30	1.02	8.40	1.01	8.90	0.978
0.003			8.40	1.01	8.40	1.03	8.50	1.03	9.05	0.980	12.5	0.833
0.01			8.50	1.00	8.60	1.02	9.70	0.944	14.0	0.796	33.6	0.747
0.03			9.00	1.01	10.1	0.943	15.8	0.771	31.7	0.600	81.0	0.375
0.5			31.2	0.520	64.0	0.397	157	0.333	370	0.295	980	0.262
0.7			38.6	0.412	79.8	0.338	196	0.286	473	0.260	1270	0.235
1			46.8	0.339	99.0	0.284	247	0.248	600	0.228	1640	0.209
3			77.4	0.172	175	0.151	465	0.143	1150	0.137	3250	0.136
10			120	0.077	290	0.074	800	0.072	2050	0.073	6000	0.077
30			172	0.036	428	0.035	1210	0.035	3150	0.036	9300	0.038
100			243	0.014	617	0.014	1760	0.015	4630	0.016	13,800	0.016
1000			448	0.004	1140	0.002	3280	0.002	8800	0.004	26,000	0.003

Pr	Nu$_{oo}$	θ_o^*	Nu$_{oo}$	θ_o^*	Nu$_{oo}$	θ_o^*	Nu$_{oo}$	θ_o^*	Nu$_{oo}$	θ_o^*	Nu$_{oo}$	θ_o^*
					Outer wall heated							
0.0	4.88	0.104	5.83	0.140	5.92	0.146	6.10	0.151	6.16	0.152		
0.001			5.83	0.140	5.92	0.144	6.10	0.151	6.30	0.154	6.35	0.157
0.003			5.83	0.140	6.00	0.146	6.22	0.150	6.90	0.150	6.92	0.153
0.01			5.95	0.141	6.20	0.146	7.40	0.144	11.4	0.131	10.2	0.135
0.03			6.22	0.140	7.55	0.140	12.7	0.125	26.3	0.098	24.6	0.089
0.5			22.5	0.071	51.5	0.064	130	0.055	310	0.049	80.0	0.086
0.7			29.4	0.063	64.3	0.055	165	0.049	397	0.044	823	0.044
1			35.5	0.051	80.0	0.046	206	0.042	504	0.039	1070	0.040
3			60.0	0.026	145	0.026	390	0.024	980	0.024	1390	0.035
10			98.0	0.013	243	0.013	680	0.012	1750	0.012	2760	0.023
30			142	0.006	360	0.006	1030	0.006	2700	0.006	4980	0.012
100			205	0.003	520	0.003	1500	0.003	4000	0.003	7850	0.006
1000			380	0.001	980	0.001	2830	0.001	7500	0.001	12,000	0.003
											22,500	0.001

TABLE 14-4
Nusselt numbers and influence coefficients for $r^* = 0.50$; fully developed turbulent flow in a circular-tube annulus; constant heat rate

	Laminar		10^4		3×10^4		10^5		3×10^5		10^6	
Pr	Nu_{ii}	θ_i^*	Nu_{ii}	θ_i^*	Nu_{ii}	θ_i^*	Nu_{ii}	θ_i^*	Nu_{ii}	θ_i^*	Nu_{ii}	θ_i^*
					Inner wall heated							
0.0	6.18	0.528	6.28	0.622	6.30	0.633	6.30	0.652	6.30	0.659	6.30	0.653
0.001			6.28	0.622	6.30	0.633	6.30	0.652	6.40	0.659	6.75	0.643
0.003			6.28	0.622	6.30	0.633	6.40	0.657	6.85	0.638	9.40	0.585
0.01			6.37	0.623	6.45	0.637	7.30	0.623	10.8	0.540	23.2	0.427
0.03			6.75	0.627	7.53	0.598	12.0	0.533	24.9	0.432	65.5	0.333
0.5			24.6	0.343	52.0	0.293	130	0.253	310	0.229	835	0.208
0.7			30.9	0.300	66.0	0.258	166	0.225	400	0.206	1080	0.185
1.0			38.2	0.247	83.5	0.218	212	0.208	520	0.182	1420	0.170
3.0			66.8	0.129	152	0.121	402	0.115	1010	0.114	2870	0.111
10.0			106	0.059	260	0.059	715	0.059	1850	0.059	5400	0.061
30.0			153	0.028	386	0.028	1080	0.028	2850	0.031	8400	0.032
100.0			220	0.006	558	0.006	1600	0.006	4250	0.007	12,600	0.007
1000.0			408	0.002	1040	0.002	3000	0.002	8000	0.002	24,000	0.002

Pr	Nu_{oo}	θ_o^*	Nu_{oo}	θ_o^*	Nu_{oo}	θ_o^*	Nu_{oo}	θ_o^*	Nu_{oo}	θ_o^*	Nu_{oo}	θ_o^*
					Outer wall heated							
0.0	5.04	0.216	5.66	0.281	5.78	0.294	5.80	0.297	5.83	0.303	5.95	0.310
0.001			5.66	0.281	5.78	0.294	5.80	0.297	5.92	0.303	6.40	0.304
0.003			5.66	0.281	5.78	0.294	5.85	0.294	6.45	0.301	9.00	0.278
0.01			5.73	0.281	5.88	0.289	6.80	0.289	10.3	0.264	22.6	0.217
0.03			6.03	0.279	7.05	0.284	11.6	0.258	24.4	0.214	64.0	0.163
0.5			22.6	0.162	49.8	0.142	125	0.123	298	0.111	795	0.098
0.7			28.3	0.137	62.0	0.119	158	0.107	380	0.097	1040	0.090
1.0			34.8	0.111	78.0	0.101	200	0.092	490	0.085	1340	0.078
3.0			60.5	0.059	144	0.058	384	0.055	960	0.054	2730	0.052
10.0			100	0.028	246	0.028	680	0.028	1750	0.028	5030	0.028
30.0			143	0.013	365	0.013	1030	0.014	2700	0.014	8000	0.015
100.0			207	0.006	530	0.006	1500	0.006	4000	0.006	12,000	0.006
1000.0			387	0.001	990	0.001	2830	0.001	7600	0.001	23,000	0.001

TABLE 14-5
Nusselt numbers and influence coefficients for fully developed turbulent flow between parallel plates; constant heat rate; one side heated and the other insulated

								Re					
	Laminar		10^4		3×10^4		10^5		3×10^5		10^6	
Pr	Nu	θ^*	Nu	θ^*	Nu	θ^*	Nu	θ^*	Nu	θ^*	Nu	θ^*
0.0	5.385	0.346	5.70	0.428	5.78	0.445	5.80	0.456	5.80	0.460	5.80	0.468
0.001			5.70	0.428	5.78	0.445	5.80	0.456	5.88	0.460	6.23	0.460
0.003			5.70	0.428	5.80	0.445	5.90	0.450	6.32	0.450	8.62	0.422
0.01			5.80	0.428	5.92	0.445	6.70	0.440	9.80	0.407	21.5	0.333
0.03			6.10	0.428	6.90	0.428	11.0	0.390	23.0	0.330	61.2	0.255
0.5			22.5	0.256	47.8	0.222	120	0.193	290	0.174	780	0.157
0.7			27.8	0.220	61.2	0.192	155	0.170	378	0.156	1030	0.142
1.0			35.0	0.182	76.8	0.162	197	0.148	486	0.138	1340	0.128
3.0			60.8	0.095	142	0.092	380	0.089	966	0.087	2700	0.084
10.0			101	0.045	214	0.045	680	0.045	1760	0.045	5080	0.046
30.0			147	0.021	367	0.022	1030	0.022	2720	0.023	8000	0.024
100.0			210	0.009	514	0.009	1520	0.010	4030	0.010	12,000	0.011
1000.0			390	0.002	997	0.002	2880	0.002	7650	0.002	23,000	0.002

TABLE 14-6
Effect of eccentricity on turbulent flow heat transfer in circular-tube annuli (experimental data for air, Leung et al.[11])

Radius ratio	e^*	$\dfrac{\mathrm{Nu}_{ii,\max}}{\mathrm{Nu}_{ii,\mathrm{conc}}}$	$\dfrac{\mathrm{Nu}_{ii,\min}}{\mathrm{Nu}_{ii,\mathrm{conc}}}$	$\dfrac{\mathrm{Nu}_{oo,\max}}{\mathrm{Nu}_{oo,\mathrm{conc}}}$	$\dfrac{\mathrm{Nu}_{oo,\min}}{\mathrm{Nu}_{oo,\mathrm{conc}}}$
0.255	0.27	0.99	0.97	1.02	0.93
	0.50	0.94	0.92	0.98	0.86
	0.77	0.92	0.88	0.93	0.77
0.500	0.54	0.96	0.87	1.01	0.78
	0.77	0.87	0.67	0.88	0.62

$e^* = e/(r_o - r_i)$, where e is the eccentricity of the tube centerlines.
$\mathrm{Nu}_{ii,\mathrm{conc}}$ and $\mathrm{Nu}_{oo,\mathrm{conc}}$ refer to the Nusselt numbers for the concentric annulus at the same Reynolds number and Prandtl number.

A variation of heat flux around the periphery of either or both surfaces of an annular passage can have a larger effect than in the simple circular tube. Employing the method of Gartner et al.[8] and using the velocity and diffusivity data of Kays and Leung,[1] Sutherland and Kays[10] present the necessary functions to solve for any arbitrary variation of heat flux on both surfaces.

Some degree of eccentricity must usually be tolerated in an annular passage. Leung, Kays, and Reynolds[11] present experimental data on the effects of eccentricity. In general, the effects of eccentricity are (1) to decrease the average Nusselt number and (2) to yield a Nusselt number variation around the periphery. Heat conduction in the tube wall in the peripheral direction can alter this variation, but these experiments were carried out with negligible peripheral conduction. Table 14-6 presents an abstract of these results for two radius ratios and for various eccentricities, for the inner surface alone heated and for the outer surface alone heated, in the Reynolds number range 30,000–80,000. Further information will be found in Kakac et al.[2]

FULLY DEVELOPED TURBULENT FLOW IN OTHER TUBE GEOMETRIES

As long as Pr is greater than about 0.5, the greater portion of the heat-transfer resistance in tube flow is in the sublayer region near the wall surfaces, and if these surfaces are at a nearly uniform temperature then the temperature distribution over most of the flow cross-sectional flow area is relatively flat. Under these circumstances, one would expect that the heat-transfer rate from the surface would be reasonably independent of the cross-sectional shape of the tube. In other words, the computed behavior for circular tubes should be applicable to tubes of any cross-sectional shape. There remains, however, the problem of finding a characteristic dimension of a noncircular tube that has the same significance as the *diameter* of the circular tube.

An examination of the Reynolds number from a different point of view provides a suggestion. Consider a circular tube and let p be the tube surface perimeter and A_c the cross-sectional area. Then

$$\text{Re} = \frac{DG}{\mu} = \frac{4A_c G}{p\mu} = \frac{4\dot{m}}{p\mu}$$

where $G = V\rho$. The Reynolds number is seen to be proportional to the ratio of the mass flow rate to the tube perimeter. If the heat-transfer resistance is primarily in the region close to the wall, it would seem reasonable that this ratio is more significant than the fact that the tube is circular and has diameter D. If this is the case, it should be possible to replace D in the circular-tube solutions and experimental correlations by $4A_c/p$; and since the latter is unambiguously defined for any cross-sectional shape, the results should be applicable to tubes of any cross-sectional shape. $4A_c/p$ is the hydraulic diameter D_h, or four times the hydraulic radius r_h:

$$D_h = 4r_h = \frac{4A_c}{p} \tag{14-18}$$

(See also a similar discussion in Chap. 12; note that this definition of D_h is identical to Eq. (12-16).)

Experimentally it is found that with this substitution for D the circular-tube results are indeed applicable quite accurately to many tubes of noncircular cross-section. It was noted in Chap. 12 that this is also the case for the friction coefficient. At very low Prandtl numbers this approximation cannot be expected to apply, because the heat-transfer resistance is no longer confined to the region near the wall surface. This is also true in laminar flow; for laminar flow in tubes the use of the hydraulic diameter does not correlate for different geometries. This approximation can also be expected to break down for tubes with very sharp corners, such as triangular tubes with one corner angle very small, because in the vicinity of the sharp corner the sublayer becomes large relative to the distance between the adjacent sides. It also does not apply for tubes with disconnected walls where the temperature is different on the different walls. With these exceptions, the approximation is very good, and the circular-tube solutions are applicable to rectangular tubes and any number of unusual shapes.

In this discussion the surface temperature is assumed everywhere uniform around the periphery, and the heat-transfer coefficient calculated is the mean coefficient around the periphery. The local coefficient around the periphery of a noncircular tube may vary considerably, but the hydraulic-diameter approximation sheds no light on this problem.

There is a considerable body of information in the heat-transfer literature, both analytic and experimental, on flow in noncircular tubes, including investigations of local coefficients around the periphery, non-uniform heating around the periphery, and many other variations. It is not practicable to summarize all of these investigations here. A comprehensive and excellent summary may be found in the *Handbook of Single-Phase Convective Heat Transfer* by Kakac et al.[2]

EXPERIMENTAL CORRELATIONS FOR FLOW IN TUBES

Until now, our major emphasis has been on analytic solutions to the energy equation, and experimental results have only been referred to when they serve to validate the assumptions used in building a mathematical model of the heat-transfer process, or where uncertainty in the information going into the mathematical models have been such that experimental data is probably more reliable. We encountered this kind of uncertainty when attempting to calculate heat transfer at very high and very low Prandtl numbers. However, experiments can and do form a primary source of convective heat-transfer data for engineering applications; if the flow geometry is complex, it is often far easier to perform experiments than to attempt to deduce heat-transfer rates by analysis. In fact, in many cases experiment is the only practicable approach.

A by-product of the analytic approach, in which we started with the applicable differential equations, is that we can now readily see how experimental results can be generalized by the use of nondimensional variables, so that it is not necessary to carry out experiments under the operating conditions of the proposed application or even with the same fluid. For example, in the absence of thermal- or hydrodynamic-entry effects, we have seen that for flow in a tube under, say, constant-heat-rate conditions, the heat-transfer performance can be expressed by a relation of the type

$$\text{Nu or St} = f(\text{Re}, \text{Pr})$$

Experiments can be used to establish this functional relationship; and, since the variables are nondimensional, it does not matter if the dimensional variables making up the nondimensional groups are not the same in the application as in the experiment. The length dimension in the Reynolds number (and Nusselt number) can be any convenient length that characterizes the size of the passage (including the hydraulic diameter, if desired, but this is not mandatory). Since the experimental results are strictly applicable to geometrically similar systems only, all length dimensions bear the same relation to one another.

Frequently it is not possible or practicable to vary the Prandtl number over a wide range in a single experimental system in order to adequately establish the functional dependence on Prandtl number. For this reason experimenters often *assume* a Prandtl number dependence, based on experience with other geometries or on the analytic solutions. Thus one often finds experimental results presented in the form

$$\text{St Pr}^{0.6} = f(\text{Re}) \quad \text{or} \quad \text{Nu Pr}^{-0.4} = f(\text{Re})$$

(It should be noted that, regardless of the length variable in Nu and Re the following relation must apply: $\text{Nu} = \text{St Pr Re}$.)

Other powers of the Prandtl number are also frequently used, depending either on what is necessary to correlate the results or on some theoretical basis. For example, in some high-performance heat-exchanger flow passages, the wall surface is interrupted at frequent intervals so that each segment of surface sees essentially a laminar external boundary layer. In Chap. 10 it was seen that for a laminar boundary layer the Prandtl number dependence can be quite accurately represented by the group $\text{St Pr}^{2/3}$ in the Prandtl number range 0.5–15. Thus there is a *theoretical* basis for using this group to generalize experimental data for such flow passages.

Obviously if there are other significant variables in the experiment, these can be included in the function, provided only that they are normalized with respect to some other variable (or variables) so as to provide an additional nondimensional group. For example, if the thermal- or hydrodynamic-entry length is significant (and this is discussed in sections to follow), the distance x from the entrance becomes significant, and an additional nondimensional parameter could be x/D. But it could just as well be xG/μ.

Such experimental correlations may be found throughout the literature for both laminar and turbulent flow in a large variety of types of flow passages and in different Prandtl number ranges. It is the intent of this book to place primary emphasis on the available analytic solutions, for it is felt that these should be thoroughly understood before the purely experimental data can be used intelligently. However, it is recognized that in engineering applications the reader will generally find it absolutely necessary to take advantage of the wealth of experimental data that exists. Typical of such available data is the compilation in *Compact Heat Exchangers*,[12] some samples of which are presented in Chap. 19. Other useful sources of data are the *Handbook of Single-Phase Convective Heat Transfer*,[2] and the *Handbook of Heat Transfer*.[13]

THERMAL-ENTRY LENGTH FOR TURBULENT FLOW IN A CIRCULAR TUBE

Consider the case of a fully developed velocity *profile*, but with a uniform fluid temperature at the point where heat transfer begins. We are concerned with the development of the temperature profile, that is, the *thermal-entry length*.

This problem can be readily solved by direct numerical integration of the energy differential equation, using the turbulence models previously discussed, for any particular problem. But it is also possible to obtain solutions in more generalized form, and this is often useful because it leads to closed-form results for certain particular cases, and provides a technique for investigating certain effects that would involve multiple direct numerical solutions.

The differential equation that must be solved for the circular tube is again Eq. (14-2); but, of course, the derivative on the right-hand side is no longer a constant or a function of x alone, as was the case for a fully developed temperature profile. The problem is essentially the same as the corresponding laminar flow problem. Equation (14-2) can be expressed in the same (x^+, r^+) coordinate system, but now, since $\varepsilon_H + \alpha$ is a function of both Reynolds number and Prandtl number, rather than a constant, it is apparent that for given boundary conditions a family of solutions is obtained rather than a single solution.

Circular-tube solutions have been obtained for both *constant surface temperature* and *constant heat rate* by Notter and Sleicher.[14] These solutions follow the same procedures as were employed for the laminar flow counterpart to the problem. Separation of variables is first employed, but then it has been necessary to use a computer to evaluate the eigenvalues and constants in the resulting series solutions. Information on ε_H and $\bar{u}$ must be supplied. Notter and Sleicher used more recent information on Pr_t than was used in the earlier Sleicher and Tribus solution to the same problem, and their results are quite consistent with the Sleicher and Rouse equations cited earlier, Eqs. (14-9), (14-10), and (14-13).

These solutions can be put in precisely the same form as for laminar flow. Thus Eqs. (9-39), (9-40), and (9-41) for *constant surface temperature*, and Eq. (9-42) for *constant heat rate*, are all directly applicable. The *constant-heat-rate* solution yields the following equation for fully developed velocity *and* temperature profiles:

$$\mathrm{Nu}_x = \left(\frac{1}{2} \sum_{m=0}^{\infty} \frac{1}{A_m \gamma_m^4} \right)^{-1}$$

The series in this equation is very slowly convergent, and there are not sufficient eigenvalues and constants to evaluate it. An alternative

TABLE 14-7
Infinite-series-solution functions for turbulent flow in a circular tube; thermal-entry length

n, m	Pr	Re	λ_m^2	G_n	γ_m^2	A_m
0	0.002	50,000	10.61	0.975		
1			58.19	0.897	29.66	0.00773
2			144.6	0.862	98.51	0.01015
3			269.7	0.841	206.5	0.00105
4					353.4	0.000607
0	0.002	100,000	11.23	1.036		
1			61.74	0.943	31.61	0.00725
2			153.6	0.903	104.7	0.00209
3			286.7	0.88	219.2	0.000977
4					374.9	0.00056
0	0.002	500,000	15.91	1.527		
1			90.19	1.246	48.83	0.00437
2			227.4	1.137	158	0.001202
3			426.8	1.08	331	0.000547
4					564.8	0.000307
0	0.01	50,000	13.39	1.277		
1			75.23	1.079	40.12	0.00431
2			188.9	0.996	131.8	0.00152
3			353.9	0.949	275	0.000696
4					469.3	0.000395
0	0.01	100,000	16.75	1.642		
1			96.15	1.29	53.11	0.00394
2			243.8	1.154	172.7	0.001068
3			458.5	1.085	358.8	0.000479
4						
0	0.01	500,000	39.31	4.2		
1			252.9	2.42	159.8	0.001152
2			665.6	1.93	504.6	0.000273
3			1271	1.752	1032	0.000112
4					1737	0.0000591
0	0.03	50,000	21.28	2.17		
1			126.7	1.532	73.68	0.00272
2			325.2	1.32	237.8	0.000705
3			614.8	1.189	491.7	0.000306
4					835.3	0.000167
0	0.03	100,000	31.06	3.28		
1			194.2	2.03	119.4	0.0016
2			506.9	1.649	380.5	0.00039
3			964.8	1.488	781.4	0.000162
4					1,320	0.000087
0	0.03	500,000	91.5	10.38		
1			678.1	4.39	477.7	0.000362
2			1,844	3.22	1,490	0.000076
3			3,562	2.9	3,011	0.000029
4					5,027	0.000015
0	0.72	10,000	64.38	7.596		

TABLE 14-7 (Continued)

n, m	Pr	Re	λ_m^2	G_n	γ_m^2	A_m
1			646.8	1.829	519.5	0.000301
2			1,870	1.217	1,624	0.0000513
3					3,202	0.0000149
0	0.72	50,000	219	26.6		
1			2,350	5.63	1,952	0.0000886
2			6,808	3.32	6,154	0.0000179
3					12,480	0.000006
0	0.72	100,000	375.9	45.8		
1			4,183	9.25	3,510	0.0000494
2			12,130	5.48	11,030	0.0000101
3					22,340	0.0000035
0	8	50,000	685.6	85.4		
		100,000	1,232	154		
		500,000	5,020	625		
1	10	50,000			2.736×10^4	5.02×10^{-6}
2					7.316×10^4	1.21×10^{-6}
3					1.373×10^5	4.36×10^{-3}
4					2.196×10^5	1.87×10^{-7}
5					3.198×10^5	8.79×10^{-8}
1	10	100,000			5.04×10^4	2.78×10^{-6}
2					1.346×10^5	6.97×10^{-7}
3					2.528×10^5	2.68×10^{-7}
4					4.046×10^5	1.26×10^{-7}
5					5.904×10^5	6.56×10^{-8}
0	50	100,000	2,570	321		
		200,000	4,778	598		
		500,000	10,800	1,350		
0	100	100,000	3,317	415		
		200,000	6,129	766		
		500,000	14,040	1,750		

equation for fully developed *constant heat rate,* based on variable-surface-temperature theory and the constant-surface-temperature functions, is Eq. (9-49).

An abstract of the Notter and Sleicher results is presented in Table 14-7. In addition, a set of functions for $Pr = 10$ for the *constant-heat-rate* problem is tabulated there, based on the calculations of Sparrow, Hallman, and Seigel.[15] The latter used $Pr_t = 1$ throughout their calculations. These functions must be used in conjunction with other data on Nu_∞ in order to be useful.

These series solutions, with the exception noted above, are very much more convergent than the corresponding laminar flow series, especially at higher Prandtl numbers, and the number of terms given is quite sufficient for most calculations.

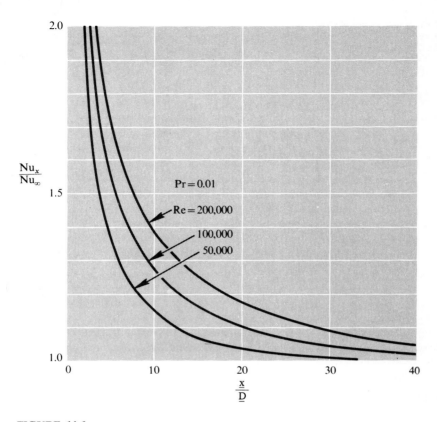

FIGURE 14-6
Nusselt numbers in the thermal-entry length of a circular tube, for constant heat rate: influence of Re at low Pr.

Figures 14-6 to 14-8 show the thermal-entry-length effect for flow in a circular tube for constant heat rate for a number of different cases. At $Pr = 0.01$ the effect is rather pronounced and increases with Reynolds number (Fig. 14-6). Figure 14-7 illustrates the strong influence of Prandtl number. At Prandtl numbers above 1 the thermal-entry-length effect becomes of decreasing importance. Note in Fig. 14-8 that at $Pr = 0.7$ there is very little influence of Reynolds number.

We can also investigate the effect of thermal-entry length on the *length-mean* Nusselt number. In heat-exchanger design this is generally of more importance than the local Nusselt number. If we move sufficiently far downstream that the local Nusselt number has closely approached its asymptotic value (which also means that only the first term in the infinite-series solution is of importance), the mean Nusselt number for

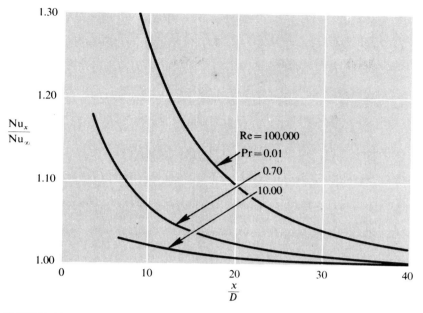

FIGURE 14-7

Nusselt numbers in the thermal-entry length of a circular tube, for constant heat rate: influence of Pr.

constant surface temperature is given by

$$\mathrm{Nu}_m = \frac{\lambda_0^2}{2} + \frac{1}{2x^+}\ln\frac{\lambda_0^2}{8G_0} \tag{14-19}$$

But $\lambda_0^2 = 2\,\mathrm{Nu}_\infty$, from Eq. (9-41). Dividing this into Eq. (14-19), we obtain

$$\frac{\mathrm{Nu}_m}{\mathrm{Nu}_\infty} = 1 + \frac{1}{x^+\lambda_0^2}\ln\frac{\lambda_0^2}{8G_0} \tag{14-20}$$

Now, for an example, consider $\mathrm{Pr} = 0.72$ and $\mathrm{Re} = 50,000$. From Table 14-7,

$$\lambda_0^2 = 219, \qquad G_0 = 26.6, \qquad x^+ = \frac{x/r_0}{\mathrm{Re}\,\mathrm{Pr}} = \frac{x/D}{17,500}$$

$$\frac{\mathrm{Nu}_m}{\mathrm{Nu}_\infty} = 1 + \frac{17,500}{219x/D}\ln\frac{219}{8 \times 26.6}$$

$$= 1 + \frac{2.14}{x/D} \tag{14-21}$$

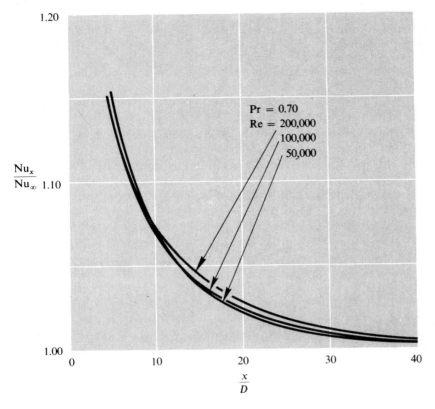

FIGURE 14-8
Nusselt numbers in the thermal-entry length of a circular tube, for constant heat rate: influence of Re at Pr = 0.7.

Thus, even for a tube 100 diameters in length, there is still a 2 percent effect of the entry length, primarily due to the high heat-transfer coefficient in the first 10 diameters.

THERMAL-ENTRY LENGTH FOR TURBULENT FLOW BETWEEN PARALLEL PLANES

Hatton and Quarmby[16] developed eigenvalue solutions for turbulent flow between parallel planes for both the case of *constant temperature* on one surface with the other surface insulated, and the case of *constant heat rate* on one surface with the other surface insulated. $Pr_t = 1$ was assumed throughout.

TABLE 14-8
Infinite-series-solution functions for turbulent flow between parallel planes; thermal entry length; one side at constant temperature and the other insulated

n	Re = 7096 λ_n^2	G_n	Re = 73,612 λ_n^2	G_n	Re = 494,576 λ_n^2	G_n
			Pr = 0.1			
0	16.7	1.86	59.4	6.87	228	27.1
1	158	1.34	611	3.78	2,685	10.9
2	450	1.03	1,788	2.54	8,200	6.47
3	893	0.859	3,535	2.14	16,210	5.37
4	1,473	0.794	5,830	2.04	26,700	5.10
5	2,201	0.762	8,660	1.95	39,820	4.85
			Pr = 1			
0	52.1	6.28	296	36.1	1,303	160
1	656	2.17	4,400	9.44	2,244	33.0
2	2,067	1.16	14,100	4.85	7,360	16.3
3	4,175	0.939	22,150	3.73	14,890	12.6
4	6,775	0.936	46,850	3.27	24,530	10.9
5	9,945	0.981	70,100	2.87	36,660	9.55
			Pr = 10			
0	153	19.1	1,038	129.4	5,180	648
1	4,880	1.81	37,200	10.2	199,600	46.9
2	17,000	0.914	132,000	4.45	696,000	20.2
3	34,250	0.865	265,600	3.26	1,427,000	14.5
4	53,800	1.01	442,000	2.73	2,368,000	11.8
5	76,700	1.18	663,000	2.35	3,556,000	9.65

The resulting eigenvalues and constants for one surface at *constant temperature* are presented in Table 14-8 and are directly applicable to Eqs. (9-38)–(9-41). For *constant heat rate* on one surface the eigenvalues and constants are also available in the original paper, but it is more convenient to have the Nusselt numbers and influence coefficients as presented in Table 14-9. This table then corresponds to Table 9-11 for laminar flow; for asymmetric heating Eq. (9-28) is applicable. Note that these results overlap the fully developed temperature-profile results of Table 14-5, but, despite the use of somewhat different data on diffusivities and velocity profiles, they are quite consistent.

It is worth noting that these parallel-planes results are applicable with only small error to the concentric circular-tube annulus for $r^* > 0.5$,

TABLE 14-9
Nusselt numbers and influence coefficients for turbulent flow between parallel planes; constant heat rate; one side heated and the other insulated

x/D_h	Re = 7096		Re = 73,612		Re = 494,576	
	Nu_{11}	θ_1^*	Nu_{11}	θ_1^*	Nu_{11}	θ_1^*
			Pr = 0.1			
1	19.7	0.056	75.2	0.018	241	0.005
3	14.3	0.122	56.2	0.146	194	0.023
10	10.7	0.267	42.4	0.115	155	0.062
30	9.44	0.352	34.8	0.233	132	0.147
100	9.34	0.359	32.1	0.290	120	0.219
			Pr = 1.0			
1	47.3	0.013	234	0.005	940	0.000
3	37.9	0.033	203	0.018	851	0.009
10	31.5	0.089	177	0.049	761	0.030
30	28.0	0.173	160	0.114	697	0.077
100	27.1	0.200	152	0.155	661	0.123
			Pr = 10			
1	102	0.004	602	0.004	2925	0.000
3	88.6	0.012	575	0.008	2829	0.003
10	81.9	0.027	550	0.018	2724	0.010
30	78.6	0.057	532	0.041	2640	0.027
100	77.5	0.070	522	0.057	2590	0.045

as can be seen by comparing the asymptotic results for the annulus and parallel planes, Tables 14-4 and 14-5.

More extensive tables of eigenvalues and constants for flow between parallel planes may be found in Sakakibara and Endo.[17]

THE EFFECTS OF AXIAL VARIATIONS OF SURFACE TEMPERATURE AND HEAT FLUX

The axially varying surface-temperature and the axially varying heat-flux problems may be handled in exactly the same manner as for laminar flow. The caveat is still that any kind of boundary conditions can be solved just as easily by direct numerical integration of Eq. (14-1), but the classical solutions allow whole classes of problems to be generalized, and in some cases closed-form results can be obtained. The equations developed in

Chap. 9 are directly applicable without alteration, but for turbulent flow the eigenvalues and constants from Tables 14-7 and 14-8 must now be used. Note also that Eqs. (9-28) and (9-29) are again applicable for the parallel-planes case, using Table 14-9.

The trends are found to be identical to those discussed for laminar flow. An increasing heat flux, or an increasing temperature difference in the flow direction, leads to higher heat-transfer coefficients, whereas the converse leads to lower heat-transfer coefficients. It is also possible to have zero or negative heat-transfer coefficients. However, the relative importance of axial variations is a strong function of Prandtl number. In the very low-Prandtl-number region (liquid metals) the effects can be very striking for a turbulent flow—even greater than those obtained for laminar flow. At Prandtl numbers near 1 and higher, the effects are usually negligible. Figure 14-5, which shows the relation between the Nusselt number for *constant heat rate* and the Nusselt number for *constant surface temperature,* provides a good indication of the regions where axial variations are of importance; it is noted that the effect at Prandtl number 0.7 is already quite small.

As an example of the significance of these results, consider the design of a nuclear reactor cooling system where the heat flux is a known function of length along the tube. Typically the heat flux is low near the entrance and exit, reaching a maximum at the midpoint. If the coolant is a liquid metal, considerable errors in the predicted surface temperatures can result unless variable-heat-flux theory is employed, that is, Eq. (9-55). On the other hand, if the coolant is a gas, or pressurized water, varying heat flux has very little influence, and it is perfectly adequate to use a Nusselt number based on *constant*-heat-flux theory or experiments to calculate the local temperature difference between the fluid and the wall surface. (Note, of course, that the *local* heat flux must be used to calculate the temperature difference, even though it is varying along the tube.)

COMBINED HYDRODYNAMIC- AND THERMAL-ENTRY LENGTH

In technical applications we are often more concerned about the development of the hydrodynamic and thermal boundary layers together than merely the thermal boundary layer alone. The turbulent tube-flow problem is further complicated by the fact that the turbulent boundary-layer development is very strongly influenced by the character of the tube entrance. If the tube has a well-designed nozzle entrance so that the velocity profile is close to uniform at the tube entrance, and if a boundary-layer trip is provided, then the combined hydrodynamic- and thermal-entry-length behavior is close to that which can be predicted by numerical integration of Eq. (14-1). However, these are laboratory conditions that are seldom encountered in real applications. If the boundary layer is not tripped, a laminar boundary layer may develop, and, depending on the Reynolds number and the free-stream turbulence, it is even possible to get fully developed *laminar* flow before a transition to turbulent flow occurs. In typical heat-exchanger applications the entrance, far from being a nozzle, is either a sharp-edged contraction or is preceded by a bend in the tube. In either case there is a separated flow at the entrance, with sufficient vorticity shed into the main stream that the heat-transfer rate is very much higher than would be obtained in a developing turbulent boundary layer where the turbulence originates from the surface.

At present we must rely upon experimental data for the turbulent flow entry region of a pipe for most applications. Some rather extensive experiments were carried out by Boelter, Young, and Iverson[18] using air flowing through a steam-heated tube in which the steam jacket was sectioned in such a way that the local heat-transfer rate at each section could be determined from a measurement of the condensate rate. An abstract of these results is plotted in Fig. 14-9. The ratio of local Nusselt number to fully developed Nusselt number is plotted as a function of x/D for a tube-diameter Reynolds number close to 50,000. The lowest curve is for the case of a hydrodynamically fully developed flow, and thus would correspond to the curves of Fig. 14-8. The other curves are for various

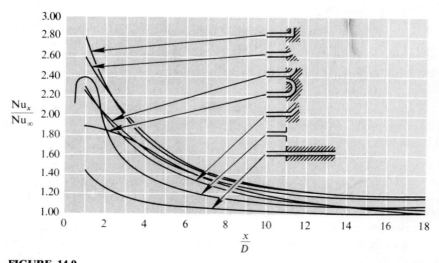

FIGURE 14-9
Measured local Nusselt numbers in the entry region of a circular tube for various entry configurations, air with constant surface temperature.[18]

typical heat-exchanger entrances: the abrupt contraction; and 45°, 90°, and 180° bends and angles.

When account is taken of the probable experimental uncertainty of data obtained in this manner, together with the fact that small changes in the geometry of the various bends may have considerable effects, a detailed quantitative comparison of the various curves is probably not warranted. What is important is the fact that all the curves for bends, and the abrupt contraction, lie substantially above the curve for the fully developed velocity profile. The interesting behavior of the abrupt-contraction curve (and this is probably the most common geometry encountered in heat exchangers) is apparently caused by the flow contraction and then re-expansion during the first diameter of tube length. The 180° bend shows similar behavior, possibly as a result of a stall region on the inside of the bend. The very high Nusselt number obtained for the 90° right-angled bend may be representative of the value that would be obtained for such a bend when preceded by a length of straight pipe; but the similar performance for the 45° round bend is probably attributable to a flow contraction rather than the bend itself, and thus would not be representative of such a bend following a section of straight pipe.

The curves of Fig. 14-9 all refer to the *local* Nusselt number. In many applications the *mean* Nusselt number with respect to tube length is of more use. For tubes with lengths greater than the entry length (and this means a greater than about 20 diameters) it should be possible to express the mean Nusselt number in the form of Eq. (14-21), that is,

$$\frac{\text{Nu}_m}{\text{Nu}_\infty} = 1 + \frac{C}{x/D} \tag{14-22}$$

The following values of C have been evaluated by integration of the curves in Fig. 14-9:

fully developed velocity profile	1.4
abrupt-contraction entrance	6
90° right-angled bend	7
180° round bend	6

Note that C for the fully developed velocity profile comes out to be 1.4 rather than the predicted 2 in Eq. (14-21), but experimental uncertainty may account for this difference.

The effect on the mean Nusselt number of the abrupt-contraction and bend entrances is seen to be very substantial. For a tube of 100 diameters length the mean Nusselt number for the abrupt contraction is still 6 percent above the asymptotic value, although most of this effect is attributable to the first 10 diameters.

All of these data were obtained with a fluid with $Pr = 0.7$. It is difficult to tell what the effect will be at very high and very low Prandtl numbers other than to say that it will be substantially greater in the liquid-metal region, and probably somewhat less at high Prandtl numbers.

THE INFLUENCE OF SURFACE ROUGHNESS

All the preceding discussion has been based on the assumption of an aerodynamically smooth surface. The effects of surface roughness have been discussed in some detail in connection with the external turbulent boundary layer (Chaps. 11 and 13) and with the hydrodynamics of flow in a tube or pipe. The effects of roughness on heat transfer for flow in a tube are essentially the same as described in Chap. 13 for the external boundary layer. The only difference arises from the fact that for flow in a tube the friction coefficient is based on the *mean* flow velocity, and the heat-transfer coefficient is based on the mixed *mean fluid* temperature.

The same mixing-length theory as discussed in Chap. 11 for rough surfaces is applicable to the tube-flow problem. The eddy diffusivity in the region far away from the wall surface differs little from that for a smooth surface. All the experiments (and these do not include the very low-Pr region) indicate a turbulent Prandtl number no different from that for a smooth surface. Thus numerical integration of Eq. (14-1) is quite feasible, and the more restricted cases of fully developed velocity profiles, Eq. (14-2), or fully developed velocity *and* temperature profiles, with constant heat rate, Eq. (14-3), all for circular tubes, can be readily handled. The surface resistance problem discussed in Chap. 13 is no different in a tube or pipe.

Dipprey and Sabersky[19] present correlations for heat transfer in tubes that take on essentially the same form as Eq. (13-48). A simpler empirical correlation is suggested by Norris:[20]

$$\frac{Nu}{Nu_{smooth}} = \left(\frac{c_f}{c_{f_{smooth}}}\right)^n \tag{14-23}$$

where $n = 0.68 \, Pr^{0.215}$. For $(c_f/c_{f_{smooth}}) > 4.0$ Norris finds that the Nusselt number no longer increases with increasing roughness.

Equation (14-23) incorporates most of the effects of roughness that one would expect. The effect of roughness on heat transfer increases with Prandtl number because the sublayer becomes of dominant influence at high Prandtl numbers. On the other hand, one would expect little effect of roughness for liquid metals. The heat-transfer coefficient is not affected as strongly as the friction coefficient because of the additional conduction heat-transfer resistance at the surface. When the roughness effects on friction become very large, no further increase in heat transfer is observed because the heat-transfer resistance has become primarily a conduction resistance at the surface in the spaces between the roughness elements.

It should be noted that artificial roughness is frequently employed as a device to increase the heat-transfer coefficient. A popular and very effective scheme is to lay a small-diameter wire along the surface transverse to the flow direction (or else to machine a similar protrusion on the surface) so as to break up the sublayer. Typically the height of the protrusions should be about $y^+ = 50$, with protrusions spaced 10–20 heights along the surface. The friction coefficient is also increased because of pressure drag on the protrusion, but if the protrusion is small enough to be primarily in the sublayer region then the increase in friction is not disproportionate to the increase in heat transfer. Therefore this scheme is more effective than so-called *turbulence promoters* that are frequently inserted inside tubes to increase heat transfer at a large cost in pumping-power requirements. This whole problem is discussed in more detail in Chap. 19 in connection with a discussion of compact and high-performance heat-exchanger surfaces.

A comprehensive review of methods for augmentation of convective heat transfer is given by Bergles.[21,22]

PROBLEMS

14-1. Consider fully developed turbulent flow between parallel planes. Let the Reynolds number (based on hydraulic diameter) be 50,000 and let the Prandtl number of the fluid be 3.

The heat flux on one plate, *into* the fluid, is constant everywhere. The heat flux on the other plate is *out* of the fluid, is also constant everywhere, and is the same in magnitude as the heat flux on the first plate. The problem is to calculate and plot the temperature distribution across the fluid from plate to plate.

Use any equations for velocity distribution and or eddy diffusivity that you feel are reasonable. For simplicity, it is sufficiently precise to assume that the eddy diffusivity is constant across the center region of the duct at the value given at $r = 0$ by Eq. (12-5), and that Eq. (12-4) is valid over the remainder of the flow area, excepting the sublayer, which can be treated as in the development leading to Eq. (11-16).

Explain the shape of the temperature profile by referring to the basic mechanisms involved. How would this profile change with Prandtl number? What would the profile be if the flow were laminar?

14-2. Consider fully developed turbulent flow in a circular tube with heat transfer to or from the fluid at a constant rate per unit of tube length. Let there also be internal heat generation (perhaps from nuclear reaction) at a rate S, W/m^3, which is everywhere constant. If the Reynolds number is 50,000 and the Prandtl number is 4, evaluate the Nusselt number as a function of the pertinent parameters. The heat-transfer coefficient in the Nusselt number should be defined in the usual manner on the basis of the heat flux *at the surface,* the surface temperature, and the mixed mean fluid temperature. Use a two-layer model to handle the sublayer [see Chap. 11 and the development leading to Eq. (11-16)] and evaluate the eddy diffusivity as described for Prob. 14-1.

14-3. Consider turbulent flow between parallel planes with Reynolds number equal to 100,000. For a fluid with $Pr = 10$, and then for a fluid with $Pr = 0.01$, evaluate the Nusselt number for fully developed constant heat rate for only one side heated and then for both sides equally heated, using the solutions given in the text. Discuss the differences between the cases of one side heated and both sides heated, in terms of the heat-transfer mechanisms and the temperature profiles. How is the Nusselt number related to the "shape" of the temperature profile?

14-4. Consider fully developed flow in a circular tube with constant heat rate per unit of tube length. Let the mean flow velocity be 8 m/s. Evaluate the heat-transfer coefficient h for the following cases, and discuss the reasons for the differences:
(*a*) air, 90°C, 1 atm pressure, 2.5 cm diameter tube;
(*b*) same with 0.6 cm diameter tube;
(*c*) hydrogen gas, 90°C, 1 atm pressure, 2.5 cm diameter tube;
(*d*) liquid oxygen, −200°C, 2.5 cm diameter tube;
(*e*) liquid water, 38°C, 2.5 cm diameter tube;
(*f*) liquid sodium, 200°C, 2.5 cm diameter tube;
(*g*) aircraft engine oil, 90°C, 2.5 cm diameter tube;
(*h*) air, 90°C, 1000 kPa pressure, 2.5 cm diameter tube.

14-5. Consider a 1.25 cm inside-diameter, 1.8 m long tube wound by an electric resistance heating element. Let the function of the tube be to heat an organic fuel from 10 to 65°C. Let the mass-flow rate of the fuel be 0.126 kg/s, and let the following average properties be treated as constant:

$$Pr = 10, \qquad \rho = 753 \text{ kg/m}^3, \qquad c = 2.100 \text{ kJ/(kg} \cdot \text{K)}$$
$$k = 0.137 \text{ W/(m} \cdot \text{K)}, \qquad \mu = 660 \text{ } \mu\text{Pa} \cdot \text{s}$$

Calculate and plot both tube surface temperature and fluid mean temperature as functions of tube length.

14-6. Liquid potassium flows in a 2.5 cm diameter tube at a mean velocity of 2.4 m/s and a mean temperature of 550°C. Suppose the tube is heated at a constant rate per unit of length but the heat flux varies around the periphery of the tube in a sinusoidal manner, with the maximum heat flux twice the minimum heat flux. If the maximum surface temperature is 700°C, evaluate the axial mean temperature gradient, °C/m, and prepare a plot of temperature around the periphery of the tube.

14-7. Consider the flow of first air ($Pr = 0.7$) and then mercury ($Pr = 0.01$) at a Reynolds number of 100,000 in a 2 cm diameter circular tube with constant heat rate per unit of length. At a distance of 1.25 m from the tube entrance the heating stops, and the tube is insulated from then on.

Using relative units of temperature, calculate and plot the decay of surface temperature as it approaches the mean fluid temperature in the insulated region. Discuss the difference in behavior of the two fluids in terms of the basic mechanisms involved.

14-8. Let air at 21°C and 1 atm pressure flow at a Reynolds number of 50,000 in a 2.5 cm diameter circular tube. The tube wall is insulated for the first 75 cm, but for the next 125 cm the tube surface temperature is constant at 40°C. Then it abruptly increases to 50°C and remains constant for another 125 cm. Plot the local heat-transfer coefficient h as a function of an axial distance. Does the abrupt increase in surface temperature to 50°C cause a significant change in the average heat-transfer coefficient over the entire 250 cm of heated length? In simple heat-exchanger theory a mean heat-transfer coefficient with respect to tube length is generally employed. In this case how much does the mean differ from the asymptotic value of h?

14-9. Starting with the constant-surface-temperature, thermal-entry-length solutions for a circular tube, calculate and compare the Nusselt numbers for constant surface temperature and for a linearly varying surface temperature, for very long tubes for a Reynolds number of 50,000 and a Prandtl number of 0.01. Repeat for a Prandtl number of 0.7 and discuss the reasons for the differences noted.

14-10. The following are the proposed specifications for the cooling tubes in a pressurized-water nuclear power reactor:

Tube configuration	Concentric circular-tube annulus, with heating from the inner tube (containing the uranium fuel), and the outer tube surface having no heat flux
Tube dimensions	Inner-tube diameter, 2.5 cm
	Outer-tube diameter, 5.0 cm
	Tube length, 5 m
Water temperature	Inlet, 275°C
	Outlet, 300°C
Water mean velocity	1 m/s
Axial heat-flux distribution	$\dot{q}''/\dot{q}''_{max} = \frac{1}{3}[1 + 2 \sin (\pi x / L)]$

Assume that the water properties may be treated as constant at 287°C.

Calculate and plot heat flux, mean water temperature, and inner and outer tube surface temperatures as functions of x. Assume that the conductance h is independent of x and that the value for fully developed constant heat rate is a reasonable approximation. (Can you justify these assumptions?) How high must the water pressure be to avoid boiling? Or is it possible to avoid boiling with these specifications? What do you think would be the effect of local boiling at the highest-temperature parts of the system?

14-11. Repeat Prob. 14-10 for the case of liquid sodium in place of high-pressure water. Let the inlet and outlet sodium temperatures be 330 and 355°C, respectively. However, this time the effect of axially varying heat flux on the heat-transfer coefficient cannot be ignored.

14-12. Helium flows in a thin-walled circular tube of 2.5 cm inside diameter. Down the center of the tube is inserted a 5 mm diameter circular electric heater. The Reynolds number of the flow, based on the hydraulic diameter of the resulting passage, is 30,000. Heat is generated and transferred through the heater surface at a rate of 30 kW/m². The outer surface of the outer tube is bare, exposed to an atmospheric environment at 21°C. Heat is transferred from the outer surface to the surroundings by both free convection and radiation. For the free convection, assume a heat-transfer coefficient h of 10 W/(m² · K). For the radiation, let the surface emissivity be 0.8.

At a particular point along the tube the mixed mean fluid temperature is 200°C. Assuming that the gas is transparent to thermal radiation and that the radiation emissivity of the two inner surfaces is 0.8, calculate the surface temperature of the heater and the temperature of the outer tube. Determine the fractions of the original heat generated in the core tube that are ultimately transferred to the helium and to the surroundings. Is radiation a major factor?

It may be assumed that the tube is sufficiently long that fully developed conditions are closely approached.

14-13. *Computer analysis of turbulent flow in a circular tube with $\dot{q}_0'' = const$ and constant fluid properties:* Use a mixing-length turbulence model with the Van Driest damping function and the variable turbulent Prandtl number model to calculate the flow in a tube at Pr = 0.01, 0.7, and 100, and for Reynolds numbers 30,000 and 150,000, from the tube entrance to a location where the Nusselt number becomes constant. Examine a temperature profile at the thermally fully developed state to evaluate the *thermal law of the wall.* Feel free to investigate any other attribute of the tube flow. You can choose how to set up the problem in terms of a suitable choice of dimensions and fluid properties (constant). For the initial conditions at the tube entrance use a flat velocity and temperature profiles.

14-14. *Computer analysis of turbulent flow in a circular tube with $t_0 = const$ and constant fluid properties:* Use a mixing-length turbulence model with the Van Driest damping function and the variable turbulent Prandtl number model to calculate the flow in a tube at Pr = 0.7 and for Reynolds numbers 30,000 and 150,000, from the tube entrance to a location where the Nusselt number becomes constant. Compare the results with the constant-heat-flux boundary-condition results. Examine a temperature profile at the thermally fully developed state to evaluate the *thermal law of the wall.* Feel free to investigate any other attribute of the tube flow. You can choose how to set up the problem in terms of a suitable choice of dimensions and fluid properties (constant). For the initial conditions at the tube entrance use a flat velocity and temperature profiles.

14-15. *Computer analysis of turbulent flow between parallel planes with $\dot{q}_0'' = const$*

and constant fluid properties: Follow Prob. 14-13 for further specification of the problem.

14-16. *Computer analysis of Prob. 14-5 with $\dot{q}_0'' = const$ and constant fluid properties:* Calculate the flow, starting with initial conditions of flat velocity and temperature profiles at the inlet. Compare the computer results with the analytical results.

14-17. *Computer analysis of Prob. 14-7 with $\dot{q}_0'' = f(x)$ and constant fluid properties:* Calculate the flow, starting with inital conditions of flat velocity and temperature profiles at the inlet. Compare the computer results with the analytical results.

14-18. *Computer analysis of Prob. 14-8 with $t_0 = f(x)$ and constant fluid properties:* Calculate the flow, starting with inital conditions of flat velocity and temperature profiles at the inlet. Compare the computer results with the analytical results.

14-19. *Computer analysis of Prob. 14-9 with $t_0 = f(x)$ and constant fluid properties:* Calculate the flow, starting with inital conditions of flat velocity and temperature profiles at the inlet. Compare the computer results with the analytical results.

REFERENCES

1. Kays, W. M., and E. Y. Leung: *Int. J. Heat Mass Transfer*, vol. 6, pp. 537–557, 1963.
2. Kakac, S., R. K. Shah, and W. Aung: *Handbook of Single-Phase Convective Heat Transfer*, Wiley, New York, 1987.
3. Gnielinski, V.: *Int. Chem. Engng*, vol. 16, pp. 359–368, 1976.
4. Sleicher, C. A., and M. W. Rouse: *Int. J. Heat Mass Transfer*, vol. 18, pp. 677–683, 1975.
5. Dittus, P. W., and L. M. K. Boelter: *Univ. Calif. Publ. Engng*, vol. 2, no. 13, pp. 443–461, October 17, 1930; reprinted in *Int. Commun. Heat Mass Transfer*, vol. 12, pp. 3–22, 1985.
6. Skupinski, E., J. Tortel, and L. Vautry: *Int. J. Heat Mass Transfer*, vol. 8, pp. 937–951, 1965.
7. Sleicher, C. A., and M. Tribus: *Proceedings of Heat Transfer and Fluid Mechanics Institute, Stanford University, 1956*, p. 59, Department of Mechanical Engineering, Stanford University, 1956.
8. Gartner, D., K. Johannsen, and H. Ramm: *Int. J. Heat Mass Transfer*, vol. 17, pp. 1003–1018, 1974.
9. Deissler, R. G.: NACA (now NASA) Report 1210, Washington, 1955.
10. Sutherland, W. A., and W. M. Kays: *Int. J. Heat Mass Transfer*, vol. 7, p. 1187, 1964.
11. Leung, E. Y., W. M. Kays, and W. C. Reynolds: Report AHT-4, Department of Mechanical Engineering, Stanford University, April 15, 1962.
12. Kays, W. M., and A. L. London: *Compact Heat Exchangers*, McGraw-Hill, New York, 1984.
13. Rohsenow, W. M., and J. P. Hartnett (eds.): *Handbook of Heat Transfer*, McGraw-Hill, New York, 1973.
14. Notter, R. H., and C. H. Sleicher: *Chem. Engng Sci.*, vol. 27, pp. 2073–2093, 1972.
15. Sparrow, E. M., T. M. Hallman, and R. Speigel: *Appl. Sci. Res.*, ser. A, vol. 7, pp. 37–52, 1957.
16. Hatton, A. P., and A. Quarmby: *Int. J. Heat Mass Transfer*, vol. 6, pp. 903–914, 1963.

17. Sakakibara, M., and K. Endo: *Int. Chem Engng,* vol. 18, pp. 728–733, 1976.
18. Boelter, L. M. K., G. Young, and H. W. Iversen: NACA (now NASA) TN 1451, Washington, July 1948.
19. Dipprey, D. F., and R. H. Sabersky: *Int. J. Heat Mass Transfer,* vol. 6, pp. 329–353, 1963.
20. Norris, R. H.: *Augmentation of Convection Heat and Mass Transfer,* American Society of Mechanical Engineers, New York, 1971.
21. Bergles, A. E.: *Applied Mechanics Reviews,* vol. 26, pp. 675–682, 1973.
22. Bergles, A. E.: Report HTL-8, Engineering Research Institute, Iowa State University, Ames, 1975.

CHAPTER
15

THE INFLUENCE OF TEMPERATURE-DEPENDENT FLUID PROPERTIES

In all the heat-transfer and flow friction solutions considered in the previous chapters it has been assumed that the fluid properties remain constant throughout the flow field. When applied to real heat-transfer problems, this assumption is obviously an idealization, since the transport properties of most fluids vary with temperature and thus vary through the boundary layer or over the flow cross section of a tube. In this chapter we examine the results of a number of analytic solutions and experiments in which this influence has been investigated, and we propose methods whereby the constant-property solutions can be corrected in a simple manner to take this influence into consideration.

The temperature-dependent-property problem is further complicated by the fact that the properties of different fluids behave differently with temperature. For *gases* the specific heat varies only slightly with temperature, but the viscosity and thermal conductivity increase to about the 0.8 power of the absolute temperature (in the moderate temperature range). Furthermore, the density varies inversely with the first power of the absolute temperature. On the other hand, the Prandtl number $\mu c/k$ does not vary significantly with temperature.

For most *liquids* the specific heat and thermal conductivity are relatively independent of temperature, but the viscosity decreases very markedly with temperature. This is especially so for oils, but even for water the viscosity is very temperature-dependent. The density of liquids, on the other hand, varies little with temperature. The Prandtl number of liquids varies with temperature in much the same manner as the viscosity.

A number of variable-property analyses are contained in the literature, as well as a considerable body of experimental data. The general effect of the variation of the transport properties with temperature is to change the velocity and temperature profiles, yielding different friction and heat-transfer coefficients than would be obtained if properties were constant.

For engineering applications it has been found convenient to employ the constant-property analytic solutions, or the experimental data obtained with small temperature differences, and then to apply some kind of correction to account for property variation. Fortunately, most of the variable-property results indicate that fairly simple corrections generally suffice over a moderate range of properties. Obviously, if the absolute value of the properties varies severalfold through the boundary layer, no simple correction scheme is going to be adequate, and thus a complete numerical integration of the applicable differential equations for each such application is required. Such cases are fortunately rare.

Two schemes for correction of the constant-property results are in common use. In the *reference temperature* method a characteristic temperature is chosen at which the properties appearing in the non-dimensional groups (Re, Pr, Nu, etc.) may be evaluated such that the constant-property results at that temperature may be used to evaluate variable-property behavior. Typically, this may be the surface temperature or a temperature part way between the surface temperature and the free-stream or mixed mean temperature; there is no general rule. In the *property ratio* method all properties are evaluated at the free-stream temperature, or the mixed mean temperature, and then all the variable-property effects are lumped into a function of a ratio of some pertinent property evaluated at the surface temperature to that property evaluated at the free-stream or mixed mean temperature.

The *reference temperature* method, although extensively used, leads to a certain awkwardness—at least in internal flow applications—that is avoided by the property ratio method. For example, in evaluating the Reynolds number it becomes necessary to split the mass velocity $G = V\rho$ so that the density may be evaluated at the reference temperature. But for internal flows, such as flows in a pipe, G is also the mass-flow rate divided by the flow cross-sectional area; and it is one parameter about which there is no ambiguity regardless of density variation over the flow area.

The *property ratio* scheme involves a different kind of correction for liquids and for gases. For *liquids,* where the viscosity variation is responsible for most of the effect, it is found that equations of the following type are often excellent approximations:

$$\frac{\text{Nu}}{\text{Nu}_{\text{CP}}} = \frac{\text{St}}{\text{St}_{\text{CP}}} = \left(\frac{\mu_0}{\mu_m}\right)^n \text{ or } \left(\frac{\mu_0}{\mu_\infty}\right)^n \tag{15-1}$$

$$\frac{c_f}{c_{f\,\text{CP}}} = \left(\frac{\mu_0}{\mu_m}\right)^m \text{ or } \left(\frac{\mu_0}{\mu_\infty}\right)^m \tag{15-2}$$

All properties in the nondimensional parameters are evaluated at local mixed mean fluid temperature (or free-stream temperature in the case of external flows), and the subscript CP refers to the appropriate constant-property solution or small-temperature-difference experimental results. The viscosity μ_0 is evaluated at the surface temperature, while μ_m is evaluated at the mixed mean temperature (or the free-stream temperature in the case of external flows). The exponents m and n are functions of geometry and the type of flow.

For *gases* the viscosity, thermal conductivity, and density are functions of *absolute* temperature. Fortunately, the absolute temperature dependence tends to be similar for different gases, although the similarity does break down at the temperature extremes. Thus it is found that the temperature-dependent-property effects can usually be adequately correlated by the equations

$$\frac{\text{Nu}}{\text{Nu}_{\text{CP}}} = \frac{\text{St}}{\text{St}_{\text{CP}}} = \left(\frac{T_0}{T_m}\right)^n \text{ or } \left(\frac{T_0}{T_\infty}\right)^n \tag{15-3}$$

$$\frac{c_f}{c_{f\,\text{CP}}} = \left(\frac{T_0}{T_m}\right)^m \text{ or } \left(\frac{T_0}{T_\infty}\right)^m \tag{15-4}$$

Again, all properties in the nondimensional parameters are evaluated at the mixed mean temperature for internal flows or at the free-stream temperature for external flows.

In the remainder of this chapter we examine the evidence, both analytic and experimental, for values of the exponents m and n; and we also indicate some appropriate reference properties. We first consider flow inside tubes and then external boundary layers. In each case we consider laminar flow and then turbulent flow, and under these headings liquids and then gases. For one case—the laminar external boundary layer with a gas—we carry through a simple approximate analytic development.

In this chapter we are concerned with only low-velocity flow where viscous dissipation may be neglected; high-velocity effects are considered in Chap. 16.

LAMINAR FLOW IN TUBES: LIQUIDS

Fully developed laminar flow of a liquid in a tube with temperature-dependent properties is a relatively simple analytic problem. Only the variation of viscosity with temperature is significant. The momentum and energy differential equations for fully developed, constant-heat-rate conditions in a circular tube follow directly from Eqs. (4-11) and (4-37):

$$\frac{dP}{dx} = \frac{1}{r}\frac{\partial}{\partial r}\left(r\mu\frac{\partial u}{\partial r}\right) \tag{15-5}$$

$$u\rho c\frac{dt_m}{dx} = \frac{1}{r}\frac{\partial}{\partial r}\left(rk\frac{\partial t}{\partial r}\right) \tag{15-6}†$$

Equations (15-5) and (15-6) can be integrated by an iterative procedure in which the temperature distribution for constant properties is used as a first approximation. Then using the appropriate viscosity variation with temperature, the momentum equation is integrated numerically to yield a second approximation for the velocity distribution. This velocity distribution is employed in a numerical integration of the energy equation to yield a second approximation for the temperature distribution. The procedure is repeated until the velocity and temperature distributions do not change. Then the mean velocity, the mixed mean temperature, the friction coefficient, and the convection conductance are evaluated as in the constant-property solutions.

Deissler[1] carried out such calculations for a circular tube for a fluid for which the viscosity varies with temperature as

$$\mu \propto T^{-1.6}$$

which corresponds approximately to the behavior of liquid metals. When the results are put in the form of Eqs. (15-1) and (15-2), we obtain

$$n = -0.14$$

which is a number that has also been used extensively to correlate experimental data for laminar flow of moderate- and high-Prandtl-number liquids.

For the friction coefficient Deissler's results indicate a slightly different exponent, depending on whether the liquid is being heated or cooled:

$$\frac{\mu_0}{\mu_m} > 1, \quad m = 0.50 \quad \text{(cooling)}$$

$$\frac{\mu_0}{\mu_m} < 1, \quad m = 0.58 \quad \text{(heating)}$$

---

† After changing to a cylindrical coordinate system.

Yang[2] considered the laminar thermal-entry-length problem in a circular tube for both a constant surface temperature and a constant heat rate. He employed an integral method that yielded results very close to the exact eigenvalue solution for constant viscosity, and he assumed that the viscosity dependence on temperature could be expressed as

$$\frac{\mu}{\mu_0} = \frac{1}{1 + \gamma\theta}$$

where $\theta = (t_0 - t)/(t_0 - t_e)$ and γ is a parameter.

Yang considered only the heat-transfer behavior, and his results for both constant surface temperature and constant heat rate can be very well approximated by

$$n = -0.11$$

and this result (which refers to the local heat-transfer coefficient) applies to both the thermal-entry length and the subsequent fully developed region. The difference in exponent, -0.11 as compared with -0.14, is evidently related to the different form of viscosity–temperature function employed. The difference is small, and in either case the influence of viscosity ratio on Nusselt number is evidently not great. Also, there appears to be little effect due to different types of thermal boundary conditions. The influence on the friction coefficient, on the other hand, is very substantial.

Calculations by the authors using a finite-difference code and the real properties of *water* yield substantially the same conclusions.

This problem has apparently not been handled for other than a circular tube.

LAMINAR FLOW IN TUBES: GASES

In this case reasonably complete analytic solutions are available as well as a considerable body of experimental data. Deissler[1] first investigated this problem analytically, assuming fully developed velocity and temperature profiles and employing an iterative procedure similar to that described for the liquid case. However, experimental data[3,4] are only in moderate agreement with these results. The difficulty apparently is that, because of density changes with temperature, there is always a radial component of velocity and under large-temperature-difference conditions it is not possible to have a fully developed velocity profile. Worsøe-Schmidt[5] has solved the applicable differential equations using a finite-difference technique for a gas with the properties of air in the temperature range 300–1700 K. A circular tube is considered, with a developed velocity profile and uniform temperature at the tube entrance. Heating and cooling with a constant surface temperature and heating with a constant heat rate are considered. In this solution the radial velocity is included.

Worsøe-Schmidt's results show a rather small effect of temperature ratio on Nusselt number but a substantial effect on friction coefficient. Near the entrance, and also well downstream, the results can be satisfactorily correlated for both heating and cooling by

$$n = 0, \quad m = 1$$

Experiments by Kays and Nicoll[3] verify the conclusion that n is approximately zero over the temperature ratio range 0.5–2. Experiments by Davenport[4] for temperature ratios from 1.0 to 2.2 yield $m = 1.35$, but this is a difficult experiment to carry out accurately and the conclusion from the analysis is probably to be preferred.

Calculations (by the authors) for flow between *parallel planes*, using the fully developed flow assumption and employing essentially the procedure of Deissler, yield substantially the same results for the friction coefficient as were obtained for the circular tube under the same assumptions. However, the results for the Nusselt number are substantially different.

TURBULENT FLOW IN TUBES: LIQUIDS

The mixing-length methods developed in Chap. 11 can be used to find solutions for the variable-property problem provided only that the variable-property forms of the momentum and energy differential equations are employed, i.e., the turbulent flow versions of Eqs. (4-11) and (4-37). It is necessary to evaluate fluid properties at the local temperature, which then requires a simultaneous solution of the momentum and energy equations. However, extensive experimental data are also available, and these are used for the recommendations to follow because the validity of any turbulent flow analytic solutions of this problem must rely heavily on experimental verification anyway.

Petukhov[6] has examined heat-transfer data covering these ranges: $Pr = 2$–140; $Re = 5000$–123,000; and $\mu_0/\mu_m = 0.025$–12.5. He finds that the data are quite well correlated by

$$\frac{\mu_0}{\mu_m} > 1, \quad n = -0.25 \quad \text{(cooling)}$$

$$\frac{\mu_0}{\mu_m} < 1, \quad n = -0.11 \quad \text{(heating)}$$

Allen and Eckert[7] conducted experiments with water at $Pr = 8$ for the Reynolds number range 13,000–111,000 for the heating case and found that the friction coefficient results are reasonably correlated by

$$\frac{\mu_0}{\mu_m} < 1, \quad m = 0.25 \quad \text{(heating)}$$

Friction data for the cooling case appear to be rare, but the very early data of Rohonczy[8] for $Re = 33,000 - 225,000$, $Pr = 1.3 - 5.8$, and $\mu_0/\mu_m = 1 - 2$ are quite well correlated by the same value of m:

$$\frac{\mu_0}{\mu_m} > 1, \quad m = 0.25 \quad \text{(cooling)}$$

TURBULENT FLOW IN TUBES: GASES

There is a considerable body of experimental data for $T_0/T_m > 1$. The data of Humble, Lowdermilk, and Desmon[9] for air yield $n = -0.55$. The extensive data of McCarthy and Wolf[10] for helium and hydrogen yield $n = -0.55$. The experiments of Barnes[11] for air, helium, and carbon dioxide yield values of n that differ for the different gases and are somewhat lower. Petukhov[6] presents an excellent summary of these and many other experiments, which yield roughly the same conclusions but which also suggest that the absolute value of n increases with temperature ratio, Sleicher and Rouse[18] examined a large amount of data and concluded that

$$\text{for } 1 < \frac{T_0}{T_m} < 5, \quad n = -\left(\log_{10} \frac{T_0}{T_m}\right)^{1/4} + 0.3 \quad \text{at } \frac{x}{D} > 40$$

Calculations (by W. M. Kays) using a finite-difference procedure and the mixing-length theory developed in Chap. 11 are in quite good agreement with this equation for $x/D > 30$ but yield smaller absolute values of n for shorter lengths, which means that there is a considerable entry-length effect. One of the difficulties with this problem is that for either a constant heating rate or a constant surface temperature there is a continuous axial variation of T_0/T_m, a continuous flow acceleration, and the radial component of velocity is never zero, so that a fully developed flow never exists. This fact is very apparent in the analytic solution referred to. It is worth noting that in evaluating the Van Driest damping function for the sublayer (see Chap. 11), *local* values of viscosity and density were used to evaluate y^+ in the term y^+/A^+, although the shear stress is the wall value. $A^+ = 26$ was used throughout. Since the major portion of the heat-transfer resistance is in the sublayer, this procedure is relatively critical.

The cooling experiment, $T_0/T_m < 1$, is more difficult to carry out, especially if local heat-transfer coefficients are desired, and few results have been reported. Humble, Lowdermilk, and Desmon[9] report a small amount of data for T_0/T_m varying from 1 to 0.46; there appears to be a negligible effect of temperature ratio; that is, $n = 0$. Nicoll and Kays[12] carried out similar experiments and could detect no effect over the range of temperature ratio 0.63 to 0.38. Brim and Eustis[19] carried out experiments with a mixture of combustion gases and nitrogen and measured local heat-transfer coefficients at $x/D = 50$ for values of T_0/T_m as low as 0.25, with the conclusion that $n = 0$. Kays' analytic solutions yield n between 0 and -0.1, that is, very close to 0.

Experimental data on the friction coefficient are not extensive, and they are difficult to interpret because of the pressure drop caused by flow acceleration or deceleration. McEligot[13] concludes that $m = -0.1$ on the basis of his own experiments and an examination of the data of Lel'chuk and Dyadyakin,[14] all for $T_0/T_m > 1$. There have apparently been few experiments for $T_0/T_m < 1$. Kays' analytic results suggest a value of m between -0.1 and -0.2 for both heating and cooling.

On the basis of this evidence, the following are recommended as simple correlations for turbulent flow of a gas in a circular tube for $x/D > 30$:

$$\text{for } T_0/T_m > 1, \quad m = -0.1, \quad n = -0.5 \quad \text{(heating)}$$

$$\text{for } T_0/T_m < 1, \quad m = -0.1, \quad n = 0.0 \quad \text{(cooling)}$$

The equation recommended by Sleicher and Rouse is probably a better approximation for n for the heating case.

It seems unlikely that tube cross-sectional shape is a significant parameter for turbulent flow, and thus these conclusions should be applicable for noncircular tubes.

THE LAMINAR EXTERNAL BOUNDARY LAYER: GASES

Consider a steady-flow, laminar, low-velocity, two-dimensional boundary layer with no pressure gradient and with constant but different surface and free-stream temperatures. The applicable differential equations under the conditions of variable-fluid properties are as follows:

Continuity, from Eq. (4-1),

$$\frac{\partial(\rho u)}{\partial x} + \frac{\partial(\rho v)}{\partial y} = 0 \tag{15-7}$$

Momentum, from Eq. (4-10),

$$\rho u \frac{\partial u}{\partial x} + \rho v \frac{\partial u}{\partial y} - \frac{\partial}{\partial y}\left(\mu \frac{\partial u}{\partial y}\right) = 0 \tag{15-8}$$

Energy, from (4-37), omitting the viscous dissipation term,

$$u\rho c \frac{\partial t}{\partial x} + v\rho c \frac{\partial t}{\partial y} - \frac{\partial}{\partial y}\left(k\frac{\partial t}{\partial y}\right) = 0 \qquad (15\text{-}9)$$

The boundary conditions are

$$u, v = 0, \qquad t = t_0 \quad \text{at } y = 0$$

$$u = u_\infty, \qquad t = t_\infty \quad \text{at } y \to \infty$$

$$u = u_\infty, \qquad t = t_\infty \quad \text{at } x = 0$$

Similarity solutions may be obtained for this system of equations. Consider the following transformation of variables. First define a stream function ψ that satisfies the continuity equation:

$$\rho u = \rho_\infty \frac{\partial \psi}{\partial y}, \qquad \rho v = -\rho_\infty \frac{\partial \psi}{\partial x} \qquad (15\text{-}10)$$

Next define

$$\eta = \sqrt{\frac{u_\infty}{\nu_\infty x}} \int_0^y \frac{\rho}{\rho_\infty}\, dy, \qquad \zeta(\eta) = \frac{\psi}{\sqrt{u_\infty \nu_\infty x}}, \qquad t = t(\eta) \qquad (15\text{-}11)$$

As in the constant-property case, we find that these changes of variables reduce the momentum and energy equations to ordinary differential equations:

$$\frac{d}{d\eta}\left(\frac{\mu\rho}{\mu_\infty\rho_\infty}\frac{d^2\zeta}{d\eta^2}\right) + \tfrac{1}{2}\zeta\frac{d^2\zeta}{d\eta^2} = 0 \qquad (15\text{-}12)$$

$$\frac{d}{d\eta}\left(k\frac{\rho}{\rho_\infty}\frac{dt}{d\eta}\right) + c\mu_\infty \frac{\zeta}{2}\frac{dt}{d\eta} = 0 \qquad (15\text{-}13)$$

Thus it is apparent that similarity solutions can be obtained regardless of how the properties vary with temperature.

Let us first consider a hypothetical gas for which Pr is a constant, c is a constant, μ varies directly as T, and ρ varies as T^{-1}. This is not very far from the room-temperature behavior of air, although the temperature dependence of μ (and by implication k) is a little large. (The properties of low-pressure steam and NH_3 correspond closely to these approximations.)

Under these conditions, Eq. (15-13) can be rewritten as

$$\frac{d}{d\eta}\left(\frac{\mu\rho}{\mu_\infty\rho_\infty}\frac{1}{\text{Pr}}\frac{dt}{d\eta}\right) + \tfrac{1}{2}\zeta\frac{dt}{d\eta} = 0$$

Since $\mu\rho$ is a constant under these assumptions, the momentum and energy equations reduce to

$$\zeta''' + \tfrac{1}{2}\zeta\zeta'' = 0$$
$$t'' + \tfrac{1}{2}\Pr\zeta t' = 0$$

These are, of course, exactly the equations obtained for the corresponding constant-property problem, Eqs. (8-8) and (10-4) [if t is replaced by τ, as in Eq. (10-1)]. We already have the desired solutions. Let us evaluate a Nusselt number and friction coefficient, evaluating all properties in the nondimensional parameters at the free-stream temperature t_∞:

$$\dot{q}_0'' = -k_0\left(\frac{\partial t}{\partial y}\right)_0 = -k_0(t_\infty - t_0)\left(\frac{\partial \tau}{\partial y}\right)_0$$

We recall that

$$\tau = \tau(\eta), \qquad \frac{\partial \tau}{\partial y} = \frac{d\tau}{d\eta}\frac{\partial \eta}{\partial y} = \tau'\sqrt{\frac{u_\infty}{\nu_\infty x}}\frac{\rho}{\rho_\infty}$$

$$\dot{q}_0'' = k_0(t_0 - t_\infty)\tau'(0)\sqrt{\frac{u_\infty}{\nu_\infty x}}\frac{\rho}{\rho_\infty}$$

$$= h(t_0 - t_\infty)$$

$$h = k_0\frac{\rho_0}{\rho_\infty}\sqrt{\frac{u_\infty}{\nu_\infty x}}\tau'(0)$$

$$\mathrm{Nu}_\infty = \frac{hx}{k_\infty} = \frac{k_0}{k_\infty}\frac{\rho_0}{\rho_\infty}\mathrm{Re}_{\infty,x}^{1/2}\tau'(0)$$

But, since $\Pr = \mu c/k$ and c are constants, k must vary as μ, that is, as T. Thus the property ratios cancel, and we obtain

$$\mathrm{Nu}_\infty = \mathrm{Re}_{\infty,x}^{1/2}\tau'(0)$$

In other words, our constant-property solutions are applicable if all properties are evaluated at the free-stream temperature. (In fact, under these property variations, $\mathrm{Nu}/\mathrm{Re}^{1/2}$ has the same value regardless of what temperature is used to evaluate properties.) In terms of Eq. (15-3), we conclude that $n = 0$.

Similar evaluation of the friction coefficient leads to an identical result if free-stream properties are used in the definition of friction coefficient, as well as in the Reynolds number. That is,

$$c_f = \frac{\tau_0}{\tfrac{1}{2}\rho_\infty u_\infty^2}$$

$$= 2\zeta''(0)\mathrm{Re}_{\infty,x}^{1/2} = 0.664\,\mathrm{Re}_{x,\infty}^{-1/2}$$

In terms of Eq. (15-4), we conclude that $m = 0$. (However, in this case, had properties been evaluated at t_0, a substantial temperature-ratio effect would have been noted.)

This approximate analysis illustrates how the temperature variation of the properties of a gas tends to compensate, so that, despite large property variations in a real gas, the constant-property solutions have a remarkable range of validity.

Solutions based on more realistic gas properties are presented by Brown and Donoughe[15] for the family of wedge solutions and the blowing and suction solutions. Reshotko and Cohen[16] include the axisymmetric stagnation point. For these solutions a slightly different transformation of variables is used, in addition to Eqs. (15-10) and (15-1):

$$\eta = y \sqrt{\frac{u_\infty}{v_0 x}}, \qquad \zeta(\eta) = \frac{\psi}{\sqrt{u_\infty v_0 x}} \qquad (15\text{-}14)$$

Again Eqs. (15-8) and (15-9) reduce to ordinary differential equations. For property variations with temperature the following relations, which approximate for air in the 600–1600 K range, are used:

$$\frac{\mu}{\mu_0} = \left(\frac{T}{T_0}\right)^{0.70}, \qquad \frac{c}{c_0} = \left(\frac{T}{T_0}\right)^{0.19}, \qquad \frac{k}{k_0} = \left(\frac{T}{T_0}\right)^{0.85}$$

The density variation is taken as $\rho/\rho_0 = (T/T_0)^{-1}$, and $Pr_0 = 0.70$. Note that, under these assumptions, Pr varies as $T^{0.04}$. With these relations substituted for the properties, the equations were integrated numerically and the results presented in tabular form. When put in the form of Eqs. (15-3) and (15-4), the data in Table 15-1 are a good approximation for $u_\infty = $ constant and for the two-dimensional stagnation point. Other cases, including blowing and suction, are found in the indicated references.

For $u_\infty = $ constant these results are very close to the results of the approximate solution described above: in fact, the only case for which there is a significant effect is for the friction coefficient for the stagnation point.

TABLE 15-1
Temperature-ratio exponents for the laminar external boundary layer with a gas

	$T_0/T_\infty > 1$ (heating)		$T_0/T_\infty < 1$ (cooling)	
	m	n	m	n
$u_\infty = $ constant	-0.10	-0.01	-0.05	0
Two-dimensional stagnation point	0.40	0.10	0.30	0.07

Experimental data appear to be lacking, but there are data for *flow normal to a cylinder.*[3] For $T_0/T_\infty > 1$ it is found that $n = 0.02$, which is quite consistent with the results in Table 15-1 since, on a cylinder, there is a two-dimensional stagnation point, followed by a laminar boundary layer approaching $u_\infty = $ constant. Thus a value between 0.10 and -0.01 would be expected. Similar results would be expected for the cylinder for $T_0/T_\infty < 1$, and for a *bank of tubes*, such as are frequently used in heat exchangers.

As an alternative to the temperature-ratio corrections suggested above, Eckert[17] suggests the use of a reference temperature scheme for gases. He finds that if the specific heat does not vary markedly through the boundary layer, the constant-property, laminar boundary-layer solutions for constant free-stream velocity can be correlated with the variable-property calculations by introducing *all* properties at a *reference temperature T_R*:

$$T_R = T_\infty + 0.5(T_0 - T_\infty) \tag{15-15}†$$

Under this scheme, the Reynolds number in both the constant-property heat-transfer and friction coefficient equations becomes

$$\mathrm{Re}_{x,R} = \frac{\rho_R u_\infty x}{\mu_R}$$

and similarly for the Nusselt, Stanton, and Prandtl numbers. The local surface shear stress is evaluated from Eq. (8-13), but with ρ_R replacing ρ.

For extremely large temperature differences the *specific heat* can no longer be treated as constant, and a formulation of the heat-transfer problem based on enthalpies rather than temperatures become more convenient. However, this problem is of more concern at high velocities and thus is discussed in Chap. 16.

THE LAMINAR EXTERNAL BOUNDARY LAYER: LIQUIDS

The laminar external boundary layer with liquids has evidently not been investigated experimentally. However, calculations by the authors using a finite-difference code and the real properties of *water* yielded the following results when comparison is made at the same *x Reynolds*

---

† Extension of this scheme to the high-velocity boundary layer is given in Chap. 16 and specifically in Eq. (16-40).

number:

$$\text{for } \mu_0/\mu_\infty > 1, \quad n = -0.25 \quad \text{(cooling)}$$
$$\text{for } \mu_0/\mu_\infty < 1, \quad n = -0.25 \quad \text{(heating)}$$
$$\text{for } \mu_0/\mu_\infty > 1, \quad m = 0.09 \quad \text{(cooling)}$$
$$\text{for } \mu_0/\mu_\infty < 1, \quad m = 0.20 \quad \text{(heating)}$$

If comparison is made at the same *enthalpy thickness* and *momentum thickness* Reynolds numbers, respectively, then the exponents become $n = -0.5$, $m = 0.18$ for cooling, and $n = -0.5$, $m = 0.4$ for heating.

THE TURBULENT EXTERNAL BOUNDARY LAYER: LIQUIDS

No experimental results for the turbulent boundary layer for liquids with large temperature differences exist. However, calculations by the authors using a finite-difference code, the real properties of water, and the mixing-length models described in Chaps 11 and 13 yielded the following results when comparison is made at the same *enthalpy thickness* (for n) and *momentum thickness* (for m) Reynolds numbers:

$$\text{for } \mu_0/\mu_\infty > 1, \quad n = -0.12 \quad \text{(cooling)}$$
$$\text{for } \mu_0/\mu_\infty < 1, \quad n = -0.23 \quad \text{(heating)}$$
$$\text{for } \mu_0/\mu_\infty > 1, \quad m = 0.26 \quad \text{(cooling)}$$
$$\text{for } \mu_0/\mu_\infty < 1, \quad m = 0.26 \quad \text{(heating)}$$

Note that these results are very close to the experimental results for turbulent flow in a tube with liquids. This is not unexpected, but it does provide additional confidence in the mixing-length models discussed in Chaps. 11 and 13.

The exponents for the case where comparison is made at the same x *Reynolds* number can be obtained by *subtracting* approximately 0.25 from each of the above.

THE TURBULENT EXTERNAL BOUNDARY LAYER: GASES

The turbulent external boundary layer with temperature-dependent properties has received little attention, either analytic or experimental. Most of the available experimental data involve high Mach numbers, which introduces the additional complication of viscous energy dissipation, a subject that is discussed in Chap. 16. However, it is difficult to see how the effect of temperature ratio on Stanton number (or Nusselt number) can be materially different from that for flow in a tube. In fact, the principal difference must be simply that the temperature ratio on which n is based for the external boundary layer, T_0/T_∞, is analogous in the case of flow in a tube to $T_0/T_{\text{centerline}}$, rather than T_0/T_m. Recall that in Chap. 14 it was suggested that $(T_0 - T_m)/(T_0 - T_{\text{centerline}})$ is about 0.90. For the heating case this suggests that the absolute value of n should be less than the value 0.5 recommended for flow in a tube. A check of some representative values of T_0 and T_m yields a value for n for the external boundary layer between -0.35 and -0.40. This is presumably the correction that should be made when the comparison between constant and variable properties is made at the same *enthalpy thickness* Reynolds number.

Calculations by the authors using a finite-difference code, the real properties of *air,* and the mixing-length theory discussed in Chaps. 11 and 13, with constant free-stream velocity, yield the following results when comparison is made at the same *momentum thickness* Reynolds number for c_f, and the same *enthalpy thickness* Reynolds number for St:

$$\text{for } T_0/T_\infty > 1, \quad m = -0.33, \quad n = -0.30 \quad \text{(heating)}$$
$$\text{for } T_0/T_\infty < 1, \quad m = -0.22, \quad n = -0.14 \quad \text{(cooling)}$$

When these results are converted to an x *Reynolds number* basis, they become

$$\text{for } T_0/T_\infty > 1, \quad m = -0.58, \quad n = -0.55 \quad \text{(heating)}$$
$$\text{for } T_0/T_\infty < 1, \quad m = -0.47, \quad n = -0.39 \quad \text{(cooling)}$$

As was seen for flow in a tube, the exponents tend to increase somewhat with temperature ratio; the results cited above are valid for $T_0/T_\infty \approx 2.00$ and 0.75.

Eckert[17] recommends the same *reference temperature* for the turbulent boundary layer for a gas as for the laminar boundary layer, i.e., Eq. (15-15). Such a procedure gives somewhat smaller values of m and n, as do the results of Spalding and Chi.[20]

PROBLEMS

15-1. Consider fully developed laminar flow between symmetrically heated parallel plates with constant heat rate per unit of duct length. The plate spacing is 1 cm, the fluid is an aircraft engine oil, and the heat flux is 1.4 kW/m^2. The mass velocity G is $600 \text{ kg/(s} \cdot \text{m}^2)$. At a particular point in question the surface temperature is $110°C$. Assuming that the viscosity is the only temperature-dependent property of significance, carry out the necessary calculations (numerical integration is necessary) to evaluate the friction coefficient and the heat-transfer conductance. Compare the results with the recommended procedures given in the text.

15-2. For air as a working substance and using the actual tabulated properties, compare the recommended temperature ratio and reference property schemes for evaluating heat-transfer coefficients and friction coefficients for a laminar boundary layer with constant free-stream velocity for both heating and cooling.

15-3. Repeat Prob. 15-2 but for a turbulent boundary layer with constant free-stream velocity.

REFERENCES

1. Deissler, R. C.: NACA (now NASA) TN 2410, Washington, July 1951.
2. Yang, K. T.: *J. Heat Transfer,* vol. 84, pp. 353–362, 1962.
3. Kays, W, M.: and W. B. Nicoll: *J. Heat Transfer,* vol. 85, pp. 329–338, 1963.
4. Davenport, M. E., PhD. dissertation, Stanford University, 1962.
5. Worsøe-Schmidt, P. M.: Technical Report 247-8. Thermosciences Division, Department of Mechanical Engineering, Stanford University, November 1964.
6. Petukhov, B. S.: *Advances in Heat Transfer,* vol. 6, Academic Press, pp. 503–564, 1970.
7. Allen, R. W., and E. R. C. Eckert: *J. Heat Transfer,* vol. 86, pp. 301–310, 1964.
8. Rohonczy, G.: *Schweitzer Arch.,* no. 5, 1939.
9. Humble, L. V., W. H. Lowdermilk, and L. G. Desmon: NACA (now NASA) Report 1020, Washington, 1951.
10. McCarthy, J. R., and H. Wolf: Research Report RR-60-12, Rocketdyne Corporation. Canoga Park, California, 1960.
11. Barnes, J. F.: National Gas Turbine Establishment Report R, 241, Pyestock, Hants, England, 1960.
12. Nicoll, W. B., and W. M. Kays: Technical Report 43, Department of Mechanical Engineering, Stanford University, 1959.
13. McEligot, D. M.: Ph.D. dissertation, Stanford University, 1963.
14. Lel'chuk, V. D., and B. V. Dyadyakin: AEC-TR-4511, Atomic Energy Commission Translation, 1962.
15. Brown, W. B., and P. L. Donoughe: NACA (now NASA) TN 2379, Washington, September 1951.
16. Reshotko, E., and C. B. Cohen: NACA (now NASA) TN 3513, Washington, July 1955.
17. Eckert, E. R. G.: *J. Aero. Sci.,* vol. 22, pp. 585–587, 1955.
18. Sleicher, C. A., and M. W. Rouse: *Int. J. Heat Mass Transfer,* vol. 18, pp. 677–683, 1975.
19. Brim, L. H., and R. H. Eustis: ASME Paper 70-HT/SpT-44, 1970.
20. Spalding, D. B., and S. W. Chi: *J. Fluid Mech.* vol. 18, p. 117, 1964.

CHAPTER
16

CONVECTIVE HEAT TRANSFER AT HIGH VELOCITIES

In all the forced-convection solutions considered up to this point the velocities and velocity gradients have been assumed sufficiently small that the effects of kinetic energy and viscous energy dissipation could be neglected. We now consider the influence of high velocities and viscous energy dissipation. Although these effects can be of importance in internal tube flow, we are primarily concerned with external boundary layers, where the more important applications appear.

High-velocity convection involves essentially two different phenomena:

1. Conversion of mechanical energy to thermal energy, resulting in temperature variations in the fluid
2. Variation of the fluid properties as a result of the temperature variation

Extremely high velocities in gases like air lead to very high temperatures, dissociation, mass concentration gradients, and thus mass diffusion, which further complicates the problem. However, at present we restrict our attention to the boundary layer with no chemical reaction; this means that for air, at least, we do not deal with temperatures greater than about 2000 K, or Mach numbers greater than about 5.

Generally the high-velocity effects are of most importance in high-speed aerodynamics. However, the effects of the mechanical-to-thermal energy conversion can actually be of considerable importance at quite moderate velocities of a liquid if the Prandtl number is sufficiently high. We restrict attention to fluids with Prandtl number near 1, with primary emphasis on gases.

Much of the literature on high-velocity convection involves a consideration of both mechanical energy conversion and the effects of temperature-dependent fluid properties, so that it is often difficult to see what are the separate effects. Here we first consider the energy conversion problem and then take up the variable-property effects.

The energy conversion process can take place either reversibly or irreversibly. Shock waves and viscous interaction (such as in a boundary layer) give rise to irreversible interchanges of mechanical and thermal energy. Inviscid velocity changes (such as the deceleration of the fluid approaching a subsonic stagnation point) yield reversible, or very nearly reversible, exchanges.

The significant differences in the two types of behavior are well illustrated by a comparison of the stagnation point with the high-speed boundary layer. At a stagnation point (Fig. 16-1) the velocity decreases, and both the pressure and the temperature rise, *outside* the boundary later. For subsonic flow the process is close to isentropic, and the viscosity of the fluid plays no part.

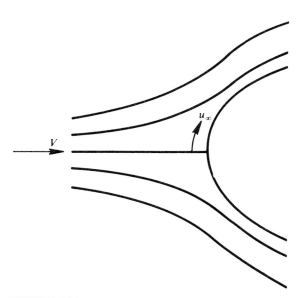

FIGURE 16-1
Flow at a stagnation point.

Regardless of whether the deceleration is reversible or irreversible, the temperature of the fluid *outside* the stagnation-region boundary layer is the *stagnation temperature*, that is, the fluid temperature resulting from an adiabatic deceleration:

$$t^*_\infty = t_\infty + \frac{V^2}{2c} \qquad (16\text{-}1)\dagger$$

Thus the *high-velocity stagnation-point problem* involves nothing that has not already been discussed. The low-velocity stagnation-point solutions are applicable (although a correction for the influence of temperature-dependent properties may be required), and it is only necessary to replace t_∞ by t^*_∞ in the convection rate equation, that is.

$$\dot{q}''_0 = h(t^*_\infty - t_0) \qquad (16\text{-}2)$$

The *high-velocity boundary-layer problem,* on the other hand, is somewhat different. If the free-stream velocity is large, there will be large velocity gradients *within* the boundary layer and a substantial conversion of mechanical energy to thermal energy by viscous shear within the boundary layer. If we consider for the moment a body that is *insulated,* as in Fig. 16-2, it is apparent that the thermal energy generated can escape from the region close to the wall surface only by the mechanism of molecular or eddy conduction. In the steady state an equilibrium is established between viscous energy dissipation and heat conduction, and the result is a temperature distribution such as shown in Fig. 16-2. The surface temperature under these conditions is known as the *adiabatic wall temperature* t_{aw}.

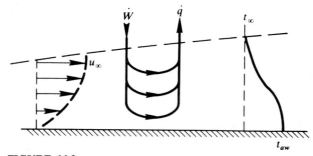

FIGURE 16-2
Velocity and adiabatic wall-temperature profile in a high-velocity boundary layer.

---

† This equation is only valid as long as the specific heat may be treated as constant.

It is apparent that determination of this adiabatic wall temperature is going to be of importance in the calculation of heat transfer, since whether heat is transferred to or from the wall must depend on whether the surface temperature is above or below t_{aw}.

It is also apparent that the Prandtl number of the fluid is going to have some bearing on the adiabatic wall temperature. The Prandtl number is the ratio of the viscosity (responsible for the energy dissipation) to the thermal diffusivity (associated with the mechanism whereby heat escapes from the boundary layer). In fact, other things being equal, this argument would suggest that a high Prandtl number should lead to a high adiabatic wall temperature, and that a low Prandtl number should lead to a low adiabatic wall temperature.

We now concentrate attention on the high-velocity *boundary-layer problem* and in particular on high-velocity *gas* flow, so that we are concerned with Prandtl numbers near 1. First we develop a particular form of the energy equation of the boundary layer, and then we consider two special cases that lend themselves to simple and direct solution, before going on to the more general problem.

THE STAGNATION ENTHALPY EQUATION

Referring to Chap. 4, note that two basic forms of the laminar energy equation of the boundary layer are developed. The first, which might be called the *total* energy equation, is Eq. (4-26). The second is Eq. (4-27), followed by some particularizations of it, namely Eqs. (4-37) and (4-38). These latter equations were developed from Eq. (4-26) by combining it with the momentum equation. Equation (4-27) is a *thermal* energy equation, every term representing thermal energy or conversion of mechanical to thermal energy. Actually, we make use of both forms of the energy equation in the developments to follow, and it is well to note that the shear-stress term has a different meaning in the two equations. In Eq. (4-26) it represents work done against viscous forces, part of which is a reversible mechanical work rate and part of which is dissipation into thermal energy. In Eq. (4-27) the viscous term represents conversion of mechanical to thermal energy resulting from the viscous work term, and this term is thus called the *viscous dissipation term.*

Let us first make use of Eq. (4-26) and develop a laminar energy equation with *stagnation enthalpy* as the dependent variable, because at Prandtl numbers at or near 1 this turns out to be a particularly convenient form. Let us neglect work against body forces, internal heat sources, and mass concentration gradients. Then we introduce the *stagnation enthalpy* i^*, the perfect-gas relation between enthalpy and temperature, and the relation between the shear stress and the viscosity coefficient:

$$i^* = i + \tfrac{1}{2}u^2$$

$$di = c\,dt$$

(c is the specific heat at constant pressure)

$$\tau_{yx} = \mu\,\frac{\partial u}{\partial y}$$

$$\rho u\,\frac{\partial i^*}{\partial x} + \rho v\,\frac{\partial i^*}{\partial y} - \frac{\partial}{\partial y}\left(\frac{k}{c}\frac{\partial i}{\partial y}\right) = \frac{\partial}{\partial y}\left(u\mu\,\frac{\partial u}{\partial y}\right) \tag{16-3}$$

But

$$\frac{\partial i^*}{\partial y} = \frac{\partial i}{\partial y} + u\,\frac{\partial u}{\partial y} \tag{16-4}$$

We solve for $\partial i/\partial y$ and substitute into Eq. (16-3). After transposing and introducing the definition of the Prandtl number, we obtain

$$\rho u\,\frac{\partial i^*}{\partial x} + \rho v\,\frac{\partial i^*}{\partial y} - \frac{\partial}{\partial y}\left(\frac{k}{c}\frac{\partial i^*}{\partial y}\right) = \frac{\partial}{\partial y}\left[\left(1-\frac{1}{\mathrm{Pr}}\right)\mu\,\frac{\partial(\tfrac{1}{2}u^2)}{\partial y}\right] \tag{16-5}$$

For *turbulent flows* the boundary-layer form of the stagnation enthalpy equation was developed in Chap. 5, and it can be recast using the same ideas leading to Eq. (16-5) for the laminar boundary layer. Neglecting the body force and source terms in Eq. (5-32), and substituting for the molecular heat flux and shear stress, leads to

$$\rho\bar{u}\,\frac{\partial \bar{i}^*}{\partial x} + \rho\bar{v}\,\frac{\partial \bar{i}^*}{\partial y} - \frac{\partial}{\partial y}\left(\frac{k}{c}\frac{\partial \bar{i}^*}{\partial y} - \overline{\rho v' i'}\right) = \frac{\partial}{\partial y}\left[\mu\left(1-\frac{1}{\mathrm{Pr}}\right)\frac{\partial(\tfrac{1}{2}\bar{u}^2)}{\partial y} - \overline{\rho u' v'}\,\bar{u}\right]$$
$$\tag{16-6}$$

The eddy diffusivity models developed in the previous turbulence chapters can now be applied to Eq. (16-6). First, the turbulent heat flux and Reynolds stress are recast using mean field closure:

$$-\overline{\rho v' i'} = \rho\varepsilon_H\,\frac{\partial \bar{i}}{\partial y} = \frac{k_t}{c}\frac{\partial \bar{i}}{\partial y} = \frac{k_t}{c}\left[\frac{\partial \bar{i}^*}{\partial y} - \frac{\partial(\tfrac{1}{2}\bar{u}^2)}{\partial y}\right]$$

$$-\overline{\rho u' v'}\,\bar{u} = \rho\varepsilon_M\,\frac{\partial(\tfrac{1}{2}\bar{u}^2)}{\partial y} = \mu_t\,\frac{\partial(\tfrac{1}{2}\bar{u}^2)}{\partial y}$$

Then the definitions of effective viscosity, conductivity, and Prandtl number are introduced:

$$k_{\mathrm{eff}} = k + k_t = k + \rho c\varepsilon_H$$

$$\mu_{\mathrm{eff}} = \mu + \mu_t = \mu + \rho\varepsilon_M$$

$$\mathrm{Pr}_{\mathrm{eff}} = \frac{\mu_{\mathrm{eff}}}{k_{\mathrm{eff}}/c} = \frac{1 + \varepsilon_M/v}{1/\mathrm{Pr} + (\varepsilon_M/v)/\mathrm{Pr}_t}$$

Using these, the resulting turbulent boundary-layer equation for the stagnation enthalpy becomes

$$\rho\bar{u}\frac{\partial\overline{i^*}}{\partial x} + \rho\bar{v}\frac{\partial\overline{i^*}}{\partial y} - \frac{\partial}{\partial y}\left(\frac{k_{eff}}{c}\frac{\partial\overline{i^*}}{\partial y}\right) = \frac{\partial}{\partial y}\left[\left(1-\frac{1}{\mathrm{Pr}_{eff}}\right)\mu_{eff}\frac{\partial(\frac{1}{2}\bar{u}^2)}{\partial y}\right] \quad (16\text{-}7)$$

Given a solution to Eq. (16-5) or Eq. (16-7), the heat flux at the surface $\dot{q}_0''$ can be evaluated as follows (for turbulent flows the time-averaged quantities replace the laminar quantities):

$$\dot{q}_0'' = -\left(k\frac{\partial t}{\partial y}\right)_0$$

From Eq. (16-4),

$$\left(\frac{\partial i^*}{\partial y}\right)_0 = \left(\frac{\partial i}{\partial y}\right)_0 = \left(\frac{di}{dt}\frac{\partial t}{\partial y}\right)_0 = \left(c\frac{\partial t}{\partial y}\right)_0$$

Thus

$$\dot{q}_0'' = -\left(\frac{k}{c}\frac{\partial i^*}{\partial y}\right)_0 \quad (16\text{-}8)$$

The advantage of this particular formulation of the energy equation for use at Pr near 1 is apparent when one notes that at $\mathrm{Pr} = 1$ the entire right-hand side goes to zero. Further, it is worth noting that, even for a low-velocity flow (where $i^* = i$), it is perfectly feasible to develop convective heat-transfer theory on the basis of enthalpy rather than temperature; and if the specific heat is not constant, this is a more natural formulation. In Chap. 20 such a formulation is introduced since it is particularly useful when mass transfer is present or when there is chemical reaction in the boundary layer. In using Eq. (16-5), or Eq. (16-7), it becomes convenient to define an *enthalpy conductance* g_i as an alternative to the heat-transfer coefficient h:

$$g_i = \frac{[(k/c)\,\partial i^*/\partial y]_0}{i_\infty^* - i_0^*}$$

from which

$$\dot{q}_0'' = g_i(i_0^* - i_\infty^*) \quad (16\text{-}9)\dagger$$

Then, if the velocity is low ($i^* = i$) and the specific heat c is constant, the following relation exists between g_i and h:

$$g_i = h/c \quad (16\text{-}10)$$

---

† See also the development of Eq. (20-33) in the chapter on mass-transfer (Chap. 20).

THE HIGH-VELOCITY THERMAL BOUNDARY LAYER FOR A FLUID WITH Pr = 1

Let us now examine some particular features of the high-velocity boundary layer by using Eq. (16-5) and considering a fluid with Pr = 1. Let us further assume a fluid with constant specific heat so that we can employ the *stagnation temperature* t^* rather than the stagnation enthalpy; this, in turn, permits some useful comparisons with previously obtained low-velocity solutions.

By definition,

$$t^* = t + \frac{u^2}{2c}$$

Then, with Pr = 1 and c constant, Eq. (16-5) becomes

$$u\rho c \frac{\partial t^*}{\partial x} + v\rho c \frac{\partial t^*}{\partial y} - \frac{\partial}{\partial y}\left(k \frac{\partial t^*}{\partial y}\right) = 0 \tag{16-11}$$

With Pr = 1 and c constant, μ and k can still be temperature-dependent, although they must be the same functions of temperature. These are reasonable approximations for most gases.

Equation (16-11) is precisely the same as Eq. (15-9), which was developed and solved for a low-velocity boundary layer. One need only replace t by t^* in the solutions. If all properties are treated as constant, Eq. (16-11) becomes identical with Eq. (4-39) or Eq. (10-2), for which solutions already exist.

The preceding argument is based on a laminar boundary layer, but if it is assumed that $\mathrm{Pr}_t = 1$ as well as Pr = 1, the same conclusion is reached for a turbulent boundary layer as for a laminar boundary layer.

Note that the momentum equation is unaffected by viscous dissipation, except insofar as the resulting static temperature distribution affects the fluid properties.

The boundary conditions for the low-velocity boundary-layer solutions under discussion (the constant-surface-temperature solutions) have been

$$u, v = 0, \quad t = t_0 \quad \text{at } y = 0$$

$$u \to u_\infty, \quad t \to t_\infty \quad \text{as } y \to \infty$$

$$u = u_\infty, \quad t = t_\infty \quad \text{at } x = 0$$

By replacing t by t^*, the corresponding boundary conditions for Eq.

(16-11) are

$$u, v = 0, \qquad t^* = t_0^* \quad \text{at } y = 0$$

$$u \to u_\infty, \qquad t^* \to t_\infty^* \quad \text{as } y \to \infty$$

$$u = u_\infty, \qquad t^* = t_\infty^* \quad \text{at } x = 0$$

These are, of course, exactly the boundary conditions desired for a constant-temperature plate in a high-velocity stream.

In the low-velocity case the solutions were reduced to a local convection conductance h such that

$$\dot{q}_0'' = -k\left(\frac{\partial t}{\partial y}\right)_0 = h(t_0 - t_\infty)$$

The corresponding solutions to Eq. (16-11) can be expressed as follows, where h is obviously the same function of the system variables as for the low-velocity case:

$$-k\left(\frac{\partial t^*}{\partial y}\right)_0 = h(t_0^* - t_\infty^*)$$

However, $(\partial t^*/\partial y)_0 = (\partial t/\partial y)_0$ and $t_0^* = t_0$ because $u = 0$ at $y = 0$. Thus

$$-k\left(\frac{\partial t}{\partial y}\right)_0 = q_0'' = h(t_0 - t_\infty^*) \qquad (16\text{-}12)$$

Evidently we can calculate surface heat-transfer rates, at least for $Pr = 1$, by merely replacing the free-stream temperature in the convection rate equation by the free-stream stagnation temperature; this, then, is the *major influence of high-velocity viscous dissipation.*

We can now evaluate the adiabatic wall temperature under this condition of $Pr = 1$ by merely letting $\dot{q}_0'' = 0$ in Eq. (16-12):

$$h(t_0 - t_\infty^*) = 0$$

$$t_{aw} = t_0 = t_\infty^* = t_\infty + \frac{u_\infty^2}{2c} \qquad (16\text{-}13)$$

From these results, a diagram of the temperature distribution in the boundary layer, shown in Fig. 16-3, can be readily constructed.

Since the Prandtl number for all gases does not differ greatly from 1, we really do not expect substantially different results for real gases.

THE LAMINAR CONSTANT-PROPERTY BOUNDARY LAYER FOR $Pr \neq 1$

Let us now consider the problem of a laminar high-velocity boundary layer on a flat plate with t_∞ and u_∞ constant but with Prandtl number

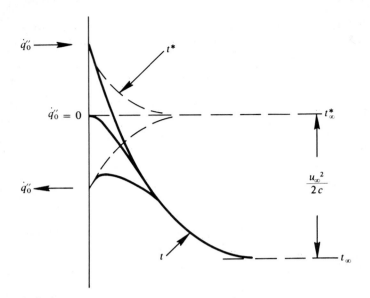

FIGURE 16-3
Temperature distribution in the high-velocity boundary layer for a fluid with Pr = 1.

different from 1. So that we may more clearly see the influence of viscous dissipation, and the effect of the Prandtl number on this influence, we consider first the rather hypothetical case of *constant* fluid properties.

The applicable differential equations are the same continuity and momentum equations employed for the corresponding low-velocity problem, Eqs. (8-1) and (8-2), but the energy equation must contain a viscous dissipation term. A convenient form of the energy equation is Eq. (4-38). Thus we have

$$\frac{\partial u}{\partial x} + \frac{\partial v}{\partial y} = 0 \qquad \text{(continuity)} \qquad (16\text{-}14)$$

$$u\frac{\partial u}{\partial x} + v\frac{\partial u}{\partial y} - v\frac{\partial^2 u}{\partial y^2} = 0 \qquad \text{(momentum)} \qquad (16\text{-}15)$$

$$u\frac{\partial t}{\partial x} + v\frac{\partial t}{\partial y} - \alpha\frac{\partial^2 t}{\partial y^2} - \frac{\alpha\,\text{Pr}}{c}\left(\frac{\partial u}{\partial y}\right)^2 = 0 \qquad \text{(energy)} \qquad (16\text{-}16)$$

Since the continuity and momentum equations are the same as for the low-velocity boundary layer discussed in Chap. 8, and the velocity boundary conditions are likewise the same, it is quite apparent that similarity solutions for the velocity field exist; in fact, the solution is simply Table 8-1.

For the energy equation one is tempted to again seek similarity solutions, employing the same similarity parameter as was used for the momentum equation. If we assume that

$$\zeta = \zeta(\eta), \qquad t = t(\eta)$$

where

$$\eta = \frac{y}{\sqrt{vx/u_\infty}}$$

and that

$$u = \frac{\partial \psi}{\partial y}, \qquad v = -\frac{\partial \psi}{\partial x}, \qquad \psi = \sqrt{vxu_\infty}\,\zeta$$

then Eq. (16-16) indeed reduces to an ordinary differential equation:

$$t'' + \frac{u_\infty^2}{c}\,\mathrm{Pr}\,\zeta''^2 + \tfrac{1}{2}\,\mathrm{Pr}\,\zeta t' = 0 \tag{16-17}$$

It is now convenient to define a new dependent variable such that

$$\theta(\eta) = \frac{t - t_\infty}{u_\infty^2/2c} \tag{16-18}$$

Then Eq. (16-17) becomes

$$\theta'' + \tfrac{1}{2}\,\mathrm{Pr}\,\zeta\theta' + 2\,\mathrm{Pr}\,\zeta''^2 = 0 \tag{16-19}$$

Our first concern is a *particular solution* to Eq. (16-19) for the case where there is an *adiabatic wall*, that is, $\dot{q}_0'' = 0$. We call this solution θ_{aw}, and the boundary conditions on Eq. (16-19) must then be

$$\theta(\infty) = 0, \qquad \theta'(0) = 0$$

Equation (16-19) is a linear equation that is readily solved for θ_{aw} by use of an integrating factor.

Of special interest is the wall surface temperature under adiabatic wall conditions, and it is a straightforward process to obtain

$$\theta_{aw}(0) = \int_0^\infty \frac{\int_0^\infty \exp\left(\int_0^\eta \tfrac{1}{2}\,\mathrm{Pr}\,\zeta\,d\eta\right)2\,\mathrm{Pr}\,(\zeta'')^{1/2}\,d\eta}{\exp\left(\int_0^\eta \tfrac{1}{2}\,\mathrm{Pr}\,\zeta\,d\eta\right)}\,d\eta \tag{16-20}$$

Equation (16-20) can be readily integrated numerically for various values of the Prandtl number, employing $\zeta(\eta)$ from Table 8-1. The results are then in the form of a table of $\theta_{aw}(0)$ as a function of Pr.[2] However, in the range of Pr from 0.5 to 10 the solution is quite well represented by

$$\theta_{aw}(0) \approx \mathrm{Pr}^{1/2} \tag{16-21}$$

Usually $\theta_{aw}(0)$ is known as the *recovery factor* r_c. The adiabatic wall temperature t_{aw} can be evaluated from

$$t_{aw} = t_\infty + r_c \frac{u_\infty^2}{2c} \qquad (16\text{-}22)$$

where $r_c \approx \mathrm{Pr}^{1/2}$. Note that at $\mathrm{Pr} = 1$, $r_c = 1$, and Eq. (16-22) reduces to Eq. (16-13).

Next we would like to obtain a complete solution to Eq. (16-19) for any surface temperature. We already have one particular solution, θ_{aw}, but this, of course, gives us only the temperature field when there is no heat transfer at the surface. We can develop a complete solution if we can find a general solution to the homogeneous part of Eq. (16-19), since then the complete solution is the sum of the general solution and the particular solution.

We already have such a solution from the low-velocity problem. Recall Eq. (10-4):

$$\tau'' + \tfrac{1}{2} \mathrm{Pr}\, \zeta \tau' = 0$$

We have $\tau(\eta)$ for boundary conditions $\tau(0) = 0$ and $\tau(\infty) = 1$. Then a complete solution to Eq. (16-19) is

$$\theta = \theta_{aw} + C_1 \tau + C_2 \qquad (16\text{-}23)$$

as substitution into Eq. (16-19) readily demonstrates.

We would now like to satisfy the boundary conditions

$$\theta(\infty) = 0, \qquad \theta(0) = \frac{t_0 - t_\infty}{u_\infty^2/2c}$$

For the first boundary condition,

$$0 = 0 + C_1 \times 1 + C_2, \qquad C_1 = -C_2$$

From the second boundary condition,

$$\frac{t_0 - t_\infty}{u_\infty^2/2c} = \theta_{aw}(0) + 0 + C_2$$

$$C_2 = \frac{t_0 - t_\infty}{u_\infty^2/2c} - \theta_{aw}(0)$$

Substituting the constants into Eq. (16-23), and rearranging,

$$\theta = \frac{t - t_\infty}{u_\infty^2/2c} = \theta_{aw} + \left(\frac{t_0 - t_\infty}{u_\infty^2/2c} - \frac{t_{0\,aw} - t_\infty}{u_\infty^2/2c} \right)(1 - \tau)$$

from which

$$t - t_\infty = \theta_{aw} \frac{u_\infty^2}{2c} + (t_0 - t_{0\,aw})(1 - \tau)$$

But $t_{0\,aw}$ is simply t_{aw}.

We can now evaluate the heat flux at the wall:

$$\dot{q}_0'' = -k\left(\frac{\partial t}{\partial y}\right)_0$$

$$\frac{\partial t}{\partial y} = \frac{\partial \theta_{aw}}{\partial y}\frac{u_\infty^2}{2c} - (t_0 - t_{aw})\frac{\partial \tau}{\partial y}$$

$$\left(\frac{\partial t}{\partial y}\right)_0 = 0 - (t_0 - t_{aw})\left(\frac{\partial \tau}{\partial y}\right)_0$$

$$\dot{q}_0'' = -k\left(\frac{\partial t}{\partial y}\right)_0 = k(t_0 - t_{aw})\frac{\tau'(0)}{\sqrt{vx/u_\infty}}$$

But, from Eqs. (10-9) and (10-10),

$$\tau'(0) = \left[\int_0^\infty \exp\left(-\tfrac{1}{2}\Pr\int_0^\eta \zeta \, d\eta\right) d\eta\right]^{-1} \approx 0.332\,\Pr^{1/3}$$

Then

$$\dot{q}_0'' = k(t_0 - t_{aw})\frac{0.332\,\Pr^{1/3}}{\sqrt{vx/u_\infty}}$$

This result immediately suggests that for a high-velocity boundary layer we should define a convection conductance h on the basis of the difference between surface temperature and *adiabatic wall temperature*. This is perfectly consistent with the $\Pr = 1$ results of the preceding section, because in that case $t_{aw} = t_\infty^*$, and it is also perfectly consistent with the low-velocity problem, since then $t_{aw} = t_\infty$. Thus let

$$\dot{q}_0'' = h(t_0 - t_{aw}) \tag{16-24}$$

Substituting,

$$h = k\frac{0.332\,\Pr^{1/3}}{\sqrt{vx/u_\infty}}$$

or

$$\mathrm{Nu}_x = 0.332\,\Pr^{1/3}\,\mathrm{Re}_x^{1/2} \tag{16-25}$$

Note that this is identical to the low-velocity result given in Eq. (10-10). We see that if fluid properties are constant, the effects of viscous dissipation are entirely taken care of by the simple expedient of substituting t_{aw} for t_∞ in the defining equation for the conductance h. Although we have demonstrated this fact only for the laminar boundary layer with constant u_∞ and t_0, it is actually a far more general conclusion and is applicable also to the turbulent boundary layer. Of course, there remains the problem of evaluating the recovery factor for other situations so that t_{aw} may be determined.

THE LAMINAR BOUNDARY LAYER FOR A GAS WITH VARIABLE PROPERTIES

We first examine an approximate solution and then discuss a more refined analysis. In both cases we consider the case of constant free-stream velocity and constant surface temperature.

If the fluid properties are variable, the three applicable differential equations, corresponding to Eqs. (16-14), (16-15), and (16-16), are

$$\frac{\partial(\rho u)}{\partial x} + \frac{\partial(\rho v)}{\partial y} = 0 \qquad \text{(continuity)} \qquad (16\text{-}26)$$

$$\rho u \frac{\partial u}{\partial x} + \rho v \frac{\partial u}{\partial y} = \frac{\partial}{\partial y}\left(\mu \frac{\partial u}{\partial y}\right) \qquad \text{(momentum)} \qquad (16\text{-}27)$$

and, from Eq. (4-37),

$$\rho u c \frac{\partial t}{\partial x} + \rho v c \frac{\partial t}{\partial y} - \frac{\partial}{\partial y}\left(k \frac{\partial t}{\partial y}\right) - \mu\left(\frac{\partial u}{\partial y}\right)^2 = 0 \qquad \text{(energy)} \qquad (16\text{-}28)$$

In Chap. 15 we found that similarity solutions are obtainable from this system of equations when the dissipation term in the energy equation is omitted. Therefore it is suggested that the same change of variables be tried here. Substituting Eqs. (15-10) and (15-11), the momentum and energy equations are found to reduce to the following ordinary differential equations:

$$\frac{d}{d\eta}\left(\frac{\mu\rho}{\mu_\infty\rho_\infty}\frac{d^2\zeta}{d\eta^2}\right) + \frac{1}{2}\zeta\frac{d^2\zeta}{d\eta^2} = 0 \qquad (16\text{-}29)$$

$$\frac{d}{d\eta}\left(k\frac{\rho}{\rho_\infty}\frac{dt}{d\eta}\right) + \frac{\mu\rho}{\rho_\infty}u_\infty^2\left(\frac{d^2\zeta}{d\eta^2}\right)^2 + c\mu_\infty\frac{\zeta}{2}\frac{dt}{d\eta} = 0 \qquad (16\text{-}30)$$

If we now make the only slightly restrictive assumption that c is a constant, Eq. (16-30) can be rewritten as

$$\frac{d}{d\eta}\left(\frac{\mu\rho}{\mu_\infty\rho_\infty}\frac{1}{\text{Pr}}\frac{dt}{d\eta}\right) + \frac{\mu\rho}{\mu_\infty\rho_\infty}\frac{u_\infty^2}{c}\left(\frac{d^2\zeta}{d\eta^2}\right)^2 + \frac{1}{2}\zeta\frac{dt}{d\eta} = 0 \qquad (16\text{-}31)$$

As a first approximation we could assume $\text{Pr} = \text{constant}$ and $\mu\rho = \text{constant}$, as was discussed in Chap. 15. This is not a bad approximation for a gas, but then Eq. (16-29) again becomes the same as Eq. (8-8), the Blasius equation obtained for the corresponding constant-property problem. Equation (16-31) becomes the same as Eqs. (16-17) and (16-19), the constant-property energy equations for high-velocity flow.

Thus, under the assumptions of $\mu\rho = $ constant and $\mathrm{Pr} = $ constant, we find that the solution for constant u_∞ and constant t_0 is identical to the corresponding constant-property solution. We can evaluate all properties at free-stream static temperature, and the high-velocity viscous dissipation effects are again entirely taken care of by merely introducing the adiabatic wall temperature into the convection rate equation.

The second approximation to this problem must obviously involve insertion of more nearly correct functions of T/T_∞ for μ/μ_∞ and ρ/ρ_∞ in Eqs. (16-29) and (16-31). Under these conditions, it is not possible to absorb the free-stream velocity in the dependent variable, as was done using Eq. (16-18).

In order to see more clearly the form that the final solution must take, let us introduce a nondimensional free-stream velocity, the Mach number M:

$$M = \frac{u_\infty}{\sqrt{\gamma R T_\infty}} \tag{16-32}$$

where γ is the ratio of the specific heats and R is the gas constant. Then letting $\tau = T/T_\infty$ and $f(\tau) = \mu\rho/\mu_\infty\rho_\infty$, Eq. (16-31) becomes

$$\frac{d}{d\eta}\left[f(\tau)\frac{1}{\mathrm{Pr}}\frac{d\tau}{d\eta}\right] + f(\tau)M^2(\gamma - 1)\zeta''^2 + \tfrac{1}{2}\zeta\tau' = 0 \tag{16-33}$$

We then seek the family of solutions $\tau(\eta)$ for the boundary conditions

$$\tau(\infty) = 1, \qquad \tau'(0) = \text{constant}$$

Again an important particular solution τ_{aw} is obtained for the condition $\tau'(0) = 0$. From this a recovery factor can be obtained, defined again by Eq. (16-22).

For the solutions for $\tau'(0)$ other than zero we can evaluate the heat flux at the surface from $\tau'(0)$, define a convection conductance again by Eq. (16-24), and evaluate a Nusselt number. If we refer all fluid properties in the Nusselt and Reynolds numbers to free-stream static temperature, it can readily be shown that

$$\frac{\mathrm{Nu}_{x,\infty}}{\mathrm{Re}_{x,\infty}^{1/2}} = f(\tau(0))\frac{\tau'(0)}{\tau_{\mathrm{aw}}(0) - \tau(0)} \tag{16-34}$$

Since the Mach number, as well as the Prandtl number, is a parameter in Eq. (16-33) and $\tau(0)$ is an additional parameter [uniquely determined by the specification of $\tau'(0)$ as a parameter], it is apparent that the final result is in the form

$$\frac{\mathrm{Nu}_{x,\infty}}{\mathrm{Re}_{x,\infty}^{1/2}} = F\left(M, \frac{T_0}{T_\infty}, \mathrm{Pr}\right) \tag{16-35}$$

Alternatively, the Mach number can be replaced by T_{aw}/T_∞, since if we combine Eqs. (16-32) and (16-22), we obtain

$$\frac{T_{aw}}{T_\infty} = 1 + \tfrac{1}{2}r_c(\gamma - 1)M^2 \qquad (16\text{-}36)$$

Thus we can also express the final result as

$$\frac{\mathrm{Nu}_{x,\infty}}{\mathrm{Re}_{x,\infty}^{1/2}} = F\left(\frac{T_{aw}}{T_\infty}, \frac{T_0}{T_\infty}, \mathrm{Pr}\right) \qquad (16\text{-}37)$$

Van Dreist[3] presents an extension of a solution to Eqs. (16-29) and (16-33) developed by Crocco. In this method the $\mu\rho = $ constant assumption is used as a first approximation, and a method of successive approximations is employed. The solution is developed for a gas with $\mathrm{Pr} = 0.75$, constant c, $\gamma = 1.40$, and the Sutherland law for viscosity.

Despite the temperature dependence of the fluid properties, it is again found that to a good approximation

$$r_c \approx \mathrm{Pr}^{1/2}$$

The remainder of the solution is shown graphically in Fig. 16-4. [Note that $\mathrm{St}\,\mathrm{Re}_{x,\infty}^{1/2} = \mathrm{Pr}^{-1}\,\mathrm{Nu}_{x,\infty}\,\mathrm{Re}_{x,\infty}^{-1/2}$.] The friction coefficient behaves in the same manner as the Stanton number, except, of course, $(c_f/2)\,\mathrm{Re}_{x,\infty}^{1/2} = 0.332$ at $M = 0$ and $T_0/T_\infty = 1$.

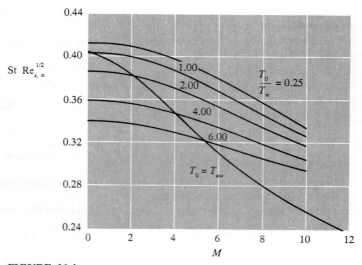

FIGURE 16-4
Heat transfer to or from the laminar high-velocity boundary layer; air with $\mathrm{Pr} = 0.75$.[3]

Note that the curve for $T_0 = T_{aw}$ is derivable from the others through Eq. (16-36). This tendency of the Stanton number (and the friction coefficient) to decrease with Mach number is quite characteristic of the high-velocity boundary layer for both laminar and turbulent flow when the fluid properties are temperature-dependent. On the other hand, there is no such influence of Mach number when the properties are constant.

These results suggest that the phenomenon under discussion here is basically the same as in the low-velocity, large-temperature-difference problem discussed in Chap. 15. In fact, if T_{aw}/T_∞ is used to replace M through Eq. (16-36), a fair approximation to Fig. 16-4 can be constructed using temperature ratios raised to a power:

$$\frac{St}{St_{CP}} = \left(\frac{T_0}{T_\infty}\right)^{-0.08}\left(\frac{T_{aw}}{T_\infty}\right)^{-0.04} \tag{16-38}$$

or, alternatively,

$$\frac{St}{St_{CP}} = \left(\frac{T_0}{T_{aw}}\right)^{-0.08}\left(\frac{T_{aw}}{T_\infty}\right)^{-0.12} \quad \text{for } T_0 > T_\infty, M \leqslant 6 \tag{16-38a}$$

For the particular case of $T_0 = T_{aw}$, Eq. (16-38) becomes

$$\frac{St}{St_{CP}} = \left(\frac{T_{aw}}{T_\infty}\right)^{-0.12} = \left(\frac{T_0}{T_\infty}\right)^{-0.12} \tag{16-39}$$

How well Eq. (16-39) fits Fig. 16-4 is illustrated in Fig. 16-5.

Note that these exponents are small, which is again consistent with the low-velocity, large-temperature-difference problem. However, note that when Eq. (16-38) is reduced to low velocities where $T_{aw} = T_\infty$, the exponent on T_0/T_∞ is -0.08, whereas in Chap. 15 exponents of -0.01 for heat transfer and -0.10 for friction were suggested. The difference is

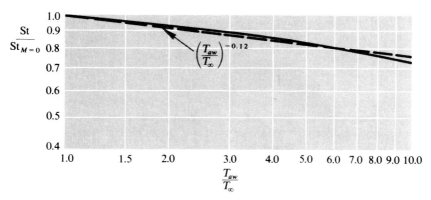

FIGURE 16-5
Correlation of the results of Fig. 16-4 on a temperature-ratio basis.

simply due to the slightly different property variation with temperature used in the two analyses. In dealing with a real gas the most important conclusion is that the effects are small, and Eq. (16-38) is probably as good an approximation as Fig. 16-4.

THE USE OF REFERENCE PROPERTIES FOR HIGH-VELOCITY LAMINAR BOUNDARY-LAYER CALCULATIONS

In the preceding section it was shown that the constant-property boundary-layer solutions can be used for approximate calculation of high-velocity boundary layers by evaluating all properties in the constant-property solutions at free-stream static temperature and then including an empirical temperature-ratio correction in the equations. Many engineers prefer to make this correction by simply evaluating the properties in the constant-property solutions at some *reference temperature* different from the free-stream temperature.

On the basis of examination of a number of "exact" laminar boundary-layer solutions for constant u_∞, t_∞, and t_0, Eckert[4] concludes that if the specific heat may be treated as constant, the following reference temperature correlates the available solutions within a few percent for Mach numbers up to 20 and over a wide range of free-stream and surface temperatures:

$$T_R = T_\infty + 0.5(T_0 - T_\infty) + 0.22(T_{aw} - T_\infty) \qquad (16\text{-}40)$$

The adiabatic wall temperature T_{aw} is again evaluated from Eq. (16-22), and the recovery factor from Eq. (16-21), but the Prandtl number in Eq. (16-21) is evaluated at T_R. Equations (16-24) and (16-25) are employed, but *all* properties in Eq. (16-25) are evaluated at T_R. In other words, the Reynolds number in Eq. (16-25) becomes

$$\mathrm{Re}_{x,R} = \frac{\rho_R u_\infty x}{\mu_R}$$

and, similarly,

$$\mathrm{Nu}_{x,R} = \frac{hx}{k_R} \quad \text{or} \quad \mathrm{St}_R = \frac{h}{u_\infty \rho_R c}$$

The friction coefficient can be evaluated from the constant-property equation (8-15), but using $\mathrm{Re}_{x,R}$ in place of Re_x. The local surface shear stress is evaluated from Eq. (8-12), but with ρ_R replacing ρ.

For very large temperature differences through the boundary layer the assumption of constant specific heat, which has been made in all the preceding solutions and discussion, is no longer valid. It was shown earlier in the development of Eq. (16-5), and the subsequent discussion, that stagnation enthalpy can be used as the dependent variable in the energy equation, and the entire problem can be formulated in terms of enthalpy rather than temperature. If the specific heat cannot be treated as constant, this is a more natural formulation. Eckert's reference property method is again applicable; and, following a development completely analogous to the previous one, we obtain

$$\dot{q}_0'' = g_i(i_0 - i_{aw}) \tag{16-41}$$

where i_{aw} is the adiabatic wall or recovery enthalpy

$$i_{aw} = i_\infty + \tfrac{1}{2}r_c u_\infty^2, \qquad r_c = Pr_R^{1/2}$$

and g_i is an enthalpy conductance evaluated from the Stanton number

$$St_R = \frac{g_i}{\rho_R u_\infty} = \frac{Nu_R}{Re_R\, Pr_R}$$

St_R is the *constant-property* Stanton number evaluated with all properties at a temperature corresponding to a *reference enthalpy* i_R:

$$i_R = i_\infty + 0.5(i_0 - i_\infty) + 0.22(i_{aw} - i_\infty) \tag{16-42}$$

This scheme is thus virtually identical to that based on temperatures.

THE TURBULENT BOUNDARY LAYER FOR A GAS WITH VARIABLE PROPERTIES

Equation (16-7) can be quite readily and satisfactorily solved for a turbulent boundary layer for any desired boundary conditions using the mixing-length theory developed in Chaps. 11 and 13, but with fluid properties *variable,* employing a finite-difference computing technique. For many applications, especially those involving complex boundary conditions, this is undoubtedly the most satisfactory procedure. However, in order to appreciate the influence of high velocity on a turbulent boundary layer, it is useful to examine some results for simple boundary conditions: constant free-stream velocity and constant surface temperature. For this purpose we could examine either some finite-difference solutions of Eq. (16-7) or some of the solutions available in the literature that involve various degrees of approximation.

The semiempirical results of Spalding and Chi[5] are a very thorough and probably accurate summary of the available data on friction coefficient. For heat transfer the results of Deissler and Loeffler[6] are less complete, but are convenient for present purposes and are in reasonably good agreement with experiment.

Deissler and Loeffler reduced the boundary-layer equations to Couette flow form and then used a mixing-length theory for their solution. The Stanton number is determined as a function of the local momentum thickness Reynolds number. The Prandtl number is considered constant at 0.73, the specific heat is held constant, and

$$\frac{\mu}{\mu_0} = \frac{k}{k_0} = \left(\frac{T}{T_0}\right)^{0.68}$$

The analysis covers the Mach number range 0–8 and momentum thickness Reynolds numbers from 10^3 to 10^5, corresponding to length Reynolds numbers from 0.5×10^6 to 0.6×10^9. Only two heat-transfer cases are considered: $T_0 \rightarrow T_{\mathrm{aw}}$ and $T_0/T_{\mathrm{aw}} = 0.5$.

Both the recovery factor and the Stanton number are determined in the analysis. For Mach numbers from 0 to 8 and Re_x from 10^6 to 10^8 the recovery factor is found to vary from about 0.87 to 0.91. Experimental values for air generally lie in this range. It has been frequently proposed that for a turbulent boundary layer.

$$r_c \approx \mathrm{Pr}^{1/3} \tag{16-43}$$

This is in reasonably good agreement with Deissler's result.

The Stanton number is strongly affected by the Mach number, but this effect is only slightly dependent on the Reynolds number. The results for $Re_{\delta_2} = 10^4$ alone are described here. The approximate nature of the analysis probably does not warrant including a Reynolds number influence in the correlations to be proposed.

In Fig. 16-6, $St/St_{M=0}$ is plotted as a function of Mach number for $T_0/T_{\mathrm{aw}} = 0.5$ and 1. Also shown are two experimental test points obtained by Pappas and Rubesin.[7] $St_{M=0}$ is the Stanton number at $M=0$ at the same *momentum thickness* Reynolds number.† The Pappas and Rubesin data are based on a comparison at the same x Reynolds number, but the difference is probably within the experimental uncertainty. Note that the correlation between analysis and experiment is quite good. Calculations by the authors using the mixing-length models discussed in Chaps. 11 and 13 and a finite-difference computer code yield substantially the same results.

Two facts are immediately evident. Although the influence of Mach number is qualitatively similar to that seen for the laminar boundary

---

† *Enthalpy thickness* Reynolds number is not a very useful parameter for compressible boundary layers because thermal energy is generated within the boundary layer in addition to that transferred to or from the boundary layer at the surface.

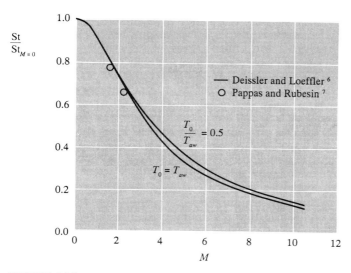

FIGURE 16-6
Heat transfer to or from the turbulent high-velocity boundary layer; air, $Pr = 0.73$, $Re_{\delta_2} = 10^4$ (analysis of Deissler and Loeffler;[6] data of Pappas and Rubesin[7]).

layer, it is at the same time very much greater. This is quite similar to the effect of temperature-dependent properties at low velocities noted in Chap. 15. Furthermore, the influence of the Mach number is approximately the same regardless of T_0/T_{aw}, although, of course, only two values of T_0/T_{aw} are considered.

The nature of the Mach number influence can be better appreciated if the Mach number is replaced by T_{aw}/T_∞, using Eq. (16-36), and then $St/St_{M=0}$ is plotted as shown in Fig. 16-7.

On logarithmic coordinates the result is very close to a straight line for T_{aw}/T_∞ up to about 7, corresponding to Mach numbers up to about 6.

Of particular interest is the exponent -0.6, which is close to the -0.4 power of T_0/T_∞ that was found in Chap. 15 to correlate most of the large-temperature-difference results for heating. In fact, some experimental *pipe* data for air yield exponents as high as -0.575. It thus appears that the Mach number effect for high-velocity convection differs little from the large-temperature-difference effect at low velocities.

The fact that the curve for $T_0/T_{aw} = 0.5$ is very close to the curve for $T_0/T_{aw} = 1$ in Fig. 16-6 suggests that the influence of T_0/T_{aw} can be separately correlated. Since only two points on the curve are available, it is not possible to tell whether a simple power of T_0/T_{aw} would be adequate. However, all our previous experience with these effects suggests that this might be the case. The Deissler and Loeffler results

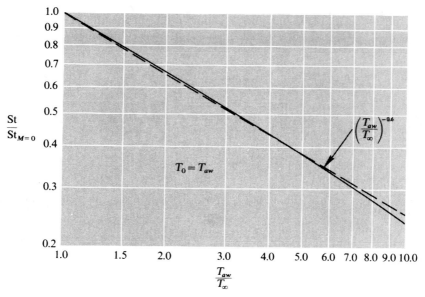

FIGURE 16-7
Correlation of the results of Fig. 16-6 on a temperature-ratio basis.

indicate that the ratio of Stanton number to Stanton number at the same Mach number but with $T_0/T_{aw} = 1$ is about 1.32 over the entire Mach number range of the analysis. Then

$$\frac{St}{St_{aw}} = \left(\frac{T_0}{T_{aw}}\right)^{-0.4}$$

Finally, putting both the Mach number and wall-temperature effects together, we obtain

$$\frac{St}{St_{CP}} = \left(\frac{T_{aw}}{T_\infty}\right)^{-0.6}\left(\frac{T_0}{T_{aw}}\right)^{-0.4} \quad (16\text{-}44)$$

where all properties in the equation used to evaluate St_{CP} are presumed to be evaluated at the free-stream static temperature.

Examination of Eq. (16-44) indicates that T_{aw} has only a minor influence, and a simplified version of Eq. (16-44) is

$$\frac{St}{St_{CP}} = \left(\frac{T_0}{T_\infty}\right)^{-0.5} \quad (16\text{-}45)$$

This is, of course, identical to the correlation recommended in Chap. 15 for a *low*-velocity gas being *heated* in a tube, as well as close to the exponents recommended for an external boundary layer.

The friction coefficient is found to behave in a quite similar manner as the Stanton number, so to a good approximation

$$\frac{c_f}{c_{f\,\text{CP}}} = \frac{\text{St}}{\text{St}_{\text{CP}}} \qquad (16\text{-}46)$$

REFERENCE PROPERTIES FOR HIGH-VELOCITY TURBULENT BOUNDARY-LAYER CALCULATIONS

As an alternative to the temperature-ratio corrections to the constant-property solutions in order to account for the influence of temperature-dependent properties, the properties in the constant-property solutions can be introduced at a reference temperature, or a reference enthalpy, as suggested for the laminar boundary layer. Eckert[4] shows that, in fact, the available data can be satisfactorily correlated with the same reference properties as given for the laminar boundary layer. In other words, Eq. (16-40) should be used where the specific heat can be treated as approximately constant, and Eq. (16-42) where it is also temperature-dependent.

MACH NUMBER AND LARGE-TEMPERATURE-DIFFERENCE CORRECTIONS FOR VARIABLE FREE-STREAM VELOCITY AND VARIABLE TEMPERATURE DIFFERENCES

We have considered only the case of a constant free-stream velocity and constant free-stream and surface temperature. In applications where these parameters vary in the flow direction, i.e., in most applications, there is no doubt that direct integration of the momentum and energy equations using finite-difference procedures provides the most direct and accurate route to an engineering solution. For the laminar boundary layer there are no real difficulties, and many computer codes are available. For the turbulent boundary layer codes are available, but their applicability depends on the adequacy of the turbulence models employed. The authors have found that the turbulence models discussed in Chaps. 11 and 13 are actually quite adequate for Mach numbers up to 3 or 4.

In the absence of finite-difference computer procedures, a reasonable approximation is to use the constant-property procedures discussed in the preceding chapters and then to apply a local correction, i.e., Eq. (16-38) or Eq. (16-44), for the compressible-flow effects. Equation (16-24) should, of course, be used as the defining equation for the heat-transfer coefficient, and Eqs. (16-21) and (16-43) are generally adequate approximations for the recovery factor. Alternatively, the reference property corrections can be used.

Such a procedure is undoubtedly not accurate for a laminar boundary layer, but fortunately we have seen that the influence of the temperature ratio for a typical gas is quite small, so that the accuracy of the correction is not of great importance.

On the other hand, the correction for a turbulent boundary layer can be substantial. But we have seen that, except for very low-Prandtl-number fluids, the turbulent boundary layer responds very quickly to axial changes in boundary conditions. Thus a local correction for the influence of the temperature dependence of the fluid properties appears justified.

Equation (13-36) for a turbulent boundary layer, when corrected as indicated above, has been found to provide a remarkably good approximation for heat transfer in supersonic nozzles.

PROBLEMS

16-1. Consider an aircraft flying at Mach 3 at an altitude of 17,500 m. Suppose the aircraft has a hemispherical nose with a radius of 30 cm. If it is desired to maintain the nose at 80°C, what heat flux must be removed at the stagnation point by internal cooling? As a fair approximation, assume that the air passes through a normal detached shock wave and then decelerates isentropically to zero at the stagnation point; then the flow near the stagnation point is approximated by low-velocity flow about a sphere.

16-2. In Prob. 16-1 it is desired to cool a particular rectangular section of the aircraft body to 65°C. The section is to be 60 cm wide by 90 cm long (in the flow direction) and is located 3 m from the nose. Estimate the total heat-transfer rate necessary to maintain the desired surface temperature. As an approximation, the boundary layer may be treated as if the free-stream velocity were constant along a flat surface for the preceding 3 m. It may also be assumed that the preceding 3 m of surface is adiabatic. To obtain the state of the air just outside the boundary layer, it is customary to assume that the air accelerates from behind the normal shock wave at the nose, isentropically to the free-stream static pressure. In this case the local Mach number then becomes 2.27, and the ratio of local absolute static temperature to free-stream stagnation temperature is 0.49. The local static pressure is the same as the free-stream, that is, the pressure at 17,500 m altitude.

16-3. Consider a laminar constant-property boundary layer on a flat plate with constant free-stream velocity. Evaluate the recovery factor for a fluid with $Pr = 100$, using Eq. (16-20). The correct answer is 7.63. Why is it difficult to obtain accuracy in solving this problem numerically?

16-4. Listed below, as functions of *axial* distance, are the radius, mass velocity, temperature, and pressure inside a supersonic nozzle. The fluid is air. It is desired to maintain the nozzle surface uniformly at 200°C. Calculate the necessary heat flux along the nozzle surface. Assume that a turbulent boundary layer originates at the start of the nozzle.

Axial distance, cm	Radius, cm	Mass velocity, kg/(s · m²)	Pressure, kPa	T, K
0	9.70	0	2068	893
1.94	8.59	98	2068	893
6.07	6.15	366	2068	891
8.14	4.95	561	2048	890
10.20	3.81	879	2013	886
12.26	2.62	1733	1841	856
13.50	2.13	2734	1262	806
14.33	2.06	2710	931	669
15.15	2.29	1855	331	518
16.39	2.84	1318	172	446
18.45	3.76	928	96.5	380
20.51	4.57	698	62.1	334
22.58	5.28	586	48.3	308
24.64	5.92	464	34.5	287
26.72	6.48	391	27.6	265
28.75	6.99	352	20.7	250
30.81	7.44	322	17.3	237
34.95	8.20	249	13.8	219

16-5. In Prob. 16-4 the nozzle is to be constructed of 0.5 cm thick stainless-steel walls. The entire nozzle is to be surrounded by a bath of water, maintained at 27°C by constant changing. The heat-transfer coefficient on the water side of the nozzle walls is estimated to be 1100 W/(m² · K) uniformly. Calculate the temperature along the inner surface of the nozzle wall and the local heat flux. Assume that heat conduction in the nozzle wall is significant in the radial direction only.

16-6. Repeat Prob. 16-4 but assume that the nozzle is constructed with a uniformly porous wall so that transpiration cooling may be used. Air is available as a coolant at 30°C. Determine the transpiration air rate, as a function of position along the surface, to maintain the surface of 200°C.

16-7. A particular rocket ascends vertically with a velocity that increases approximately linearly with altitude, reaching 3000 m/s at 60,000 m. Consider a point on the cylindrical shell of the rocket 5 m from the nose. Calculate and plot, as functions of altitude, the adiabatic wall temperature, the local convection conductance, and the internal heat flux necessary to prevent the skin temperature from exceeding 50°C (see Prob.

16-2 for remarks about the state of the air just outside the boundary layer in such a situation).

16-8. Consider again Prob. 16-7, but let the skin be of 3 mm thick stainless steel, insulated on the inner side. Treating the skin as a single element of capacitance, calculate the skin temperature as a function of altitude. [The specific heat of stainless steel is $0.46\,kJ/(kg \cdot K)$.]

16-9. Consider the gas turbine blade on which Prob. 11-8 is based (see Fig. 11-19). It is desired to maintain the blade surface uniformly at 650°C by internal cooling. Calculate the necessary heat flux around the periphery of the blade.

16-10. In Prob. 16-9 the turbine blade is 15 cm long, and for present purposes it may be assumed that blade dimensions and operating conditions are the same along the entire blade length. It appears feasible to allocate up to 0.03 kg/s of air at 200°C to cool the blade. Assuming a hollow blade of mild-steel construction, a minimum wall thickness of 0.75 mm, and any kind of internal inserts as desired (a solid core leaving a narrow passage just inside the outer wall could be used, for example), make a study of the feasibility of internally cooling this blade. Assume that the cooling air can be introduced at the blade root and discharged through the tip. Treat the blade as a simple heat exchanger.

16-11. The gas turbine blade of Prob. 11-8 is to be cooled to a uniform surface temperature of 650°C by transpiration of air through a porous surface. If the cooling air is available at 200°C, calculate the necessary cooling-air mass-transfer rate per unit of surface area as a function of position along the blade surface. Discuss the probable surface temperature distribution if it is only mechanically feasible to provide a transpiration rate that is uniform along the surface.

16-12. Consider the rocket nozzle described in Prob. 11-6 (Fig. 11-18). Calculate the heat flux along the nozzle surface necessary to maintain the surface at 1100°C. If the convergent part of the nozzle is exposed to black-body radiation at 2200°C (the reactor core) and the nozzle surface itself is a black body, will thermal radiation contribute significantly to the heat flux through the nozzle walls?

16-13. Suppose the nozzle of Prob. 16-12 is constructed of 6 mm thick molybdenum. Copper cooling tubes are then to be wrapped around the nozzle and bonded to the surface. Room-temperature water is available as a coolant. Make a study of the feasibililty of water cooling in this manner, after first choosing a tubing size, flow arrangement, and reasonable water velocity.

16-14. The helium rocket nozzle of Probs. 11-6 and 16-12 is to be cooled to a uniform surface temperature of 800°C by transpiration of additional helium that is available at 38°C. Calculate the necessary local transpiration rates if this turns out to be a feasible scheme. What is the total necessary coolant rate? How does this compare with the total hot-helium rate passing through the nozzle?

16-15. Consider again Prob. 13-11, but let the aircraft speed be 300 m/s. At what airspeed does direct cooling become impossible?

REFERENCES

1. Cebeci, T., and A. M. O. Smith: *Analysis of Turbulent Boundary Layers,* pp. 47–79, Academic Press, New York, 1974.
2. Pohlhausen, E.: *Z. Angew. Math. Mech.,* vol. 1, pp. 115–121, 1921.
3. Van Dreist, E. R.: NACA (now NASA) TN 2597, Washington, January 1952.
4. Eckert, E. R. G.: *J. Aero. Sci.,* vol. 22, pp. 585–587, 1955.
5. Spalding, D. B., and S. W. Chi: *J. Fluid Mech.* vol. 18, pp. 117–143, 1964.
6. Deissler, R. G., and A. L. Loeffler, Jr: NACA (now NASA) TN 4262, Washington, April 1958.
7. Pappas, C. C., and M. W. Rubesin: *Heat Transfer and Fluid Mechanics Institute Papers,* pp. 19–28, Stanford University, California, 1953.

CHAPTER
17

FREE-CONVECTION BOUNDARY LAYERS

In Chap. 1 convection was defined as the transport of mass and energy by potential gradients and by gross fluid motion. If the fluid motion is induced by some external means such as fluid machinery or vehicle motion, the process is generally called *forced convection*. In contrast, fluid motion can also be induced by body forces such as gravitational, centrifugal, or Coriolis forces, and this process is generally called *natural convection*. Natural flows occur in atmospheric and oceanic circulation, electric machinery and nuclear reactor cooling systems, heated or cooled enclosures, electronic power supplies, and so forth. The flow is a buoyancy-induced motion resulting from body forces acting on density gradients, which, in turn, arise from mass concentration and/or temperature gradients in the fluid. In this chapter we consider only temperature-induced buoyancy from a single surface within quiescent surroundings. Natural flow on such a surface, unbounded by other surfaces, is usually called *free convection,* and it has many of the characteristics of forced-convection boundary-layer flows.

In this chapter we deal with only the simplest free-convection solutions. More in-depth presentations can be found in Ostrach[1] or Ede.[2] An introduction to the various aspects of external natural convection flows is given by Gebhart.[3]

We start with the mass, momentum, and energy equations that govern laminar free-convection boundary-layer flows. We then develop similarity solutions to the equations for constant-wall-temperature, constant-wall-heat-flux, and transpiration boundary conditions, and an approximate integral solution for the constant-wall-temperature case. The effects of variable properties and the free-convection flow regimes are then discussed. For turbulent free convection we present two approximate solutions and recommended correlations based on experimental data. Finally, heat-transfer correlations for other geometries are presented, and mixed free and forced convection is discussed.

BOUNDARY-LAYER EQUATIONS FOR FREE CONVECTION

The equations governing external laminar-free convection are essentially the same as those for forced convection. We assume the fluid to be of one species and Newtonian, and the flow field two-dimensional and steady. The applicable equations from Chap. 4 are *continuity,* from Eq. (4-1),

$$\frac{\partial(\rho u)}{\partial x} + \frac{\partial(\rho v)}{\partial y} = 0 \tag{17-1}$$

momentum, from Eq. (4-10) and including $X = -\rho g$ (a gravitational body force oriented in the negative x direction),

$$\rho u \frac{\partial u}{\partial x} + \rho v \frac{\partial u}{\partial y} + \frac{dP}{dx} = \frac{\partial}{\partial y}\left(\mu \frac{\partial u}{\partial y}\right) - \rho g \tag{17-2}$$

and *energy,* from Eq. (4-37), assuming constant specific heat. Note that work done against external force fields uX does not appear, because this equation is the combination of the total energy equation and the momentum equation. We also neglect viscous dissipation and the pressure-gradient term:

$$\rho u c \frac{\partial t}{\partial x} + \rho v c \frac{\partial t}{\partial y} - \frac{\partial}{\partial t}\left(k \frac{\partial t}{\partial y}\right) = 0 \tag{17-3}$$

The pressure-gradient term in the momentum equation is due to the hydrostatic pressure field outside the boundary layer.† From hydrostatics,

---

† The pressure is assumed to be a function of x. If the gravitational field is not aligned with the principal direction of motion or if the surface is curved or rotating, the pressure field is two-dimensional. For surface inclinations at a small angle θ to the gravitational field, g can usually be replaced by $g \cos \theta$, the component along the surface.

the gradient is

$$\frac{dP}{dx} = -\rho_\infty g$$

Thus the body force and pressure-gradient terms combine, and the momentum equation becomes

$$\rho u \frac{\partial u}{\partial x} + \rho v \frac{\partial u}{\partial y} = g(\rho_\infty - \rho) + \frac{\partial}{\partial y}\left(\mu \frac{\partial u}{\partial y}\right)$$

These equations can be further simplified by making the *Boussinesq approximation* for free-convection flows. It is a two-part approximation: (1) it neglects all variable-property effects in the three equations, except for density in the momentum equation; and (2) it approximates the density difference term with a simplified equation of state:

$$\rho_\infty - \rho = \rho\beta(t - t_\infty)$$

where β is the volumetric coefficient of thermal expansion. Note that for ideal gases $\beta = 1/T_\infty$, where T is the absolute temperature in Kelvin, and for incompressible liquids $\rho\beta$ is nearly constant. This approximation is good for gravity-induced flows over nearly vertical surfaces with negligible temperature variation of t_∞ with x. For further discussion of these restrictions see Gebhart.[4]

If we invoke the Boussinesq approximation, the free-convection equations become

$$\frac{\partial u}{\partial x} + \frac{\partial v}{\partial y} = 0 \tag{17-4}$$

$$u \frac{\partial u}{\partial x} + v \frac{\partial u}{\partial y} = g\beta(t - t_\infty) + v \frac{\partial^2 u}{\partial y^2} \tag{17-5}$$

$$u \frac{\partial t}{\partial x} + v \frac{\partial t}{\partial y} - \alpha \frac{\partial^2 t}{\partial y^2} = 0 \tag{17-6}$$

with boundary conditions

$$
\begin{aligned}
u, v &= 0, & t &= t_0(x) & &\text{at } y = 0 \\
u &= 0, & t &= t_\infty & &\text{as } x = 0 \\
u &\to 0, & t &\to t_\infty & &\text{as } y \to \infty
\end{aligned}
$$

SIMILARITY SOLUTIONS: LAMINAR FLOW ON A CONSTANT-TEMPERATURE, VERTICAL, AND SEMI-INFINITE FLAT PLATE

One of the earliest analytic and experimental investigations of free convection over isothermal vertical plates was by Schmidt and Beckmann

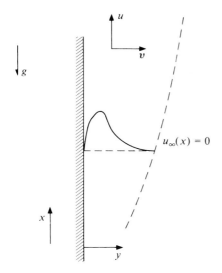

FIGURE 17-1
Coordinate system and velocity components for a free-convection momentum boundary layer.

(see Ref. 5). Their experimental results confirm the boundary-layer character of the flow. Figures 17-1 and 17-2 depict the shapes of the velocity and temperature profiles obtained from the experiments.†

The experimental profiles suggests the possibility of profile similarity and the possibility that similarity solutions might exist for the governing equations. There are several methods for obtaining these solutions if they exist, and the one outlined here is similar to that used in Chap. 8 leading to the Blasius solution.

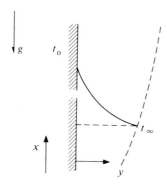

FIGURE 17-2
A free-convection thermal boundary layer.

---

† Note that Figs. 17-1 and 17-2 depict the case $t_0 > t_\infty$, which induces a vertically upward flow. The leading edge of the plate ($x = 0$) is at the bottom and g is in the $-x$ direction. For the case of $t_0 < t_\infty$ the buoyancy-induced flow would be downward, with the leading edge at the top. For either case the heat-transfer solutions are identical.

Because the continuity, momentum, and energy equations contain the v component of velocity, it becomes convenient to transform the boundary-layer equations by defining a stream function. It is also convenient to nondimensionalize the temperature. Let

$$u = \frac{\partial \psi}{\partial y}, \qquad v = -\frac{\partial \psi}{\partial x}, \qquad \theta = \frac{t - t_\infty}{t_0 - t_\infty}$$

Then the momentum equation becomes

$$\frac{\partial \psi}{\partial y} \frac{\partial^2 \psi}{\partial x \, \partial y} - \frac{\partial \psi}{\partial x} \frac{\partial^2 \psi}{\partial y^2} = g\beta(t_0 - t_\infty)\theta + v \frac{\partial^3 \psi}{\partial y^3} \tag{17-7}$$

and the energy equation becomes

$$\frac{\partial \psi}{\partial y} \frac{\partial \theta}{\partial x} - \frac{\partial \psi}{\partial x} \frac{\partial \theta}{\partial y} - \alpha \frac{\partial^2 \theta}{\partial y^2} + \theta \frac{\partial \psi}{\partial y} \frac{dt_0}{dx} = 0 \tag{17-8}$$

The premise that the velocity and temperature profiles are geometrically similar, differing only by stretching factors in the x direction, suggests that we define a similarity variable as

$$\eta = yH(x) \tag{17-9}$$

Under the transformation, two partial differential equations for $\psi(x, \eta)$ and $\theta(x, \eta)$ are obtained. The objective now is to reduce these partial differential equations to ordinary differential equations, and a classical technique is to separate variables. Thus we assume

$$\psi(x, \eta) = vF(\eta)G(x) \tag{17-10}$$

The resulting set of equations then reduces to ordinary differential equations in η (see Gebhart[4]) when

$$G(x) = 4(\tfrac{1}{4} \, \mathrm{Gr}_x)^{1/4}$$

$$H(x) = \frac{1}{x}(\tfrac{1}{4} \, \mathrm{Gr}_x)^{1/4}$$

where

$$\mathrm{Gr}_x = \frac{g\beta x^3(t_0 - t_\infty)}{v^2}$$

is the *Grashof number*, which is often interpreted as the parameter describing the ratio of buoyancy to viscous forces.

The ordinary differential equations for $t_0(x) = \text{constant}$ become

$$F''' + \theta + 3FF'' - 2F'^2 = 0 \tag{17-11}$$

$$\theta'' + 3 \, \mathrm{Pr} \, F\theta' = 0 \tag{17-12}$$

The boundary conditions on these equations transform as follows:

$$\left.\begin{array}{ll} u = 0 & \text{at } y = 0 \\ y = 0 & \text{at } y = 0 \\ u \to 0 & \text{as } y \to \infty \\ t = t_0 & \text{at } y = 0 \\ t \to t_\infty & \text{as } y \to \infty \end{array}\right\} \text{ giving } \left\{\begin{array}{l} F'(0) = 0 \\ F(0) = 0 \\ F'(\infty) = 0 \\ \theta(0) = 1 \\ \theta(\infty) = 0 \end{array}\right.$$

Equations (17-11) and (17-12) are coupled, nonlinear ordinary differential equations that were initially solved by approximate methods. The first complete numerical solution was carried out by Ostrach[6] for a range of Prandtl numbers from 0.01 to 1000. Heat-transfer results from the numerical solution can be defined in terms of a local heat-transfer coefficient

$$h_x = \frac{\dot{q}_0''}{t_0 - t_\infty} \qquad \text{(positive } \dot{q}'' \text{ in positive } y \text{ direction)}$$

where

$$\dot{q}_0'' = -k\left(\frac{\partial t}{\partial y}\right)_0 = -k(t_0 - t_\infty)\left(\frac{\partial \theta}{\partial y}\right)_0 = -k(t_0 - t_\infty)\left(\frac{d\theta}{d\eta}\frac{\partial \eta}{\partial y}\right)_0$$

$$\dot{q}_0'' = -k(t_0 - t_\infty)\theta'(0)H(x)$$

A local Nusselt number is then defined to put h_x into nondimensional form:

$$\mathrm{Nu}_x = \frac{xh_x}{k} = -\theta'(0)xH(x)$$

$$= -\frac{\theta'(0)}{\sqrt{2}}\mathrm{Gr}_x^{1/4} \tag{17-13}$$

The local Nusselt number is thus found to be a function of Pr and Gr_x. Solutions for $-\theta'(0)/\sqrt{2}$ from Ostrach and others are presented in Table 17-1. Ede[2] also gives a function that is a very satisfactory correlation to the entries in this table:

$$\mathrm{Nu}_x = \frac{3}{4}\left[\frac{2\,\mathrm{Pr}}{5(1 + 2\,\mathrm{Pr}^{1/2} + 2\,\mathrm{Pr})}\right]^{1/4}(\mathrm{Gr}_x\,\mathrm{Pr})^{1/4} \tag{17-14}$$

Solutions also exist for the limiting cases of $\mathrm{Pr} \to 0$ and $\mathrm{Pr} \to \infty$. Le Fevre[8] obtained these solutions by transforming Eqs. (17-11) and (17-12):

$$\mathrm{Pr}\,(F_1''' + \theta_1) + \theta_1 + F_1 F_1'' = \tfrac{2}{3}F_1'^2 \tag{17-15}$$

$$\theta_1'' + \theta_1' F_1 = 0 \tag{17-16}$$

TABLE 17-1
Values of $Nu_x \, Gr_x^{-1/4}$ for various Prandtl numbers; laminar heat transfer to a vertical free-convection boundary layer with $t_0 =$ constant, adapted from Refs. 2, 4, 6, and 7

Pr	0.01	0.1	0.72	1.0	10	100	1000
$Nu_x \, Gr_x^{-1/4}$	0.0570	0.164	0.357	0.401	0.827	1.55	2.80

In this form the differential equations can be simplified for $Pr \to 0$ and $Pr \to \infty$. The results are

$$Nu_x = \begin{cases} 0.600(Gr_x \, Pr^2)^{1/4} & \text{as } Pr \to 0 \quad\quad (17\text{-}17) \\ 0.503(Gr_x \, Pr)^{1/4} & \text{as } Pr \to \infty \quad\quad (17\text{-}18) \end{cases}$$

These asymptotic results compare favorably with the $Pr = 0.01$ and $Pr = 1000$ entries in Table 17-1. They also indicate a variation in the Prandtl number exponent. By plotting $-\theta'(0)/\sqrt{2}$ versus Prandtl number, it can be shown that the $\frac{1}{2}$-power dependence holds to about $Pr = 0.1$.

Frequently the total plate heat transfer is of interest rather than the local heat transfer. To calculate the total rate, we need a mean Nusselt number with respect to x. Thus

$$h = \frac{1}{x} \int_0^x h_x \, dx = \frac{1}{x} \int_0^x \frac{k}{x} \left[\frac{-\theta'(0)}{\sqrt{2}} Gr_x^{1/4} \right] dx$$

$$= \frac{1}{x} \int_0^x \frac{k}{x} C x^{3/4} \, dx = \frac{4}{3} \frac{k}{x} C x^{3/4} = \frac{4}{3} h_x$$

or

$$Nu = \frac{4}{3} Nu_x \quad\quad (17\text{-}19)$$

The coefficient $\frac{4}{3}$ in Eq. (17-19) for free convection on a vertical surface compares with a coefficient of 2 in Eq. (10-14) for forced convection over a flat plate. This reflects the fact that h varies as $x^{-1/4}$ for free convection, compared with $x^{-1/2}$ for forced convection.

SIMILARITY SOLUTIONS WITH VARIABLE SURFACE TEMPERATURE

Similarity solutions can also be obtained for certain $t_0(x)$ functions. Again the requirement is to have the governing equations reduce to ordinary differential equations. Yang[9] and Sparrow and Gregg[10] have shown that the power-law form of $t_0(x)$ leads to similarity solutions.

Consider the power-law variation

$$t_0 - t_\infty = Ax^n$$

Using the same method and variables as outlined in the previous section, Eqs. (17-11) and (17-12) become

$$F''' + \theta + (n+3)FF'' - (2n+2)F'^2 = 0 \qquad (17\text{-}20)$$
$$\theta'' + \text{Pr}\left[(n+3)F\theta' - 4nF'\theta\right] = 0 \qquad (17\text{-}21)$$

with the same boundary conditions

$$F(0) = F'(0) = 0, \qquad \theta(0) = 1$$
$$F'(\infty) = 0, \qquad \theta(\infty) = 0$$

Recall that the wall heat flux is

$$\dot{q}_0'' = -k(t_0 - t_\infty)\theta'(0)H(x) = -k(t_0 - t_\infty)\frac{\theta'(0)}{\sqrt{2}}\frac{1}{x}\text{Gr}_x^{1/4}$$

$$= k\left(\frac{g\beta}{v^2}\right)^{1/4}\frac{-\theta'(0)}{\sqrt{2}}(t_0 - t_\infty)^{5/4}\frac{x^{3/4}}{x}$$

$$= C(t_0 - t_\infty)^{5/4}x^{-1/4}$$

If $t_0 - t_\infty$ now varies as $x^{1/5}$, a constant-heat-flux boundary condition obtains. The heat flux can be recast in terms of a heat-transfer coefficient and Nusselt number, and Eq. (17-13) is again obtained. Actually, the constant wall temperature is a special case where $n = 0$.

Table 17-2 gives solutions for $-\theta'(0)/\sqrt{2}$ for the constant-heat-flux case. Comparison of the constant-heat-flux values with the constant-wall-temperature values in Table 17-1 reveals that the constant-heat-flux Nusselt number is about 15 percent higher. This compares with ratios of about 36 percent for laminar forced-convection boundary layers. A correlating equation for the entries in Table 17-2 is given by Fujii and Fujii[11] in a form similar to Eq. (17-14):

$$\text{Nu}_x = \left(\frac{\text{Pr}}{4 + 9\,\text{Pr}^{1/2} + 10\,\text{Pr}}\right)^{1/5}(\text{Gr}_x^* \,\text{Pr})^{1/5} \qquad (17\text{-}22)$$

TABLE 17-2
Values of $\text{Nu}_x \,\text{Gr}_x^{-1/4}$ for various Prandtl numbers; laminar heat transfer to a vertical free-convection boundary layer with $\dot{q}_0'' =$ constant, adapted from Refs. 2, 4, 6, 7, and 11

Pr	0.01	0.1	0.72	1.0	10	100	1000
$\text{Nu}_x \,\text{Gr}_x^{-1/4}$	0.0669	0.189	0.406	0.457	0.931	1.74	3.14

In Eq. (17-22) heat-flux data are correlated using a modified Grashof number containing $\dot{q}_0''$ explicitly:

$$\text{Gr}_x^* = \text{Gr}_x \, \text{Nu}_x = \frac{g\beta}{kv^2} \, \dot{q}_0'' x^4$$

SIMILARITY SOLUTIONS WITH WALL SUCTION OR BLOWING

Heat-transfer solutions exist for the case of vertical free-convection flow over a porous surface with mass transfer at the wall. As with the variable-surface-temperature problems, the transpired fluid velocity v_0 needs to be a particular function of x for similar velocity and temperature profiles. Eichhorn[12] obtained similarity solutions for prescribed $v_0(x)$ and $t_0(x)$. The applicable equations are Eqs. (17-20) and (17-21), with

$$t_0 - t_\infty = Ax^n$$

Eichhorn's $v_0(x)$ function that leads to similarity can be derived using the same procedures as in the previous sections. From the stream function definition for v, Eq. (17-10) for ψ, and the previously determined $G(x)$ and $H(x)$ functions, we have

$$v = -\frac{\partial \psi}{\partial x} = -vF\frac{dG}{dx} - vG\frac{dF}{d\eta}\frac{\partial \eta}{\partial x} = -vF\frac{dG}{dx} - vGF'y\frac{dH}{dx}$$

At the wall, $y = 0$ and $t_0 - t_\infty = Ax^n$:

$$v_0 = -vF(0)\frac{dG}{dx} = -vF(0)\frac{d}{dx}\left[4\left(\frac{\text{Gr}_x}{4}\right)^{1/4}\right]$$

$$= -\frac{v}{x}\left(\frac{\text{Gr}_x}{4}\right)^{1/4}(n+3)F(0) \qquad (17\text{-}23)$$

For similarity solutions to exist, the ordinary differential equations and boundary conditions must have constant coefficients. Similarity results if, in Eq. (17-23),

$$F(0) = \text{constant}$$

Therefore

$$v_0 \propto \frac{1}{x}\left(\frac{\text{Gr}_x}{4}\right)^{1/4} \propto x^{(n-1)/4}$$

The requirement $F(0) = \text{constant}$ becomes the boundary condition corresponding to $v = v_0$ at $y = 0$. For $F(0) > 0$ there will be wall suction, while $F(0) < 0$ corresponds to blowing. The Nusselt number is exactly the

TABLE 17-3
Values of $\mathrm{Nu}_x \mathrm{Gr}_x^{-1/4}$ for $\mathrm{Pr} = 0.73$ and wall suction and blowing; laminar heat transfer to a vertical free-convection boundary layer with $t_0 =$ constant, adapted from Ref. 12

	Suction				Blowing		
$F(0)$	$+1.0$	$+0.8$	$+0.4$	0	-0.4	-0.8	-1.0
$\mathrm{Nu}_x \mathrm{Gr}_x^{1/4}$	1.55	1.27	0.758	0.359	0.115	0.0187	0.00529

same as for the variable-temperature case:

$$\mathrm{Nu}_x = \frac{-\theta'(0)}{\sqrt{2}} \mathrm{Gr}_x^{1/4}$$

Table 17-3 gives solutions for $-\theta'(0)/\sqrt{2}$ for the constant-wall-temperature boundary condition, $n = 0$, and for $\mathrm{Pr} = 0.73$.

In Table 17-3 we see that blowing reduces heat transfer, while suction has the opposite effect. The blowing limit is $\theta'(0) \to 0$, corresponding to a zero wall temperature gradient. The suction limit is the case where the wall heat transfer equals the change in enthalpy of the fluid being sucked. These limits correspond to those for the laminar forced-convection case in Chap. 10. However, for the free-convection case, the $\theta'(0) = 0$ limit does not result in a zero wall shear stress because the blowing increases the temperature gradient, and this increases the buoyant driving force, which accelerates and stabilizes the flow. For more discussion see Eichhorn.[12]

APPROXIMATE INTEGRAL SOLUTIONS: LAMINAR FLOW ON A CONSTANT-TEMPERATURE, VERTICAL, AND SEMI-INFINITE FLAT PLATE

The free-convection integral solution technique can be developed by following the same procedures outlined in Chap. 8 for the forced-convection laminar boundary layer. The idea is to solve the integral momentum and energy equations by assuming velocity and temperature profile shapes and laws for wall shear stress and heat transfer. The resulting differential equations are coupled through the density, and their solutions follow that outlined by Eckert.[13]

The momentum integral equation to be used is Eq. (6-4) with an added body force term, $v_0 = u_\infty = 0$, R constant, and the Boussinesq approximation:

$$-\tau_0 = \frac{d}{dx}\int_0^Y \rho u^2 \, dy + Y\frac{dP}{dx} + \int_0^Y \rho g \, dy$$

$$= \frac{d}{dx}\int_0^Y \rho u^2 \, dy - g\beta\int_0^Y \rho(t - t_\infty) \, dy \qquad (17\text{-}24)$$

The applicable energy integral equation is Eq. (6-13) with $i_s = c(t - t_\infty)$, $v_0 = 0$, R constant, and the Boussinesq approximation:

$$\dot{q}_0'' = \frac{d}{dx}\int_0^Y \rho u c(t - t_\infty) \, dy = \rho c\frac{d}{dx}\int_0^Y u(t - t_\infty) \, dy \qquad (17\text{-}25)$$

To solve these equations, we require profile shapes that approximate the velocity and temperature profiles in Figs 17-1 and 17-2. Eckert assumed the following shapes, along with the assumption that the momentum and thermal boundary layers are of equal thickness:

$$\frac{u}{U} = \frac{y}{\delta}\left(1 - \frac{y}{\delta}\right)^2$$

$$\frac{t - t_\infty}{t_0 - t_\infty} = \left(1 - \frac{y}{\delta}\right)^2$$

These profiles are substituted into the integral equations, along with Newton's law of viscosity and Fourier's conduction law for τ_0 and $\dot{q}_0''$. This results in a set of coupled ordinary differential equations involving the characteristic velocity U and δ. Eckert solved these equations by assuming them to be power-law functions in x. The resulting expression for the boundary-layer thickness is

$$\frac{\delta}{x} = 3.93\left(\frac{0.952 + \mathrm{Pr}}{\mathrm{Pr}^2}\right)^{1/4} \mathrm{Gr}_x^{-1/4} \qquad (17\text{-}26)$$

The expression for δ can now be substituted into the temperature profile to evaluate the wall heat flux, and the resulting Nusselt number is

$$\mathrm{Nu}_x = 0.508\left(\frac{\mathrm{Pr}^2}{0.952 + \mathrm{Pr}}\right)^{1/4} \mathrm{Gr}_x^{1/4} \qquad (17\text{-}27)$$

For $\mathrm{Pr} = 0.72$, Eq. (17-27) predicts $\mathrm{Nu}_x\,\mathrm{Gr}_x^{-1/4} = 0.379$. This result is within 6 percent of the exact similarity solution in Table 17-1.

THE EFFECT OF VARIABLE PROPERTIES

Up to this point, we have considered heat-transfer solutions derived using the Boussinesq approximation, which assumed, in part, that properties except for the density in the momentum equation were constant. From

Chap. 15 we know that for gases μ, k, and ρ are functions of the *absolute temperature*. Since the free-convection motion itself is derived from a temperature gradient across the boundary layer, we expect variable properties to influence the Nusselt number.

Clearly the influence of variable properties depends on how μ and k vary with T. We can assume ρ to obey the ideal-gas equation of state, and $\beta = 1/T_\infty$. The simplest case for analysis is the hypothetical ideal gas in which c is constant and μ and k vary directly as T. Sparrow and Gregg[14] studied this problem and showed that the governing equations reduce to the constant-property equations. Thus the $\text{Nu}_x = f(\text{Pr}, \text{Gr}_x)$ relations are exactly the same as those we obtained from the constant-property solutions. An analogous conclusion was obtained for the laminar forced-convection boundary layer in Chap. 15.

Sparrow and Gregg obtained numerical solutions for a number of other hypothetical fluid-property variations. They included $\mu \propto T^{3/4}$ and $k \propto T^{3/4}$, which is close to the 0.8-power law for gases in the moderate temperature range. The major conclusion from their work was that a *reference temperature* could satisfactorily be used to correct the constant-property solutions. They proposed that all properties in the constant-property correlations be evaluated at a reference temperature

$$T_R = T_0 - 0.38(T_0 - T_\infty) \qquad (17\text{-}28)$$

For gases they recommend $\beta = 1/T_\infty$ and for low-Pr fluids $\beta = 1/T_R$. *However*, almost all Nu_x data and data correlation reported in the open literature, *and in this chapter*, are formulated using $T_R = 0.5(T_0 + T_\infty)$, the average film temperature. For the heat-flux boundary condition, T_R is an unknown. To evaluate Gr_x and Nu_x, the wall temperature must be guessed and iterated until the T_0 used in T_R is the same as that calculated from $\dot{q}_0''$ and Nu_x.

FREE-CONVECTION FLOW REGIMES

In forced convection we find that a length Reynolds number arises when the Navier–Stokes equations are nondimensionalized. For $\text{Re} \to 0$ the inertial terms are no longer important and the flow is termed *slow-viscous* or "creeping" flow. At moderate Re the flow is still laminar, but viscous and inertial effects are important, and for high Re the flow is turbulent. These same flow regimes are observed in free convection. To define a Reynolds number for free convection, we need a characteristic velocity to describe the inertial forces within the boundary layer. Consider the region of the boundary layer where the velocity profile reaches a maximum. If we ignore the effects of diffusion and cross-stream convection in this region and assume that t can be approximated by t_0, the momentum equation reduces to

$$\rho U \frac{dU}{dx} = \rho g \beta (t_0 - t_\infty)$$

This equation relates the convective acceleration of the fluid to the buoyancy driving force. By integrating the equation, we obtain a characteristic velocity for the flow:

$$U \propto \sqrt{g \beta (t_0 - t_\infty)x}$$

A Reynolds number for the flow then follows:

$$\mathrm{Re}_x = \frac{Ux}{\nu} = \sqrt{\frac{g\beta(t_0 - t_\infty)x^3}{\nu^2}}$$

$$\mathrm{Re}_x = \mathrm{Gr}_x^{1/2}$$

From this approximation we see that the Grashof number serves as the length Reynolds number for free convection from vertical flat surfaces. For gases and moderate-Pr liquids the slow-viscous flow regime is generally found below about $\mathrm{Gr}_x = 10^4$. Transition to turbulent flow occurs between $\mathrm{Gr}_x = 10^9$ and 10^{10}, with a tendency toward higher Gr_x for higher Pr. Surface inclination and surface heating rate also affect transition. A discussion of transition phenomena is given by Gebhart.[3]

Frequently the *Rayleigh number* $\mathrm{Ra}_x = \mathrm{Gr}_x \mathrm{Pr}$ is used in place of Gr_x for defining these flow regimes. It also appears in many heat-transfer correlations. However, most experimental evidence indicates that Ra_x does not entirely reflect the Pr dependence. Le Fevre[15] points out that $\mathrm{Gr}_x \mathrm{Pr}^2$, the *Boussinesq number,* is the proper functional dependence for large Gr_x or small Pr, which eliminates the effect of viscosity.

TURBULENT FLOW ON A VERTICAL AND SEMI-INFINITE FLAT PLATE

A turbulent integral analysis was carried out by Eckert and Jackson[16] using precisely the same technique that Eckert[13] used for laminar free-convection flow. Equations (17-24) and (17-25) are again applicable. Based on available experimental velocity and temperature profile data, the profile shapes used by Eckert and Jackson were

$$\frac{u}{U} = \left(\frac{y}{\delta}\right)^{1/7}\left(1 - \frac{y}{\delta}\right)^4$$

$$\frac{t - t_\infty}{t_0 - t_\infty} = 1 - \left(\frac{y}{\delta}\right)^{1/7}$$

In these profile shapes U is a characteristic velocity, t_0 is constant, and δ is the outer edge of the boundary layer. Note that differentiating the velocity profile gives $u_{max} = 0.537U$. Thus U is defined to eliminate the constant obtained in curve-fitting the experimental velocity profile data.

To solve the integral equations, the laws for wall shear stress and heat flux need to be modeled. Eckert and Jackson argued that in the near-wall region τ_0 would be similar to that for forced convection. The shear-stress law they used follows from the power-law form of the wall law for the near-wall region, Eq. (11-17). Because $u = 0$ at $y = \delta$, Eckert and Jackson used the characteristic velocity U in place of u_∞ at $y = \delta$. The resulting equation for the shear stress becomes

$$\tau_0 = 0.0225\rho U^2\left(\frac{\nu}{U\delta}\right)^{1/4}$$

The wall heat flux is obtained by using an analogy between momentum and heat transfer with a Prandtl number function for fluids with Prandtl numbers near unity:

$$St = \frac{\dot{q}_0''}{\rho c U(t_0 - t_\infty)} = 0.0225\left(\frac{\nu}{U\delta}\right)^{1/4}Pr^{-2/3}$$

Solution for the turbulent case follows exactly the solution procedure for the laminar case. The profiles and expressions for τ_0 and $\dot{q}_0''$ are substituted into Eqs. (17-24) and (17-25). This results in two ordinary differential equations for U and δ. The equations are solved by assuming the power-law functions

$$U = C_u x^m, \qquad \delta = C_\delta x^n$$

These functions are substituted into the differential equations, and the exponents and constants are obtained directly. The resulting Nusselt number is found to be

$$Nu_x = 0.0295\left[\frac{Pr^7}{(1 + 0.494\,Pr^{2/3})^6}\right]^{1/15} Gr_x^{2/5} \qquad (17\text{-}29)$$

This analysis suggests that h varies as $x^{0.2}$. To obtain an average heat-transfer coefficient, we follow the same procedure used with Eq. (17-19) for the laminar solution, which leads to

$$Nu = 0.834\,Nu_x$$

Bailey[17] carried out a turbulent integral analysis using Eqs. (17-24) and (17-25) and the same velocity profile and τ_0 expression as Eckert and Jackson. For the temperature profile, and therefore $\dot{q}_0''$, Bailey used a two-layer temperature model with a laminar sublayer out to $y^+ = 15$ and then a constant-diffusivity turbulent region. He found that

$$\begin{aligned} Pr &= 0.73, \quad Nu_x = 0.10(Gr_x\,Pr)^{1/3} \\ Pr &= 0.01, \quad Nu_x = 0.060\,Gr_x^{1/4} \end{aligned} \qquad (17\text{-}30)$$

This analysis suggests that h is independent of x for gases and that $\text{Nu}_x = \text{Nu}$.

By comparing the Eckert and Jackson solution with the Bailey solution for air ($\text{Pr} = 0.73$), we find the Bailey solution to consistently predict a lower Nusselt number. Experimental data can be found in the literature to support both analyses. The Eckert and Jackson analysis is supported by Cheesewright,[18] and the data of Warner and Arpaci[19] fit the Bailey correlation. Numerical predictions of turbulent free convection have been reported in the literature, and the prediction trend is for the Nusselt number to lie between the two correlations.

Figure 17-3 shows the experimental data of Seibers, Schwind, and Moffat,[37] together with the Bailey equation for air. These data also show the transition from a laminar free-convection boundary layer to a turbulent free-convection boundary layer, and in addition provide some idea of the effects of temperature-dependent fluid properties on a turbulent free-convection boundary layer. The two sets of data were obtained at wall-to-free-stream temperature ratios of 1.15 and 1.39.

Experimental heat-transfer data for turbulent free convection with a constant wall heat flux have been reported by Vliet and Ross.[20] They found that their air data correlate using

$$\text{Nu}_x = 0.0942(\text{Gr}_x \, \text{Pr})^{1/3} \tag{17-31}$$

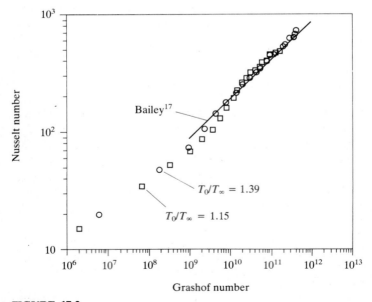

FIGURE 17-3
Turbulent free convection; air, constant surface temperature (data of Seibers, Schwind, and Moffat[37]).

This equation is nearly identical to the Bailey equation and suggests that the turbulent Nusselt numbers for constant-heat-flux and constant-temperature boundary conditions are similar for gases. The same conclusion is found in forced convection.

A study of numerous sets of experimental data has been carried out by Churchill and Chu.[21] They propose the following general correlation for turbulent free convection on an isothermal vertical surface:

$$\text{Nu} = \frac{0.15(\text{Gr Pr})^{1/3}}{[1 + (0.492/\text{Pr})^{9/16}]^{16/27}} \tag{17-32}$$

The correlation is valid for all Pr and for Gr_x to 10^{12}. For a constant-heat-flux surface they suggest changing the coefficient 0.492 to 0.437.

HEAT-TRANSFER SOLUTIONS FOR OTHER GEOMETRIES

When heated plates are inclined from the vertical, the buoyancy force can be decomposed into $\sin\theta$ and $\cos\theta$ components. Here θ is the angle of inclination from the vertical; $+\theta$ refers to an upward-facing heated surface (unstable with a buoyancy component away from the surface) and $-\theta$ is downward-facing (stable). Moran and Lloyd,[22] and others, indicate that by replacing g with $g\cos\theta$ in the Grashof number, vertical-surface Nusselt number correlations can be used in the range from $+20°$ to $-60°$ relative to the vertical. This applies to laminar flow on constant-surface-temperature and constant-heat-flux surfaces for the case where the surface span is wide enough to neglect edge effects. Vliet and Ross[20] indicate that the Nusselt number is independent of angle for turbulent flow with $0° < \theta < +30°$. For the stable condition, $0° < \theta < -80°$, they indicate that the Grashof number should be computed using $g\cos^2\theta$.

For completeness, we include results from available experimental data correlations for horizontal surfaces and horizontal and vertical cylinders. The reader is referred to Raithby and Hollands[23] for an approximate integral method for predicting free-convection coefficients for these and numerous other geometries. The concept of local similarity and local nonsimilarity can also be applied to obtain approximate heat-transfer solutions for nonsimilar free-convection boundary layers. For reference to the method (similar to that outlined in Chap. 8) see Minkowycz and Sparrow.[24]

Average Nusselt number correlations for gases over isothermal *horizontal surfaces* with the heated surface facing *upward* are given by Fishenden and Saunders.[25] The Grashof numbers in the correlation are based on the length of a square plate:

Laminar

$$\text{Nu} = 0.54(\text{Gr}_L\,\text{Pr})^{1/4} \quad \text{for } 10^5 < \text{Gr}_L\,\text{Pr} < 2 \times 10^7$$

Turbulent

$$\text{Nu} = 0.14(\text{Gr}_L\,\text{Pr})^{1/3} \quad \text{for } 2 \times 10^7 < \text{Gr}_L\,\text{Pr} < 3 \times 10^{10} \quad (17\text{-}33)$$

Lloyd and Moran[26] have also studied the horizontal-surface-free convection problem using the heat–mass-transfer analogy, and they find about the same results as those given in Eq. (17-33). The differences are as follows: in the laminar equation the limits are $2.2 \times 10^4 \leqslant \text{Gr}_L\,\text{Pr} \leqslant 8 \times 10^8$; in the turbulent equation the coefficient is 0.15 and the limits are $8 \times 10^6 \leqslant \text{Gr}_L\,\text{Pr} \leqslant 1.6 \times 10^9$. The length scale in their Grashof number is that developed by Goldstein, Sparrow, and Jones:[27] $L = A/P$, where A is the surface area and P is the surface perimeter. Lloyd and Moran showed that with L, Eq. (17-33) correlates data for circular, square, rectangular, and right-angled triangular surfaces.

For average Nusselt numbers on isothermal *horizontal cylinders* Kuehn and Goldstein[28] recommend a slightly modified form of the laminar correlation by Churchill and Chu[29] and the turbulent correlation by Fishenden and Saunders[25] for $\text{Gr}_D\,\text{Pr} > 10^9$:

Laminar

$$\text{Nu} = 0.518\left[1 + \left(\frac{0.599}{\text{Pr}}\right)^{3/5}\right]^{-5/12}(\text{Gr}_D\,\text{Pr})^{1/4}$$

Turbulent

$$\text{Nu} = 0.1(\text{Gr}_D\,\text{Pr})^{1/3} \qquad (17\text{-}34)$$

To account for curvature effects or low Gr_D when the boundary layer is thick compared with the cylinder diameter, the Nusselt number correlation is modified using

$$\overline{\text{Nu}} = \frac{2}{\ln(1 + 2/\text{Nu})} \qquad (17\text{-}35)$$

An extensive survey of free-convection data from horizontal cylinders can be found in a study by Fand, Morris, and Lum[30] for $0.7 < \text{Pr} < 3090$. They also discuss the effects of reference temperature on correlations.

Vertical circular cylinders have average Nusselt numbers that depend on the ratio D/L for the surface and on the number $\text{Gr}_D\,\text{Pr}$. In a laminar study Sparrow and Gregg[31] conclude that the vertical flat-plate solution is acceptable for air when

$$(\text{Gr}_D\,\text{Pr})^{1/4}\frac{D}{L} > 38 \qquad (17\text{-}36)$$

Raithby and Hollands[23] recommend Eq. (17-35) to correct the flat-surface Nusselt number for curvature effects or low Gr_D when the boundary layer is thick compared with the cylinder diameter. Note that Eq. (17-35) yields a larger Nusselt number, so the flat-surface value will always be conservative. Ede[2] presents data for turbulent flow on vertical cylinders that indicate a marked increase in Nusselt number for $D/L <$ 0.05.

MIXED FREE AND FORCED CONVECTION

The major effect of buoyancy is to alter the velocity and temperature fields in the forced-convection flow, and this in turn alters the Nusselt number and friction coefficient. Consider the case of upward forced convection over a vertical surface (or vertical tube). If the surface is heated such that $t_0 > t_\infty$, the resulting buoyancy force aids the convective motion, especially in the near-wall region, and we would expect the Nusselt number to be greater than the forced-convection value at that Reynolds number. Likewise, for $t_0 < t_\infty$ the buoyancy force would oppose the flow, and the Nusselt number would decrease. For flow over a horizontal surface a similar enhancement or decrease of the Nusselt number occurs. In this case the buoyancy effect appears as a pressure gradient because the isotherms are not parallel to the surface. Forced convection in heated horizontal tubes can result in enhanced Nusselt numbers if buoyancy sets up a secondary flow.

Fully established turbulent flows and laminar flows with low Gr have negligible buoyancy effects. *Mixed* convection primarily occurs with laminar and transitional flows and moderate to large Gr, and its onset depends on the relative magnitudes of Gr and Re. In the following paragraphs we present some analytic results to illustrate the onset of mixed convection. A review of mixed convection in internal flows is given by Kays and Perkins;[32] for external mixed convection flows see Gebhart.[4]

Lloyd and Sparrow[33] have carried out a local similarity analysis for forced convection over isothermal vertical surfaces with aided buoyancy. They compared $\mathrm{Nu}_x/\mathrm{Re}_x^{1/2}$ with the ratio $\mathrm{Gr}_x/\mathrm{Re}_x^2$ for various Prandtl number fluids to show the effects of buoyancy. For a 5 percent increase in Nusselt number the $\mathrm{Gr}_x/\mathrm{Re}_x^2$ values were 0.24, 0.13, 0.08, and 0.056 at $\mathrm{Pr} = 100$, 10, 0.72, and 0.08. The paper also presents the effects of buoyancy on velocity and temperature profiles.

Mixed-convection flow on horizontal plates with $\mathrm{Pr} = 0.7$ was studied by Chen et al.[34] using local similarity and nonsimilarity analyses. They found that for a heated surface facing upward a 5 percent increase in Nusselt number occurred when $\mathrm{Gr}_x/\mathrm{Re}_x^{2.5}$ was about 0.05. The paper also presents the effects of buoyancy on velocity and temperature profiles.

An analytical and experimental study of convection heat transfer with buoyancy inside heated horizontal tubes was carried out by Mori and Futagami.[35] For air with $Pr = 0.72$ they found that the Nusselt number for forced convection begins to increase when the product $Re\,Gr\,Pr$ exceeds 10^3, with a 50 percent increase by 10^5. The Grashof number is defined using the tube radius, while the Reynolds number is defined in terms of the tube diameter. The main cause of the increased heat transfer is the secondary flow, similar to that found in curved-tube flow. Mixed convection on the external surface of horizontal tubes is discussed by Morgan.[36]

PROBLEMS

17-1. Consider a flat plate surrounded by a fluid at rest (at rest outside the boundary layer) and oriented vertically to a gravity field of strength g. If the plate is heated to a temperature above that of the fluid, the fluid immediately adjacent to the plate will be heated, its density will decrease below that of the surrounding fluid, the resulting buoyancy force will put the fluid in motion, and a free-convection boundary layer will form. If the thermal expansion coefficient for the fluid is defined as

$$\beta = -\frac{1}{\rho}\left(\frac{\partial \rho}{\partial T}\right)_P$$

develop the applicable momentum integral equation of the boundary layer under conditions where the density variation through the boundary layer is small relative to the free-fluid density.

17-2. For laminar flow over a constant-heat-flux surface develop two expressions for the mean Nusselt number. In the first case base the heat-transfer coefficient on the temperature difference between the average surface temperature and the ambient; in the second case let the temperature difference be the surface minus the ambient at $\frac{1}{2}L$. Compare the two mean Nusselt numbers with those for a constant-wall-temperature surface, Eq. (17-19), at Prandtl numbers of 0.1, 0.72, 1.0, and 100.0, and discuss.

17-3. Develop an analytic solution for the laminar temperature profile and Nusselt number for the case of large wall suction. Start with Eq. (17-21). The final expression will contain n, $F(0)$, and Pr as parameters. Compare this asymptotic expression for Nusselt number with the results in Table 17-3.

17-4. A flat ribbon heat strip is oriented vertically on an insulating substrate. Let the ribbon be 1 m wide by 3 m long. Its energy dissipation is $0.5\ W/cm^2$ to air at 25°C. What are the average heat-transfer coefficient and surface temperature of the ribbon? Where will transition to turbulent flow occur? Would you be justified in neglecting the laminar contribution to the heat transfer?

17-5. Using the approximate integral solution method for turbulent free convection over a vertical and constant-temperature surface, develop an equation for δ/x. Compare and discuss how the boundary-layer thickness varies with x for laminar and turbulent free and forced convection.

17-6. Consider the free-convection cooling of a thick, square plate of copper with one surface exposed to air and the other surfaces insulated. Let the air temperature be 25°C and the copper temperature be 45°C. The copper is 10 cm on a side. Compare the average heat-transfer coefficients for three exposed face orientations: vertical, inclined 45° to the vertical, and horizontal

17-7. Consider the wire on a constant-temperature hot-wire anemometer sensor. The wire diameter is 0.00038 cm and its length is 580 diameters. Let the wire temperature be 260°C and the air temperature be 25°C. Compare the heat-transfer coefficients for the wire placed in the horizontal and vertical positions. Note that the effects of the wire support prongs are neglected.

REFERENCES

1. Ostrach, S.: *Theory of Laminar Flows,* pp. 528–615, Princeton University Press, Princeton, NJ, 1964.
2. Ede, A. J.: *Advances in Heat Transfer,* vol. 4, pp. 1–64. Academic Press, New York, 1967.
3. Gebhart, B.: *J. Fluids Engng,* vol. 101, pp. 5–28, 1979.
4. Gebhart, B.: *Heat Transfer,* 2d ed., pp. 316–402, McGraw-Hill, New York, 1971.
5. Eckert, E. R. G., and E. E. Soehngen: Tech. Rep. 5747, ATI 44580, Air Material Company, Dayton, Ohio, December 1948.
6. Ostrach, S.: NASA Report 1111, Washington, 1953.
7. Sparrow, E. M., and J. L., Gregg: *Trans. ASME,* vol. 78, pp. 435–440, 1956.
8. Le Fevre, E. J.: *Proceedings of 9th International Congress of Applied Mechanics, Brussels, 1956,* vol. 4, pp. 168–174.
9. Yang, K.-T.: *J. Appl. Mech.,* vol. 27, p. 230, 1960.
10. Sparrow, E. M., and J. L. Gregg: *Trans. ASME,* vol. 80, pp. 379–386, 1958.
11. Fujii, T., and M. Fujii: *Int. J. Heat Mass Transfer,* vol. 19, pp. 121–122, 1976.
12. Eichhorn, R.: *J. Heat Transfer,* vol. 82, pp. 260–263, 1960.
13. Eckert, E. R. G.: *Introduction to the Transfer of Heat and Mass,* pp. 158–164. McGraw-Hill, New York, 1950.
14. Sparrow, E. M., and J. K. Gregg: *Trans. ASME,* vol. 80, pp. 879–886, 1958.
15. Le Fevre, E. J.: *Int. J. Heat Mass Transfer,* vol. 19, pp. 1215–1216, 1976.
16. Eckert, E. R. G., and T. W. Jackson: NACA (now NASA) TN 2207, Washington, 1950.
17. Bailey, F. J.: *Proc. Inst. Mech. Engrs,* vol. 169, pp. 361–368, 1955.
18. Cheesewright, R.: *J. Heat Transfer,* vol. 89, pp. 1–8, 1968.
19. Warner, C. Y., and V. S. Arpaci: *Int. J. Heat Mass Transfer,* vol. 11, pp. 397–406, 1968.
20. Vliet, G. C., and D. C. Ross: *J. Heat Transfer,* vol. 97, pp. 549–555, 1975.
21. Churchill, S. W., and H. H. S. Chu: *Int. J. Heat Mass Transfer,* vol. 18, pp. 1323–1329, 1975.
22. Moran, W. R., and J. R. Lloyd: *J. Heat Transfer,* vol. 97, pp. 472–474, 1975.
23. Raithby, G. D., and K. G. T. Hollands: *J. Heat Transfer,* vol. 98, pp. 72–80, 1976.
24. Minkowycz, W. J., and E. M. Sparrow: *J. Heat Transfer,* vol. 96, pp. 178–183, 1974.

25. Fishenden, M., and O. A. Saunders: *An Introduction to Heat Transfer,* Oxford University Press, London, 1950.
26. Lloyd, J. R., and W. R. Moran: *J. Heat Transfer,* vol. 96, pp. 443–447, 1974.
27. Goldstein, R. J., Sparrow, E. M., and D. C. Jones: *Int. J. Heat Mass Transfer,* vol. 16, pp. 1025–1034, 1973.
28. Kuehn, T. H., and R. J. Goldstein: *Int. J. Heat Mass Transfer,* vol. 19, pp. 1127–1134, 1976.
29. Churchill, S. W., and H. H. S. Chu: *Int. J. Heat Mass Transfer,* vol. 18, pp. 1049–1053, 1975.
30. Fand, R. M., Morris, E. W., and M. Lum: *Int. J. Heat Mass Transfer,* vol. 20, pp. 1173–1184, 1977.
31. Sparrow, E. M., and J. L. Gregg: *Trans. ASME,* vol. 78, pp. 1823–1828, 1956.
32. Kays, W. M., and H. C. Perkins: *Handbook of Heat Transfer,* pp. 7-37–7-46, McGraw-Hill, New York, 1973.
33. Lloyd, J. R., and E. M. Sparrow: *Int. J. Heat Mass Transfer,* vol. 13, pp. 434–438, 1970.
34. Chen, T. S., E. M. Sparrow, and A. Mucoglu: *J. Heat Transfer,* vol. 99, pp. 66–71, 1977.
35. Mori, Y., and K. Futagami: *Int. J. Heat Mass Transfer,* vol. 10, pp. 1801–1813, 1967.
36. Morgan, V. T.: *Advances in Heat Transfer,* vol. 11, pp. 199–264, Academic Press, New York, 1975.
37. Seibers, D. L., R. G. Schwind, and R. J. Moffat: Report HMT-36, Thermosciences Division, Department of Mechanical Engineering, Stanford University, February 1983.

CHAPTER
18

HEAT-EXCHANGER ANALYSIS AND DESIGN

One of the most important applications of convective heat-transfer theory is in the analysis and design of *heat exchangers*—devices to transfer heat from one fluid stream to another. These are found in abundance in thermal power systems, refrigeration systems, cooling systems, and throughout the process industries. They are designed in an almost infinite variety of sizes, shapes, and configurations.

Perhaps the simplest conceivable heat exchanger would be a length of circular tube through which one fluid flows, while the other fluid flows either normally to it, or axially along the outside of the tube. But most heat exchangers involve a multiplicity of tubes arranged in various ways, and often the tubes have flow cross-sections other than circular. In any case, however, the typical scheme is to transfer heat by convection from one fluid to a separating wall, through that wall by conduction, and then by convection to the other fluid.

In heat-exchanger analysis we will generally be considering fluids flowing inside long tubes, or flowing normally to banks of a large number of tubes. In either case it is the convection theory developed in the tube-flow chapters, rather than the external-boundary-layer chapters, that will most often be called upon. At any point within the heat exchanger we will calculate the local rate of heat transfer between the fluid and wall using an equation of the type

$$\dot{q}'' = h(t_{fluid} - t_{surface}) \tag{18-1}$$

We will presume that we know how to evaluate the heat-transfer coefficient h from our previous study. The question now is the way in which t_{fluid} and $t_{surface}$ vary throughout the heat exchanger. The answer to that question is the fundamental one in heat-exchanger analysis; it is the feature that distinguishes the latter from the subject of convective heat transfer in general.

Entire volumes have been devoted to heat-exchanger design, so obviously only an introduction to the subject is feasible in a two-chapter treatment. For brevity, consideration will be limited to two-fluid heat exchangers (some heat exchangers involve several fluids), steady-state performance, single-phase fluids, and only a few of the many possible flow arrangements. Particular emphasis will be placed on heat exchangers involving *gases,* and Chap. 19 is entirely devoted to the particular problems posed by gas-flow heat exchangers.

The heat-transfer coefficient h in Eq. (18-1) will be assumed to be invariant in the direction of flow, or at least it will be assumed that an average heat-transfer coefficient can be defined that is invariant. For those cases where the variation of fluid properties in the flow direction is substantial, so that h varies substantially, special procedures will be described.

There are two methods of heat exchanger analysis that are commonly used. One, the *log-mean-temperature-difference* method, is extensively used in the process industries; the other, the *effectiveness–NTU* method is more often employed for gas-flow heat exchangers, especially in thermal power systems and cooling applications. The authors believe that the *effectiveness–NTU* scheme provides a superior way to look at heat exchangers in terms of nondimensional variables. Since the emphasis will in any case be primarily on gas-flow heat exchangers, only the *effectiveness–NTU* scheme will be described. However, it should be clear that, despite this emphasis, the *effectiveness–NTU* scheme is applicable to any kind of heat exchanger.

We will first introduce the variables that enter into the analysis of a heat exchanger, and then examine some of the various flow arrangements that might be used. Finally we will discuss some of the particular heat-exchanger mathematical solutions. The actual design of heat exchangers using these solutions will be considered in Chap. 19.

EXCHANGER VARIABLES

A surface heat exchanger will be defined as a steady-state device in which heat is transferred from one fluid stream to another without any direct contact or mixing of the fluids. The heat must therefore be transferred

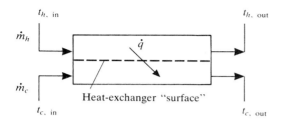

FIGURE 18-1
Schematic of a two-fluid heat exchanger.

through some kind of separating wall or partition, which is known as the *heat-transfer surface*; see Fig. 18-1.

Analysis of a heat exchanger involves an analysis of its thermal performance and its pressure-drop performance. Both will generally be of importance in the design of the heat exchanger, especially if the fluids are gases.

The thermal, or heat-transfer, performance analysis of a heat exchanger generally involves one of two general types of problem:

1. Given the complete operating conditions, to design a heat exchanger, i.e., specify the surface area requirement, after having chosen a type of surface and a flow arrangement (the *design problem*).
2. Given a particular heat exchanger, and fluid inlet conditions, to determine the outlet states of both fluids (the *performance prediction problem*).

In either case a convenient expression for the thermal performance of the heat exchanger is the *heat-transfer effectiveness* (which has no relation to the *efficiency*).

The effectiveness will be defined as the ratio of the actual rate of heat transfer in the exchanger to the maximum possible rate of heat transfer, as limited by thermodynamic considerations:

$$\varepsilon = \dot{q}_{\text{actual}} / \dot{q}_{\text{max.possible}} \qquad (18\text{-}2)$$

If we make an energy balance on each of the two fluid sides, we obtain

$$\dot{q}_{\text{actual}} = \dot{m}_c c_c (t_{c,\text{out}} - t_{c,\text{in}}) = \dot{m}_h c_h (t_{h,\text{in}} - t_{h,\text{out}})$$

Note that these expressions are exact for a perfect gas (if $c = c_p$, and c_p may be treated as constant), but for any other fluid they are exact only for a constant-pressure process. For an incompressible fluid (i.e., a liquid) the assumption is implicit that the change in flow work (Pv) involved is negligible relative to the change in internal thermal energy of the fluid, and this idealization is usually perfectly valid.

For convenience, we will lump together the flow rates and specific heats, and term the products the *capacity rates*:

$$C_c = \dot{m}_c c_c$$
$$C_h = \dot{m}_h c_h$$

The thermodynamic limit on the maximum possible heat-transfer rate would obtain if one of the two fluids left the heat exchanger at the entering temperature of the other fluid (a greater heat transfer rate would violate the second law of thermodynamics). If $C_h < C_c$,

$$\dot{q}_{max} = C_h(t_{h,in} - t_{c,in})$$

otherwise

$$\dot{q}_{max} = C_c(t_{h,in} - t_{c,in})$$

Thus

$$\dot{q}_{max} = C_{min}(t_{h,in} - t_{c,in})$$

Then the *effectiveness* is defined:

$$\varepsilon = \frac{C_c(t_{c,out} - t_{c,in})}{C_{min}(t_{h,in} - t_{c,in})}$$

$$= \frac{C_h(t_{h,in} - t_{h,out})}{C_{min}(t_{h,in} - t_{c,in})} \tag{18-3}$$

Note that the effectiveness is fundamentally a heat-transfer-rate ratio, rather than a temperature-difference ratio, although it reduces to the latter for the unique case of equal capacity rates, i.e.,

$$C_{min} = C_{max}, \quad \text{or} \quad C_c = C_h$$

The heat-transfer thermal analysis is one of determining the effectiveness as a function of the relevant heat-exchanger variables. In general it will be found that

$$\varepsilon = \varepsilon(U, A, C_h, C_c, \text{flow arrangement})$$

Let us now examine these variables:

1. U, $\text{W}/(\text{m}^2 \cdot \text{K})$: the unit overall conductance for heat transfer from the hot fluid to the cold fluid. We will examine this variable in more detail shortly.
2. A, m^2: the heat-transfer surface area upon which U is based. We will say more about this later.
3. C_c, W/K: the cold-fluid capacity rate.
4. C_h, W/K: the hot-fluid capacity rate.
5. Flow arrangement: i.e., counterflow, parallel flow, cross-flow, etc.

For any of the flow arrangements that will be discussed we find that ε can be expressed as a function of two nondimensional parameters. Any two containing all of the exchanger variables could be used, but we will settle upon a particular pair that happen to be convenient:

1. AU/C_{min}, where C_{min} is the lesser of C_c and C_h.
2. C_{min}/C_{max}.

We will call AU/C the NTU, the *number of transfer units* (sometimes written N_{tu}). Thus

$$AU/C_{min} = \text{NTU}_{max} \qquad (18\text{-}4)$$

C_{min}/C_{max} will simply be called the *capacity rate ratio*.
Therefore, in general,

$$\varepsilon = \varepsilon(\text{NTU}_{max}, C_{min}/C_{max}, \text{ flow arrangement})$$

EVALUATION OF THE OVERALL CONDUCTANCE

Examine a small section of the wall separating the two fluids in a heat exchanger; Fig. 18-2. All of the terms in Fig. 18-2 are based on 1 m^2 of wall area. Note that the fluid flow directions can be various.

U is obviously built up from the two local convective conductances (heat-transfer coefficients) on either side of the wall, plus the conduction resistance of the wall material, and any other resistances that might be present. In this case

$$U = \frac{1}{\sum_i R_i} = \frac{1}{R_h + R_{wall} + R_c} = \frac{1}{1/h_h + \delta/k + 1/h_c}$$

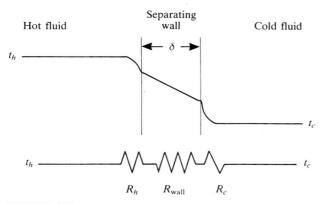

FIGURE 18-2
A heat-exchanger wall section.

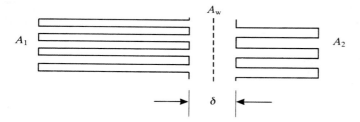

FIGURE 18-3
A_1 and A_2 include both fin and base area.

For the more general case where there are fins (extended surface) on both sides, and the total surface areas are unequal, Fig. 18-3, we can readily deduce

$$U_1 = \frac{1}{1/(\eta_{0,1}h_1) + (\delta/k)A_1/A_w + (A_1/A_2)/(\eta_{0,2}h_2) + \text{other resistances}}$$

(18-5)

Here

$$\eta_0 = 1 - \frac{A_f}{A}(1 - \eta_f)$$

is the *overall surface effectiveness*, where A_f is the fin area and η_f is the *fin heat-transfer effectiveness*:

$$\eta_f = \dot{q}/\dot{q}_{max}$$

where $\dot{q}_{max}$ is what the heat-transfer rate through the fin base would be were the entire fin at base temperature. If the fin conductivity is indefinitely large then $\eta_f = 1$.

Analytic expressions for η_f can be developed for various fin geometries. The most important one is for the simple rectangular-section fin, Fig. 18-4. For this case

$$\eta_f = \frac{\tanh ml}{ml}$$

(18-6)

where $m = (2h/k\delta)^{1/2}$.

Equation (18-6) was developed assuming that the heat-transfer coefficient h is constant along the fin, which may not be accurate. However, if the effectiveness is near 1 this will not matter much.

Note that U_1 is based on a particular surface area (A_1 in this case). However, it can readily be shown that

$$A_1 U_1 = A_2 U_2$$

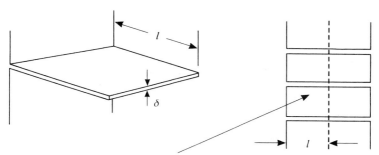

Typical heat-exchanger fin arrangement

FIGURE 18-4
A simple rectangular fin.

U may be a constant, or it may vary through the heat exchanger if the hs vary. The heat-exchanger theory that we will develop is based on a constant U, or a mean U with respect to surface area.

Finally we need to be able to evaluate the heat-transfer coefficients. These can be obtained from analytic solutions for the simple geometries, or from correlations of experimental data. In either case, for fully developed flow in tubes we can generally express the heat-transfer

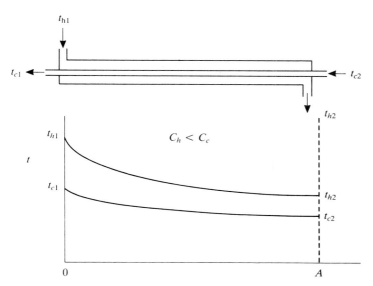

FIGURE 18-5
A counterflow heat exchanger.

coefficients in nondimensional form:

$$\text{Nu or St} = \phi(\text{Re}, \text{Pr})$$

If the tube length is relevant, the ratio L/D can be included, although for most heat exchangers the length effect is negligible.

Once the surface and flow-passage configuration has been chosen (geometry fixed), and the material is specified (k fixed), then it is apparent that

$$U_1 = \phi(G_1, G_2)$$

where $G = \dot{m}/A_c$, and A_c is the total flow cross-sectional area and $\dot{m}$ the total mass-flow rate.

FLOW ARRANGEMENTS

A few of the most common flow arrangements will now be illustrated.

Counterflow (see Fig. 18-5)

Although only single tubes are illustrated, the heat exchanger could be made up of any number of tubes or passages of any cross-sectional shape.

By inspection, it is apparent that for the temperature diagram shown, $C_h < C_c$. If $C_h = C_c$, $C_{min}/C_{max} = 1$, and the two temperature lines would be parallel straight lines.

Note that for the case shown the heat flux is greatest near the left end; for $C_{min}/C_{max} = 1$ the heat flux would be constant along the length of the heat exchanger.

For the capacity-rate ratio shown, if the heat-transfer area is increased indefinitely (Fig. 18-6), t_{h2} will approach t_{c2}, but t_{c1} cannot approach t_{h1}. For this reason the effectiveness is based on the maximum possible change of the *minimum* capacity rate fluid.

For $C_{min}/C_{max} = 1$, if the area is increased indefinitely, both terminal temperature differences will approach zero. This, then, is the only case where heat-transfer irreversibility can approach zero.

Parallel Flow (see Fig. 18-7)

Note that if the capacity rates of the two fluids are equal, the final temperatures, t_{c2} and t_{h2}, can only approach a temperature halfway between the two initial temperatures. Thus ε approaches 0.5 as A increases indefinitely.

This is a flow arrangement that is generally used only when the desired effectiveness is very low, or when the capacity rate ratio is close to zero so that the flow arrangement becomes immaterial.

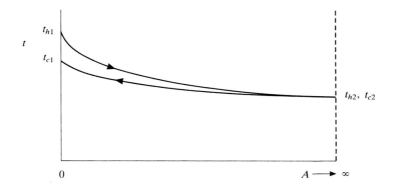

FIGURE 18-6
Effect of infinite heat-transfer area; $C_h = C_{min}$.

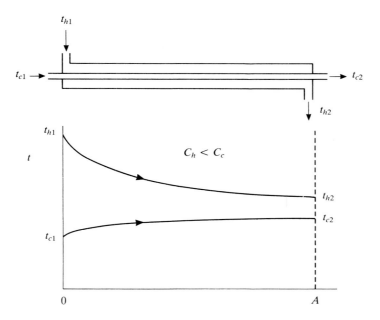

FIGURE 18-7
A parallel-flow heat exchanger.

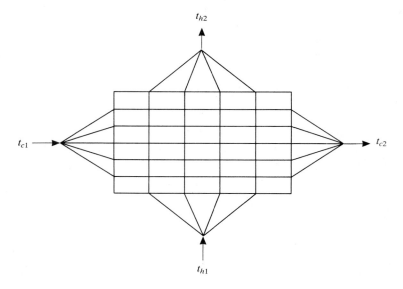

FIGURE 18-8
A cross-flow heat exchanger.

Cross-Flow, Both Fluids Unmixed (see Fig. 18-8)

This is the most common type of cross-flow heat exchanger. The fluid temperatures within the heat exchanger are both functions of two spatial coordinates rather than a single coordinate as in the previous case. A basic assumption in this case is that there is no lateral mixing of the fluids, so that each "flow tube" stays independent of its neighbors. There is thus a lateral temperature gradient in each fluid at the plane where it emerges. When we speak of the effectiveness of this heat exchanger, we are implying that each fluid mixes with itself completely after emerging, and the effectiveness is defined on the basis of mixed-mean temperatures.

The simplest version of a cross-flow heat exchanger would be a bank of circular tubes with one fluid flowing inside the tubes and the other flowing normally to them. Unless there are baffles, a certain amount of lateral mixing will occur in the fluid flowing normally to the tubes. However, this is seldom sufficient to invalidate the assumption of no mixing.

Counterflow with Cross-Flow Headers (see Fig. 18-9)

A common method for constructing a heat exchanger is employ a stack of plates, with each fluid flowing between alternate plates. Fins (extended

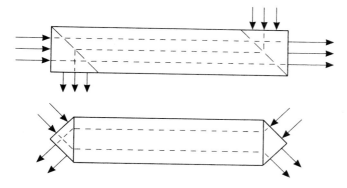

FIGURE 18-9
Counterflow heat exchanger with cross-flow headers.

surface) might be added to increase the amount of surface area, in which case this would be called a *plate-fin* heat exchanger. Such a scheme lends itself very well to a cross-flow design because alternate flow passages can be sealed off at the outer edge of the heat exchanger. A counterflow design becomes more difficult, however, because the two fluids must be somehow separated at the entrances and exits. As will be seen, a counterflow design will always provide a higher effectiveness for a given amount of surface area, so it would often be desirable to use counterflow with plate-fin construction. One way to at least partially accomplish this is to use counterflow with cross-flow headers, as is illustrated in Fig. 18-9.

Periodic-Flow Heat Exchangers (see Fig. 18-10)

This kind of heat exchanger makes use of a rotating matrix of tubes or channels so that each flow tube passes back and forth from the hot fluid

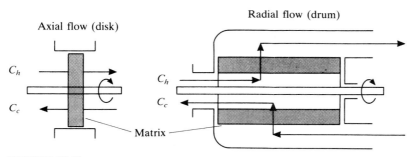

FIGURE 18-10
Two types of periodic-flow heat exchanger.

to the cold fluid. It then depends upon the "heat capacity" of the matrix material to store thermal energy while it carries it from one fluid to the other. Seals must be provided to separate the two fluids, and it is thus limited to moderate pressure differences and to fluids for which a small amount of mixing is not objectionable. Regenerative air heaters for combustion systems fulfill these specifications.

The major virtue of this arrangement is that it allows the use of very compact heat-transfer surfaces (large surface area per unit of volume), and tube matrices made from materials that would not be practicable for direct-transfer heat exchangers (ceramics, for example.)

Multi-Pass Heat Exchangers (see Fig. 18-11)

As will be seen, the counterflow arrangement will always yield a higher effectiveness for a given value of NTU. It is possible to approach counterflow performance with any component flow arrangement by multi-passing.

Figure 18-11 shows a two-pass system where each of the passes is a cross-flow heat exchanger. This scheme can be extended to three, four, five, etc. passes, and a still closer approach to counterflow performance will obtain.

Whether one uses a counterflow arrangement in a given application, or a multi-pass cross-flow arrangement, is generally dictated by the choice of heat-transfer surfaces. If it is desired to use a bank of circular tubes as the basic heat-transfer surface, with one fluid inside the tubes and the

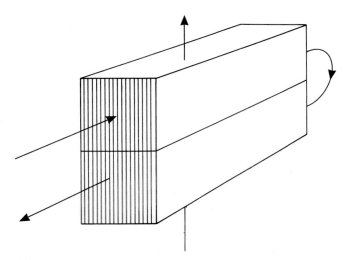

FIGURE 18-11
A two-pass cross-counterflow heat exchanger.

other flowing normally to them, then a multi-pass arrangement is the only way counterflow performance can be approached. Of course, the fluid on the outside of the tubes could be made to flow axially along the tubes, but it turns out to be possible to obtain much higher heat-transfer coefficients if the flow is normal. If plate-fin surfaces are used, again a cross-flow arrangement is more convenient, so multi-passing provides a route to approaching counterflow performance. Alternatively, the counterflow arrangement with cross-flow headers could be used.

ANALYSIS OF A SIMPLE HEAT EXCHANGER

If one capacity rate is very much larger than the other, C_{min}/C_{max} approaches 0, and the flow arrangement becomes immaterial. This is also true if one of the fluids is evaporating or condensing; the essential feature is that one of the fluids is at a constant temperature throughout the heat exchanger.

The thermal analysis for this case is very simple, and it serves nicely to introduce the methodology of heat-exchanger analysis, and the variables, and it provides an effective way to get a "feel" for the results and for the order of magnitude of the nondimensional variables.

Figure 18-12 shows how the temperatures will vary for this case.

Note that by using the surface area A rather than tube length as the independent variable, the result will apply to any number of tubes in parallel, and to tubes of any shape.

We now cut out an elemental section of tube, apply a heat-transfer rate equation to it, and make an energy balance. The rate equation is

$$\delta \dot{q} = U \, \delta A \, (t_h - t_c)$$

An energy balance gives

$$\delta \dot{q} = C_c \frac{dt_c}{dA} \, \delta A$$

Combining these, we have

$$C_c \frac{dt_c}{dA} \, \delta A = U \, \delta A \, (t_h - t_c)$$

Integrating,

$$\int_{t_{c1}}^{t_{c2}} \frac{dt_c}{t_h - t_c} = \int_0^A \frac{U \, dA}{C_c}$$

that is,

$$-\ln \left(\frac{t_h - t_{c2}}{t_h - t_{c1}} \right) = NTU_{max}$$

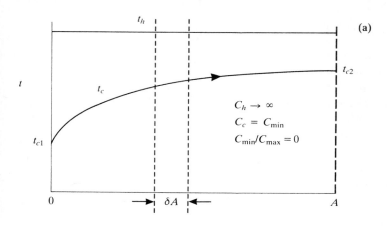

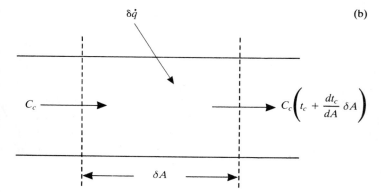

FIGURE 18-12
A heat exchanger with $C_{min}/C_{max} = 0$.

provided that U and C_c are constant. Then

$$\frac{t_h - t_{c2}}{t_h - t_{c1}} = e^{-NTU_{max}}$$

$$\frac{t_h - t_{c1}}{t_h - t_{c1}} - \frac{t_h - t_{c2}}{t_h - t_{c1}} = 1 - e^{-NTU_{max}}$$

and finally

$$\frac{t_{c2} - t_{c1}}{t_h - t_{c1}} = 1 - e^{-NTU_{max}}$$

that is,

$$\varepsilon = 1 - e^{-NTU_{max}} \tag{18-7}$$

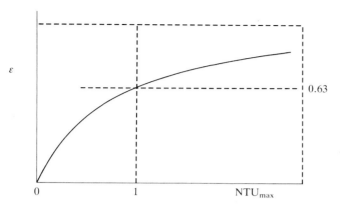

FIGURE 18-13
The solution for $C_{min}/C_{max} = 0$.

Note the form of the solution, i.e., $\varepsilon = \varepsilon(\text{NTU}_{max})$. When C_{min}/C_{max} is introduced as an additional variable, we merely get more complex functions that differ for each flow arrangement.

This result can be represented graphically as in Fig. 18-13.

From this simple result one can begin to get a "feel" for the physical significance of NTU. A value of NTU = 1 yields an effectiveness of 0.63. Most often, heat exchangers have NTU values ranging from about 0.3 to 5.0. Since NTU = AU/C, we see that effectiveness is increased by increasing A or U, or by decreasing C.

ANALYSIS FOR COUNTERFLOW

The procedure is very similar to that just used, except that now we must deal with two fluid streams. The final results can be put into the form

$$\varepsilon = \frac{1 - e^{-\text{NTU}_{max}(1 - C_{min}/C_{max})}}{1 - (C_{min}/C_{max})e^{-\text{NTU}_{max}(1 - C_{min}/C_{max})}} \qquad (18\text{-}8)$$

Note that if $C_{min}/C_{max} = 0$, Eq. (18-8) reduces to the previous case. The other extreme is $C_{min}/C_{max} = 1$. The equation is indeterminate, but use of L'Hôpital's rule reduces it to

$$\varepsilon = \frac{\text{NTU}}{1 + \text{NTU}} \quad \text{for} \quad \frac{C_{min}}{C_{max}} = 1 \qquad (18\text{-}9)$$

The complete results are given graphically in Fig. 18-14. Note that for $C_{min}/C_{max} > 0$ the effectivness is reduced for any given value of NTU.

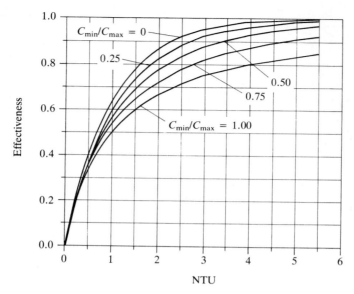

FIGURE 18-14
Effectiveness of a counterflow heat exchanger.

ANALYSIS FOR CROSS-FLOW, BOTH FLUIDS UNMIXED

This flow arrangement can best be solved numerically, although a series solution to the resulting partial differential equation is possible. The numerical scheme can be visualized on the diagram shown in Fig. 18-15.

Let M and N be the number of discrete flow passages for the hot and cold fluids, respectively. For each cell in the resulting matrix we write an energy balance for each fluid, and a rate equation connecting the two. Accuracy is greatly improved if central differences are used for the rate equation, i.e., the arithmetic average of inflow and outflow temperatures. If we start at the upper left-hand corner, there is now sufficient information to solve for the outlet temperatures of each cell. The whole procedure can obviously be easily set up on a computer, and it can be done very quickly using a spreadsheet program.

$N = M = 10$ will give close to three-figure accuracy for ε.

Note that the outlet temperatures will vary over the outlet planes, so it is necessary to average the outlet temperatures before defining the effectiveness.

The final results are shown graphically on Fig. 18-16. Note that for $C_{min}/C_{max} = 0$ the results are identical to the counterflow case, i.e., Eq. (18-7), but for $C_{min}/C_{max} = 1$ the effectiveness is substantially lower, especially at high NTU. For example, to obtain $\varepsilon = 0.75$ for counterflow

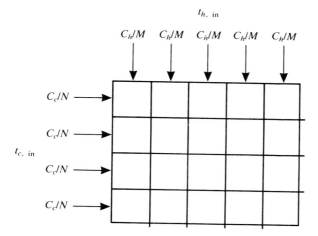

FIGURE 18-15
Numerical layout for cross-flow.

requires NTU = 3, while for cross-flow the requirement is about 4.75, i.e., 58 percent more area, which generally means 58 percent more volume and weight.

It is often useful to have an algebraic equation for ε rather than a graph or table. The following is an empirical equation that provides a fair

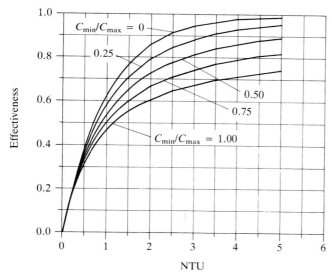

FIGURE 18-16
Effectiveness of a cross-flow heat exchanger with both fluids unmixed.

fit to Fig. 18-16 except at the extremes of the variables:

$$\varepsilon = 1 - \exp\left[\frac{\exp\left(-NTU^{0.78}C_{min}/C_{max}\right) - 1}{NTU^{-0.22}C_{min}/C_{max}}\right]$$

MULTI-PASS HEAT EXCHANGERS

Consider the multi-pass counterflow system shown in Fig. 18-17, where all of the passes are identical, and where the fluids are completely mixed between the passes (the illustration suggests cross-flow passes, but actually any flow arrangement could be used for the individual passes.)

Let ε_p be the effectiveness of each identical pass. Let n be the number of passes, and ε_0 the overall effectiveness of the entire system considered as a single heat exchanger. The following equations can be readily developed from simple energy balances:

$$\varepsilon_0 = \frac{\left(\dfrac{1 - \varepsilon_p C_{min}/C_{max}}{1 - \varepsilon_p}\right)^n - 1}{\left(\dfrac{1 - \varepsilon_p C_{min}/C_{max}}{1 - \varepsilon_p}\right)^n - \dfrac{C_{min}}{C_{max}}} \tag{18-10}$$

$$\varepsilon_p = \frac{\left(\dfrac{1 - \varepsilon_0}{1 - \varepsilon_0 C_{min}/C_{max}}\right)^{1/n} - 1}{\dfrac{C_{min}}{C_{max}}\left(\dfrac{1 - \varepsilon_0}{1 - \varepsilon_0 C_{min}/C_{max}}\right)^{1/n} - 1} \tag{18-11}$$

and, for $C_{min}/C_{max} = 1$,

$$\varepsilon_0 = \frac{n\varepsilon_p}{1 + (n - 1)\varepsilon_p} \tag{18-12}$$

A plot of Eq. (18-12), for simple cross-flow passes, is shown on Fig. 18-18. The NTU on the diagram is the total of the NTUs of the various

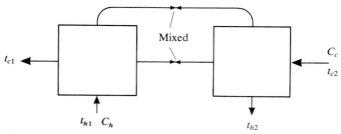

FIGURE 18-17
A multi-pass heat exchanger with both fluids mixed between passes.

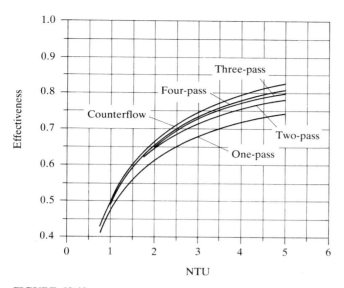

FIGURE 18-18
Effectiveness of a multi-pass heat exchanger with cross-flow passes; $C_{min}/C_{max} = 1$; fluid mixed between passes.

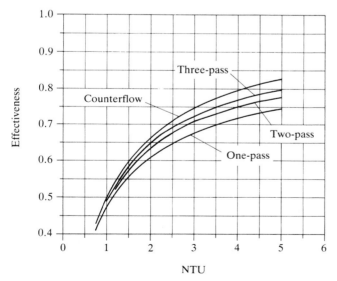

FIGURE 18-19
Effectiveness of a multi-pass heat exchanger with cross-flow passes, with no mixing between the passes for one of the fluids; $C_{min}/C_{max} = 1$.

passes. Note that pure counterflow performance is approached as the number of passes in increased.

Figure 18-19 shows the performance of a multi-pass heat exchanger with cross-flow passes where mixing between passes has *not* been allowed for one of the two fluids. This result was obtained by numerical integration, similar to that described for a simple cross-flow heat exchanger. Note that mixing between passes has only a very small effect, so that Eqs. (18-10)–(18-12) can evidently be used with only small error.

ANALYSIS OF A COUNTERFLOW HEAT EXCHANGER WITH CROSS-FLOW HEADERS

Numerical solutions for this arrangement can be obtained in the same way as for simple cross-flow. Figure 18-20 shows some results for $C_{min}/C_{max} = 1$.

To use this figure, one calculates the NTU for the main counterflow section, and then the NTUs for the two headers separately. If fluid properties are assumed constant and the surface configuration is the same throughout the heat exchanger then U/C_{min} is the same in the cross-flow headers as in the counterflow part. Then the NTUs for the parts differ only by the amount of surface area in each.

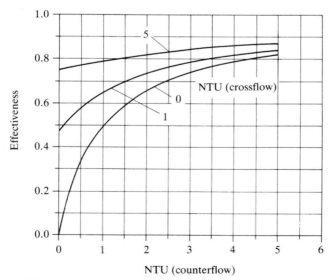

FIGURE 18-20
Effectiveness of a counterflow heat exchanger with cross-flow headers; $C_{min}/C_{max} = 1$.

As an example, consider a heat exchanger with a total NTU of 5. If the entire heat exchanger were counterflow then for $C_{min}/C_{max} = 1$ the effectiveness would be 0.834. On the other hand, if the heat exchanger were simple cross-flow, the effectiveness would be 0.752.

Now consider a counterflow arrangement with cross-flow headers for which NTU = 4 in the counterflow section, and NTU = 0.5 in each of the cross-flow headers. Thus the total NTU = 4 + 0.5 + 0.5 = 5. From Fig. 18-20 the effectiveness is approximately 0.816. This is not quite counterflow performance, but it is substantially better than cross-flow alone.

ANALYSIS FOR PERIODIC-FLOW HEAT EXCHANGERS

Figure 18-21 shows some results, obtained numerically, for a periodic-flow heat exchanger for $C_{min}/C_{max} = 1.00$ and 0.90. Since these heat exchangers find most of their use in regenerative combustion systems, these two values of capacity rate ratio cover the really useful range.

Note that a new variable, C_r, appears. This is the *rotor capacity rate*, i.e., the rate at which the solid heat capacity of the rotor material is rotated:

$$C_r = \text{mass of rotor} \times \text{specific heat} \times \text{rev/sec}$$

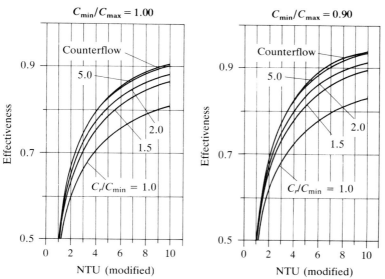

FIGURE 18-21
Effectiveness of a periodic-flow heat exchanger.

The larger the value of C_r/C_{min}, the closer is counterflow performance approached, but the greater is the fluid "carryover", which detracts from performance. Fortunately, high values for C_r/C_{min} can usually be obtained with quite small rotating speed (like 30 rpm). A modified NTU definition is used:

$$NTU_0 = \frac{1}{C_{min}} \left[\frac{1}{(1/hA)_c + (1/hA)_h} \right]$$ (18-13)

where $0.25 < (hA)^* < 4.0$, with

$$(hA)^* = (hA)_c/(hA)_h$$

This representation is an approximation [note the limitations on $(hA)^*$], but it is a quite good one.

Note that ε approaches the counterflow ε as C_r/C_{min} approaches infinity.

One of the problems with periodic-flow heat exchangers, especially at high ε, is *longitudinal conduction*, which tends to reduce ε. The same problem exists in counterflow heat exchangers, but it tends to be much more acute in periodic-flow heat exchangers. Figure 18-22 shows some

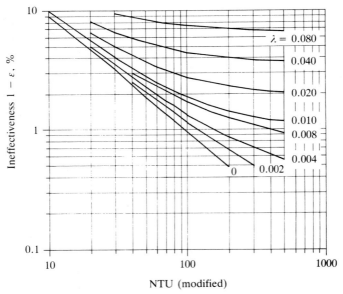

FIGURE 18-22
Effect of longitudinal conduction on the performance of a periodic-flow heat exchanger; $C_{min}/C_{max} = 1$.

numerical solutions for this effect. The parameter λ is defined by

$$\lambda = k_s A_s / C_{min} L \qquad (18\text{-}14)$$

where A_s is the conduction cross-sectional area, k_s is the thermal conductivity of the matrix material, and L is the heat-exchanger flow length.

CALCULATION OF HEAT EXCHANGERS BY USE OF A SPREADSHEET

The description of a method for numerical solution for a simple cross-flow heat exchanger suggests that it ought to be practicable to solve such a heat exchanger by use of a computer spreadsheet program such as Lotus123† or Excel.‡ In fact, it is an eminently practicable method for dealing with much more complex heat exchangers such as multi-pass arrangements, and heat exchangers in which the fluid properties vary significantly with temperature. Even the simple counterflow heat exchanger for a case where fluid properties vary can be easily handled in this manner.

Basically the idea is to break the heat exchanger up into a number of cells and then to use the matrix layout of the spreadsheet to perform the calculations for each cell. Any number of spreadsheet cells can be used to represent one heat-exchanger cell; the number depends on the number of calculations that must be made. For example, as many as 15 spreadsheet cells might be used for one heat-exchanger cell if for each cell it is necessary or desirable to separately calculate Reynolds numbers, heat-transfer coefficients, friction coefficients, etc. If the overall heat exchanger is represented by a 10×10 matrix of cells then a total of 1500 spreadsheet cells will be required—a quite practicable number. But only a few cells need to be programmed, since the spreadsheet can automatically copy and re-reference multiple groups of cells. The spreadsheet program is then set for iterative calculations, and calculations are carried out until the results converge on a solution.

CALCULATION OF PRESSURE DROP IN A HEAT EXCHANGER

In gas-flow heat exchangers the pressure drop on both fluid sides is virtually always as important as the thermal performance. (This fact will

---

† Lotus123 is a trademark of Lotus Development Corporation.
‡ Excel is a trademark of Microsoft Corporation.

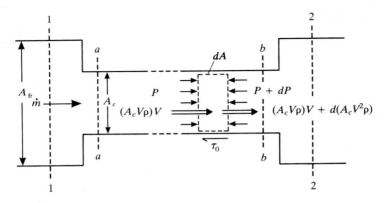

FIGURE 18-23
Calculation of pressure drop.

be discussed in much greater detail in Chap. 19.) In liquid heat exchangers the pressure drop is seldom of controlling influence in the design, but nevertheless it usually must be evaluated. The equation developed below provides a useful way to make this calculation.

Consider a tube of any cross-sectional shape, or a bank of tubes, Fig. 18-23. In the usual situation the fluid approaches the heat exchanger in a large inlet duct, and then the flow subdivides into a large number of small heat exchanger passages. The cross-sectional area of the inlet duct is denoted by A_{fr} (frontal area), while the total cross-sectional area of the heat-exchanger flow passages is denoted by A_c. At the exit of the heat exchanger the flow expands from the small passages to a single large duct that is presumed in this case to have the same flow area as the inlet duct. The following definitions will be used:

$$G = \frac{\dot{m}}{A_c} = V\rho, \qquad \sigma = \frac{A_c}{A_{\mathrm{fr}}}, \qquad \tau_0 = \tfrac{1}{2}c_f \rho V^2$$

K_c and K_e are empirical entrance and exit loss coefficients that are associated with the presumed abrupt contraction of the flow area at the entrance and exit to the heat exchanger and the flow separations that inevitably occur. If flow-turning headers are employed, the entrance and exit problems are more complex, but those cases will not be considered here.

We apply the momentum theorem to the elemental control volume shown in the tube:

$$dP = \frac{c_f G^2 \, dA}{2A_c \rho} + \frac{G^2}{\rho^2} \, d\rho$$

Integrating from a to b,

$$P_a - P_b = \frac{G^2}{2}\left[\frac{c_f A}{\rho_m A_c} + 2\left(\frac{1}{\rho_b} - \frac{1}{\rho_a}\right)\right]$$

If we now add in the pressure changes (and losses) at the entrance and exit, and rearrange the equation, we obtain

$$\left(\frac{\Delta P}{P_1}\right) = \frac{G^2}{2}\frac{1}{P_1\rho_1}\left[(K_c + 1 - \sigma^2) + 2\left(\frac{\rho_1}{\rho_2} - 1\right)\right.$$

$$\left. + c_f\frac{A}{A_c}\frac{\rho_1}{\rho_m} - (1 - \sigma^2 - K_e)\frac{\rho_1}{\rho_2}\right] \qquad (18\text{-}15)$$

where subscript 1 refers to conditions in the inlet duct, and 2 to conditions in the outlet duct. ρ_m is the inverse of the mean specific volume with respect to tube length. Usually a simple arithmetical average will suffice.

Figure 18-24 provides some values for K_c and K_e for a typical plate-fin heat exchanger, but the values of K_c and K_e do not differ greatly for other entrance and exit geometries.

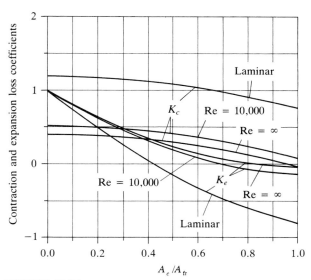

FIGURE 18-24
Entrance and exit pressure-loss coefficients; two-dimensional contraction and expansion; multiple square tubes.

PROBLEMS

18-1. Carry out the development of Eq. (18-6).

18-2. Carry out the development of Eqs. (18-8) and (18-9).

18-3. Develop a computer program to numerically determine the effectiveness of a single-pass cross-flow heat exchanger as a function of NTU and C_{min}/C_{max}. Use a 10×10 cell grid. Compare your results with Fig. 18-16.

18-4. Use a spreadsheet computer program to determine the effectiveness of a two-pass cross-flow heat exchanger as a function of NTU and C_{min}/C_{max} for the case where one of the fluids is completely mixed between passes but the other is unmixed. Use a 10×10 cell grid for each pass. Compare your results with Fig. 18-19.

CHAPTER
19

COMPACT HEAT-EXCHANGER SURFACES

Compact heat-exchanger surfaces are, by definition, surface configurations that have a characteristic of large surface area per unit of volume. A bundle of very small-diameter circular tubes would be a compact heat-exchanger surface, but large area per unit volume can also be attained by attaching extended surface, or fins, to a relatively larger tube. So one commonly used compact surface is constructed by starting with a bank of circular tubes and attaching circular fins to the outside of each tube. A variation of this surface uses continuous sheet fins instead of individual circular fins. But the tubes need not be circular, and a common variation uses flat tubes.

The compact surfaces described above (finned-tube surfaces) are adequate if a large amount of surface area is needed on only one fluid side. If large surface area is desired on *both* fluid sides, an effective scheme is to build up the heat exchanger from a stack of flat plates, with the two fluids flowing between alternate pairs of plates. If the plates are placed very close together, the area per unit volume can be very large, but an alternative arrangement is to place the plates relatively far apart and to connect them by an extended surface, or fins. This is what is known as *plate-fin* construction.

In either the finned-tube surfaces or the plate-fin surfaces the fins are frequently cut into small segments, or otherwise interrupted in various ways. Or sometimes indentations of various shapes are made in the fins. These are all schemes for increasing the heat-transfer coefficient so as to minimize the amount of surface area required.

The number of possible surface configurations, when all of these variations are considered, is very large indeed. One of the purposes of this chapter is to attempt to develop an understanding of these surfaces and their characteristics so that surfaces for a particular application can be chosen or designed in a rational way.

Compact heat-exchanger surfaces find most of their applications in heat exchangers involving gases rather than liquids. This is because heat-transfer coefficients tend to be much lower with gases, and therefore larger surface area is required to transfer a given amount of heat. But the reason that gases yield low heat-transfer coefficients is related to the mechanical power necessary to pump a gas through a heat exchanger. So we are going to start our investigation of these surfaces by an analysis of this pumping-power problem.

One of the conclusions that will be reached is that in the design of gas-flow heat exchangers the pressure-drop calculations (again the pumping-power problem) are just as important as the heat-transfer calculations, and, in fact, if care is not taken, it is very easy to expend in mechanical pumping power as much as is gained by the heat-transfer function of the heat exchanger. The situation is quite different for liquids, where the pressure drop is seldom of controlling influence in the design. In this sense the design of gas-flow heat exchangers is a more complex problem than that of liquid-flow heat exchangers.

A THEORY OF COMPACT HEAT-EXCHANGER SURFACES

Let us consider flow in any kind of long passage, be it a circular tube, a rectangular tube, a tube with interrupted fins, or even flow normal to a bank of tubes. In any case the heat-transfer and fluid friction characteristics of the surface may be represented by functions of the type

$$St\, Pr^{2/3} = \phi_h(Re) \qquad (19\text{-}1)$$

$$c_f = \phi_f(Re) \qquad (19\text{-}2)$$

where

$$Re = \frac{4r_h G}{\mu}, \qquad 4r_h = \frac{4A_c L}{A}, \qquad G = \frac{\dot{m}}{A_c} = V\rho$$

A_c is the *minimum* total free-flow cross-sectional area, L is the tube length, and A is the total surface area.

The two-thirds power of the Prandtl number is only an approximation, but it will suffice for present purposes. It is a fair approximation for turbulent flow in long continuous tubes of any cross-sectional shape over a wide range of Prandtl number, and it is a very good approximation for the laminar boundary layer that is found on the fins of some of the interrupted fin surfaces that will be considered. For fully developed laminar flow in long tubes the power should be 1, but again it matters little in the present context. Note also that the difference between *constant surface temperature* and *constant heat rate* is being ignored, as is the effect of tube length.

These functions can be used to correlate data for any kind of long "tube," and the data could be either analytic or experimental.

Since $St = h^{-1}(Gc/Pr^{2/3})\phi_h$, we can solve for h in terms of ϕ_h. Thus

$$h = \frac{\mu c}{Pr^{2/3}} \frac{1}{4r_h} \text{Re } \phi_h, \quad W/(m^2 \cdot K) \qquad (19\text{-}3)$$

Now, instead of looking upon h as a heat-transfer coefficient, let us look upon it as representing *"heat transfer power per unit of surface area"* for one degree of temperature difference.

We would like to compare this with the corresponding *"friction power per unit of surface area,"* which we shall call E. Let dP_f be the friction pressure drop associated with the heat-transfer area dA. Then

$$E = -\frac{-dP_f}{\rho} GA_c \frac{1}{dA} \quad \begin{array}{l}\text{("loss of mechanical energy}\\ \text{due to pressure drop")}\end{array}$$

Then, using Eq. (18-15) (ignoring entrance and exit effects),

$$\frac{dP_f}{\rho} = -\frac{c_f G^2}{2 \rho^2} \frac{dA}{A_c}$$

Combining and rearranging,

$$E = \frac{1}{2} \frac{\mu^3}{\rho^2} \left(\frac{1}{4r_h}\right)^3 \text{Re}^3 \phi_f \quad \text{("friction power per unit of } A\text{," } W/m^2)$$

$$(19\text{-}4)$$

Equations (19-3) and (19-4) provide a rational basis for comparing the performance of one surface configuration against another. Assuming that the functions ϕ_h and ϕ_f are known for a number of surfaces, and the hydraulic diameter and the fluid properties are specified, a plot could be prepared for h as a function of E for each surface. A surface that plotted "high" on this diagram would be able to transfer a given amount of heat for less friction power (smaller pressure drop), or for a given expenditure of friction power that surface would transfer more heat. As it will turn out, there are quite large differences in the heat-transfer–friction-power characteristics of different surface configurations. We will see later why these differences occur.

However, for the moment we would like to use Eqs. (19-3) and (19-4) for a different purpose.

Let us examine a particular very simple case. Consider fully developed *turbulent flow* in a long smooth-walled tube of any cross-sectional shape (circular, rectangular, etc.) The following are reasonable approximations for ϕ_f, and for ϕ_h over a range of Prandtl number from the gases to the very viscous liquids (i.e., virtually all fluids except the liquid metals):

$$\phi_h = 0.023 \, \text{Re}^{-0.2} \tag{19-5}$$

$$\phi_c = 0.046 \, \text{Re}^{-0.2} \tag{19-6}$$

We now substitute Eqs. (19-5) and (19-6) into Eqs. (19-3) and (19-4), and combine the results to eliminate Re:

$$E = \frac{12{,}465 h^{3.5} \mu^{1.83} (4r_h)^{0.5}}{k^{2.33} c^{1.77} \rho^2}, \quad \text{W/m}^2 \tag{19-7}$$

This equation first demonstrates that, at least for a turbulent flow, mechanical energy must be expended in order to transfer heat, and this expenditure is a very strong function of the heat transfer rate. But it also shows that the amount of mechanical energy expended is heavily dependent upon the fluid properties.

A parameter to this equation, which does not appear directly, is the fluid velocity. Note that from Eqs. (19-3) and (19-5)

$$h \text{ varies as } V^{0.8}, \quad \text{and thus } E \text{ varies as } V^{2.8}$$

In a turbulent flow h can be increased as much as desired by increasing the velocity, but a large price is paid in E. In the design of a heat exchanger for a particular application V is usually an independent variable. It is inversely proportional to the frontal area of the heat exchanger, which the designer can often vary at will.

Whether or not E is an important variable in the design of a heat exchanger obviously depends upon whether E is large or small. In this regard, notice the influence of the fluid properties in Eq. (19-7). In particular, note the density ρ. For water, for example, ρ is typically 1000 times that for air. Thus (other things being equal, which they usually are not) air would require 10^6 more friction power for the same rate of heat transfer.

All of this suggests that the ratio E/h in a heat exchanger might be an important parameter. We obviously would want E/h to be small, but the question is how small? We next want to demonstrate that in a given

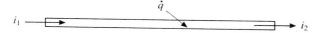

FIGURE 19-1
A single heat-exchanger tube.

application the ratio E/h, or at least the maximum permissible value of E/h, is essentially a pure *thermodynamic function* and is independent of fluid properties, velocities, and surface configuration. This fact will then place limitations on the values of heat-transfer coefficient attainable in a heat exchanger, and will explain why gas heat exchangers must operate with relatively very low heat-transfer coefficients.

Consider steady flow in one of the tubes of a heat exchanger, Fig. 19-1. If the fluid in the tube is heated, as indicated, its thermodynamic *potential* for producing useful work in any kind of interaction with an environment at T_0 is *increased*. We can calculate this increase by using the *steady-flow availability function b*:

$$b = i + T_0 s$$

where i is the enthalpy and s the entropy. Then the *increase in work potential* is

$$\Delta b = b_2 - b_1 = (i_2 - T_0 s_2) - (i_1 - T_0 s_1) = (i_2 - i_1) - T_0(s_2 - s_1)$$

For a *constant-property* liquid

$$i_2 - i_1 = c(T_2 - T_1) + \frac{P_2 - P_1}{\rho}$$

$$s_2 - s_1 = c \ln \left(\frac{T_2}{T_1} \right)$$

Thus

$$\Delta b = c(T_2 - T_1) + \frac{P_2 - P_1}{\rho} - T_0 c \ln \left(\frac{T_2}{T_1} \right)$$

(A similar argument can be developed for a *gas*, and will lead to slightly different equations but to the same general conclusions.)

Now consider the extreme case where $\Delta b = 0$. Even if $T_2 > T_1$, we are evidently losing, through *pressure drop*, everything that has been gained by heating the fluid. In other words, the increase in work potential resulting from heating the fluid has been lost in the mechanical energy needed to overcome the pressure drop. Therefore the pressure drop for this *break-even* case is

$$\left(\frac{\Delta P}{\rho} \right)_{\Delta b = 0} = \frac{P_1 - P_2}{\rho} = c(T_2 - T_1) - T_0 c \ln \left(\frac{T_2}{T_1} \right)$$

It is then evident that we must design for ΔP very much lower than this if the design is going to be useful. Let F be a design factor that represents what fraction of the heat-transfer work potential we are willing to put back into the system as mechanical pumping work. Obviously, F is going to vary from one application to another, depending on many optimization criteria, but one can guess that typically F will have a practicable maximum value in the range 0.01–0.02. So the maximum tolerable pressure drop in a heat exchanger would be

$$\frac{\Delta P}{\rho} = F\left[c(T_2 - T_1) - T_0 c \ln\left(\frac{T_2}{T_1}\right)\right] \qquad (19\text{-}8)$$

From Eq. (19-8) we can now calculate a typical acceptable value for E/h:

$$E = \frac{\dot{m}}{A}\frac{\Delta P}{\rho}$$

(where A is the heat-transfer area), and

$$\dot{q} = hA\,\Delta T$$

where ΔT is a mean temperature difference, fluid to wall. Thus

$$h = \frac{\dot{q}}{A\,\Delta T}$$

But

$$\dot{q} = \dot{m}(i_2 - i_1) = \dot{m}c(T_2 - T_1) \qquad \text{(if } \Delta P \text{ is small)}$$

Then

$$h = \frac{\dot{m}c(T_2 - T_1)}{A\,\Delta T}$$

Forming the ratio E/h, we obtain

$$\frac{E}{h} = F\,\Delta T\left[1 - \frac{T_0 \ln(T_2/T_1)}{T_2 - T_1}\right] \qquad (19\text{-}9)$$

Note that E/h is then a function only of temperatures, and has nothing to do with fluid properties or the characteristics of the heat exchanger, except only insofar as those parameters might affect the optimimization criteria that lead to a value for F.

Now let us return to Eq. (19-7) and see how this result affects the permissible magnitude of the heat-transfer coefficient. Equation (19-7) can be rearranged to the following:

$$h = \frac{k^{0.93}c^{0.47}\rho^{0.8}(E/h)^{0.4}}{43.48\mu^{0.73}(4r_h)^{0.2}} \qquad (19\text{-}10)$$

A numerical example will be instructive. Consider one of the fluids in a simple heat exchanger. Let the fluid be heated from 300 K to 550 K. Let the environmental temperature be 293 K, and let the mean temperature difference between the fluid and the wall be 20 K. Choose F to be 0.01. Then from Eq. (19-9), $E/h = 0.06$.

Let the tube have a hydraulic diameter of 1 cm and suppose the fluid is air at 1 atm pressure. Inserting the properties of air into Eq. (19-10), along with $E/h = 0.06$, we obtain

$$h = 53 \text{ W}/(\text{m}^2 \cdot \text{K})$$

(Note that E/h is raised to the power 0.4, so a precise value for E/h is not necessary to the argument. The same can be said for $4r_h$.)

This is a very *small* heat-transfer coefficient. But if we attempt to increase h by increasing the fluid velocity, we greatly increase E.

Next we use Eqs. (19-1) and (19-5) to solve for the Reynolds number Re:

$$\text{Re} = 5500$$

This is a very *low* Reynolds number. Figure 19-2 shows a typical plot of St $\text{Pr}^{2/3}$ versus Re for fully developed flow in a smooth-walled tube, and it can be seen that this Reynolds number is in the middle of the *transition region* between laminar and turbulent flow, where h would be even lower.

The conclusion that can be drawn is that the Reynolds number in typical gas-flow heat exchangers is going to be not too far removed from the transition region, and that h is going to be small. Higher pressures (high ρ) will increase h and Re somewhat. Varying tube size has only a

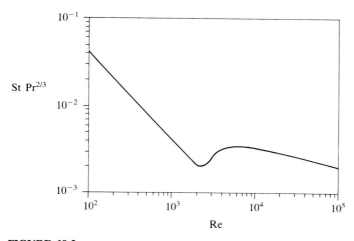

FIGURE 19-2
Heat-transfer characteristics for flow in a typical tube.

small effect on h, but very small tubes may move the Reynolds number operating point down into the *laminar flow* region.

These conclusions are of course based on long continuous-walled tubes. We will find that considerably higher heat-transfer coefficients are obtainable by various modifications of the wall surface, but, because such surfaces have higher friction coefficients, the operating Reynolds numbers will turn out to be still lower.

Let us now consider the same operating conditions, but change the fluid to water. Now we obtain

$$h = 27,500 \text{ W}/(\text{m}^2 \cdot \text{K})$$
$$\text{Re} = 112,600$$

This is a *high* heat-transfer coefficient and a *high* Reynolds number. But now, from the Reynolds number, let us calculate the fluid velocity:

$$V = 9 \text{ m/s}$$

This is a much higher velocity than can be tolerated in a water system; cavitation at flow separation points would limit velocities to a small fraction of this figure.

The conclusion then is that when using water, or other moderate-viscosity *liquids*, the friction power expenditure (pressure drop) is seldom of controlling influence in the design of a heat exchanger. Other factors cause the design to be restricted to low velocities, while at the same time the heat-transfer coefficients are still very much higher than is possible with gases. The possible exceptions to this rule would be oils or other high-viscosity liquids.

But for *gases* friction power expenditure, or pressure drop, is virtually always a critical design consideration, and this has a large influence on the way that gas heat exchangers are designed.

Because h tends to be low in gas-flow systems, in order to minimize friction power, the surface-area requirement tends to be large. Unless we are to have very large heat exchangers, we must try to use heat-exchanger surface configurations having large area per unit volume, i.e., *compact heat-exchanger surfaces*.

As we implied earlier, some surface configurations are more favorable than others from the point of view of their *heat-transfer–friction-power characteristics*. The reasons for this will be examined shortly.

HEAT-TRANSFER AND FRICTION CHARACTERISTICS OF A SIMPLE COMPACT SURFACE

For illustrative purposes, we will discuss *plate-fin* surfaces, although most of our conclusions will also apply to finned-tube surfaces. The simplest version of a plate-fin surface, along with its experimentally determined heat transfer and friction characteristics,[1] is shown in Fig. 19-3, where the particular surface is designated as Surface #6.2. The constant β is the total heat-transfer area per unit volume between the plates.

The flow tubes of Surface #6.2 are essentially long rectangular tubes with slight variation at the corners. The length-to-diameter ratio of the tubes for this particular test core was 55, which suggests that there is a small length effect in the turbulent region, and a larger effect in the laminar region. However, in using these data the tube-length effects can usually be neglected, especially in the turbulent region. The experiments were carried out using air and a *constant surface temperature*, and it will be recalled that this has little influence on heat transfer in the turbulent flow region, but can have a considerable influence in laminar flow.

Note that there is a distinct transition from a laminar flow to a turbulent flow, starting at about the classical transition Re = 2300. The transition region extends over a considerable range of Re, partially because we are dealing with a multiplicity of parallel tubes, each slightly different from its neighbor. Close examination of these data will reveal that they are very close to the equations developed in Chapters 7, 9, 12, and 14 for this particular geometry. Recall that for laminar flow the shape of the flow passage has a large influence, but in turbulent flow it is virtually negligible, at least where there are no acute angle corners.

Note further that the data extend over a Reynolds number range from about 800 to 12,000, which is precisely the range discussed above for gas-flow heat exchangers.

HEAT-TRANSFER AND FRICTION CHARACTERISTICS OF A STRIP-FIN SURFACE

Next examine Surface #1/8-19.86 shown in Fig. 19-4. Here the fins have been cut and offset at $\frac{1}{8}$ inch (3.175 mm) intervals. This is a more *compact* surface than the previous one (more surface area per unit volume, larger β), but both $\mathrm{St}\,\mathrm{Pr}^{2/3}$ and c_f are very much higher at any given Reynolds number. For example, at Re = 4000, $\mathrm{St}\,\mathrm{Pr}^{2/3}$ is higher by a factor of 2.3, while c_f has increased by a factor of 3.2. Since these are all nondimensional parameters, the differences must be caused by the geometry and not the size.

We will shortly examine the *reasons* for these very substantial differences, but first let us look at the consequences for the heat-transfer–friction-power characteristics. On Fig. 19-5, h is plotted as a function of E using the properties of air at 300 K and 1 atm, and using a common hydraulic diameter, 1.54 mm.

We now consider an example. If a heat-transfer coefficient $h = 100$ W/(m$^2 \cdot$ K) is desired, Surface #1/8-19.86 requires $E = 27$ W/m^2, while

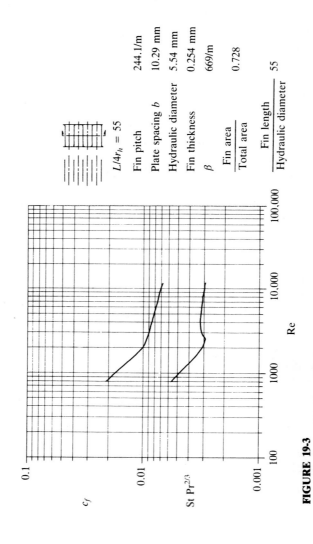

$L/4r_h = 55$	
Fin pitch	244.1/m
Plate spacing b	10.29 mm
Hydraulic diameter	5.54 mm
Fin thickness	0.254 mm
β	669/m
$\dfrac{\text{Fin area}}{\text{Total area}}$	0.728
$\dfrac{\text{Fin length}}{\text{Hydraulic diameter}}$	55

FIGURE 19-3
Surface #6.2 (a plain-fin surface).

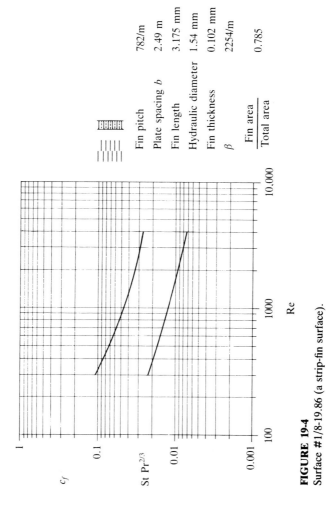

Fin pitch	782/m
Plate spacing b	2.49 m
Fin length	3.175 mm
Hydraulic diameter	1.54 mm
Fin thickness	0.102 mm
β	2254/m
$\dfrac{\text{Fin area}}{\text{Total area}}$	0.785

FIGURE 19-4
Surface #1/8-19.86 (a strip-fin surface).

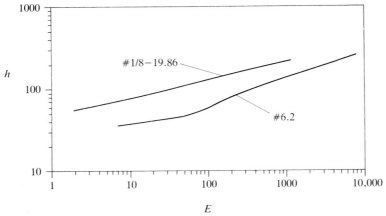

FIGURE 19-5
A comparison of $h-E$ characteristics.

Surface #6.2 requires $E = 270 \, \text{W/m}^2$. Or, looking at it the other way, for $E = 100 \, \text{W/m}^2$, Surface #1/8-19.86 will give $h = 140 \, \text{W/(m}^2 \cdot \text{K)}$ while Surface #6.2 will yield $h = 63 \, \text{W/(m}^2 \cdot \text{K)}$. Either way, there is a large difference. The reason for these large differences is that at any given flow velocity the interrupted fin surface yields a substantially higher heat-transfer coefficient. When the comparison is made at the same value of the heat-transfer coefficient, the interrupted fin surface is operating at a lower velocity, and, since the friction power varies at about the cube of the velocity, the differences can be very large. Note that the friction coefficient c_f is considerably higher for the interrupted fin surface, but the velocity-cubed effect can more than compensate for a very much higher friction coefficient.

A consequence of this behavior is that *high-performance interrupted-fin surfaces* tend to operate at lower fluid velocities. The designer obtains lower velocities by using a larger total flow area, and thus a larger frontal area. This may or may not be a disadvantage in any particular application.

Let us now see how Surface #1/8-19.86 attains such a high heat-transfer coefficient. Consider the flow over one of the fin segments, Fig. 19-6.

A new boundary layer forms on each fin segment. If we examine the x Reynolds number for this boundary layer, we will find that over the Reynolds range of the data on Fig. 19-4 the x Reynolds number has a maximum value of about 8000. This is well down into the range of very stable *laminar* boundary layers. Despite the fact that the fin segment has a finite thickness, and thus a blunt leading edge and undoubtedly a

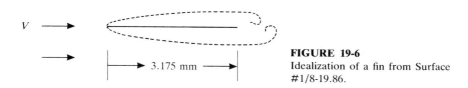

V

$\longrightarrow$ 3.175 mm $\longrightarrow$

FIGURE 19-6
Idealization of a fin from Surface
#1/8-19.86.

separation region, the boundary layer quickly re-attaches and a simple laminar boundary layer with no appreciable pressure gradient is established. To test this hypothesis, let us calculate $St\,Pr^{2/3}$ using the laminar boundary-layer equations (10-15) and (8-15):

$$St\,Pr^{2/3} = 0.664(xG/\mu)^{-0.5} \tag{19-11}$$

$$c_f = 1.328(xG/\mu)^{-0.5} \tag{19-12}$$

(Note that the laminar boundary-layer equations for the *average* coefficients are used, rather than *local* coefficients.)

For this particular surface, $x/4r_h = 0.00318/0.00154$. Substituting into Eqs. (19-11) and (19-12), we obtain

$$St\,Pr^{2/3} = 0.463(4r_h G/\mu)^{-0.5} = 0.463\,Re^{-0.5} \tag{19-13}$$

$$c_f = 0.926\,Re^{-0.5} \tag{19-14}$$

where Re is based on $4r_h$.

We now test Eq. (19-13) at, say, Re = 1000:

$$St\,Pr^{2/3} = 0.0146$$

This corresponds closely to the experimental result for Surface #18-19.86 as shown on Fig. 19-4, and in fact it is a fair approximation over the entire test range. It is surprising that it is so close, because the interrupted fins comprise only 79 percent of the surface area, but evidently the basic area is similarly affected.

However, the corresponding computed value of c_f is 0.0293, which does not compare at all well with the experimental value of 0.049. Obviously something is missing. Figure 19-7 shows a diagram of the fin, and the forces acting on the fin, when the *finite thickness* of the fin (somewhat exaggerated) is included.

There will be flow stagnation at the leading edge, and the total force on the leading edge, denoted by F_u, will be the sum of the static pressure and the dynamic pressure resulting from the stagnation. The forces acting on the two sides, the viscous shear forces, will be denoted by F_{sf}. On the trailing edge there will be only the static pressure force, denoted by F_d, since the streamlines from the boundary layer on the sides will leave the body tangentially to the side surfaces. Let A be the surface area on the sides of the fin, and A_δ the area at each of the leading and trailing edges.

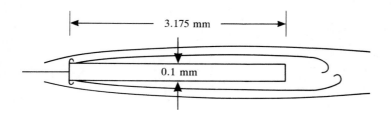

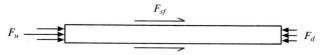

FIGURE 19-7
A fin with finite thickness.

Then

$$F_u = A_\delta (P + \tfrac{1}{2} C_D \rho V^2)$$

where C_D is a *drag coefficient*,

$$F_{sf} = \tfrac{1}{2} A c_f \rho V^2$$

where c_f can be calculated from Eq. (19-14), and

$$F_d = A_\delta P$$

assuming that the pressure change across the very small fin is negligible. We now define a total friction coefficient based on A:

$$F_{\text{total}} = \tfrac{1}{2} A c_{f\,\text{total}} \rho V^2$$

where $F_{\text{total}} = F_u + F_{sf} - F_d$. We substitute and solve for $c_{f\,\text{total}}$:

$$c_{f\,\text{total}} = c_f + \frac{A_\delta}{A} C_D$$

C_D must be less than unity, and 0.9 would be a reasonable guess. Then, using the dimensions of the fin to evaluate A_δ/A, and Eq. (19-14), we obtain

$$c_{f\,\text{total}} = 0.926 \, \text{Re}^{-0.5} + 0.0144 \tag{19-15}$$

This equation is in quite good agreement with the experimental data, Fig. 19-4. Note that approximately half the pressure drop for this surface is apparently attributable to *pressure drag*, and that proportion increases at higher Reynolds numbers. It would appear, at least from this analysis, that the pressure drag is contributing nothing to the heat-

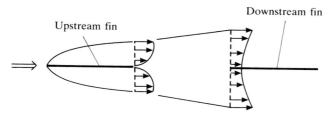

FIGURE 19-8
The effect of the wakes of fins.

transfer performance, and thus is parasitic. However, this needs further examination.

The heat-transfer analysis was based on two assumptions: (1) that there exists a simple laminar boundary layer on the fins; and (2) that the velocity and temperature of the fluid approaching a fin is at the local mixed mean conditions. Under these assumptions, the entire heat-transfer resistance is in the laminar boundary layer on each fin segment.

In reality, if the fins were infinitely thin, any one fin would lie in the *wake* of the preceding fin, or fins. Then the velocity and temperature profiles approaching each fin would appear somewhat as shown on Fig. 19-8.

The downstream fin would experience a velocity lower than the mean velocity, and a temperature difference smaller than the difference between surface and mixed mean temperatures. The momentum boundary layer would be thicker than otherwise, and both effects would tend to *decrease* the heat-transfer coefficient. The fact that this does not seem to occur, at least for this particular surface, suggests that there is indeed a mechanism present to mix the fluid in the wakes.

If the x Reynolds number is sufficiently large, the wake will become unstable and break down on its own, although in the Reynolds number range of these types of heat-exchanger surfaces this is unlikely. However, the blunt leading and trailing edges of the fins cause flow separations, and vorticity is shed into the main stream. This *turbulence* then provides a mixing mechanism to apparently erase the velocity and temperature wakes completely. The blunt leading and trailing edges were seen to approximately double the friction power expenditure, but this analysis suggests that this extra friction power is not necessarily parasitic. Some amount of vorticity shed into the main stream is needed if the surface is to realize its full heat-transfer potential.

What is not known is *how much* vorticity is needed. Examination of the test performance of strip-fin surfaces that are nearly identical except for thicker fins shows the expected higher friction coefficients, and the heat-transfer comparison suggests that there probably exists an optimum fin thickness—but this optimum has apparently not been systematically explored. The problem is further complicated by the need to optimize fin thickness for any given application because of the influence of fin thermal conductivity and thickness on fin effectiveness. A conclusion that seems inescapable, however, is that any attempt to *streamline* strip-fins to reduce friction power would probably not be worth the costs.

OTHER METHODS FOR OBTAINING HIGH PERFORMANCE

Virtually any scheme for increasing the heat-transfer coefficient without increasing the velocity will result in a surface with a higher $h-E$ characteristic than a continuous smooth tube. In *liquid* systems inserts of various shapes are frequently placed inside tubes. These are generally known as *turbulence promoters*, since they do in fact increase the turbulence level. For liquids they frequently provide a useful way to increase the heat-transfer rate. However, for *gas* applications turbulence promoters increase the friction power to an unnecessary extent, and the friction power problem is of course critical for gases. The difficulty is that for a turbulent flow the heat-transfer resistance is predominantly in the sublayer, and turbulence promoters only affect the sublayer in a secondary way. The key to efficiently increasing heat transfer is to destroy the sublayer.

Surface roughness decreases sublayer thickness, and a *fully rough* surface is one for which the sublayer is completely destroyed. Therefore any kind of surface roughness that leads to a fully rough surface provides an efficient way to obtain high performance.

Another commonly used scheme is to provide indentations or dimples, either along the walls of a tube or in fins. In this case boundary-layer separation is induced on the downstream side of the dimple, along with a certain amount of shedding of vorticity, and a new boundary layer is forced to form. Another scheme is to employ a wavy surface so that the fluid is forced to change direction at frequent intervals. Again boundary-layer separation is induced. Holes cut into fins provide another way to interrupt the growth of the boundary layer, although there is a disadvantage in removing heat-transfer surface area. A very common scheme in finned surfaces is to cut the fin at frequent intervals, and then to bend the fin material out into the flow-stream. This is usually known as a *louvered* surface. All of these schemes can be used in plate-fin heat-exchanger surfaces, or they can be used on the fins of finned-tube surfaces. Finally, the simple geometry formed when a fluid flows normally to a bank of circular tubes is a very high-performance configuration because a new boundary layer must form on each tube, although, because of boundary-layer separation and the large exposed frontal area of each tube, the friction coefficient is extremely high and is caused almost entirely by pressure drag.

At the present time it appears that for *gas-flow* applications the simple *strip-fin* surface probably represents about the optimum surface configuration. Its great virtue relative to the other schemes discussed is that the flow is not turned through an angle. Turning induces large pressures on the surface, which are not recovered after flow separation occurs, and thus results in unnecessarily high friction coefficients.

The approximate analysis leading to Eqs. (19-13) and (19-15) suggests that still better performance could be obtained by cutting the fins into still shorter segments. This is true, but ultimately a point of diminishing returns is approached because fins must have a finite thickness in order to conduct heat, and decreasing fin flow length while maintaining fin thickness leads to ever higher friction coefficients. Sometimes circular pins are used for fins, and these would represent the limit of short-segment strip-fins.

THE GENERAL $h-E$ PERFORMANCE OF STRIP-FINS

Consider the performance of a long smooth tube, such as is represented in Fig. 19-2. If the tube is sufficiently long that fully developed conditions prevail, the heat-transfer coefficient in the laminar flow region becomes a constant (the constant Nusselt number for fully developed flow). The $h-E$ plot for such a surface would be as shown in Fig. 19-9, where both the laminar and turbulent flow regions are shown. Also shown are the characteristics of a *strip-fin* surface, which tend to plot as close to a straight line on logarithmic coordinates, although at high and low

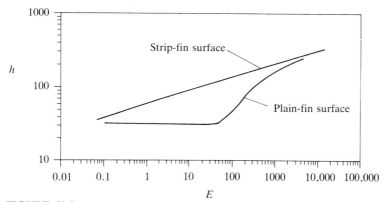

FIGURE 19-9
Comparison of $h-E$ characteristics of different kinds of surfaces.

Reynolds numbers there will be a slight tendency to curve upwards. Different strip-fin surfaces, or other high-performance surfaces, will have similar characteristics, although some will be higher or lower than the one illustrated. But the significant point is that the differences between the strip-fin surfaces and a smooth-walled surface diminish and can become small at both the low and high ends of the chart. The strip-fin and other high-performance surfaces show a significant advantage over only a limited range of operating conditions (although that advantage can be spectacular). Further examination reveals that the Reynolds number range (based on hydraulic diameter) over which such surfaces excel is approximately 400–15,000. In other words, it is the range from a little below transition to a bit above transition. But it was seen earlier this is precisely the Reynolds number range of most gas-flow applications.

In high-pressure systems, if the resulting high density leads to Reynolds numbers much above 15,000, there is probably not much to be gained from high-performance heat-transfer surfaces. Very low pressures and densities lead to very low Reynolds numbers that result in laminar flow in smooth tubes, and again little advantage for high-performance surfaces. Another factor that forces the operating range toward the low end of Fig. 19-9 is small hydraulic diameter. If the surface area density is increased (more area per unit of volume), the hydraulic diameter is inevitably decreased. Therefore extremely compact heat-exchanger surfaces tend to operate in this region. Because of their requirement for very large surface area, the rotating periodic-flow regenerators frequently used in automotive gas turbines are usually constructed of very compact matrices with very small-hydraulic-diameter flow passages. Significantly, they employ smooth noninterrupted flow passages with laminar flow because interrupted surfaces show little advantage. Additionally, very small-passage interrupted surfaces would also be more expensive to manufacture and might easily be fouled.

EXPERIMENTAL DATA FOR COMPACT HEAT-EXCHANGER SURFACES

Figures 19-3 and 19-4 show experimentally obtained heat transfer and flow friction characteristics for two compact surfaces. Figures 19-10 to 19-17 provide the same information for a representative variety of other compact surfaces. These include four tube-bank surfaces, one with bare tubes, and three with various types of fins. Three more plate-fin surfaces are presented: one with plain fins, one with strip-fins, and one a modern automobile radiator surface. Finally, a very compact matrix for use in rotating periodic-flow heat exchangers is included.

Here α is the total heat-transfer area per unit heat-exchanger volume, β is the total heat-transfer area per unit volume between the

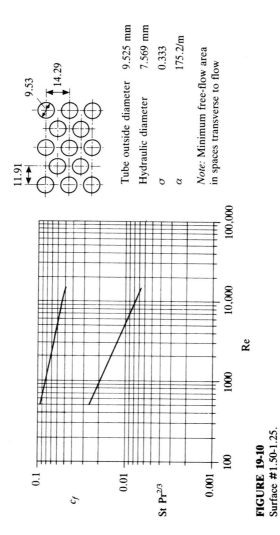

FIGURE 19-10

Surface #1.50-1.25.

Tube outside diameter   9.525 mm

Hydraulic diameter   7.569 mm

σ 0.333

α 175.2/m

Note: Minimum free-flow area in spaces transverse to flow

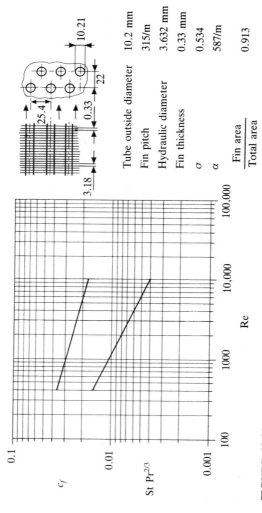

Tube outside diameter	10.2 mm
Fin pitch	315/m
Hydraulic diameter	3.632 mm
Fin thickness	0.33 mm
σ	0.534
α	587/m
$\dfrac{\text{Fin area}}{\text{Total area}}$	0.913

FIGURE 19-11
Surface #8.0-3/8T.

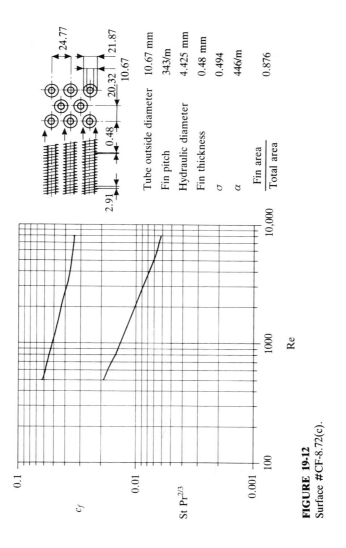

Tube outside diameter	10.67 mm
Fin pitch	343/m
Hydraulic diameter	4.425 mm
Fin thickness	0.48 mm
σ	0.494
α	446/m
$\dfrac{\text{Fin area}}{\text{Total area}}$	0.876

FIGURE 19-12
Surface #CF-8.72(c).

463

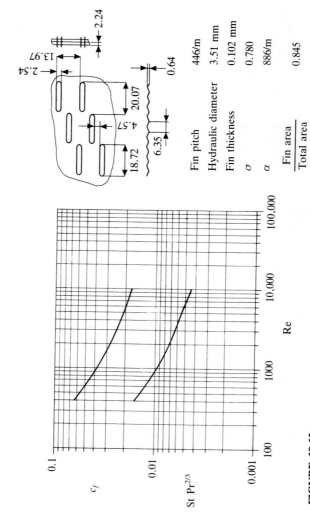

Fin pitch   446/m
Hydraulic diameter   3.51 mm
Fin thickness   0.102 mm
σ 0.780
α 886/m

$\dfrac{\text{Fin area}}{\text{Total area}}$ 0.845

FIGURE 19-13
Surface #11.32-0.737-SR.

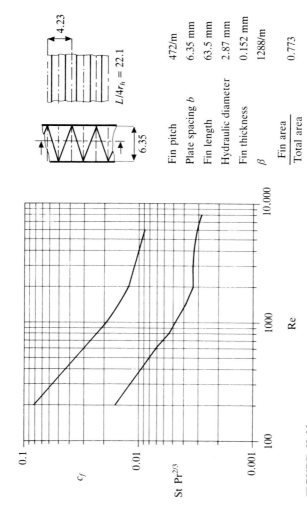

Fin pitch	472/m
Plate spacing b	6.35 mm
Fin length	63.5 mm
Hydraulic diameter	2.87 mm
Fin thickness	0.152 mm
β	1288/m
$\dfrac{\text{Fin area}}{\text{Total area}}$	0.773

$L/4r_h = 22.1$

4.23

6.35

FIGURE 19-14
Surface #12.00T.

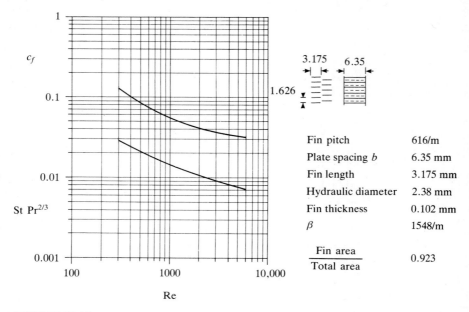

Fin pitch	616/m
Plate spacing b	6.35 mm
Fin length	3.175 mm
Hydraulic diameter	2.38 mm
Fin thickness	0.102 mm
β	1548/m
$\dfrac{\text{Fin area}}{\text{Total area}}$	0.923

FIGURE 19-15
Surface #1/8-15.61.

plates of one fluid side, and σ is the ratio of free-flow to frontal area, A_c/A_{fr}.

These data should provide enough information to investigate what can be accomplished with compact heat exchangers. The test results are taken from Kays and London,[1] where data from more than 130 such surfaces are tabulated.

Note that the Reynolds number in all of these charts is consistently defined on the basis of a hydraulic diameter defined by Eq. (7-17), and the friction coefficient is consistent with Eq. (7-13).

DESIGN OF COMPACT HEAT EXCHANGERS

In principle, the design of compact heat exchangers is not significantly different from the design of any other heat exchanger. However, some different things tend to be important.

The most important thing to bear in mind is that the pressure-drop design is just as important as the heat-transfer design. Another point is that the *shape* of the resulting heat exchanger often presents an awkward problem.

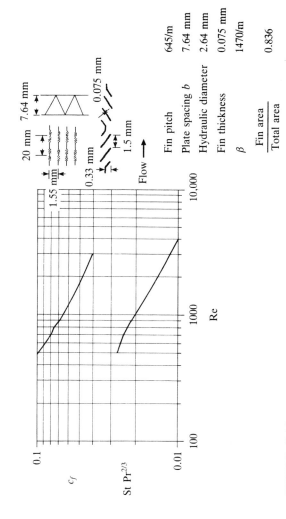

Fin pitch	645/m
Plate spacing b	7.64 mm
Hydraulic diameter	2.64 mm
Fin thickness	0.075 mm
β	1470/m
$\dfrac{\text{Fin area}}{\text{Total area}}$	0.836

FIGURE 19-16
Surface Radiator-H.

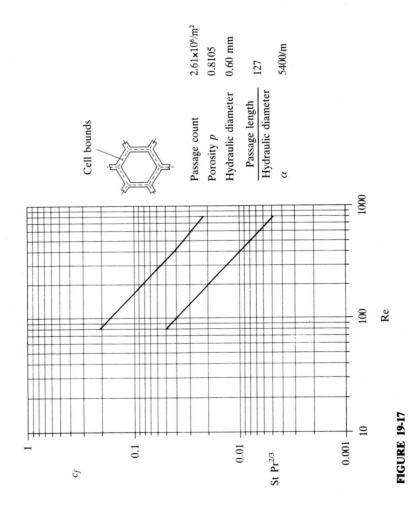

Cell bounds

Passage count	$2.61 \times 10^6 / m^2$
Porosity p	0.8105
Hydraulic diameter	0.60 mm
$\dfrac{\text{Passage length}}{\text{Hydraulic diameter}}$	127
α	5400/m

Re

c_f

St $Pr^{2/3}$

FIGURE 19-17
Surface #519 (matrix).

The *design problem* to be posed is as follows:

Given both fluid flow rates,
all fluid properties,
the desired effectiveness, and
the desired pressure drop for both fluids
Find the dimensions of the heat exchanger

One of the things that will be shown is that, regardless of the particular pair of surfaces chosen, there does exist a *unique design* of *cross-flow* heat exchanger (or multi-pass cross-flow heat exchanger) that will meet all of the above specifications. This is *not true* for the *counterflow* heat exchanger. For the rotating *periodic-flow* heat exchanger there is again a *unique design* that will meet all specifications.

Choice of Surface Pair

For a cross-flow or a counterflow design a surface configuration must be chosen for each fluid side (although there are some limitations discussed below). For a periodic-flow design only one surface need be chosen, since both fluid sides use the same surface.

The designer's freedom in the choice of a pair of surfaces is quite variable. If the heat exchanger is to be of plate-fin construction, Fig. 19-18, there is a great deal of choice. In principle, virtually any arbitrarily chosen pair of surfaces could be used for either a cross-flow or counterflow design.

Fin-tube surfaces, or simple bare tube banks, Fig. 19-19, do not allow so much freedom, since once the surface configuration for the outside of the tubes has been chosen, the possibilities for the other side are severely limited.

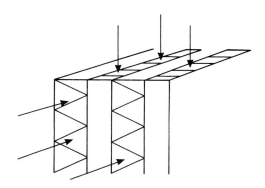

FIGURE 19-18
A cross-flow plate-fin heat exchanger.

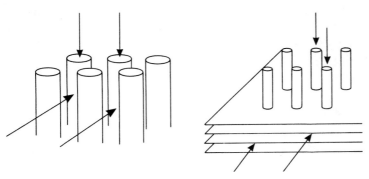

FIGURE 19-19
Tube-bank heat exchangers.

Note also that the fin orientation in the plate-fin construction allows the possibility of either a cross-flow or a counterflow arrangement, but the tube-bank surfaces are limited to a cross-flow arrangement (unless one is willing to have one of the fluids flow parallel to the outsides of the tubes). Obviously, availability, manufacturing costs, and many other considerations dictate the choice of surface pair.

The following are some approximate "rules of thumb" that will assist in choosing surfaces for particular applications:

1. The optimum heat-exchanger design generally results if the heat-transfer *resistances* on the two fluid sides are approximately *equal.*
2. Low-density fluids tend to yield low heat-transfer coefficients, for reasons discussed, so if the densities of the two fluids differ markedly then a surface pair should be chosen with *more surface area* on the *low-density side.* In a liquid-to-gas heat exchanger the surface-area ratio should be very large for optimum conditions, and this usually leads to the use of fin-tube surfaces.
3. Higher heat-transfer coefficients can usually be obtained by *surface interruptions,* and these are then generally more useful on the *low-density side.*
4. If there is a choice, it would be well to allow a *larger flow area* on the *low-density side,* because low-density fluids tend to give a large pressure drop, which can only be decreased by lower velocities, i.e., large flow areas. Otherwise, the heat exchanger may assume a very strange shape.
5. If the *fin effectiveness* in a particular design turns out to be low, say less than 70–80 percent, it is apparent that the heat exchanger is carrying a lot of useless surface area, which costs in volume and weight. On the other hand, if η_f is closs to 1, the fins are probably

unnecessarily thick, which costs in weight. Fin thickness obviously presents an optimization problem. One must be careful, however, with strip-fin surfaces, where fin thickness has a definite influence on the friction coefficient, and to a lesser extent on the heat-transfer coefficient.

Design of a Cross-Flow Heat Exchanger

Once that a pair of surfaces has been chosen, certain geometrical variables are fixed. Designating the two sides of the heat exchanger as 1 and 2, these are

$(4r_h)_1$, $(4r_h)_2$	hydraulic diameters
$\alpha_1 = A_1/V$, $\alpha_2 = A_2/V$	where V is the total volume
$\sigma_1 = A_{c1}/A_{fr1}$, $\sigma_2 = A_{c2}/A_{fr2}$	free-to-frontal area ratios
$(A_f/A)_1$, $(A_f/A)_2$	where A_f is the fin area
$(A/A_w)_1$, $(A/A_w)_2$	where A_w is the wall or direct area

(note that A is the total surface area, $A_f + A_w$).

α_1 and α_2 are the ratios of total heat-transfer area to total heat-exchanger volume for each of the two fluid sides. It will be noted that on the figures giving the basic heat-transfer and friction characteristics for various plate-fin surfaces a parameter β is given. This is the ratio of total heat-transfer area to the *volume between the plates* for that particular surface. The *total volume* of a plate-fin heat exchanger includes the volume occupied by the flow passages of both sides, plus the small volume occupied by the separating plates. α_1 and α_2 are related to β_1 and β_2 by the following equation, which can be readily developed from the geometry:

$$\alpha_1 = \frac{b_1\beta_1}{b_1 + b_2 + 2a} \tag{19-16}$$

where b_1 is the plate spacing for side 1, b_2 is the plate spacing for side 2, and a is the thickness of the separating plate.

σ_1 and σ_2 relate the *free-flow area* on each fluid side to the corresponding *frontal area*. Again a simple relation can be developed from the geometry and the definition of the hydraulic radius:

$$\sigma_1 = \frac{b_1\beta_1 r_{h1}}{b_1 + b_2 + 2a} = (\alpha r_h)_1 \tag{19-17}$$

where $r_h = A_c L/A$ is the ratio of the free-flow area to the passage perimeter.

Note that the αs and σs can only be evaluated after the surface configurations for both fluid sides have been chosen.

With the surfaces specified, the geometrical variables fixed, and the fluid properties determined, Eqs. (19-1) and (19-2) can now be solved for h and c_f on both sides of the heat exchanger as functions of the mass velocities:

$$h = h(G), \qquad c_f = c_f(G)$$

where

$$G = \frac{\dot{m}}{A_c} = \frac{\dot{m}}{A_{\text{fr}}\sigma}$$

The fin effectiveness η_f and surface effectiveness η_0 for each of the two sides now become functions of the Gs.

The equation for the overall conductance U, Eq. (18-5), can now be expressed in terms of G_1 and G_2.

The cross-flow heat exchanger will appear as in Fig. 19-20. The three dimensions a, b, and c are what we seek to establish. Note that c is the flow length in the direction of fluid 1, b is the flow length in the direction of fluid 2, and a is the nonflow dimension.

The following relations follow from the geometry:

$$V = abc \quad \text{(volume)}$$

$$A_{\text{fr}1} = ab, \qquad A_{c1} = \sigma_1 A_{\text{fr}1} = ab\sigma_1$$

$$A_{\text{fr}2} = ac, \qquad A_{c2} = \sigma_2 A_{\text{fr}2} = ac\sigma_2$$

The calculational procedure is inevitably an iterative one. The two mass velocities G_1 and G_2 will turn out to be the most convenient iteration variables. The procedure requires a first estimate of G_1 and G_2, and then subsequent corrections until all of the specifications are satisfied.

A reasonable first estimate of G_1 and G_2 can be made using the

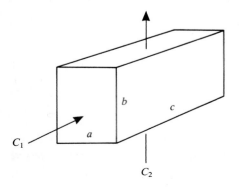

FIGURE 19-20
A cross-flow heat exchanger.

following equation, which can be derived, after a number of simplifying assumptions, from a combination of the pressure-drop equation (18-15) and the equation for U, Eq. (18-5):

$$G = \sqrt{\frac{St\, Pr^{2/3}}{c_f} \frac{\Delta P}{NTU} \frac{\rho}{Pr^{2/3}}}$$

The ratio $St\, Pr^{2/3}/c_f$ is reasonably constant for most heat-transfer surfaces over a considerable range of Reynolds number, and does not differ much for a wide range of surfaces. A value of 0.3 for this ratio is usually quite adequate because the subsequent iteration procedure is rapidly convergent.

By choosing G_1 and G_2 as the variables, we are, in effect, estimating the two frontal areas, $A_{fr1} = ab$ and $A_{fr2} = ac$. We then solve for the total volume $V = abc$, which establishes a, b, and c separately. A suggested step-by-step procedure follows.

We calculate the following in order using the various equations that have been developed in this chapter and in the preceding one:

1. h_1, h_2 from G_1, G_2
2. η_{o1}, η_{o2}
3. U
4. $NTU_{max} = f(\varepsilon, C_{min}/C_{max})$
5. $A = NTU_{max}C_{min}/U$
6. $V = A/\alpha$
7. a, b, c from V, ab, ac
8. c_{f1}, c_{f2} from G_1, G_2
9. $(\Delta P/P_1)_1$, $(\Delta P/P_1)_2$

The two pressure-drop ratios are then compared with the desired values. If either differs more than some predetermined amount, the values of G_1 and G_2 are adjusted according to the following relation, which is merely a simplification of the pressure-drop equation:

$$G \propto (\Delta P)^{1/2}$$

Then steps 1–9 are repeated. The iteration usually converges very rapidly, so it *can* be done by hand. However, it should be apparent that it is a fairly simple procedure to program on a computer. If any kind of optimization is being attempted, there is really no alternative to using a computer.

Regardless of the original choice of surface pair, and no matter how nonoptimal the pair may be, this procedure will yield a single unique design of cross-flow heat exchanger. However, the resulting design may involve a very strange shape of heat exchanger—a "pancake," or a long thin strip. These geometric troubles, which frequently occur when very compact surfaces are used with large mass flow rates, and where the densities of the two fluids differ markedly, can only be remedied by a new and different choice of surface pair, or perhaps by breaking up the heat exchanger into a number of separate units.

Design of a Multi-Pass Cross-Flow Heat Exchanger

Although a multi-pass heat exchanger can be designed directly, it is probably simpler to design a single-pass heat exchanger, and then *transform* it into a multi-pass design. Compare the single-pass and two-pass heat exchangers in Fig. 19-21.

If the surfaces are the same then the volume and weight of both heat exchangers are the same. If the mass flow rates and fluid properties are the same then the mass velocities are the same, and thus the Us and NTUs are the same. Since the total flow lengths are the same, the pressure drops are the same for both fluids (ignoring for the moment the turning losses in the two-pass design).

The only thing that has changed is the effectiveness. ε is *higher* for the two-pass design. One can thus design the two-pass heat exchanger as if it were a single-pass cross-flow heat exchanger, but use the $\varepsilon(\text{NTU})$ relation for the two-pass case. Then the dimensions can be altered as indicated in the figure.

Higher ε has been obtained at the price of a doubling of the nonflow dimension a. This may or may not be a disadvantage, but it is often inconvenient if mass flow rates are large and very compact surfaces are used.

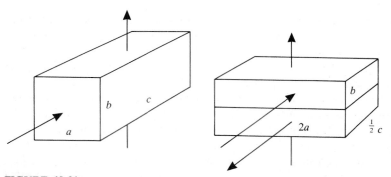

FIGURE 19-21
Transformation of a single-pass cross-flow heat exchanger to a two-pass heat exchanger.

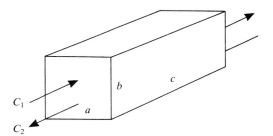

FIGURE 19-22
A counterflow heat exchanger.

Three-pass, four-pass, etc. heat exchangers are handled in exactly the same way. But for very high ε it is often preferable to go to the straight counterflow design, perhaps with cross-flow headers, although this may not be practicable if tubular surfaces are used.

Design of a Counterflow Heat Exchanger

A simple counterflow arrangement is shown in Fig. 19-22. This design differs from that of the cross-flow heat exchanger in that the mass velocities G_1 and G_2 cannot be varied separately once the surface pair is chosen. The relation between G_1 and G_2 is fixed, since they both share the same frontal area, A_{fr}.

Since G_1 and G_2 are not independent of one another, it is apparent that the pressure drops on the two sides cannot be varied independently. If an unfavorable balance of pressure drop is obtained, it can only be changed by a change in the choice of surface pairs (perhaps by scaling one of them up or down).

On the other hand, the counterflow design often has definite *shape* advantages over cross-flow, because volume and frontal area are the only geometrical parameters fixed by the design. The frontal shape can be varied at will.

Of course counterflow has a large advantage in NTU for a given ε, especially at high ε, and this translates into smaller volume and weight.

Outside of these remarks, the procedure for designing a counterflow heat exchanger differs little from that for a cross-flow heat exchanger, except that G_1 and G_2 are not independent of one another.

The design of a counterflow heat exchanger with cross-flow headers is not quite so straightforward. The easiest procedure is probably to design as if it were a simple counterflow heat exchanger, but using the ε(NTU) relations for the mixed case. But then the cross-flow headers will have to be examined in more detail (for example, the velocities may be different, and the NTU distribution may not be as originally planned), and more iterations will be required before converging on a final design.

Design of a Rotating Periodic-Flow Heat Exchanger

The design of this system is simpler than for the previous types because there is only one surface configuration to deal with rather than two.

First one has to estimate what the effects of leakage and carryover are going to be. This will lead to new *effective* values for the mass-flow rates and the terminal temperatures.

Next the necessary NTU for the heat exchanger is determined from the previously described charts. It will be necessary to arbitrarily choose a value for C_r/C_{min}. After the design is complete, this will establish RPM, from which carryover can be checked. Some iteration may be involved to find the optimum values for C_r/C_{min} and carryover rate.

From here one proceeds as in the design of a cross-flow heat exchanger. Initial estimates of G_1 and G_2 can be made using the same relation. This establishes an initial estimate for both frontal areas. Figure 19-23 illustrates what is actually being determined if the heat exchanger has a *disk* configuration.

At this point the only unknown is the flow length L, which is the same for both fluids. One then proceeds as with a cross-flow design up to step 5. Note then that the overall NTU of the periodic-flow heat exchanger is defined by

$$\text{NTU}_0 = \frac{1}{C_{min}[(1/hA)_1 + (1/hA)_2]}$$

but, by definition, $r_h = A_c L/A$ and $A = A_c L/r_h$. We substitute for A in this equation, and then factor L out of the denominator:

$$\text{NTU}_0 = \frac{L}{C_{min}r_h[(1/hA_c)_1 + (1/hA_c)_2]}$$

There is now enough information to solve for L, and the complete dimensions of the heat exchanger are fixed.

Next we calculate the pressure drops for both fluids. If they differ from what is desired, the Gs can then be adjusted as before, and the procedure is repeated until sufficient convergence is obtained.

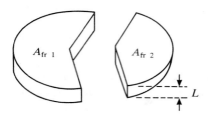

FIGURE 19-23
A periodic-flow heat-exchanger disk.

Note that for this type of heat exchanger a unique design is obtainable for all of the specifications, as was the case for a cross-flow heat exchanger. At the same time, it is possible to attain close to counterflow performance.

PROBLEMS

19-1. A single-pass cross-flow heat exchanger is constructed using the circular-fine surface designated as #CF-8.72(c), Fig. 19-12. The dimensions of the heat exchanger on the face that exposes the fins are 0.6 m × 0.6 m (i.e., a square cross-section), while the dimensions on the face exposing the ends of the circular tubes are 0.6 m × 0.5 m. Air at 1 atm pressure and 18°C flows into the finned side at a rate of 2 kg/s, while water at 80°C flows through the tubes at a rate of 5 kg/s. The tube wall thickness is 0.5 mm. The construction material is copper.

Determine the outlet temperatures of the two fluids and the pumping-power requirements for both.

19-2. It is desired to specify the dimensions of a heat exchanger whose function is to heat 3 kg/s of air at 1 atm from 18°C to 90°C using an available supply of CO_2 at 4 kg/s, 3 atm pressure, and 125°C. The pressure drop for both of the fluids is to be 3 percent. That is, the pressure drop through the heat exchanger for the air will be 3 percent of 1 atm, and that for the CO_2 will be 3 percent of 3 atm.

Plate-fin construction is to be used, and the heat exchanger is to have a single-pass cross-flow arrangement. Any two of the plate-fin surfaces described in this chapter may be used (or both sides could be made from the same surface geometry). The thickness of the plates separating the surfaces should be 0.005 mm. The material should be aluminum.

An iterative solution will be necessary in order to satisfy both the thermal requirements and the pressure drops.

One of the objectives of the problem should be to find which pair of surfaces provides a minimum-volume heat exchanger, and which pair provides a design having the most "compact" design, i.e., the design that minimizes the largest dimension.

19-3. Repeat Prob. 19-2 but use a two-pass cross–counterflow arrangement, with the CO_2 making the two passes.

19-4. In the design of an energy-efficient house, a difficulty arises when the house is sealed up so well that various types of air pollution become both unpleasant and dangerous. The obvious solution is to provide ventilation, but this destroys the reason for sealing, i.e., to prevent the leak of thermal energy. A way to address this problem is to provide forced draft ventilation through a heat exchanger in such a way that the exhaust air (warm) transfers heat to the fresh air (cool).

The objective of this problem is to design a heat exchanger to accomplish this. We will specify that the heat exchanger have a single-pass cross-flow arrangement. It will be of plate-fin construction and must use one of the surfaces for which data are provided in this chapter.

We would like to make the effectiveness as high as "practicable," although we do not really know what that term implies without heat-exchanger cost information. Lacking this, let us use 0.85.

We cannot afford very large pressure drops on the two sides because the electric power to blow the air through might otherwise be used more profitably to provide house heating. But let us assume that a further constraint is that air velocity in the ducts leading to and from the heat exchanger must be held down to no more than 4 ft/s (1.2 m/s) to keep sound levels low. Pumping power should be checked, but this constraint in itself may keep the pumping power level low. Let us further specify that the frontal areas of the two fluid faces of the heat exchanger are to be the same as the duct cross-sectional areas, so that if rectangular ducts are used, no transition sections will be necessary.

The system specifications are that the ventilation rate must be one house volume per hour. The house has 2000 ft^2 (186 m^2) of floor area with an 8 ft (2.4 m) ceiling.

The inside temperature of the house will be 70°F (21°C). The outside air temperature will be 35°F (2°C).

19-5. A gas turbine regenerator (sometimes called a recuperator) is to be designed for a 550 shaft horsepower plant (truck or bus) with the following overall specifications:

Compressor pressure ratio	4:1
Air-flow rate	3.98 kg/s
Regenerator effectiveness	80 percent
Total percentage pressure drop (air side plus gas side)	6 percent
Air temperature entering regenerator	453 K
Gas temperature entering regenerator	844 K
Specific fuel consumption	0.51 lb$_m$ fuel/(SHP h)

In a gas turbine regenerator a percent of pressure drop on one fluid side has approximately the same adverse effect on output power and specific fuel consumption as a percent of pressure drop on the other side. Thus the designer can, if desired, shift percentage pressure drop from one side to the other to seek an optimum solution.

Since the air:fuel ratio is very high, the combustion products on the hot-gas side may be assumed to have essentially the properties of pure air.

The objective of the problem is to design (i.e., to specify the dimensions for, and the weight of) a heat exchanger that will meet specifications. The machinery and ducting arrangements suggest that a two-pass cross–counterflow arrangement be used, with the high-pressure air making the turns and the low-pressure exhaust gas going straight through.

The design may use any of the heat-exchanger surfaces described in this chapter. If plate-fin construction is employed then a pair of surfaces must be chosen. Both the shape and size of the resulting design will depend heavily upon these surfaces, so care should be taken in choosing them, noting that the low-density gas side is favored by large flow areas as well as large amounts of surface area. A design that will meet specifications can be accomplished with any of the surfaces, but the resulting heat exchanger may be large and awkwardly shaped. Note that it will have to fit into a typical truck engine compartment.

The heat-exchanger material should be stainless steel, or an equivalent type of material.

A modern gasoline internal combustion engine weighs 1.3–2.5 kg per horsepower. A similar diesel engine weighs 2.5–4.5 kg/HP. The compressor–turbine–combustion-chamber of a small gas turbine probably weighs less than 0.2 kg/HP. How does the regenerative gas turbine look in the light of these figures?

19-6. Repeat Problem 19-5, but this time let the heat exchanger be a rotating periodic-flow system using the ceramic matrix designated #519, Fig. 19-17.

REFERENCE

1. Kays, W. M., and A. L. London: *Compact Heat Exchangers,* McGraw-Hill, New York, 1984.

CHAPTER
20

MASS TRANSFER: FORMULATION OF A SIMPLIFIED THEORY

The next three chapters are concerned with fluid systems in which concentration gradients of two or more components of the fluid exist, so that mass diffusion takes place within the fluid. We restrict consideration entirely to those situations where the mass-diffusion rate is proportional to the concentration gradient of the diffusing substance alone (Fick's law) and is thus independent of temperature gradients, pressure gradients, and the concentration gradients of other components of a mixture. This is an idealization of the behavior of real systems, which is exact only for a binary mixture in the absence of other gradients. There are some applications where the coupling between heat and mass transfer cannot be ignored, and others where the exact expression for mass diffusion in a multicomponent mixture must be used, but there are many where our idealization is perfectly adequate for engineering purposes (see Chap. 3). The primary rationale for developing a simplified theory for convective mass transfer, rather than starting with a complete theory, is that simple Fick's law mass transfer forms a very logical extension to the momentum and thermal boundary-layer theories developed in previous chapters, and thus the reader is traversing rather familiar ground. Mass transfer in all its ramifications is an extremely large and complex subject, and it is

480

presumed that the interested reader will regard these three chapters as merely a convenient introductory route into the subject.

Our primary concern is the calculation of mass- and energy-transfer rates at a phase interface, and the theory to be developed permits at least approximate solutions to such diverse applications as arise in psychrometry, drying, evaporative cooling, transpiration cooling, diffusion-controlled combustion, and ablation, to name a few.

The objective of this chapter is to reduce the convective mass-transfer problem to the form of Eq. (1-2),† starting with the differential equations of the boundary layer as developed in Chap. 4. It is shown that a rather impressive list of alternative simplifications to the boundary-layer equations leads to this formulation. This chapter is devoted primarily to a definition of the mass-transfer conductance g and the mass-transfer potential or driving force B.

Chapter 21 is devoted to the solution of the simplified diffusion equation under a number of conditions, for the sole purpose of evaluating the conductance g. Here we find that some of the heat-transfer boundary-layer solutions developed earlier are directly applicable, with only a change of variables.

Chapter 22 is concerned with evaluation of the driving force B, and thus the final calculation of mass-transfer rates and accompanying heat-transfer rates. In that chapter a variety of illustrative examples are considered.

DEFINITIONS

The following definitions overlap to some extent those given in Chaps. 1–4. Since an understanding of the simplifications to the diffusion equations depends largely on a complete grasp of these definitions, it is worthwhile to group them.

General Definitions

Our ultimate concern is with the total rate of mass transfer $\dot{m}''$, kg/(s · m²), across a phase boundary under steady-flow conditions, as shown on Fig. 20-1. However, in any particular application we may be more interested in a heat-transfer rate than in a mass-transfer rate; or we may want to consider the heat transfer at an interface where there is mass diffusion within the boundary layer but where $\dot{m}'' = 0$. The formulation

---

† The formulation presented here is based largely on the work of D. B. Spalding; see, for example, Ref. 1.

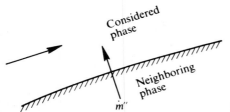

FIGURE 20-1

that we develop is sufficiently general that such problems are merely special cases.

It should be pointed out that the formulation of the mass-transfer problem to be presented is completely consistent with the more conventional formulation where the mass-transfer rate is asserted to be proportional to the difference between the concentration of the transferred substance at the interface and the concentration at the free-stream. In fact, if there is no chemical reaction and concentrations are very small (relative to 0) this formulation reduces to precisely that. However, the present formulation is additionally valid where concentrations are large and where chemical reaction is taking place within the boundary layer.

The primary variables within the boundary layer (in addition to the spatial coordinates) are

m_j mass concentration of component j, kg_j/kg mix. Thus

$$\sum_j m_j = 1$$

$\mathbf{G}$ *net* mass-flux vector, $kg/(s \cdot m^2)$. Note that this includes all components of the mixture, some of which may be moving at different velocities than others because of diffusion. Note further that

$\mathbf{V} = \mathbf{G}/\rho$ where $\mathbf{V}$ is the "mixture velocity" vector and ρ is the density of the mixture, kg/m^3. Also, $u = G_x/\rho$, $v = G_y/\rho$, etc.

$\dot{m}'' = G_{y,0}$ the normal component of the net mass flux vector at the interface; $\dot{m}''$ is then the *net mass-transfer* rate at the interface, including all components, and is defined as *positive* into the considered phase.

$m_j \mathbf{G}$ *convected* component of the flux of component j.

$\dot{m}_j'' = m_j \dot{m}''$ *convected* flux of component j at or below the interface (this is further subscripted later).

$\dot{m}_j'''$ mass-creation rate of component j at some point in the boundary layer due to chemical reaction,

$kg/(s \cdot m^3)$. Note that

$$\sum_j \dot{m}_j''' = 0$$

n_α — mass concentration of element α, kg of element α per kg of total mixture.

$n_{\alpha,j}$ — a constant fixed by the chemical formula for compound j, (kg of element α)/(kg of substance j). Thus

$$n_\alpha = \sum_j n_{\alpha,j} m_j$$

$$\sum_\alpha n_\alpha = 1$$

$$\sum_j n_{\alpha,j} \dot{m}_j''' = 0$$

i — specific enthalpy of mixture, J/kg.

i_j — partial enthalpy of component j, J/kg_j; thus

$$i = \sum_j m_j i_j$$

$c_j = \left(\dfrac{\partial i_j}{\partial t} \right)_{P, m_j, m_k}$ — specific heat at constant pressure for component j. For perfect-gas mixtures the subscripts may be dropped, and this is also a reasonable approximation for the liquid applications considered. Under these conditions,

$$c_j = \frac{di_j}{dt}$$

$c = \sum_j m_j c_j$ — specific heat of the mixture. Note that $c \neq di/dt$, *except* where composition is uniform (no mass diffusion) or where the specific heats of all the components are equal and there is no chemical reaction (see below).

Definitions Concerned with the States of the Transferred Substance

Here we consider some definitions associated with mass concentration and then some definitions associated with energy transfer. Consider first the control volumes indicated in Fig. 20-2 and the steady-flow transfer of component j across the indicated control surfaces.

Let the 0 surface be an infinitesimal distance inside the considered phase (that is, inside the boundary layer) and the L surface be an infinitesimal distance inside the neighboring phase. The T surface is

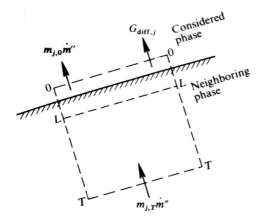

FIGURE 20-2
Control volumes at the phase interface for the mass-diffusion equation.

considered to be some finite distance inside the neighboring phase, and the fluid state at this surface is termed the *transferred-substance state*. The principal characteristic of the T state is that there are no gradients either of concentration or temperature at the T-state location. All mass and energy transports across the T-state boundary are by convection. In a particular application the "T state" may in fact be totally fictitious.

The mass flux of component j across the T surface is indicated to be

$$\dot{m}''_{j,T} = m_{j,T}\dot{m}''$$

This relation is used to *define* the T-state concentration. Thus

$$m_{j,T} = \dot{m}''_{j,T}/\dot{m}'' \qquad (20\text{-}1)$$

If the component j is the *only* transferred substance, as is the case in most simple applications, $m_{j,T} = 1$. Both $\dot{m}''$ and $\dot{m}''_{j,T}$ are described as *positive* if flowing toward the considered phase; *either or both can be negative, and $\dot{m}''_{j,T}$ can be greater in magnitude than $\dot{m}''$*, the net mass transfer at the interface. Thus $m_{j,T}$ can be greater than unity, and it can be negative. Although it is treated as a "concentration," it is seen that it is of a different nature from m_j in the considered phase, which is always between 0 and 1.

In a similar manner the "concentration" of some chemical element α in the transferred-substance state is defined as

$$n_{\alpha,T} = \dot{m}''_{\alpha,T}/\dot{m}'' \qquad (20\text{-}2)$$

where

$$\dot{m}''_{\alpha,T} = \sum_j (\dot{m}''_j n_{\alpha,j})_T$$

Consider next the control volume indicated in Fig. 20-3 and the steady-flow *energy fluxes* flowing across the L and T surfaces. The

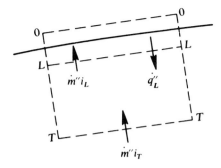

FIGURE 20-3
Control volumes at the phase interface for the energy equation.

indicated heat flux $\dot{q}_L''$ is assumed to be *conduction* heat transfer, and can be conduction within the transferred substance, fluid or solid, or within any solid structural materials present. For the moment we neglect thermal radiation and modify the equations later to include radiation.

Note that there is no temperature gradient and thus no conduction heat transfer across the T surface. If the conservation of energy principle is now applied to the control volume bounded by the L and T surfaces, a formal definition of the transferred-substance enthalpy i_T is obtained:

$$\dot{m}''i_L - \dot{q}_L'' - \dot{m}''i_T = 0$$

$$i_T = i_L - \frac{\dot{q}_L''}{\dot{m}''} \tag{20-3}$$

In actual applications there may, in fact, be no adiabatic surface, in which case i_T is simply a quantity defined by Eq. (20-3).

THE DIFFERENTIAL EQUATIONS OF THE CONCENTRATION BOUNDARY LAYER

Consider a two-dimensional boundary layer in (x, y) coordinates within the considered phase. Let us restrict consideration to steady flow in the absence of body forces, and let velocities be sufficiently low that we can neglect viscous energy dissipation and the influence of pressure gradients on the energy equation. Then, from the differential equations developed in Chap. 4, the following must be the differential equations to be satisfied within the laminar boundary layer:

diffusion of component j, Eq. (4-18),

$$G_x \frac{\partial m_j}{\partial x} + G_y \frac{\partial m_j}{\partial y} - \frac{\partial}{\partial y}\left(\gamma_j \frac{\partial m_j}{\partial y}\right) = \dot{m}_j''' \tag{20-4}$$

diffusion of chemical elements α, Eq. (4-24),

$$G_x \frac{\partial n_\alpha}{\partial x} + G_y \frac{\partial n_\alpha}{\partial y} - \frac{\partial}{\partial y} \left[\sum_j \gamma_j \frac{\partial}{\partial y} (n_{\alpha,j} m_j) \right] = 0 \qquad (20\text{-}5)$$

energy, from Eq. (4-27) or (4-26),

$$G_x \frac{\partial i}{\partial x} + G_y \frac{\partial i}{\partial y} - \frac{\partial}{\partial y} \left(\sum_j \gamma_j \frac{\partial m_j}{\partial y} i_j \right) - \frac{\partial}{\partial y} \left(k \frac{\partial t}{\partial y} \right) = 0 \qquad (20\text{-}6)$$

continuity, from Eq. (4-1),

$$\frac{\partial G_x}{\partial x} + \frac{\partial G_y}{\partial y} = 0 \qquad (20\text{-}7)$$

momentum, from Eq. (4-10),

$$G_x \frac{\partial u}{\partial x} + G_y \frac{\partial u}{\partial y} - \frac{\partial}{\partial y} \left(\mu \frac{\partial u}{\partial y} \right) + \frac{dP}{dx} = 0 \qquad (20\text{-}8)$$

These five equations then form the basis for solution of the steady-flow, low-velocity, mass-transfer, laminar boundary-layer problem. Next we introduce various simplifications to the first three equations so as to reduce them to a single standard form. After introducing the same simplifications into the boundary conditions, we can seek a single family of solutions applicable to a considerable variety of problems.

The *turbulent* boundary layer is handled in much the same way. It is only necessary to stipulate that the viscosity and the diffusion coefficients are now the corresponding turbulent quantities.

In the process of simplification we are obviously going to lose some generality. In applications where we are not satisfied with the validity of the simplifications we have no recourse but to return to the original five equations and attempt a solution by other means. If, in a multicomponent system, the binary diffusion coefficients differ sufficiently to invalidate the use of Fick's law, or if the Dufour and Soret effects are deemed to be of importance (see Chap. 3), we must obviously retreat even further before attempting a solution. Whether the increased accuracy is worth the time and effort is the classic problem of engineering analysis. At any rate in this introductory treatment we are concerned only with solution of the simplified equations.

Simplified Mass-Diffusion Equations of the Boundary Layer

DIFFUSION OF AN INERT SUBSTANCE. In this case there is no chemical reaction, and $\dot{m}_j''' = 0$. Thus Eq. (20-4) becomes

$$G_x \frac{\partial m_j}{\partial x} + G_y \frac{\partial m_j}{\partial y} - \frac{\partial}{\partial y} \left(\gamma_j \frac{\partial m_j}{\partial y} \right) = 0 \qquad (20\text{-}9)$$

Equation (20-9) is the standard form that we seek for all the other simplifications.

SIMPLE CHEMICAL REACTION WITH $\gamma_f = \gamma_o = \gamma$. Consider a simple reaction taking place within the considered phase in which a *fuel* combines with an *oxidant* to form a single *product*, without intermediaries. Then, from stoichiometric considerations, we can write

$$1 \text{ kg fuel} + r \text{ kg oxidant} \rightarrow (1 + r) \text{ kg product}$$

We can now apply Eq. (20-4) independently to the *fuel* and then the *oxidant* (or to the *product* if we prefer):

$$G_x \frac{\partial m_f}{\partial x} + G_y \frac{\partial m_f}{\partial y} - \frac{\partial}{\partial y} \left(\gamma_f \frac{\partial m_f}{\partial y} \right) = \dot{m}_f'''$$

$$G_x \frac{\partial m_o}{\partial x} + G_y \frac{\partial m_o}{\partial y} - \frac{\partial}{\partial y} \left(\gamma_o \frac{\partial m_o}{\partial y} \right) = \dot{m}_o'''$$

From the stoichiometric relation,

$$\dot{m}_f''' = \dot{m}_o'''/r$$

Then, combining these equations,

$$G_x \frac{\partial m_f}{\partial x} + G_y \frac{\partial m_f}{\partial y} - \frac{\partial}{\partial y} \left(\gamma_f \frac{\partial m_f}{\partial y} \right) = \frac{1}{r} \left[G_x \frac{\partial m_o}{\partial x} + G_y \frac{\partial m_o}{\partial y} - \frac{\partial}{\partial y} \left(\gamma_o \frac{\partial m_o}{\partial y} \right) \right]$$

Rearranging,

$$G_x \frac{\partial}{\partial x} \left(m_f - \frac{m_o}{r} \right) + G_y \frac{\partial}{\partial y} \left(m_f - \frac{m_o}{r} \right) - \frac{\partial}{\partial y} \left(\gamma_f \frac{\partial m_f}{\partial y} - \frac{\gamma_o}{r} \frac{\partial m_o}{\partial y} \right) = 0$$

Now suppose $\gamma_f = \gamma_o = \gamma$ (and this is the first of our major simplifications). Then

$$G_x \frac{\partial}{\partial x} \left(m_f - \frac{m_o}{r} \right) + G_y \frac{\partial}{\partial y} \left(m_f - \frac{m_o}{r} \right) - \frac{\partial}{\partial y} \left[\gamma \frac{\partial}{\partial y} \left(m_f - \frac{m_o}{r} \right) \right] = 0$$

$$(20\text{-}10)$$

Similar equations can obviously be developed for

$$m_f + \frac{m_p}{1 + r} \quad \text{if } \gamma_f = \gamma_p$$

$$\frac{m_o}{r} + \frac{m_p}{1 + r} \quad \text{if } \gamma_o = \gamma_p$$

Note that in assuming equality of the diffusion coefficients, we are not introducing significantly more of an approximation than is already incorporated in the differential equations. In the development of the latter, Fick's law was assumed to be valid, and from Chap. 3 Fick's law is accurately applicable to the components of a multicomponent mixture only if the binary diffusion coefficients for each pair of components in the mixture are the same.

DIFFUSION OF CHEMICAL ELEMENT α WITH $\gamma_j = \gamma_k = \gamma$. In a chemically reacting boundary layer in which the simple reaction is not applicable we can still consider conservation of the separate elements. Consider Eq. (20-5) under the assumption that $\gamma_j = \gamma_k = \gamma$ (note that the diffusion coefficients refer to the compounds m_j and not the elements α). Then

$$G_x \frac{\partial n_\alpha}{\partial x} + G_y \frac{\partial n_\alpha}{\partial y} - \frac{\partial}{\partial y}\left[\gamma \frac{\partial}{\partial y}\sum_j (n_{\alpha,j} m_j)\right] = 0$$

But

$$\sum_j (n_{\alpha,j} m_j) = n_\alpha$$

Thus

$$G_x \frac{\partial n_\alpha}{\partial x} + G_y \frac{\partial n_\alpha}{\partial y} - \frac{\partial}{\partial y}\left(\gamma \frac{\partial n_\alpha}{\partial y}\right) = 0 \tag{20-11}$$

Simplification of the Energy Equation

We now develop four different simplifications to the energy equation, all of which lead to the same standard form.

THE UNIT-LEWIS-NUMBER SIMPLIFICATION. First let us write the energy equation (20-6) in a slightly different form. The last term may be written

$$\frac{\partial}{\partial y}\left(k \frac{\partial t}{\partial y}\right) = \frac{\partial}{\partial y}\left(\Gamma c \frac{\partial t}{\partial y}\right) = \frac{\partial}{\partial y}\left(\Gamma \sum_j m_j c_j \frac{\partial t}{\partial y}\right)$$

But for a perfect gas $c_j \, dt = di_j$, and we have previously indicated an intention to use this expression as an approximation for a liquid. Thus

$$\frac{\partial}{\partial y}\left(k \frac{\partial t}{\partial y}\right) = \frac{\partial}{\partial y}\left(\Gamma \sum_j m_j \frac{\partial i_j}{\partial y}\right)$$

and Eq. (20-6) becomes

$$G_x \frac{\partial i}{\partial x} + G_y \frac{\partial i}{\partial y} - \frac{\partial}{\partial y}\left(\sum_j \gamma_j \frac{\partial m_j}{\partial y} i_j\right) - \frac{\partial}{\partial y}\left(\Gamma \sum_j m_j \frac{\partial i_j}{\partial y}\right) = 0 \tag{20-12}$$

It is apparent that a great simplification will result if we can set $\gamma_j = \Gamma$. In Chap. 3 the Lewis number was defined as $\mathrm{Le}_j = \gamma_j/\Gamma = \mathrm{Pr}/\mathrm{Sc}_j$. Thus our proposed simplification is valid if the Lewis number for each of the various fluids in the mixture is equal to unity at all points in the boundary layer. (This does not preclude γ_j and Γ varying through the boundary layer.)

For gases and gas mixtures both Pr and Sc are of the order of magnitude of unity, so the assumption of $\mathrm{Le} = 1$ for a gas mixture is frequently reasonably valid. Some representative data on binary gas mixtures are contained in App. A and can be examined to verify this assertion.

In a turbulent flow the Reynolds analogy between turbulent heat and momentum transfer (see Chap. 13) can be readily extended to turbulent mass transfer. It is found that if γ_j and Γ are defined as turbulent quantities, the analogy leads directly to $\gamma_j = \Gamma$; thus, for a turbulent boundary layer with gases, the unit-Lewis-number assumption may be even more valid than for a laminar boundary layer. It must be remembered, however, that in a turbulent boundary layer a substantial part of the resistance to heat transfer, and also to mass diffusion, is in the sublayers where the mechanism is still primarily molecular.

For liquids the situation is not nearly so favorable (see App. A), but a second simplification to the energy equation, discussed below, often is as valid for liquid mixtures as is the unit-Lewis-number assumption for gases; either simplification leads to the same standard form of the energy equation.

Note that under the unit-Lewis-number assumption no restriction is placed on possible chemical reaction within the boundary layer, so that the resulting simplified differential equation is applicable to the chemically reacting boundary layer.

Under the assumption that $\gamma_j = \Gamma$ for all components, Eq. (20-12) becomes

$$G_x \frac{\partial i}{\partial x} + G_y \frac{\partial i}{\partial y} - \frac{\partial}{\partial y}\left(\Gamma \sum_j i_j \frac{\partial m_j}{\partial y} + \Gamma \sum_j m_j \frac{\partial i_j}{\partial y} \right) = 0$$

The last term then simplifies as follows:

$$\Gamma \sum_j \left(i_j \frac{\partial m_j}{\partial y} + m_j \frac{\partial i_j}{\partial y} \right) = \Gamma \sum_j \frac{\partial}{\partial y}(m_j i_j) = \Gamma \frac{\partial}{\partial y} \sum_j m_j i_j = \Gamma \frac{\partial i}{\partial y}$$

Thus Eq. (20-12) reduces finally to

$$G_x \frac{\partial i}{\partial x} + G_y \frac{\partial i}{\partial y} - \frac{\partial}{\partial y}\left(\Gamma \frac{\partial i}{\partial y} \right) = 0 \qquad (20\text{-}13)$$

NO CHEMICAL REACTION, WITH EQUAL SPECIFIC HEATS FOR ALL COMPONENTS. We express the enthalpy of a component of a mixture by integrating the defining expression for the specific heat at constant pressure:

$$i_j = i_{j,0} + \int_{t_0}^{t} c_j \, dt$$

The temperature t_0 is an arbitrary datum temperature associated with $i_{j,0}$, the enthalpy of the j component at the datum temperature. If no chemical reaction is permitted then all the $i_{j,0}$ can be arbitrarily made zero. If, in addition, the specific heats are all equal, that is,

$$c_j = c_k = c$$

then

$$i_j = \int_{t_0}^{t} c \, dt$$

The third term of Eq. (20-6) then becomes

$$\frac{\partial}{\partial y} \left(\sum_j \gamma_j \frac{\partial m_j}{\partial y} i_j \right) = \frac{\partial}{\partial y} \left(\int_{t_0}^{t} c \, dt \sum_j \gamma_j \frac{\partial m_j}{\partial y} \right)$$

But, from Eq. (4-29), $\sum_j \gamma_j \, \partial m_j / \partial y = 0$. Thus the entire mass-diffusion term in Eq. (20-6) is equal to zero.

Also, under these conditions,

$$i = \sum_j m_j i_j = \sum_j \left(m_j \int_{t_0}^{t} c \, dt \right) = \int_{t_0}^{t} c \, dt \sum_j m_j = \int_{t_0}^{t} c \, dt$$

and $di = c \, dt$. Thus Eq. (20-6) can be simplified to

$$G_x \frac{\partial i}{\partial x} + G_y \frac{\partial i}{\partial y} - \frac{\partial}{\partial y} \left(\Gamma \frac{\partial i}{\partial y} \right) = 0 \tag{20-14}$$

SIMPLE CHEMICAL REACTION, WITH EQUAL SPECIFIC HEATS. Consider a boundary layer in which the same simple reaction is taking place as discussed above under the simplifications of the mass diffusion equation. Employing the subscripts f, o, p, and oc for the *fuel*, *oxidant*, *products*, and all *other components*, respectively, the enthalpy of the mixture can be expressed as

$$i = \sum_j m_j i_j = m_f i_f + m_o i_o + m_p i_p + m_{oc} i_{oc}$$

The enthalpy of each component can be expressed in various ways, depending on the chemical state of the system chosen for enthalpy datum. A frequently used datum scheme is to let the enthalpy of the elements be zero at some reference temperature, and then the enthalpy of a compound at the same temperature is the *change* of enthalpy associated with the reaction forming the compound from its elements, the so-called "heat of formation" of a compound. Such a scheme could be employed here, but in simple combustion reactions it is frequently more convenient to define and derive a heat of combustion H_o, J/(kg of fuel), which is actually the negative of the enthalpy difference between the products of combustion and fuel and oxidant entering into the reaction, all at t_0.

For calculation purposes H_o can be attached to the fuel, oxidant, or product, since it is a property of a chemical system and not merely a property of the fuel. It cancels out of calculations unless a reaction actually takes place, in which case all three components are involved in a fixed stoichiometric ratio. Thus if we employ H_o, there are three possible schemes for expressing the enthalpy of the components, and thereby the enthalpy of the mixture, as follows:

1.
$$i_f = \int_{t_0}^{t} c_f \, dt + H_o, \qquad i_o = \int_{t_0}^{t} c_o \, dt$$

$$i_p = \int_{t_0}^{t} c_p \, dt, \qquad i_{oc} = \int_{t_0}^{t} c_{oc} \, dt$$

2.
$$i_f = \int_{t_0}^{t} c_f \, dt, \qquad i_o = \int_{t_0}^{t} c_o \, dt + \frac{H_o}{r}$$

$$i_p = \int_{t_0}^{t} c_p \, dt, \qquad i_{oc} = \int_{t_0}^{t} c_{oc} \, dt$$

3.
$$i_f = \int_{t_0}^{t} c_f \, dt, \qquad i_o = \int_{t_0}^{t} c_o \, dt$$

$$i_p = \int_{t_0}^{t} c_p \, dt - \frac{H_o}{1+r}, \qquad i_{oc} = \int_{t_0}^{t} c_{oc} \, dt$$

For present purposes let us employ scheme 2. Then the enthalpy of the mixture becomes

$$i = m_f \int_{t_0}^{t} c_f \, dt + m_o \int_{t_0}^{t} c_o \, dt + m_o \frac{H_o}{r} + m_p \int_{t_0}^{t} c_p \, dt + m_{oc} \int_{t_0}^{t} c_{oc} \, dt$$

But now let us consider a case where $c_f = c_o = c_p = c_{oc} = c$. Then

$$i = m_o \frac{H_o}{r} + \int_{t_0}^{t} c \, dt \qquad (20\text{-}15)$$

The next-to-last term in Eq. (20-6) can now be simplified, employing scheme 2 and equal specific heats, as follows:

$$\sum_j \gamma_j \frac{\partial m_j}{\partial y} i_j = \gamma_f \frac{\partial m_f}{\partial y} \int_{t_0}^{t} c\, dt + \gamma_o \frac{\partial m_o}{\partial y} \int_{t_0}^{t} c\, dt + \gamma_o \frac{\partial m_o}{\partial y} \frac{H_o}{r}$$

$$+ \gamma_p \frac{\partial m_p}{\partial y} \int_{t_0}^{t} c\, dt + \gamma_{oc} \frac{\partial m_{oc}}{\partial y} \int_{t_0}^{t} c\, dt$$

Since the same integral is common to four of the terms, we can introduce Eq. (4-29) to eliminate all these terms. Thus we conclude that

$$\sum_j \gamma_j \frac{\partial m_j}{\partial y} i_j = \gamma_o \frac{\partial m_o}{\partial y} \frac{H_o}{r}$$

Finally, the last term in Eq. (20-6) may be developed as follows. From Eq. (20-15),

$$\frac{\partial i}{\partial y} = \frac{H_o}{r} \frac{\partial m_o}{\partial y} + c \frac{\partial t}{\partial y}$$

and

$$\Gamma \frac{\partial i}{\partial y} = \Gamma \frac{H_o}{r} \frac{\partial m_o}{\partial y} + \Gamma c \frac{\partial t}{\partial y} = \Gamma \frac{H_o}{r} \frac{\partial m_o}{\partial y} + k \frac{\partial t}{\partial y}$$

Then

$$k \frac{\partial t}{\partial y} = \Gamma \frac{\partial i}{\partial y} - \Gamma \frac{H_o}{r} \frac{\partial m_o}{\partial y}$$

Now substituting for the last two terms of Eq. (20-6),

$$G_x \frac{\partial i}{\partial x} + G_y \frac{\partial i}{\partial y} - \frac{\partial}{\partial y}\left(\gamma_o \frac{\partial m_o}{\partial y} \frac{H_o}{r} + \Gamma \frac{\partial i}{\partial y} - \Gamma \frac{H_o}{r} \frac{\partial m_o}{\partial y} \right) = 0$$

But suppose $\gamma_o = \Gamma$. Then the equation reduces to

$$G_x \frac{\partial i}{\partial x} + G_y \frac{\partial i}{\partial y} - \frac{\partial}{\partial y}\left(\Gamma \frac{\partial i}{\partial y} \right) = 0 \qquad (20\text{-}16)$$

In some cases this result may be less restricting than the unit-Lewis-number simplification, since only the diffusion coefficient for the oxidant must be equal to the heat-diffusion coefficient, although, of course, the equality of specific heats must now be assumed. Note that the unit-Lewis-number simplification includes the possibility of "simple" chemical reactions, as well as multiple reactions.

Equation (20-16) is obviously also derivable under enthalpy schemes 1 and 3. The expression for the enthalpy of the mixture and the diffusion coefficient assumption are then

(scheme 1) $\qquad i = m_f H_o + \int_{t_0}^{t} c \, dt, \qquad$ with $\gamma_f = \Gamma \qquad$ (20-17)

(scheme 3) $\qquad i = -m_p \dfrac{H_o}{1+r} + \int_{t_0}^{t} c \, dt, \qquad$ with $\gamma_p = \Gamma \qquad$ (20-18)

UNIFORM COMPOSITION. The final simplification of the energy equation to the standard form results when the composition is everywhere uniform. There is thus no mass diffusion, since $\partial m_j / \partial y = 0$ everywhere. This, of course, simply brings us back to the single-component boundary-layer problem, which was the subject matter of Chaps 10 and 13. However, this can still be a *mass-transfer* problem, since fluid of the same composition as the mainstream fluid can flow through the interface (boundary-layer blowing or suction).

Under these conditions, the third term in Eq. (20-6) drops out, and we can also express $di = c \, dt$. Then noting that $k = c\Gamma$, Eq. (20-6) becomes

$$G_x \frac{\partial i}{\partial x} + G_y \frac{\partial i}{\partial y} - \frac{\partial}{\partial y}\left(\Gamma \frac{\partial i}{\partial y}\right) = 0 \qquad (20\text{-}19)$$

An alternative form of Eq. (20-19) is

$$G_x c \frac{\partial t}{\partial x} + G_y \frac{\partial t}{\partial y} - \frac{\partial}{\partial y}\left(c\Gamma \frac{\partial t}{\partial y}\right) = 0$$

Then if we consider the case of *constant specific heat*, c drops out and we obtain another equation in the standard form

$$G_x \frac{\partial t}{\partial x} + G_y \frac{\partial t}{\partial y} - \frac{\partial}{\partial y}\left(\Gamma \frac{\partial t}{\partial y}\right) = 0 \qquad (20\text{-}20)$$

General Form of the Simplified Equations

All the simplifications of the mass diffusion and energy equations discussed above have led to the same differential equation. Let us now summarize by defining a variable $\mathscr{P}$ as follows. Let

$$\mathscr{P} = m_j, \; m_f - \frac{m_o}{r}, \; \ldots, \; n_\alpha, \; i, \; t$$

under the indicated restrictions. Let $\lambda = \Gamma$ or γ_j, as appropriate.. Then $\mathscr{P}$ must satisfy the differential equation of the boundary layer:

$$G_x \frac{\partial \mathscr{P}}{\partial x} + G_y \frac{\partial \mathscr{P}}{\partial y} - \frac{\partial}{\partial y}\left(\lambda \frac{\partial \mathscr{P}}{\partial y}\right) = 0 \qquad (20\text{-}21)$$

If we now generalize this equation to three-dimensional space, we obtain

$$\mathbf{G} \cdot \nabla \mathscr{P} - \nabla \cdot \lambda \nabla \mathscr{P} = 0 \qquad (20\text{-}22)$$

We refer to this equation as the $\mathscr{P}$ equation.

The Conserved Property of the Second Kind†

Any property that satisfies the $\mathscr{P}$ equation is defined as a *conserved property of the second kind*. A useful feature of this conserved property is contained in the following theorem.

> If, in any steady flow, there are two conserved properties of the second kind $\mathscr{P}_I$ and $\mathscr{P}_{II}$, and at each point $\lambda_I = \lambda_{II}$, then any linear combination of $\mathscr{P}_I$ and $\mathscr{P}_{II}$ is a conserved property of the second kind.

The proof is simply to substitute the linear combination into the $\mathscr{P}$ equation.

This theorem allows us to build up new conserved properties and thus additional solutions by combining properties that we already know.

BOUNDARY CONDITIONS AT THE INTERFACE

Here we develop a general form of the boundary condition at the interface by considering in turn each of the simplifications of the boundary-layer equations discussed.

Mass-Diffusion Equation: General

Consider a control volume enclosing the phase interface and bounded by the 0 and T surfaces, as shown in Fig. 20-4.

Consider the steady flow of some component j across these surfaces, and apply conservation of mass for component j:

$$\dot{m}'' m_{j,0} - \left(\gamma_j \frac{\partial m_j}{\partial y} \right)_0 - \dot{m}'' m_{j,T} = 0$$

Thus

$$\dot{m}'' = \frac{(\gamma_j \, \partial m_j / \partial y)_0}{m_{j,0} - m_{j,T}} \qquad (20\text{-}23)$$

---

† After D. B. Spalding.[1]

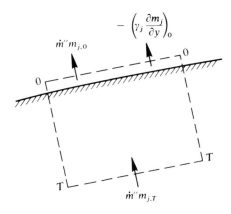

FIGURE 20-4
Control volume at the phase interface for
the mass-diffusion equation.

This is a general form of boundary condition that we seek for all
cases. It is also, of course, the boundary condition for diffusion of an
inert substance corresponding to Eq. (20-9).

Mass-Diffusion Equation: Simple Chemical Reaction

Considering the same control volume as above, and applying conservation of mass separately for the "fuel" and the "oxidant," we obtain for
the fuel

$$\dot{m}'' m_{f,0} - \left(\gamma_f \frac{\partial m_f}{\partial y} \right)_0 - \dot{m}'' m_{f,T} = 0$$

and for the oxidant

$$\dot{m}'' m_{o,0} - \left(\gamma_o \frac{\partial m_o}{\partial y} \right)_0 - \dot{m}'' m_{o,T} = 0$$

Dividing the second equation by r and subtracting from the first,

$$\dot{m}'' \left(m_f - \frac{m_o}{r} \right)_0 - \left[\gamma_f \frac{\partial m_f}{\partial y} - \gamma_o \frac{\partial}{\partial y} \left(\frac{m_o}{r} \right) \right]_0 - \dot{m}'' \left(m_f - \frac{m_o}{r} \right)_T = 0$$

We now let $\gamma_f = \gamma_o = \gamma$, and obtain

$$\dot{m}'' = \frac{[\gamma(\partial/\partial y)(m_f - m_o/r)]_0}{(m_f - m_o/r)_0 - (m_f - m_o/r)_T} \tag{20-24}$$

Note that in an application there need not necessarily be any
oxidant (or fuel) in the transferred substance.

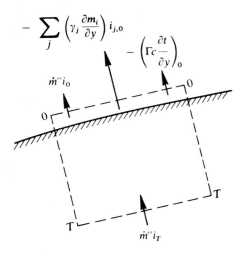

FIGURE 20-5
Control volume at the phase interface for the energy equation.

Mass-Diffusion Equation: Chemical Element α

If we apply conservation of chemical element α to the same control volume and let $\gamma_j = \gamma_k = \gamma$, we obtain

$$\dot{m}'' = \frac{(\gamma \, \partial n_\alpha / \partial y)_0}{n_{\alpha,0} - n_{\alpha,T}} \tag{20-25}$$

Energy Equation: General

Consider a similar control volume but with steady flow of energy across the 0 and T surfaces, Fig. 20-5, neglecting thermal radiation.

Applying conservation of energy,

$$\dot{m}'' i_0 - \sum_j \left(\gamma_j \frac{\partial m_j}{\partial y} \right)_0 i_{j,0} - \left(\Gamma c \frac{\partial t}{\partial y} \right)_0 - \dot{m}'' i_T = 0$$

from which,

$$\dot{m}'' = \frac{\sum_j [\gamma_j (\partial m_j / \partial y)]_0 i_{j,0} + [\Gamma c (\partial t / \partial y)]_0}{i_0 - i_T} \tag{20-26}$$

Equation (20-26) is not yet in the standard form, but the various simplifications to the energy equation lead to the form of Eq. (20-23).

The Unit-Lewis-Number Simplification

The second term in the numerator of Eq. (20-26) may be rewritten as follows:

$$\left(\Gamma c \frac{\partial t}{\partial y} \right)_0 = \left(\Gamma \frac{\partial t}{\partial y} \right)_0 \left(\sum_j m_j c_j \right)_0 = \left(\Gamma \sum_j m_j c_j \frac{\partial t}{\partial y} \right)_0 = \left(\Gamma \sum_j m_j \frac{\partial i_j}{\partial y} \right)_0$$

Under the unit-Lewis-number assumption, $\gamma_j = \gamma_k = \ldots = \Gamma$. Then Eq. (20-26) may be written

$$\dot{m}'' = \frac{\left[\Gamma \sum_j \left(i_j \frac{\partial m_j}{\partial y} + m_j \frac{\partial i_j}{\partial y}\right)\right]_0}{i_0 - i_T} = \frac{\left[\Gamma \sum_j \frac{\partial (m_j i_j)}{\partial y}\right]_0}{i_0 - i_T} = \frac{\left(\Gamma \frac{\partial}{\partial y} \sum_j m_j i_j\right)_0}{i_0 - i_T}$$

Thus

$$\dot{m}'' = \frac{(\Gamma \, \partial i / \partial y)_0}{i_0 - i_T} \qquad (20\text{-}27)$$

No Chemical Reaction, with Equal Specific Heats for All Components

We have already seen that, under these conditions,

$$\sum_j \gamma_j \frac{\partial m_j}{\partial y} i_j = 0$$

Also

$$\Gamma c \frac{\partial t}{\partial y} = \Gamma \frac{\partial i}{\partial y}$$

for equal specific heats. Thus Eq. (20-26) becomes

$$\dot{m}'' = \frac{(\Gamma \, \partial i / \partial y)_0}{i_0 - i_T} \qquad (20\text{-}28)$$

Simple Chemical Reaction, with Equal Specific Heats

In applying this simplification to the energy differential equation, it was shown that if the enthalpy of the mixture is expressed by Eq. (20-15) then

$$\sum_j \gamma_j \frac{\partial m_j}{\partial y} i_j = \gamma_o \frac{\partial m_o}{\partial y} \frac{H_o}{r}$$

It was similarly shown that

$$\Gamma c \frac{\partial t}{\partial y} = k \frac{\partial t}{\partial y} = \Gamma \frac{\partial i}{\partial y} - \Gamma \frac{H_o}{r} \frac{\partial m_o}{\partial y}$$

If these two expressions are substituted into Eq. (20-26) and we make the additional stipulation that $\gamma_o = \Gamma$, we again obtain

$$\dot{m}'' = \frac{(\Gamma \, \partial i / \partial y)_0}{i_0 - i_T} \qquad (20\text{-}29)$$

Obviously the other enthalpy schemes lead to the same result

Uniform Composition

If the composition is uniform, the first term in the numerator of Eq. (20-26) is zero, and $dt = di/c$. Thus Eq. (20-26) reduces to

$$\dot{m}'' = \frac{(\Gamma \, \partial i/\partial y)_0}{i_0 - i_T} \tag{20-30}$$

Finally if c is everywhere the same (and this includes the transferred substance):

$$\dot{m}'' = \frac{(\Gamma \, \partial t/\partial y)_0}{t_0 - t_T} \tag{20-31}$$

General Form of the Boundary Conditions

If we now introduce the concept of a conserved property $\mathcal{P}$, Eqs. (20-23)–(20-31) can be written generally as

$$\dot{m}'' = \frac{(\lambda \, \partial \mathcal{P}/\partial y)_0}{\mathcal{P}_0 - \mathcal{P}_T} \tag{20-32}$$

DEFINITION OF THE MASS-TRANSFER CONDUCTANCE AND DRIVING FORCE

Our mathematical problem now becomes one of obtaining solutions to the $\mathcal{P}$ equation for various boundary conditions on $\mathcal{P}$.

In a boundary layer let us denote by $\mathcal{P}_\infty$ the value of $\mathcal{P}$ in the free stream. Then at any point along the surface there is a $\mathcal{P}$ profile as shown in Fig. 20-6.

Examination of the $\mathcal{P}$ equation indicates that it is *almost* linear in $\mathcal{P}$. The nonlinearity arises in two ways. First, λ may be (and usually is) a function of $\mathcal{P}$. This is the same kind of nonlinear effect that arises in the

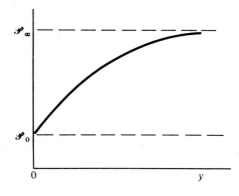

FIGURE 20-6
Conserved-property profile in a boundary layer.

simple heat-transfer problem resulting from the influence of temperature on the transport properties. However, we now have to contend with the possible influence of composition.

The other source of nonlinearity is due to the fact that the normal velocity component at the wall, v_0, is directly proportional to the mass-transfer rate $\dot{m}''$, which is, in turn, functionally related to $\mathscr{P}_\infty - \mathscr{P}_0$. Thus the whole velocity field, and more specifically G_x and G_y (which must be obtained by solution of the momentum equation), are functions of $\mathscr{P}_\infty - \mathscr{P}_0$.

Nevertheless, the $\mathscr{P}$ equation *is* linear in the limit as $\dot{m}'' \to 0$. Thus, at this limit, from Fig. 20-6,

$$\left(\lambda \frac{\partial \mathscr{P}}{\partial y} \right)_0 \propto (\mathscr{P}_\infty - \mathscr{P}_0)$$

Let us then *define* a mass-transfer conductance g such that

$$\left(\lambda \frac{\partial \mathscr{P}}{\partial y} \right)_0 = g(\mathscr{P}_\infty - \mathscr{P}_0), \quad \text{or} \quad g = \frac{(\lambda \, \partial \mathscr{P}/\partial y)_0}{\mathscr{P}_\infty - \mathscr{P}_0} \qquad (20\text{-}33)$$

The conductance g is then independent of $\dot{m}''$ for $\dot{m}''$ near zero, but may become a strong function of $\dot{m}''$ as $\dot{m}''$ becomes large. However, this does not invalidate the definition of g.

If we now combine Eqs. (20-33) and (20-32),

$$\dot{m}'' = g \frac{\mathscr{P}_\infty - \mathscr{P}_0}{\mathscr{P}_0 - \mathscr{P}_T}$$

Let us now define a mass-transfer driving force B such that

$$B = \frac{\mathscr{P}_\infty - \mathscr{P}_0}{\mathscr{P}_0 - \mathscr{P}_T} \qquad (20\text{-}34)$$

and then

$$\dot{m}'' = gB \qquad (20\text{-}35)$$

Thus, to calculate mass-transfer rates, we evaluate B from given values of $\mathscr{P}$ at the 0, T, and ∞ states, and g from an appropriate solution to the $\mathscr{P}$ equation. B can be calculated from mass concentrations, but it can also be calculated from any of the other conserved properties under suitably restricted conditions.

Note that B is a nondimensional parameter, so that g is dimensionally the same as $\dot{m}''$, that is, it has dimensions kg/(s $\cdot$ m^2). The driving force can be positive or negative, which determines the direction of $\dot{m}''$ (positive B corresponds to mass transfer *into* the boundary layer from the surface). The conductance g is always positive.

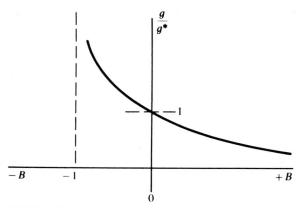

FIGURE 20-7
Typical variation of g with B.

In Chapter 21 we find it convenient to employ a parameter g^*, defined as follows:

$$g^* = \lim_{B \to 0} \frac{\dot{m}''}{B} \qquad (20\text{-}36)$$

The ratio g/g^* is frequently useful because it tends to be primarily a function of B alone. That is, for a given class of boundary layers, g/g^* is often moderately, or completely, independent of such parameters as Reynolds number. The relationship between g/g^* and B is generally, qualitatively, as shown in Fig. 20-7.

The nature of the conductance g can perhaps be better appreciated if we consider the last simplification of the energy equation, that of uniform composition, and the case where specific heat is a constant so that the temperature t is a conserved property, Eq. (20-20). Then the definition of g becomes

$$g = \frac{(\Gamma\, \partial t/\partial y)_0}{t_\infty - t_0}$$

Previously we defined the heat-transfer conductance h such that

$$\dot{q}_o'' = h(t_0 - t_\infty) = \left(-k\frac{\partial t}{\partial y}\right)_0 = \left(-\Gamma c\frac{\partial t}{\partial y}\right)_0$$

Combining these relations,

$$g = h/c$$

We can nondimensionalize this relation by dividing by G_∞ to yield

$$\frac{g}{G_\infty} = \frac{h}{G_\infty c} = \text{St}$$

Thus we see that g/G_∞ is a mass-transfer "Stanton number." Of course, it is the same as the heat-transfer Stanton number only under the restrictions indicated.

We can nondimensionalize this relation in an alternative manner by multiplying through by x/Γ:

$$\frac{gx}{\Gamma} = \frac{hx}{c\Gamma} = \frac{hxc}{ck} = \frac{hx}{k} = \text{Nu}_x$$

Thus in this form we have a mass-transfer "Nusselt number," sometimes referred to as the *Sherwood number*.

The parameter B has been developed here as a driving force, or potential, in terms of the values of conserved properties in the free stream and at the interface, but it can also be recognized as the boundary-layer "blowing" parameter B_h discussed in Chap. 13.

Consider the same simplification of the energy equation as above, and let the density be everywhere uniform. Then, starting with Eq. (20-35),

$$B = \frac{\dot{m}''}{g} = \frac{\dot{m}''}{G_\infty g/G_\infty} = \frac{\rho v_0}{\rho u_\infty} \frac{1}{\text{St}} = \frac{v_0}{u_\infty} \frac{1}{\text{St}} \qquad (20\text{-}37)$$

PROBLEMS

20-1. Consider a binary-mixture gas mass-transfer system in which the considered phase is a mixture of gases 1 and 2. Let gas 2 be injected into the mixture at the interface; that is, gas 2 is the only transferred substance. Show that if there is no phase change from the L state to the 0 state then $\dot{q}_0'' = \dot{q}_L''$, regardless of the mass-transfer aspects of the problem. [Start with Eq. (20-26).]

20-2. Consider the same system as in Prob. 20-1. In addition to no phase change at the interface, let the Lewis number be everywhere 1. At the outer edge of the boundary layer let the mixture be exclusively component 1. Show that, under these conditions,

$$\dot{q}_L'' = \dot{q}_0'' = g(i_{1,\infty} - i_{1,0})$$

If the specific heat of component 1 may be considered as constant, what is the implied relationship between St and g/G_∞?

20-3. Show that for the case of $\dot{m}'' = 0$ the heat flux at the interface is proportional to the *enthalpy* differences between the 0 and ∞ states, for any case for which enthalpy is a conserved property.

20-4. Demonstrate, with the aid of a suitable model, that in a *turbulent* flow, under the assumption of the Reynolds analogy, the "turbulent" Lewis number is equal to 1.

20-5. Starting with the energy equation of the boundary layer in the form of Eq. (4-26), show that for the case of a fluid with Pr = 1 the *stagnation enthalpy* is a conserved property of the second kind.

REFERENCE

1. Spalding, D. B.: *Int. J. Heat Mass Transfer,* vol. 1, pp. 192–207, 1960.

CHAPTER
21

MASS TRANSFER: SOME SOLUTIONS TO THE CONSERVED-PROPERTY EQUATION

This chapter is devoted to a description of some solutions to the conserved-property equation, (20-21) or (20-22). The primary present objective is to evaluate the mass-transfer conductance g.

The laminar *constant-property* boundary layer is fairly completely handled. Most of the available turbulent boundary-layer solutions to the $\mathcal{P}$ equation are primarily restricted to constant fluid properties. A major difficulty in obtaining useful general solutions to the $\mathcal{P}$ equation is the rather substantial influence of mixture ratio on the transport properties in a multicomponent system. In heat transfer without mass transfer the constant-property solutions are found to be applicable with only slight modification to many real problems; in mass transfer the number of kinds of property variation is larger, the effects can be large, and the constant-property solutions are less directly useful. Nevertheless, the constant-property boundary-layer solutions must still form the basic starting point, and thus constitute our primary concern here.

Of course, one always has the option to solve the $\mathscr{P}$ equation for any desired variable properties and boundary conditions using finite-difference techniques and machine computation; indeed, this may turn out to be the only practicable procedure for complex problems. However, one should also then consider the validity of the various approximations and simplifications that went into the development of the $\mathscr{P}$ equation, for it may not be significantly more difficult to solve the applicable differential equations without simplifications. Recall, for example, that Fick's law is accurately applicable only to a two-component mixture or that the Lewis number may, in fact, depart considerably from 1. Our present purpose, however, is to seek those applications for which a simplified approach is satisfactory for engineering work, because the rewards in time and labor are very considerable.

THE LAMINAR CONSTANT-PROPERTY BOUNDARY LAYER, SIMILARITY SOLUTIONS

Under the stipulation of constant properties, Eq. (20-21) becomes

$$G_x \frac{\partial \mathscr{P}}{\partial x} + G_y \frac{\partial \mathscr{P}}{\partial y} - \lambda \frac{\partial^2 \mathscr{P}}{\partial y^2} = 0$$

or

$$u \frac{\partial \mathscr{P}}{\partial x} + v \frac{\partial \mathscr{P}}{\partial y} - \frac{\lambda}{\rho} \frac{\partial^2 \mathscr{P}}{\partial y^2} = 0 \qquad (21\text{-}1)$$

The velocity components u and v must come from a solution to the corresponding *momentum* and *continuity* equations:

$$u \frac{\partial u}{\partial x} + v \frac{\partial u}{\partial y} - \frac{\mu}{\rho} \frac{\partial^2 u}{\partial y^2} = -\frac{1}{\rho} \frac{dP}{dx} \qquad (21\text{-}2)$$

$$\frac{\partial u}{\partial x} + \frac{\partial v}{\partial y} = 0 \qquad (21\text{-}3)$$

Consider the boundary conditions:

$$\mathscr{P} = \mathscr{P}_0 \quad \text{at} \quad \begin{cases} y = 0 \\ u = 0 \\ v = v_0 \end{cases}$$

$$\mathscr{P} \to \mathscr{P}_\infty \quad \text{as} \quad \begin{cases} y \to \infty \\ u \to u_\infty \end{cases}$$

$$\mathscr{P} = \mathscr{P}_\infty \quad \text{at} \quad \begin{cases} x = 0 \\ u = u_\infty \end{cases}$$

We already have a group of exact solutions to Eq. (21-1) for certain boundary conditions, namely the *similarity solutions* for $u_\infty = Cx^m$, in Chap. 10, where the boundary conditions include the additional specifications that

$$\mathcal{P}_0 \text{ and } \mathcal{P}_\infty \text{ are constant along the surface}$$

$$\frac{v_0}{u_\infty} \text{Re}_x^{1/2} \text{ is constant along the surface}$$

However, the similarity solutions in Chap. 10 employ temperature t as the dependent variable, rather than $\mathcal{P}$. These solutions were presented in nondimensional form (see Tables 10–3 to 10–5) as

$$\frac{\text{Nu}}{\text{Re}_x^{1/2}} = \phi\left(\frac{\mu}{\Gamma}, m, \frac{v_0}{u_\infty}\text{Re}_x^{1/2}\right) \tag{21-4}$$

(note that $\text{Pr} = \mu/\Gamma$).

Let us now convert these parameters to the corresponding mass-transfer parameters associated with the $\mathcal{P}$ equation. We first note that Re_x and m remain unchanged. The heat-transfer Nusselt number describes the dimensionless gradient at the wall; that is, in heat-transfer solutions

$$\text{Nu} = \frac{x(\partial t/\partial y)_0}{t_\infty - t_0}$$

Substituting $\mathcal{P}$ for t, and regarding Nu now as a dimensionless quantity not necessarily associated with heat transfer but merely describing the behavior of the dependent variable at the wall,

$$\text{Nu} = \frac{x(\partial \mathcal{P}/\partial y)_0}{\mathcal{P}_\infty - \mathcal{P}_0}$$

By definition, from our mass-transfer development,

$$g = \frac{\lambda(\partial \mathcal{P}/\partial y)_0}{\mathcal{P}_\infty - \mathcal{P}_0}$$

This is not dimensionless; but if both sides are multiplied by x/λ, we get a dimensionless quantity that describes the gradient of $\mathcal{P}$ at the wall:

$$\frac{xg}{\lambda} = \frac{x(\partial \mathcal{P}/\partial y)_0}{\mathcal{P}_\infty - \mathcal{P}_0}$$

The values listed in Tables 10-3 to 10-5 are dimensionless wall gradients. In heat-transfer problems we identify them with values of the Nusselt number hx/k. These tabulated values can also be interpreted as the values for the mass-transfer Nusselt number, that is,

$$xg/\lambda = \text{Nu}$$

or, similarly,

$$g/G_\infty = \text{St}$$

The Prandtl number in Eq. (21-4), $\text{Pr} = \mu/\Gamma$, now becomes the corresponding parameter μ/λ. Depending on which simplification of the mass-diffusion or energy equations is being used as a basis for the $\mathscr{P}$ equation, $\mu/\lambda = \mu/\gamma_j = Sc_j$, the Schmidt number; or $\mu/\lambda = \mu/\Gamma$, the Prandtl number.

Finally, the blowing parameter must be interpreted in terms of mass-transfer parameters:

$$\frac{v_0}{u_\infty} \text{Re}_x^{1/2} = \frac{\dot{m}''}{G_\infty} \text{Re}_x^{1/2} \quad \text{(since the density is constant)}$$

But, since $\dot{m}'' = gB$,

$$\frac{v_0}{u_\infty} \text{Re}_x^{1/2} = \frac{gB}{G_\infty} \text{Re}_x^{1/2} = \frac{xg}{\lambda} \frac{\mu}{xG_\infty} \frac{\lambda}{\mu} B \, \text{Re}_x^{1/2} = \frac{B}{\mu/\lambda} \frac{\text{Nu}}{\text{Re}_x^{1/2}}$$

or

$$B = \frac{v_0}{u_\infty} \text{Re}_x^{1/2} \frac{\mu/\lambda}{\text{Nu} \, \text{Re}_x^{-1/2}}$$

Thus we see that we can express the similarity solutions in the form [from (21-4)]

$$\frac{gx/\lambda}{\text{Re}_x^{1/2}} = \phi\left(\frac{\mu}{\lambda}, m, B\right) \qquad (21\text{-}5)$$

where the functions ϕ in Eqs. (21-4) and (21-5) are identical.

An interesting and perhaps unforeseen result is that the similarity solutions yield a constant value of B along the surface. Since $\mathscr{P}_0$ and $\mathscr{P}_\infty$ have already been assumed constant in the above development, it is apparent that in this case constant B implies constant $\mathscr{P}_T$.

Using the relations derived above, we can now convert all the similarity solutions presented in Chap. 10 to the form of Eq. (21-5). For example, consider Table 10-3 for $m = 0$ and $\text{Pr} = 0.7$. With the indicated change of parameters, it becomes Table 21-1, a family of solutions to the $\mathscr{P}$ equation for $\mu/\lambda = 0.7$.

The other results in Tables 10-3 to 10-5 can be similarly modified. Spalding and Evans[1] present much more extensive tabulations of these solutions.

On Figs. 21-1 and 21-2 some of the constant-property similarity solutions for $m = 0$ (constant u_∞) are plotted in the form of g/g^* as a

TABLE 21-1
Solutions to the $\mathscr{P}$ equation for a laminar constant-property boundary-layer for $m = 0$, $\mu/\lambda = 0.7$, and B = constant

$\dfrac{v_0}{u_\infty} \mathrm{Re}_x^{1/2}$	$\dfrac{gx/\lambda}{\mathrm{Re}_x^{1/2}}$	B
-2.500	1.850	-0.946
-0.750	0.722	-0.728
-0.250	0.429	-0.408
0	0.292	0
0.250	0.166	1.054
0.375	0.107	2.455
0.500	0.0517	6.78

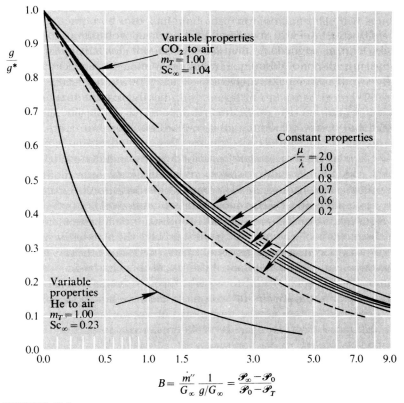

$$B = \frac{\dot{m}''}{G_\infty}\frac{1}{g/G_\infty} = \frac{\mathscr{P}_\infty - \mathscr{P}_0}{\mathscr{P}_0 - \mathscr{P}_T}$$

FIGURE 21-1
Solution of the $\mathscr{P}$ equation for a laminar boundary layer with constant free-stream velocity.

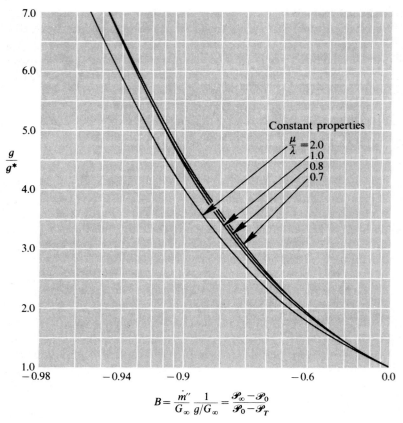

$$B = \frac{\dot{m}''}{G_\infty} \frac{1}{g/G_\infty} = \frac{\mathscr{P}_\infty - \mathscr{P}_0}{\mathscr{P}_0 - \mathscr{P}_T}$$

FIGURE 21-2
Solution of the $\mathscr{P}$ equation for a laminar boundary layer with constant free-stream velocity.

function of B, where g^* is the limiting value of g as B approaches zero. Of particular interest in these plots is the fact that g/g^* is not only independent of Re_x but is also only mildly dependent on μ/λ.

A word regarding the useful applicability of the constant-property solutions is in order here. Obviously, in no real problem are properties precisely constant. The constant-property mass-transfer solutions do provide, however, at least a fair approximation for many applications, and are accurately applicable under any one of the following three conditions:

1. Applications where λ, μ, and ρ are truly close to constant through the boundary layer because of the peculiar combination of fluids involved.

2. Small mass-transfer rates $(B \to 0)$ where the transferred-substance concentration is never large enough to change properties substantially.
3. Large mass-transfer rates where most of the fluid transferred across the interface is the same as the free-stream fluid, but there is a small concentration of the fluid of interest in the transferred fluid. In other words, if $\mathscr{P} = m_j$, this would be the case of $m_{j,T} \ll 1$.

THE LAMINAR BOUNDARY LAYER: SOME VARIABLE-PROPERTY SOLUTIONS

In Fig. 21-1 there are also plotted some calculated results by Baron[2] for which helium and carbon dioxide, respectively, are injected from the surface into a laminar boundary layer with air as the free-stream fluid. The actual properties of the binary mixtures He–air and CO_2–air are employed, and the concentration of the helium or the carbon dioxide varies from zero in the free stream to a value at the 0 state that is a function of the mass-transfer rate. [In fact, $m_0 = B/(1 + B)$.] In either case mass concentration is the conserved property $\mathscr{P}$, and $\mu/\lambda = \mu/\gamma =$ Sc. At very small values of B the concentration of the He or CO_2 is small throughout the boundary layer, the constant-property situation is approached, and g^* can be taken as the result corresponding to the constant-property solution for the value of Sc for a very dilute mixture. At high values of B the Schmidt number varies markedly through the boundary layer, especially in the case of He–air; the results of the property variation are seen very graphically in Fig. 21-1.

For engineering calculations it would be very convenient if constant-property solutions could be applied to variable-property applications by making some empirical correction to the former in much the same manner as was suggested in Chap. 15. Here, however, we are discussing variable properties caused by variation of mixture ratio through the boundary layer, whereas in Chap. 15 the variable properties were caused by temperature variations.

As was the case for temperature-dependent property effects, two empirical schemes appear possible. A *reference-property* scheme can be used whereby the appropriate constant-property solution is employed, and the fluid properties are all evaluated at some defined reference composition. Or, alternatively, the constant-property solution can be employed at the free-stream state, and then the result is corrected by multiplying by a function of the ratio of some appropriate property evaluated at the 0 state and the ∞ state; this is the *property ratio* scheme.

Knuth[3] has investigated the use of reference compositions for binary gas mixtures, employing, as a basis, an analysis of variable-property mass transfer to a Couette flow. Knuth finds that the few available variable-property laminar boundary-layer solutions can be satisfactorily correlated with the constant-property solutions if the fluid properties are all evaluated at a reference composition defined as follows:

Let component a be injected into a boundary layer in which the fluid is a mixture of component a and component b. Then the desired *mean* composition for correlation, expressed in terms of component b, is m_b^* and is given by

$$m_b^* = \frac{\mathfrak{M}_b}{\mathfrak{M}_b - \mathfrak{M}_a} \frac{\ln(\mathfrak{M}_\infty/\mathfrak{M}_0)}{\ln[m_{b,\infty}\mathfrak{M}_\infty/(m_{b,0}\mathfrak{M}_0)]} \tag{21-6}$$

where $\mathfrak{M}_a$ and $\mathfrak{M}_b$ are the molecular weights of components a and b, respectively, and $\mathfrak{M}_\infty$ and $\mathfrak{M}_0$ are the molecular weights of the mixture at the free-stream and 0 states, respectively. The molecular weight of a binary mixture of a and b can be expressed as

$$\mathfrak{M} = \frac{\mathfrak{M}_a\mathfrak{M}_b}{\mathfrak{M}_b + (\mathfrak{M}_a - \mathfrak{M}_b)m_b}$$

It is evident that molecular weight is the dominant variable here, and an alternative scheme that empirically correlates the data in Fig. 21-1 equally satisfactorily is to replace the ordinate of Fig. 21-1 (and also presumably Fig. 21-2) by

$$\frac{g}{g^*}\left(\frac{\mathfrak{M}_\infty}{\mathfrak{M}_0}\right)^{2/3}$$

In this scheme all properties are otherwise introduced at the free-stream, or ∞, state. This is, of course, a *property ratio* scheme.

Note that for a gas the *density* of the mixture varies directly with the molecular weight. In a binary gas mixture the diffusion coefficient D_j is virtually independent of mixture ratio, so that $\gamma_j = \rho D_j$ varies directly with density and thus directly with mixture molecular weight. The viscosity also varies with mixture ratio, but not nearly so much as ρ and γ_j. Thus the density variation is primarily responsible for the variable-property influence, and this is why correlation can be approximately obtained by use of a molecular-weight ratio.

THE LAMINAR CONSTANT-PROPERTY BOUNDARY LAYER WITH ARBITRARILY VARYING FREE-STREAM VELOCITY

In Chap. 10 an approximate scheme was described for extending the exact wedge-flow solutions to flow over a body of revolution with any arbitrarily varying free-stream velocity. It was shown in Fig. 10-6 that the necessary functions could be approximated by a linear relation, which then yields a particularly simple computing equation for the local Stanton number. Precisely the same technique can be used to extend the exact similarity solutions for blowing and suction to the variable-free-stream-velocity problem. In terms of the mass-transfer variables, the computing equation becomes [note the same form as Eq. (10-52)]

$$\frac{g}{G_\infty} = \frac{C_1 \mu^{1/2} R G_\infty^{c_2}}{\left(\int_0^x G_\infty^{C_3} R^2 \, dx \right)^{1/2}} \tag{21-7}$$

The constants C_1, C_2, and C_3, as computed by Spalding and Chi,[4] are functions of μ/λ and B and are given in Table 21-2.

Equation (21-7) and Table 21-2 then comprise a laminar flow constant-property solution to the $\mathscr{P}$ equation. Actually everything in

TABLE 21-2
Constants for Eq. (21-7) for various values of μ/λ for a laminar constant-property boundary layer

μ/λ	B	C_1	C_2	C_3
0.7	-0.9	1.85	0.05	1.1
	-0.6	0.812	0.15	1.3
	0.0	0.418	0.435	1.87
	1.0	0.244	0.65	2.3
	3.0	0.136	1.15	3.3
	9.0	0.060	1.90	4.8
1.0	-0.9	1.43	0.15	1.3
	-0.6	0.633	0.25	1.5
	0.0	0.332	0.475	1.95
	1.0	0.200	0.65	2.3
	3.0	0.113	1.00	3.0
	9.0	0.052	1.45	3.9
5.0	-0.9	0.431	0.45	1.9
	-0.6	0.205	0.50	2.0
	0.0	0.117	0.595	2.19
	1.0	0.073	0.65	2.3
	3.0	0.045	0.75	2.5
	9.0	0.023	0.90	2.8
>5.0	-0.9	$1.037(\mu/\lambda)^{-2/3}$	0.90	2.8
	-0.6	$0.568(\mu/\lambda)^{-2/3}$	0.90	2.8
	0.0	$0.339(\mu/\lambda)^{-2/3}$	0.90	2.8
	1.0	$0.230(\mu/\lambda)^{-2/3}$	0.90	2.8
	3.0	$0.145(\mu/\lambda)^{-2/3}$	0.90	2.8
	9.0	$0.077(\mu/\lambda)^{-2/3}$	0.90	2.8

Table 21-1 and Figs. 21-1 and 21-2, as well as in the earlier Tables 10-3 to 10-6 is incorporated in this result. These results are exact for the family of wedge-flow solutions and are presented as a useful approximation for the arbitrarily varying free-stream velocity problem.

THE TURBULENT CONSTANT-PROPERTY BOUNDARY LAYER WITH CONSTANT FREE-STREAM VELOCITY

The process whereby momentum and thermal energy are diffused in a direction transverse to the main-stream flow by the motion of turbulent eddies has been discussed in Chaps. 11 and 13. If there exists a *concentration gradient*, the same mechanism will cause the diffusion of a particular fluid species. An analysis exactly paralleling that for the Reynolds analogy in Chap. 13 leads to the definition of an *eddy diffusivity* for *mass transfer* and to the equality of the diffusivities for mass and momentum transfer. A *turbulent Schmidt number* can be defined, analogous to the turbulent Prandtl number, and all the arguments concerning the behavior of the latter seem equally applicable. Experimental information on the turbulent Schmidt number is very scarce, but what little there is seems to suggest a considerable similarity to the turbulent Prandtl number. For present purposes we assume complete similarity, and this allows us to make use of the turbulent boundary-layer heat-transfer solutions of Chap. 13 with only a change of names of the variables and parameters.

If the free-stream velocity is constant and the mass-transfer rate is small (small B) then $g \to g^*$, and Eq. (13-18) should be directly applicable for a *gas* mixture—at least for a Schmidt number near unity. That is,

$$\frac{g^*}{G_\infty} \left(\frac{\mu}{\lambda}\right)^{0.4} = 0.0287 \, \mathrm{Re}_x^{-0.2} \tag{21-8}$$

A larger rate of mass transfer to or from the surface leads to the *transpired* turbulent boundary-layer problem discussed in Chaps. 11 and 13. If all fluid properties are assumed constant, it makes no difference if the fluid crossing the surface is chemically the same or different from the free-stream fluid. Thus Eqs. (13-37) and (13-38) should be applicable. The blowing parameter B_h is, of course, the same as B when g/G_∞ is substituted for St. Thus

$$\frac{g}{g^*} = \frac{\ln(1+B)}{B} \tag{21-9}$$

and

$$\frac{g}{G_\infty}\left(\frac{\mu}{\lambda}\right)^{0.4} = 0.0287 \, \text{Re}_x^{-0.2} \frac{\ln\,(1+B)}{B} \tag{21-10}$$

THE TURBULENT BOUNDARY LAYER: SOME VARIABLE-PROPERTY SOLUTIONS

The effects of property variations that result from large concentration differences through the boundary layer can be readily investigated if the $\mathscr{P}$ equation is solved by finite-difference procedures: but, of course, the possible variations are enormous and little attempt has been made to generalize results. Before modern procedures were available, Rubesin and Pappas[6]† presented two variable-property solutions for a constant free-stream velocity, employing a Couette flow approximation. In one, helium is injected into an air boundary layer; in the other, hydrogen is injected. The conserved property is the concentration of helium, or hydrogen, in a binary mixture with air. In both cases $m_T = 1$.

The effect of the property variations proves to be very substantial, and serious error is introduced if it is ignored. However, a fairly satisfactory correlation is obtained if the left-hand side of Eq. (21-9) is replaced by

$$\frac{g}{g^*}\left(\frac{\mathfrak{M}_\infty}{\mathfrak{M}_0}\right)^{0.4}$$

and all properties are otherwise evaluated at the free-stream state.

Knuth and Dershin[7] have investigated the use of *reference compositions* to correlate variable-property turbulent boundary-layer solutions for binary gas mixtures with the constant-property solutions. They conclude that the same reference composition can be used as was suggested for the laminar boundary layer, that is, Eq. (21-6).

MASS-TRANSFER COEFFICIENTS FROM HEAT-TRANSFER DATA

On several occasions we have already taken advantage of the analogous form of the boundary-layer differential equations and boundary condi-

---

† Reference 6 is primarily concerned with heat transfer and only secondarily with the mass-transfer solution. However, the reported mass concentrations constitute a solution to the $\mathscr{P}$ equation, whereas the heat-transfer results do not, because the energy equation does not reduce to the $\mathscr{P}$ equation under the conditions of the analysis. This case is considered in Chap. 22.

tions to deduce solutions to the $\mathscr{P}$ equation from heat-transfer solutions. It should now be apparent that we can also make use of *experimental* heat-transfer data to deduce mass-transfer coefficients in exactly the same manner.

If we have heat-transfer data only for the case of no blowing or suction at the surface (and this is usually the case) then, of course, the analogy provides us only with g^*. Suppose we denote the heat-transfer Stanton number for no blowing by St^*. Experimental heat-transfer data are frequently presented in the following form, where Re is based on any convenient length parameter:

$$\text{St}^* \text{Pr}^a = C \, \text{Re}^b \tag{21-11}$$

Note that $\text{Nu} = \text{St Pr Re}$, and the Nusselt number can be used just as well as a basis for the analogy as Stanton number. From this, by analogy,

$$\frac{g^*}{G_\infty} \left(\frac{\mu}{\lambda}\right)^a = C \, \text{Re}^b \tag{21-12}$$

Thus if we know a, b, and c from heat-transfer experiments, we can evaluate g^* provided that Re and μ/λ in the proposed mass-transfer application are in the same range as Re and Pr in the heat-transfer experiments from which the data were taken.

As an approximation for large mass-transfer rates, we can combine Eq. (21-12) with Eq. (21-9) to yield

$$\frac{g}{G_\infty} \left(\frac{\mu}{\lambda}\right)^a \frac{B}{\ln(1+B)} = C \, \text{Re}^b \tag{21-13}$$

If the value of a is taken as $\frac{2}{3}$, the result is essentially the Chilton–Colburn analogy. This is a fair approximation over a very wide range of Pr or μ/λ for laminar or turbulent boundary layers, turbulent flow in pipes, flow normal to banks of tubes, and so on.

If an experimental heat-transfer coefficient has been measured and it is desired to determine the mass-transfer coefficient that would have resulted under the same flow conditions, on the same system, with the same free-stream fluid, Eqs. (21-11) and (21-12) can be combined to yield

$$g^* = \frac{h}{c} \left(\frac{\text{Pr}}{\mu/\lambda}\right)^a \tag{21-14}$$

This result can obviously be combined with Eq. (21-9) to yield a value for g for high mass-transfer rates.

It is this ability to determine mass-transfer data from heat-transfer experiments that so greatly extends the present, rather meager, supply of experimental mass-transfer data.

PROBLEMS

21-1. Consider a laminar Couette flow with mass transfer through the 0 surface. Let the transferred substance be CO_2 alone $(m_T = 1)$, and let the considered phase be a binary mixture of CO_2 and air, all at 1 atm pressure and 16°C. Let the thickness of the layer be such that the concentration of CO_2 at the ∞ state is always effectively zero. Using actual properties that vary with concentration, evaluate g/g^* as a function of B. Compare with the results given in Fig. 21-1. (Alternatively, let the transferred substance be He or H_2).

21-2. Consider again Prob. 21-1, but let $m_T \ll 1$ so that γ can be treated as constant. Let the remainder of the transferred substance be air. Compare the results of both problems. Can you suggest an empirical correlation of the variable-property results with the constant-property results?

21-3. Consider mass transfer from a sphere (it could be a liquid droplet of fuel). Let there be no effects of transverse flow; that is, the mass flows out symmetrically in all directions. Solve the $\mathscr{P}$ equation under the assumption of *constant* λ to obtain an expression for g. What is the result for g/g^*?

21-4. Let the transferred substance of Prob. 21-3 be benzene and the surrounding fluid be air. Using the concentration of the benzene as the conserved property, calculate g/g^* as a function of B using the actual variable properties, assuming that the system is isothermal. Let the temperature be 150°C. (Note that the Schmidt number for the dilute mixture is relatively independent of temperature and pressure, that D_j is relatively independent of composition, and that γ_j is relatively independent of pressure. On this basis, γ_j can be estimated as a function of composition.) Compare the results with those of Prob. 21-3.

21-5. Prepare a plot of g/g^* as a function of B for a laminar constant-property boundary layer at an axisymmetric stagnation point, and at a two-dimensional stagnation point, for $Sc = 0.5$. Compare with the similar results for the flat plate, $m = 0$.

21-6. It is required to estimate the average mass-transfer rate per unit area of benzene evaporating from the outer surface of a circular cylinder across which air is flowing at 6 m/s. The driving force B has been evaluated to be 0.90. It is found that if a cylinder of the same diameter is placed in the same airstream, the heat-transfer coefficient is $h = 85 \ W/(m^2 \cdot K)$. Estimate the mass-transfer rate explaining in detail the reasons for any assumptions that you build into the analysis. Evaluate the concentration of benzene in the 0 state, assuming no chemical reaction.

REFERENCES

1. Spalding, D. B., and H. L. Evans: *Int. J. Heat Mass Transfer,* vol. 2, pp. 314–341, 1960.
2. Baron, J. R.: Technical Report 160, Naval Supersonic Laboratory, MIT, Cambridge, Mass., 1956.
3. Knuth, E. L.: *Int. J. Heat Mass Transfer,* vol. 6, pp. 1–22, 1963.

4. Spalding, D. B., and S. W. Chi: *Int. J. Heat Mass Transfer,* vol. 6, May 1963. pp. 363–385.
5. Rubesin, M. W.: NACA (now NASA) TN 3341, December 1954; NACA (now NASA) RM A55 L13, Washington, February 1956.
6. Rubesin, M. W., and C. C. Pappas: NACA (now NASA) TN 4149, Washington, February 1958.
7. Knuth, E. L., and H. Dershin: *Int. J. Heat Mass Transfer,* vol. 6, pp. 999–1018, 1963.

CHAPTER

22

MASS TRANSFER: SOME EXAMPLES OF EVALUATION OF THE DRIVING FORCE

We have defined a driving force for mass transfer B by Eq. (20-34) and have seen that the conserved property in this definition can be any one of a number of different system properties, depending on the particular simplifications employed to obtain the $\mathcal{P}$ equation. The mere definition of B does not indicate the several ways that this parameter can be employed. In some problems B is found to be sufficiently small that g^* may be used, while in others B is large enough to require more elaborate treatment. The objective of this chapter is to examine several applications that are illustrative of this and other points.

THERMODYNAMICS OF THE AIR–WATER-VAPOR SYSTEM

The thermodynamics of the air–water-vapor mixture system are simple and provide a particularly convenient vehicle for illustrating mass-transfer

517

problems where the L state is a liquid phase while the 0 and ∞ states are gas phases. We first consider some examples involving this system, then some examples involving chemical reaction, and finally some special cases.

An adequate description of the thermodynamic state of this system can be obtained by assuming perfect-gas behavior and using only the Gibbs–Dalton law for gas mixtures, a table of vapor pressure of H_2O as a function of temperature, the latent heat of vaporization of H_2O at some reference temperature, and specific heat data for the liquid H_2O, H_2O vapor, and air.

The following mixture relations are based on the definitions of the quantities involved.

The mole fraction of component a in the mixture is

$$X_a = \frac{m_a/\mathfrak{M}_a}{m_a/\mathfrak{M}_a + m_b/\mathfrak{M}_b + \ldots} = \frac{m_a\mathfrak{M}}{\mathfrak{M}_a} \tag{22-1}$$

The equivalent molecular weight of the mixture is

$$\mathfrak{M} = X_a\mathfrak{M}_a + X_b\mathfrak{M}_b + \ldots = \frac{1}{m_a/\mathfrak{M}_a + m_b/\mathfrak{M}_b + \ldots} \tag{22-2}$$

For a perfect-gas mixture the following relations exist among the mole fraction, the volume composition, and the partial pressure:

$$X_a = \frac{V_a}{V} = \frac{P_a}{P} \tag{22-3}$$

Note that V_a is the volume that substance a alone would occupy if at the temperature and pressure of the mixture.

For an air–water-vapor mixture these equations combine to give the following relation for the mass concentration of water vapor in the mixture:

$$m_{H_2O} = \frac{P_{H_2O}}{1.61P - 0.61P_{H_2O}} \tag{22-4}$$

where P is the total pressure of the mixture and P_{H_2O} is the partial pressure of the water vapor. If in a boundary layer there is an interface between liquid H_2O and an air–water-vapor mixture, and if thermodynamic equilibrium is assumed to exist at the interface, then P_{H_2O} in Eq. (22-4) is the saturated vapor pressure of H_2O at the interface temperature $t_0 = t_L$.

In psychrometric work it is frequently preferred to work with an "absolute humidity" ω, kg H_2O/kg dry air. Obviously ω is related to

m_{H_2O} by

$$m_{H_2O} = 1 - \frac{1}{1 + \omega} \tag{22-5}$$

Another frequently used parameter in psychrometry is the "relative humidity" ϕ:

$$\phi = \frac{P_{H_2O}}{P_{H_2O,\,sat}} \tag{22-6}$$

where $P_{H_2O,\,sat}$ is the vapor pressure of H_2O at the mixture temperature.

There are a number of different ways in which we could express the enthalpy of an air–water-vapor mixture, depending on the datum states chosen. The following scheme is adequate for present purposes, but otherwise has no advantages over any other scheme.

Air Express t in °C, and let 0°C be the datum. Then

$$i_a = c_a t = 1.005t, \quad kJ/kg$$

H_2O Let the datum be *liquid* H_2O at 0°C. Then for H_2O vapor

$$i_{H_2O} = i_{fg,0} + c_{H_2O} t$$
$$= 2503 + 1.88t, \quad kJ/kg$$

Thus for the air–water-vapor mixture

$$i = \sum_j m_j i_j = m_a i_a + m_{H_2O} i_{H_2O} = (1 - m_{H_2O}) i_a + m_{H_2O} i_{H_2O}$$

Combining,

$$i = (1.005 + 0.88m_{H_2O})t + 2503m_{H_2O}, \quad kJ/kg \text{ mixture} \tag{22-7}$$

Also, for the pure *liquid water phase*

$$i_L = c_L t = 4.2t, \quad kJ/kg \tag{22-8}$$

ANALYSIS OF THE WET-BULB PSYCHROMETER

Consider a thermometer, the bulb of which is surrounded by a cloth wick that has been dipped in water. Then allow moist air, the humidity of which is to be determined, to flow over the wet bulb (Fig. 22-1). If the system is allowed to come to equilibrium and if the wick has not yet dried out, the thermometer indicates a temperature t_{wb} lower than the free-stream, or dry-bulb, temperature t_{db}, which can be measured independently. At equilibrium, $\dot{q}_L'' = 0$, and in terms of our formulation of the mass-transfer problem $t_{wb} = t_0 = t_L$ and $t_{db} = t_\infty$.

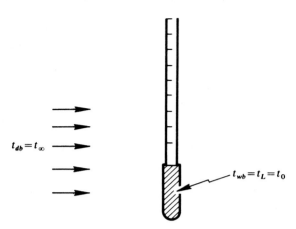

FIGURE 22-1
The wet-bulb psychrometer.

The concentration of H_2O in the mixture is a conserved property, since this problem involves the mass diffusion of an inert substance (H_2O). Equations (20-35) and (20-34) yield

$$\dot{m}'' = gB$$

where

$$B = \frac{m_{H_2O,\infty} - m_{H_2O,0}}{m_{H_2O,0} - m_{H_2O,T}}$$

We know that $m_{H_2O,T} = 1$, because only one substance is transferred. Thus

$$B = \frac{m_{H_2O,\infty} - m_{H_2O,0}}{m_{H_2O,0} - 1} \tag{22-9}$$

Let us now examine the energy equation. The air–water-vapor mixture system is close to a unit-Lewis-number system (see App. A). Thus the simplified form of the energy equation is a reasonable approximation, and enthalpy is also a conserved property.† Furthermore, if $Le = 1$ then $\Gamma = \gamma_j$, and it follows that the same value of g applies to both the energy- and mass-diffusion problems. The Bs are thus the same, because $\dot{m}''$ is common to both equations. We thus can write, from the

---

† It is shown later that, because of the very small concentrations of water vapor involved in this particular example, and the small mass-transfer rates, the unit-Lewis-number assumption is more restrictive than need be. However, at high mass-transfer rates, unit-Lewis-number provides a very useful simplification.

energy equation,

$$B = \frac{i_\infty - i_0}{i_0 - i_T}$$

From the T-state definition, Eq. (20-3),

$$i_T = i_L - \frac{\dot{q}_L''}{\dot{m}''}$$

and it is clear that $\dot{q}_L'' = 0$ at equilibrium. Then

$$B = \frac{i_\infty - i_0}{i_0 - i_L} \tag{22-10}$$

Equating Eqs. (22-9) and (22-10),

$$\frac{m_{H_2O,\infty} - m_{H_2O,0}}{m_{H_2O,0} - 1} = \frac{i_\infty - i_0}{i_0 - i_L}$$

where

$$i_\infty = (1.005 + 0.88 m_{H_2O,\infty}) t_{db} + 2503 m_{H_2O,\infty}$$
$$i_0 = (1.005 + 0.88 m_{H_2O,0}) t_{wb} + 2503 m_{H_2O,0}$$
$$i_L = 4.2t$$

If we know t_{wb}, we can determine $m_{H_2O,0}$ by assuming thermo-dynamic equilibrium at the interface, since $P_{H_2O,0}$ is then the saturation pressure of H_2O at t_{wb}; and we can evaluate $m_{H_2O,0}$ from Eq. (22-4) (see App. A). If we then measure t_{wb}, the only remaining unknown in the equation is $m_{H_2O,\infty}$; thus the wet-bulb psychrometer can be used to determine $m_{H_2O,\infty}$.

This is obviously a tedious calculation, so graphical procedures, for example the psychrometric chart, are frequently used.

Note that in this particular example it was never necessary to evaluate g. However, we would need to evaluate g if we wanted to evaluate the *rate* of mass transfer from the wick.

To appreciate the order of magnitude of the numbers involved, consider the case of $t_{wb} = 16°C$, $t_{db} = 27°C$, and $P = 1$ atm. Then, from App. A, $m_{H_2O,0} = 0.011$, and we would calculate

$$m_{H_2O,\infty} = 0.0065$$

Thus

$$B = \frac{0.0065 - 0.011}{0.011 - 1} = 0.00455$$

Reference to our boundary-layer solutions in Chap. 21 indicates that this is a very small value of B, and thus for this type of problem $g \approx g^*$.

The very small H_2O concentrations characteristic of this example suggest an alternative approach that is a little more accurate. Although the unit-Lewis-number simplification of the energy equation is a fair approximation for the air–water-vapor system, it is an approximation that need not be made in this case. If we reexamine the general form of the boundary condition for the energy equation (20-26), we note that the second term in the numerator is simply the convection heat-transfer rate $\dot{q}_0''$. If the concentration of H_2O is everywhere small, $\dot{q}_0''$ is virtually unaffected by the presence of the H_2O vapor and can be calculated as if there were no mass transfer. The first term in the numerator can be obtained from the solution to the mass-diffusion equation. If we now let i_{H_2O} be zero at the 0 state, so that $i_L = i_{fg,0}$, and express i_T in terms of i_L and $\dot{q}_L''/\dot{m}''$, we can readily obtain the following expression for $\dot{q}_L''$:

$$\dot{q}_L'' = h(t_\infty - t_0) - \dot{m}'' i_{fg,0} \tag{22-11}$$

where h is the conductance for heat transfer in the absence of mass transfer and $\dot{m}''$ is obtained from Eq. (22-9), i.e., $\dot{m}'' = gB$.

Note further that in this particular example $m_{H_2O,0} \ll m_{H_2O,T}$. (The latter of course is 1 because there is only one transferred substance.) In this case $m_{H_2O,0}$ in the denominator of Eq. (22-9) could be neglected, and the driving force B becomes simply the difference between the concentration of water vapor at the interface and in the free stream, i.e.,

$$B = m_{H_2O,0} - m_{H_2O,\infty}$$

Thus the driving force for mass transfer becomes analogous to the driving force for heat transfer, namely the temperature difference between the surface and the free stream. Although this is often the case, it is important to note that this represents only a very special case, as subsequent examples will demonstrate.

In this particular psychrometer example the combination of these equations now results in the expression

$$m_{H_2O,\infty} = m_{H_2O,0} - \frac{h}{g i_{fg,0}}(t_{db} - t_{wb})$$

An iterative solution is still required, but h and g can be evaluated independently without assuming Le = 1.

MODIFICATION OF THE DEFINITION OF B TO ACCOUNT FOR THERMAL RADIATION

In the preceding example we ignored the possible influence of thermal radiation. If g is large, radiation could be negligible, but there are obviously many applications where it is not negligible.

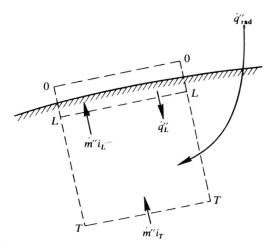

FIGURE 22-2
Modification of the energy balance at a phase interface to account for thermal radiation.

If it is assumed that radiation is absorbed and emitted at least an infinitesimal distance inside the L surface and that the considered phase fluid is transparent then the boundary layer is totally unaffected by radiation. However, the energy balance on the $L-T$ control volume (Fig. 20-3) that leads to the definition of B must be modified.

Consider the control volume shown in Fig. 22-2, in which a net radiant heat flux is included. Without radiation, the original definition of B from the simplified energy equation is

$$B = \frac{i_\infty - i_0}{i_0 - i_T}$$

If we retain the original definition of i_T, Eq. (20-3), we obtain

$$B = \frac{i_\infty - i_0}{i_0 - [i_L - (\dot{q}''_L/\dot{m}'')]}$$

But this definition of B is still valid *with* radiation, if the radiant flux is assumed to act below the L surface. Then, if we apply the conservation-of-energy principle to the $L-T$ control volume,

$$\dot{m}'' i_L - \dot{m}'' i_T - \dot{q}''_L - \dot{q}''_{\text{rad}} = 0$$

$$i_L - \frac{\dot{q}''_L}{\dot{m}''} = i_T + \frac{\dot{q}''_{\text{rad}}}{\dot{m}''}$$

Thus a modified form of the definition of B for applications of the energy equation involving thermal radiation is

$$B = \frac{i_\infty - i_0}{i_0 - i_T - \dot{q}''_{\text{rad}}/\dot{m}''} \tag{22-12}$$

In applications where it is more convenient to work with i_L than i_T the original definition is, of course, still valid, and radiation is handled as a separate effect.

DRYING

Suppose we have a wet surface and want to dry it by blowing air over it while at the same time supplying heat from its rear. (Alternatively, the rear could be insulated and heat could be supplied from the air side by thermal radiation, as in the previous section.)

Let $t_\infty = 27°C$ and $m_{H_2O,\infty} = 0.02$. Suppose we supply heat sufficient to hold the surface at 82°C, that is, $t_0 = t_L = 82°C$. Let $P = 1$ atm.

What is the drying rate $\dot{m}''$ during the period the surface is completely wet? What is the heat-transfer rate $\dot{q}''_L$ (or $\dot{q}''_{rad}$) necessary to hold the surface temperature at 82°C?

From App. A, $m_{H_2O,0} = 0.393$. For an inert substance m_{H_2O} is a conserved property and

$$B = \frac{m_{H_2O,\infty} - m_{H_2O,0}}{m_{H_2O,0} - m_{H_2O,T}} = \frac{0.02 - 0.393}{0.393 - 1} = 0.615$$

Thus

$$\dot{m}'' = gB$$

Note that in this example the H_2O concentration is no longer small and B is sufficiently large that g/g^* is no longer near unity.

To evaluate the necessary heat-transfer rate, we turn to the energy equation and the unit-Lewis-number simplification. Then enthalpy is a conserved property and

$$B = \frac{i_\infty - i_0}{i_0 - i_T} = \frac{i_\infty - i_0}{i_0 - i_L + \dot{q}''_L/\dot{m}''} = 0.615$$

Solving for $\dot{q}''_L$,

$$\dot{q}''_L = \dot{m}''(1.63i_\infty - 2.63i_0 + i_L)$$

where the enthalpies can be evaluated from Eqs. (22-7) and (22-8).

EVAPORATIVE COOLING

Suppose we have a porous surface exposed to a hot dry gas and we would like to keep the surface cool by forcing water through at a sufficient rate to keep the surface wet. Consider the example shown in Fig. 22-3.

What is the necessary cooling water rate, and what will be the equilibrium surface temperature $t_0 = t_L$?

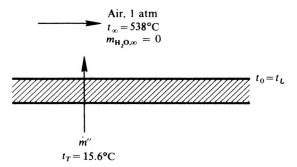

FIGURE 22-3
Example of evaporative cooling.

For an inert substance

$$B = \frac{m_\infty - m_0}{m_0 - m_T} = \frac{-m_0}{m_0 - 1}$$

For Lewis number equal to 1

$$B = \frac{i_\infty - i_0}{i_0 - i_T}$$

In this case we are not interested in $\dot{q}''_L$, but we do know the T-state temperature, 15.6°C. Equating the two expressions for B and substituting Eqs. (22-7) and (22-8) for the enthalpies, we obtain an equation with $m_{H_2O,0}$ and t_0 as unknowns. But these are related by the equation of state and can be separately evaluated. The solution is an iterative one, yielding

$$t_0 = 64.4°C, \quad m_{H_2O,0} = 0.16$$

Finally,

$$B = \frac{-0.16}{0.16 - 1} = 0.190$$

and

$$\dot{m}'' = gB = 0.190g$$

BURNING OF A VOLATILE FUEL IN AIR

Let us consider the rate of burning of a droplet of ethyl alcohol (C_2H_5OH) in air. Alternatively, this could be the burning of an alcohol-wetted wick in air. In either case we are concerned with the rate $\dot{m}''$ at which alcohol leaves the surface.

As a first approximation, let us assume that the rate of chemical reaction is so much greater than the rate at which components are diffused through the boundary layer that the reaction can be assumed to be instantaneous, and the burning rate is then *diffusion-controlled*. This assumption allows us to specify rather simple boundary conditions, since there can then be no oxygen at the 0 state and no fuel at the ∞ state.

Next we assume that the "simple chemical reaction" is a valid approximation, that all specific heats are equal and $\gamma_{O_2} = \Gamma$, and that for the mass-diffusion equation $\gamma_{O_2} = \gamma_{C_2H_5OH}$. These are admittedly rather gross approximations, but the alternative is complete solution of the basic differential equations of the boundary layer. Our objective here is an approximate solution, obtained by reducing the problem to a solution of the $\mathscr{P}$ equation.

Complete solution of this problem requires both the energy equation and the mass-diffusion equation, but we find it most convenient to use the former. To a close approximation, the energy equation alone gives us the driving force B.

The various constants for the problem are as follows:

Fuel: C_2H_5OH

$r = 2.09$, for reaction to CO_2 and H_2O

$m_{O_2,\infty} = 0.232$, atmospheric air

$t_\infty = 16°C$

$H_o = 28,145 \text{ kJ/(kg fuel vapor)}$, heat of combustion of fuel at t_0

$i_{fg,0} = 854 \text{ kJ/(kg fuel)}$, latent heat of vaporization of fuel at t_0

$c = 1.005 \text{ kJ/(kg} \cdot \text{K)}$, specific heat of air

$P = 1 \text{ atm}$

As a first step, let us assume that the entire droplet (or wick) is uniformly at the same temperature $t_0 = t_L = t_T = 78°C$, the boiling point of C_2H_5OH at 1 atm pressure. The actual surface temperature is somewhat lower and must be established by iteration. However, we will see that we do not need to know this temperature very accurately to evaluate B.

For a simple chemical reaction we have previously seen that we can express the enthalpy of the system by Eq. (20-15):

$$i = m_{O_2}\frac{H_o}{r} + \int_{t_0}^{t} c\, dt \approx m_{O_2}\frac{H_o}{r} + c(t - t_0)$$

For convenience, let us set the 0 state as the datum for enthalpy.

Since C_2H_5OH is a volatile fuel, the reaction is a gas-phase reaction and takes place somewhere between the 0 state and the ∞ state. Under the assumption of instantaneous reaction, we can conclude that

$$m_{O_2,0} = 0$$

Then

$$i_0 = 0$$

$$i_\infty = 0.232 \frac{28,145}{2.09} + 1.005(15.6 - 78)$$

$$= 3124 - 62.7 = 3061$$

$$i_T = i_L \quad \text{(since } \dot{q}_L'' = 0 \text{ by the conditions of the problem)}$$

$$= -i_{fg,0} = -854$$

Then, from the energy equation,

$$B = \frac{i_\infty - i_0}{i_0 - i_T} = \frac{3061 - 0}{0 - (-854)} = 3.59$$

Finally, $\dot{m}'' = gB$, provided that we can obtain an appropriate value for g.

Strictly speaking, to properly evaluate the surface temperature, we need to know the mixture composition at the 0 state, so that the partial pressure of the fuel at the 0 state may be evaluated. To do this, we can turn to the mass-diffusion equation for a simple chemical reaction, but we can see from the above calculations that the result for B is not very sensitive to small changes in the value of t_0.

Note that B is very much greater than for the preceding inert-substance problems.

SIMPLE GRAPHITE BURNING IN AIR

Let us consider the rate of burning of carbon from a graphite surface in an atmospheric-air environment. The surface could be a lump of coke in a furnace, or it could be a graphite heat shield designed for protection against a high-temperature air environment. In either case it is assumed that the only chemical reaction taking place in the vicinity of the surface is

$$C + \tfrac{1}{2}O_2 = CO, \quad r = 1.333$$

The vapor pressure of carbon is so low at any reasonable surface temperature that we are forced to conclude that reaction must take place at the surface. Unlike with the volatile substance just considered, it appears that the surface temperature is going to be very high and will contribute materially to the enthalpy evaluation. In the absence of any initial basis for estimating this temperature, it appears that the mass-diffusion equation for a simple chemical reaction provides a more direct route to the evaluation of the driving force B. The expected high surface temperature also provides the basis for the assumption of a reaction to CO rather than CO_2. This obviously is related to the thermodynamic equilibrium state of the carbon–oxygen system at the temperature and pressure at the surface.

On the assumption of the above simple chemical reaction, the quantity $m_C - m_{O_2}/r$ is a conserved property, and thus from Eq. (20-34) a possible formulation for the driving force is

$$B = \frac{(m_C - m_{O_2}/r)_\infty - (m_C - m_{O_2}/r)_0}{(m_C - m_{O_2}/r)_0 - (m_C - m_{O_2}/r)_T}$$

Whether this expression will be adequate depends on whether values of $m_C - m_{O_2}/r$ can be found at the ∞, 0, and T states.

If carbon is the only transferred substance,

$$\left(m_C - \frac{m_{O_2}}{r}\right)_T = 1$$

At the ∞ state there is obviously no carbon and the concentration of oxygen in atmospheric air is 0.232. Thus

$$\left(m_C - \frac{m_{O_2}}{r}\right)_\infty = 0 - \frac{0.232}{1.333} = -0.174$$

The critical state is the 0 state. It has already been assumed that the surface temperature is such that the vapor pressure of carbon is negligible, and thus $m_{C,0} = 0$. The concentration of oxygen presents a further problem. If thermodynamic equilibrium obtains, we must conclude that at an elevated temperature uncombined oxygen cannot exist at the 0 surface, and thus

$$\left(m_C - \frac{m_{O_2}}{r}\right)_0 = 0$$

This result assumes, of course, that reaction rates are sufficiently high that thermodynamic equilibrium obtains, and it also assumes that no other reactions are involved. These are simply the approximations that we make in order to obtain a simple solution. Obviously a more precise result is obtained if the reaction rates of all possible reactions are considered.

On the basis of the various approximations introduced, we then conclude that

$$B = \frac{-0.174 - 0}{0 - 1} = 0.174$$

Thus

$$\dot{m}'' = gB = 0.174g$$

Note the quite small magnitude of B as compared with the volatile fuel.

GRAPHITE ABLATION WITH MORE THAN ONE REACTION

Let us consider a chemically reacting boundary layer where more than one reaction is of importance, so that the simple-chemical-reaction idealization is inadequate. In this case we find it necessary to consider the diffusion of each element. If we can accept the assumption that the diffusion coefficients for all components are the same then the applicable diffusion equation is Eq. (20-11), where the concentration of chemical element α is a conserved property.

Consider a combustion-products mixture containing, by mass, 40 percent CO_2, 20 percent H_2O, and the remainder inerts. Let this mixture flow over a graphite surface at 1650°C. The gas mixture could be the products from a rocket motor, and the graphite surface could be a nozzle. The problem is to compute the rate of erosion of the graphite surface.

Again, because of the low vapor pressure of carbon, any reaction between the gas and the graphite can be expected to take place at the surface. However, there is more than one possible reaction at the surface. Let us examine the equilibrium constants K_p, at 1650°C, for some of these reactions.

For a reaction

$$aA + bB = cC + dD$$

we have

$$K_p = \frac{P_C^c P_D^d}{P_A^a P_B^b}$$

where P is the partial pressure of the component of the mixture, expressed in atmospheres.

The following are possible reactions, with the corresponding equilibrium constants at 1650°C:

$$C(s) + CO_2 = 2CO, \qquad K_p = 4000$$
$$C(s) + H_2O = CO + H_2, \qquad K_p = 1234$$
$$C(s) + 2H_2 = CH_4, \qquad K_p = 1/790$$
$$CO_2 = CO + \tfrac{1}{2}O_2, \qquad K_p = 1/18{,}000$$
$$H_2O = H_2 + \tfrac{1}{2}O_2, \qquad K_p = 1/40{,}000$$
$$C(s) + \tfrac{1}{2}O_2 = CO, \qquad K_p \to \infty$$

A large value of K_p means that thermodynamic equilibrium lies overwhelmingly on the right-hand side of the equation, while a very small value means that equilibrium lies on the left-hand side. With these equations (and two of them are redundant), we could calculate the equilibrium composition at the 0 state, but merely by inspection it is apparent that at the 0 state the mixture can contain only CO and H_2, plus the inerts. On the other hand, at the ∞ state only CO_2 and H_2O are present. The T state is entirely carbon. The complete mixture under consideration then contains (excluding CH_4, which evidently cannot exist anywhere at this temperature)

$$CO_2, \quad H_2O, \quad CO, \quad H_2, \quad C(s)$$

Let us now express the concentration of the elements carbon, oxygen, and hydrogen at any point in the system, employing the definition of n_α from Chap. 4:

$$n_\alpha = \sum_j n_{\alpha,j} m_j$$

$$n_C = \tfrac{12}{44} m_{CO_2} + \tfrac{12}{28} m_{CO} + m_C$$

$$n_O = \tfrac{32}{44} m_{CO_2} + \tfrac{16}{18} m_{H_2O} + \tfrac{16}{28} m_{CO}$$

$$n_H = \tfrac{2}{18} m_{H_2O} + m_{H_2}$$

Any one of these element concentrations could be used for a conserved property in the $\mathscr{P}$ equation. However, we need a conserved property that can be evaluated from presently known data at the 0, ∞, and T states, and none of these qualifies. Specifically, we do not know the concentration of CO and H_2 at the 0 state. But it is recalled from Chap. 20 that any linear combination of conserved properties is again a conserved property, provided that diffusion coefficients are the same. This suggests that a combination of several different n_α may be found containing only quantities that are known at the 0, ∞, and T states. Both n_C and n_O contain the unknown m_{CO}, but the following combination does not:

$$n_C - \tfrac{3}{4} n_O = m_C - \tfrac{3}{11} m_{CO_2} - \tfrac{2}{3} m_{H_2O}$$

This quantity is a conserved property that can be evaluated at all the necessary states. Thus

$$\mathscr{P}_0 = (n_C - \tfrac{3}{4} n_O)_0 = 0$$

$$\mathscr{P}_\infty = 0 - \tfrac{3}{11}(0.40) - \tfrac{2}{3}(0.20) = -0.242$$

$$\mathscr{P}_T = 1$$

Then

$$B = \frac{\mathscr{P}_\infty - \mathscr{P}_0}{\mathscr{P}_0 - \mathscr{P}_T} = \frac{-0.242 - 0}{0 - 1} = 0.242$$

and

$$\dot{m}'' = gB = 0.242g$$

It should be evident that the simple chemical reaction is merely a special case of this more general procedure.

THE HIGH-TEMPERATURE BOUNDARY LAYER WITH DISSOCIATION

Consider a boundary layer in which there is chemical reaction, and thus concentration gradients and mass diffusion, but in which there is no mass transfer across the 0 and L surfaces. An example might be the nose cone surface of a body entering the Earth's atmosphere, but cooled so that there is no ablation. The air behind the bow shock wave is at such a high temperature that dissociation occurs. But if the body is cooled to a moderately low temperature, recombination tends to occur within the boundary layer, and the equilibrum state at the 0 surface tends toward the state of the original free-stream air (although the process can occur sufficiently fast that thermodynamic equilibrium is not established). If the unit-Lewis-number approximation, or any of the other energy-equation simplifications, is applicable, the boundary condition to the energy equation becomes that of the $\mathscr{P}$ equation:

$$\dot{m}'' = gB = g\frac{i_\infty - i_0}{i_0 - i_T} = g\frac{i_\infty - i_0}{i_0 - i_L + \dot{q}_L''/\dot{m}''}$$

Transposing,

$$\dot{m}''(i_0 - i_L) + \dot{q}_L'' = g(i_\infty - i_0)$$

But if $\dot{m}'' = 0$, the first term drops out, and $g = g^*$. Thus

$$\dot{q}_L'' = g^*(i_\infty - i_0) \qquad (22\text{-}13)$$

The surface heat flux to a chemically reacting boundary layer is then seen to be proportional to the *enthalpy difference* across the boundary layer.

Note, however, that for the nose of a high-velocity vehicle the free-stream enthalpy i_∞ must include the free-stream kinetic energy $\frac{1}{2}u_\infty^2$. This term may be so large that the other enthalpy terms may be neglected.

TRANSPIRATION COOLING BY GAS INJECTION†

An effective way to cool a surface over which a hot gas is flowing is to construct it from a porous material and force a cool gas through it from the rear.

The injected cooling gas can be the same as the hot mainstream fluid, in which case the boundary-layer problem is that of heat transfer with simple blowing; or it can be of a *different* composition, in which case

---

† See also Chap 13, where the transpired turbulent boundary layer is discussed.

we have combined heat and mass diffusion. The two types of problem are, however, closely related, since they both involve mass transfer at the surface.

If the injected gas is the same as the mainstream gas, or if it is different but one of the other simplifications to the energy equation is applicable, enthalpy is a conserved property and the complete solution is contained in the standard formulation

$$\dot{m}'' = g\,\frac{i_\infty - i_0}{i_0 - i_T}$$

where i is a function of temperature (and composition in the second case), and the problem is generally to evaluate $\dot{m}''$ to attain a given t_0 or vice versa.

It is sometimes preferred, however, to express the solution in terms of the *heat flux* $\dot{q}_0''$ rather than use the simplified mass-transfer formulation, because if the properties of the injected gas and the mainstream gas are such that none of the simplifications to the energy equation are approximated, enthalpy is no longer a conserved property of the second kind and the standard formulation is no longer valid. In fact, we are forced all the way back to Eq. (20-26) and Fig. 20-5 for a valid expression of the boundary condition at the surface.

In either case, it can be readily demonstrated that if there is no phase change between the L and 0 surfaces then $\dot{q}_L'' = \dot{q}_0''$. And, of course,

$$\dot{q}_0'' = -\left(\Gamma c\,\frac{\partial t}{\partial y}\right)_0$$

Consider first the case where the injected gas is the same as the mainstream gas. Then the standard formulation yields

$$\dot{m}'' = g\,\frac{i_\infty - i_0}{i_0 - i_L + \dot{q}_L''/\dot{m}''} = g\,\frac{i_\infty - i_0}{i_0 - i_L + \dot{q}_0''/\dot{m}''}$$

But in this case $i_0 = i_L$. Thus

$$\dot{q}_0'' = g(i_\infty - i_0)$$

If the specific heat may be treated as constant in the considered phase, we can then define a heat-transfer convection conductance in the conventional manner and evaluate a heat-transfer Stanton number:

$$\dot{q}_0'' = gc(t_\infty - t_0)$$
$$= h(t_\infty - t_0)$$
$$h = gc$$
$$\frac{h}{G_\infty c} = \text{St} = \frac{g}{G_\infty}$$

If we denote the Stanton number for $B \to 0$ by St* then

$$\frac{St}{St^*} = \frac{g}{g^*} = \phi(B, \text{ other parameters})$$

For the constant-property laminar boundary layer for $u_\infty = $ constant, this result is plotted in Fig. 22-4 for Pr = 1 and 0.7, and the curves are simply taken from Fig. 21-1. Similarly, a corresponding turbulent boundary-layer constant-property result is plotted in Fig. 22-5. Note, however, that g/G_∞ in the definition of B has been replaced by its equivalent for this case, St, and the resulting parameter has been denoted by B_h, the transpiration parameter introduced in Chap. 13.

A few solutions, based on the "exact" energy equation rather than the simplified energy equation, have been developed for transpiration

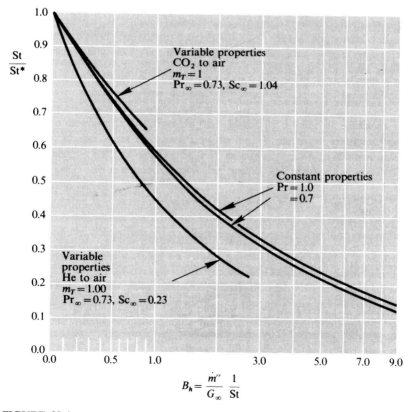

FIGURE 22-4
Heat transfer to a laminar boundary layer with transpiration cooling and constant free-stream velocity.

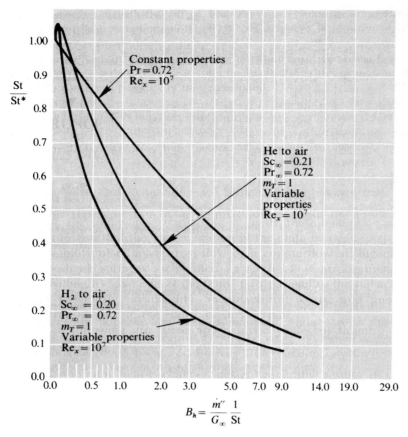

$$B_h = \frac{\dot{m}''}{G_\infty} \frac{1}{St}$$

FIGURE 22-5
Heat transfer to a turbulent boundary layer with transpiration cooling and constant free-stream velocity.

cooling in which some other gas is injected into an air boundary layer. Not only do these solutions involve simultaneous heat and mass diffusion through the boundary layer, but the transport properties of the mixture (including the Lewis number) vary markedly through the boundary layer, especially at high blowing rates where the concentration of the injected gas is high at the 0 state. (Note that as $B \to 0$, the Stanton number must approach its value for the simple constant-property air boundary layer, regardless of whether the injected fluid is the same as or different from the free-stream fluid.) The results of these variable-property solutions can be presented in graphical form on the same coordinate system as employed in Figs. 22-4 and 22-5. But note that B_h is now no longer related to conserved properties.

Results for helium injection and CO_2 injection, both into a laminar air boundary layer,[1] are plotted in Fig. 22-4. Similar results for helium and hydrogen injection into a turbulent air boundary layer[2] are plotted in Fig. 22-5.

The variable-property effects are seen to be quite substantial, especially in the cases of light-gas injection. A correlation of the variable-property results with the constant-property solutions, using the *reference property* concept, is presented by Knuth[3] and by Knuth and Dershin.[4] The latter authors employ a Couette flow model as a basis for determining a reference composition that correlates the variable-property results with constant-property solutions. Analysis of the Couette flow heat-transfer problem with foreign-gas injection indicates that the property most responsible for the large variable-property effects shown in Figs. 22-4 and 22-5 is the specific heat of the injected fluid, c_T (if different from that of the mainstream fluid, c_∞). This suggests that an alternative approximate correlation procedure might be based on a property ratio scheme, with c_T/c_∞ as the additional variable.

A fairly good correlation of the data in Fig. 22-4 is indeed obtained if all properties are based on the ∞ state and B_h is replaced by $B_h(c_T/c_\infty)^{1/3}$.

Similarly, the turbulent flow data in Fig. 22-5 can be approximately correlated if B_h is replaced by $B_h(c_T/c_\infty)^{0.6}$. This correlation does not work well at very low blowing rates, where the hydrogen and helium curves are seen to rise before they decrease. This effect is due to the fact that the very high thermal conductivity of the injected fluid tends to substantially increase the mixture thermal conductivity in the sublayers, and initially offsets the normal effect of blowing to decrease heat-transfer rate.

In applying these solutions, where the problem is to determine the surface temperature for a given mass-transfer rate, or the mass-transfer rate for a given surface temperature, an energy balance on the $L-T$ control volume yields the desired relation. Thus, referring to Fig. 20-3,

$$\dot{m}''i_L - \dot{m}''i_T = \dot{q}''_L = \dot{q}''_0 = h(t_\infty - t_0)$$

$$i_L - i_T = c_T(t_L - t_T) = c_T(t_0 - t_T)$$

$$\dot{m}'' = \frac{h(t_\infty - t_0)}{c_T(t_0 - t_T)}$$

Thermal radiation can, of course, be included in this energy balance as desired.

PROBLEMS†

22-1. Consider the burning of a droplet of ethyl alcohol (or a wet wick) in an atmosphere of pure oxygen. How does the burning rate compare with the same droplet in atmospheric air, assuming that the logarithmic equation for g/g^* is applicable, other things being equal?

22-2. Suppose you dip the wick of a wet-bulb thermometer in ethyl ether and expose it to atmospheric air at 20°C. What equilibrium temperature will the thermometer record under the unit-Lewis-number assumption?

22-3. Dry air at 315°C flows along a flat wet surface (water) at 9 m/s. The pressure is 1 atm. If the surface is insulated (except for exposure to the air) and radiation is neglected, what equilibrium temperature will the surface assume? If the boundary layer is turbulent, what is the mass-transfer rate at a point 1.5 m from the virtual origin of the boundary layer?

22-4. Consider flow of air normal to a cylinder as described in Prob. 21-6. Let the air be at 20°C, 1 atm pressure, and 60 percent relative humidity. Inside the cylinder there is water at 90°C. The surface is porous so that the water can be forced out at a rate sufficient to keep the outer surface moist. What is the average surface temperature and the average mass-transfer rate per unit of surface area?

22-5. A thin plate of solid salt (NaCl), 15 cm by 15 cm, is to be dragged through seawater (edgewise) at 20°C at a velocity of 5 m/s. Seawater has a salt concentration of about 3 percent by weight. Saturated NaCl–H_2O at 20°C has a concentration of 36 g NaCl/100 g H_2O. Estimate the total rate at which salt goes into solution.

22-6. Consider an axisymmetric stagnation point on a missile traveling through the atmosphere at 5500 m/s where the static air temperature is near zero degrees absolute. It is desired to maintain the surface at 1200°C by transpiration of hydrogen through the wall. The hydrogen is available from a reservoir at 38°C. If the "heat of combustion" of hydrogen is approximately 116,000 kJ/kg at 1200°C and the mean specific heats of hydrogen and air are, respectively, 15 kJ/(kg · K) and 1.1 kJ/(kg · K), evaluate the driving force B from the energy equation under the unit-Lewis-number assumption for the conditions of (1) reaction of H_2 with O_2, and (2) no reaction.

    Suggestion: Use the surface temperature as the temperature datum for enthalpy. Note that the free-stream enthalpy must include the very large contribution of the stagnation enthalpy.

22-7. In Prob. 16-9 the cooling of a gas turbine blade is considered. Considering

---

† Since many of the following problems are based on the air–water-vapor system, it is suggested that students first prepare a graph of m_{H_2O} versus T_{sat} from Table A-20, or else write a computer program based on Table A-20.

only the stagnation point at the leading edge of the blade, evaluate the surface temperature and the necessary water rate per square foot of stagnation area to cool by water injection through a porous surface. Assume that the water is available at 38°C and that the mass concentration of water vapor in the products of combustion is 0.01.

22-8. Continue Prob. 22-7 to include the entire blade surface, using what you feel to be the best available approximations for various parts of the surface.

22-9. A rocket nozzle is to be constructed with dimensions as shown in Fig. 11-18. The following table gives the composition of the exhaust products:

Compound	m_j, kg/(kg mixture)
CO_2	0.278
CO	0.279
H_2O	0.209
H_2	0.022
N_2	0.212

The stagnation pressure is 3400 kPa and the stagnation temperature is 3000°C. The nozzle is to be constructed of graphite. The objective of the problem is to estimate the rate of erosion (ablation) of the graphite at the throat, and particularly to estimate the time elapsed for a 10 percent increase in throat diameter.

22-10. Consider an axisymmetric blunt-nosed vehicle entering the Earth's atmosphere. At 60,000 m altitude the velocity is 6100 m/s. The radius of curvature of the nose is 1.8 m. Calculate the convective heat flux to the vehicle at the stagnation point. (Note that this calculation neglects radiation from the very high-temperature dissociated gases behind the shock wave.) If the surface is insulated, calculate the equilibrium surface temperature assuming that the surface radiates as a black body to the surroundings. Suppose the surface is a graphite heat shield; calculate the rate of ablation of the graphite in millimeters per minute.

22-11. Let a piece of ice uniformly at 0°C be immersed in water at 27°C. Evaluate the rate at which melting occurs in terms of the conductance g. How would you evaluate g?

22-12. Consider a 1.2 m square wet towel hanging from a clothes line. Let sunlight fall on one side at an angle of 45°. The normal solar flux is 946 W/m², and the absorptivity and emissivity of the towel are 1. The surrounding air is at 21°C, with a relative humidity of 65 percent. At the equilibrium temperature of the system it is determined that an equivalent system with no mass transfer would have a free-convection coefficient of 8.5 W/(m² · K). What is the drying rate during the period when the towel is sufficiently wet that the surface can be idealized as a liquid surface? (Note that, in addition to the high-frequency solar flux, the 21°C surroundings radiate to *both* sides of the towel, and the towel radiates to the surroundings. Assume that the towel is at a uniform temperature throughout its thickness.)

22-13. In a diesel engine the fuel is injected as small droplets, and, after an initial ignition delay, during which time part of the fuel vaporizes and then burns rather abruptly, the remainder of the fuel burns as fast as the fuel vaporizes from each individual droplet. The objective of this problem is to estimate the time required for complete consumption of a spherical droplet 5 μm in diameter.

During this period, let it be assumed that the combustion chamber pressure is 6200 kPa and the air temperature is 800°C. The fuel may be idealized as $C_{12}H_{26}$, with "heat of combustion" of 44,000 kJ/kg and latent heat of vaporization 358 kJ/kg. It may be assumed that these values are relatively independent of temperature. The boiling point of the fuel is approximately 427°C at this pressure.

22-14. Calculate the evaporation (sublimation) rate from a snow bank if the ambient air temperature is -23°C, solar energy is falling on the bank at a rate of 600 W/m^2, the underside of the bank is effectively insulated, and the heat-transfer coefficient for free convection from an equivalent surface with no mass transfer is 6 W/(m$^2 \cdot$ K). The absorptivity *and* emissivity of snow may be assumed to be 1. The absolute humidity of the air is 0.0003 kg H_2O/kg dry air.

22-15. At a particular region in a steam condenser the fluid is a mixture of 80 percent steam (by mass) and 20 percent air. The total pressure is 6.75 kPa, and the temperature is that of saturated steam at that pressure. What is the necessary condition at the surface of the tubes for condensation to take place? In simple condensation the resistance to heat transfer is usually considered to be entirely that of the liquid film that forms on the surface. How does the present problem differ? Suppose the liquid film has an average thickness of 0.1 mm; investigate the effect of the air on the overall resistance to heat transfer between the tube metal surfaces and the free-stream mixture.

22-16. A wet surface is to be dried by blowing dry air at 32°C over it, while at the same time electrically heating it from the rear with a fixed heat flux of 25,000 W/m^2. Investigate the variation of the drying rate $\dot{m}''$ as the conductance g is varied (presumably by varying the air velocity).

22-17. In Prob. 22-16 the heat is to be supplied from the rear by condensing steam. Let the steam be available at 110°C, and let it be assumed that the *overall conductance* for heat transfer from the steam to the L surface (including the condensing resistance, the wall resistance, and the resistance of the material being dried) is 280 W/(m$^2 \cdot$ K). Investigate the influence of the mass-transfer conductance g on the drying rate.

22-18. Consider a laundry convective clothes dryer for which there is available dry air at 20°C and 1 atm pressure. There is also available superheated steam at 1 atm pressure and 260°C. Considering only the "constant-drying-rate" period during which the cloth surface is totally wet, investigate whether it is possible to increase the drying rate by mixing some of the steam with the air. That is, how does drying rate vary with steam–air ratio?

22-19. In a part of a solar-operated seawater desalting plant, air saturated with water vapor at 1 atm and 55°C passes into a condenser where the fresh water is recovered. The condenser is built up of a bank of circular tubes with cooling seawater flowing inside the tubes and the saturated air flowing normally. The heat-transfer coefficient on the inside of the tubes is estimated to be 1400 W/(m² · K). The tube wall is 0.75 mm thick and has a thermal conductivity of 100 W/(m · K). On the outside of the tubes, it is estimated that if there were no mass transfer, the average heat-transfer coefficient would be 70 W/(m² · K). If the cooling-water temperature in the first row of tubes (the tubes over which the 55°C saturated air is flowing) is 46°C, calculate the rate of condensation per unit of surface area and the rate of heat transfer to the coolant. Neglect the resistance of the condensate film on the tubes. (Would it be significant if the film thickness was, say, 0.1 mm?)

22-20. Air, saturated with water vapor, at 1 atm and 77°C, flows downward along a flat, smooth plate at a velocity of 3 m/s. The plate is 30 cm high (flow direction) and 1.8 m wide. It is cooled by circulating water on its back side. The cooling water is circulated so that its average temperature is effectively 18°C over the entire back surfaces. The heat-transfer coefficient between the coolant and the back surface is estimated to be 1100 W/(m² · K). The wall is 1.3 mm thick and has a thermal conductivity of 26 W/(m · K). Estimate the total rate of condensation of water and the total heat-transfer rate to the coolant, assuming that the condensate on the plate surface runs off fast enough that its heat-transfer resistance may be neglected.

Is B a constant along the surface? What are the necessary conditions for B to be a constant? Describe how you would propose to analyze the problem, taking into consideration the liquid condensate film on the plate surface.

22-21. Air at 1100°C and 1013 kPa pressure flows along a flat surface at a uniform velocity of 30 m/s. Investigate the problem of cooling this surface by transpiration cooling. At a point 1 m from the leading edge of the surface, consider the use of first air, then helium, as coolants, both of which are assumed to be available at 20°C. Prepare a plot of surface temperature as a function of coolant mass-flow rate per unit of surface area. Also calculate the surface temperature as a function of coolant mass-flow rate if the coolant were used merely to absorb heat rather than to pass into the boundary layer, assuming that the coolant leaves the system at surface temperature.

22-22. An example is worked in the text for the evaluation of B for the flow of a mixture of CO_2 and H_2O at 1650°C over a graphite surface. Investigate the effect of (1) introducing O_2 into the mixture in various amounts and (2) introducing N_2 (treated as an inert) into the mixture in various amounts.

REFERENCES

1. Baron, J. R.: Technical Report 160, Naval Supersonic Laboratory, MIT, Cambridge, Mass., 1956.
2. Rubesin, M. W., and C. C. Pappas: NACA (now NASA) TN 4149, Washington, February 1958.
3. Knuth, E. L.: *Int. J. Heat Mass Transfer,* vol. 6, pp. 1–22, 1963.
4. Knuth, E. L., and H. Dershin: *Int. J. Heat Mass Transfer,* vol. 6, pp. 999–1018, 1963.

APPENDIX
A

PROPERTY
VALUES

In this appendix the transport properties of a variety of gaseous and liquid fluids are presented. The primary purpose of this compilation is to support the problems at the end of each chapter and to provide material for additional problems.

The data are presented in the International System of Units (SI), and the units should be compatible with the various presently accepted national and international standards for SI. The units in the data tables include the base units of meter (length), kilogram (mass), second (time), mole (amount of substance), and kelvin (temperature). The algebraically derived units used in the tables are the joule (energy), pascal (pressure), and watt (power). For convenience in some of the tables, the SI prefixes *kilo* ($k = 10^3$), *milli* ($m = 10^{-3}$), and *micro* ($\mu = 10^{-6}$) are used. The balance of the tables use an abbreviated scientific notation (for powers of 10) that has been found useful with hand calculators (for example, $-07 = 10^{-7}$, $+00 = 10^0$, and $+04 = 10^4$). Conversion of the table entries from SI to the English Engineering System of Units can be easily made using the conversion factors on the inside covers of the textbook.

Tables A-1 to A-7 contain the properties of a representative group of gases at standard atmospheric pressure, $P = 101.325$ kPa. For many applications these gases may be treated as perfect gases for which the

541

TABLE A-1
Properties of air at $P = 101.325$ kPa; $\mathfrak{M} = 28.966$

T, K	ρ, kg/m³	μ, Pa·s		ν, m²/s		c_p, kJ/(kg·K)	k, W/(m·K)		Pr	$g\beta/\nu^2$, 1/(K·m³)	
100	3.5985	70.60	−07	19.62	−07	1.028	92.20	−04	0.787	25.5	+09
150	2.3673	10.38	−06	43.85	−07	1.011	13.75	−03	0.763	34.0	+08
200	1.7690	13.36	−06	75.52	−07	1.006	18.10	−03	0.743	85.9	+07
250	1.4119	16.06	−06	11.37	−06	1.003	22.26	−03	0.724	30.3	+07
263	1.3421	16.70	−06	12.44	−06	1.003	23.28	−03	0.720	24.1	+07
273	1.2930	17.20	−06	13.30	−06	1.004	24.07	−03	0.717	20.3	+07
283	1.2473	17.69	−06	14.18	−06	1.004	24.86	−03	0.714	17.2	+07
293	1.2047	18.17	−06	15.08	−06	1.004	25.63	−03	0.712	14.7	+07
300	1.1766	18.53	−06	15.75	−06	1.005	26.14	−03	0.711	13.2	+07
303	1.1650	18.64	−06	16.00	−06	1.005	26.37	−03	0.710	12.6	+07
313	1.1277	19.11	−06	16.95	−06	1.005	27.09	−03	0.709	10.9	+07
323	1.0928	19.57	−06	17.91	−06	1.006	27.80	−03	0.708	94.6	+06
333	1.0600	20.02	−06	18.89	−06	1.007	28.51	−03	0.707	82.5	+06
343	1.0291	20.47	−06	19.89	−06	1.008	29.21	−03	0.706	72.2	+06
350	1.0085	20.81	−06	20.63	−06	1.008	29.70	−03	0.706	65.8	+06
353	1.0000	20.91	−06	20.91	−06	1.008	29.89	−03	0.705	63.5	+06
363	0.9724	21.34	−06	21.95	−06	1.009	30.58	−03	0.704	56.1	+06
373	0.9463	21.77	−06	23.01	−06	1.010	31.26	−03	0.703	49.6	+06
400	0.8825	22.94	−06	26.00	−06	1.013	33.05	−03	0.703	36.3	+06
450	0.7844	24.93	−06	31.78	−06	1.020	36.33	−03	0.700	21.6	+06
500	0.7060	26.82	−06	37.99	−06	1.029	39.51	−03	0.699	13.6	+06

550	0.6418	28.60 −06	44.56 −06	1.039	42.60 −03	0.698	89.7	+05
600	0.5883	30.30 −06	51.50 −06	1.051	45.60 −03	0.699	61.6	+05
650	0.5431	31.93 −06	58.80 −06	1.063	48.40 −03	0.701	43.6	+05
700	0.5043	33.49 −06	66.41 −06	1.075	51.30 −03	0.702	31.7	+05
750	0.4706	34.98 −06	74.32 −06	1.087	54.10 −03	0.703	23.7	+05
800	0.4412	36.43 −06	82.56 −06	1.099	56.90 −03	0.703	18.0	+05
850	0.4153	37.83 −06	91.10 −06	1.110	59.70 −03	0.703	13.9	+05
900	0.3922	39.18 −06	99.90 −06	1.121	62.50 −03	0.702	10.9	+05
950	0.3716	40.49 −06	10.90 −05	1.131	64.90 −03	0.705	86.9	+04
1000	0.3530	41.77 −06	11.83 −05	1.141	67.20 −03	0.709	70.0	+04
1100	0.3209	44.4 −06	13.8 −05	1.160	73.2 −03	0.705		
1200	0.2942	46.9 −06	15.9 −05	1.177	78.2 −03	0.705		
1300	0.2715	49.3 −06	18.2 −05	1.195	83.7 −03	0.705		
1400	0.2521	51.7 −06	20.5 −05	1.212	89.1 −03	0.704		
1500	0.2353	54.0 −06	22.9 −05	1.230	94.6 −03	0.704		
1600	0.2206	56.3 −06	25.5 −05	1.248	10.0 −02	0.703		
1700	0.2076	58.5 −06	28.2 −05	1.266	10.5 −02	0.702		
1800	0.1961	60.7 −06	31.0 −05	1.286	11.1 −02	0.701		
1900	0.1858	62.9 −06	33.9 −05	1.307	11.7 −02	0.700		
2000	0.1765	65.0 −06	36.8 −05	1.331	12.4 −02	0.699		
2100	0.1681	67.2 −06	40.4 −05	1.359	13.1 −02	0.696		
2200	0.1605	69.3 −06	43.2 −05	1.392	13.9 −02	0.693		
2300	0.1535	71.4 −06	46.5 −05	1.434	14.9 −02	0.688		
2400	0.1471	73.5 −06	50.0 −05	1.487	16.1 −02	0.681		
2500	0.1412	75.7 −06	53.6 −05	1.556	17.5 −02	0.673		

TABLE A-2
Properties of O_2 at $P = 101.325$ kPa; $\mathfrak{M} = 31.999$

T, K	ρ, kg/m³	μ, Pa·s		ν, m²/s		c_p, kJ/(kg·K)	k, W/(m·K)		Pr	$g\beta/\nu^2$, 1/(K·m³)	
100	3.9906	76.80	−07	19.25	−07	0.947	90.50	−04	0.804	26.5	+09
150	2.6186	11.27	−06	43.04	−07	0.918	13.76	−03	0.754	35.3	+08
200	1.9557	14.65	−06	74.91	−07	0.914	18.24	−03	0.735	87.3	+07
250	1.5598	17.77	−06	11.39	−06	0.912	22.54	−03	0.718	30.2	+07
300	1.2998	20.67	−06	15.90	−06	0.918	26.74	−03	0.710	12.9	+07
350	1.1141	23.37	−06	20.98	−06	0.929	30.56	−03	0.710	63.6	+06
400	0.9749	25.89	−06	26.56	−06	0.942	34.20	−03	0.713	34.7	+06
450	0.8665	28.28	−06	32.64	−06	0.956	37.70	−03	0.717	20.4	+06
500	0.7799	30.54	−06	39.16	−06	0.971	41.20	−03	0.720	12.8	+06
600	0.6499	34.70	−06	53.39	−06	1.002	48.00	−03	0.725	57.3	+05
700	0.5571	38.50	−06	69.11	−06	1.031	54.40	−03	0.730	29.3	+05
800	0.4874	42.10	−06	86.37	−06	1.054	60.30	−03	0.736	16.4	+05
900	0.4333	45.40	−06	10.48	−05	1.074	66.10	−03	0.737	99.2	+04
1000	0.3899	48.50	−06	12.44	−05	1.090	71.70	−03	0.737	63.3	+04
1100	0.3545	51.40	−06	14.50	−05	1.103	77.10	−03	0.736	42.4	+04
1200	0.3250	54.20	−06	16.68	−05	1.115	82.10	−03	0.736	29.4	+04
1300	0.3000	56.90	−06	18.97	−05	1.125	87.10	−03	0.735	20.9	+04
1400	0.2785	59.50	−06	21.36	−05	1.134	92.10	−03	0.733	15.3	+04

TABLE A-3
Properties of N_2 at $P = 101.325$ kPa; $\mathfrak{M} = 28.018$

T, K	ρ, kg/m³	μ, Pa·s		ν, m²/s		c_p, kJ/(kg·K)	k, W/(m·K)		Pr	$g\beta/\nu^2$, 1/(K·m³)	
100	3.4800	68.70	−07	19.74	−07	1.072	94.10	−04	0.783	25.1	+09
150	2.2890	10.00	−06	43.69	−07	1.047	13.85	−03	0.756	34.2	+08
200	1.7107	12.86	−06	75.20	−07	1.043	18.26	−03	0.734	86.7	+07
250	1.3657	15.46	−06	11.32	−06	1.039	22.22	−03	0.724	30.6	+07
300	1.1381	17.86	−06	15.69	−06	1.039	25.98	−03	0.714	13.3	+07
350	0.9755	20.08	−06	20.58	−06	1.040	29.39	−03	0.711	66.1	+06
400	0.8536	22.14	−06	25.94	−06	1.044	32.52	−03	0.711	36.4	+06
450	0.7587	24.08	−06	31.74	−06	1.049	35.64	−03	0.709	21.6	+06
500	0.6829	25.90	−06	37.93	−06	1.056	38.64	−03	0.708	13.6	+06
600	0.5691	29.27	−06	51.44	−06	1.075	44.10	−03	0.713	61.7	+05
700	0.4878	32.35	−06	66.32	−06	1.097	49.30	−03	0.720	31.8	+05
800	0.4268	35.20	−06	82.48	−06	1.122	54.10	−03	0.730	18.0	+05
900	0.3794	37.86	−06	99.80	−06	1.145	58.70	−03	0.739	10.9	+05
1000	0.3414	40.36	−06	11.82	−05	1.167	63.10	−03	0.746	70.1	+04
1100	0.3104	42.70	−06	13.76	−05	1.186	67.20	−03	0.754	47.1	+04
1200	0.2845	45.00	−06	15.82	−05	1.204	71.30	−03	0.760	32.6	+04
1300	0.2626	47.10	−06	17.93	−05	1.219	75.40	−03	0.761	23.4	+04
1400	0.2439	49.20	−06	20.17	−05	1.232	79.70	−03	0.761	17.2	+04

TABLE A-4
Properties of CO_2 at $P = 101.325$ kPa; $\mathfrak{M} = 44.010$

T, K	ρ, kg/m³	μ, Pa·s		ν, m²/s		c_p, kJ/(kg·K)	k, W/(m·K)		Pr	$g\beta/\nu^2$, 1/(K·m³)	
220	2.4728	11.19	−06	45.25	−07	0.778	10.83	−03	0.804	21.8	+08
250	2.1652	12.63	−06	58.33	−07	0.806	12.89	−03	0.790	11.5	+08
300	1.7967	14.99	−06	83.43	−07	0.852	16.62	−03	0.768	46.9	+07
350	1.5369	17.26	−06	11.23	−06	0.897	20.50	−03	0.755	22.2	+07
400	1.3408	19.42	−06	14.48	−06	0.938	24.41	−03	0.746	11.7	+07
450	1.1918	21.50	−06	18.04	−06	0.977	28.34	−03	0.741	66.9	+06
500	1.0726	23.48	−06	21.89	−06	1.013	32.28	−03	0.737	40.9	+06
600	0.8939	27.20	−06	30.43	−06	1.075	40.30	−03	0.726	17.6	+06
700	0.7662	30.60	−06	39.94	−06	1.125	48.70	−03	0.707	87.8	+05
800	0.6704	33.90	−06	50.57	−06	1.168	56.00	−03	0.707	47.9	+05
900	0.5959	36.90	−06	61.92	−06	1.204	62.10	−03	0.715	28.4	+05
1000	0.5363	39.70	−06	74.02	−06	1.234	68.00	−03	0.720	17.9	+05
1100	0.4876	42.40	−06	86.96	−06	1.259	73.30	−03	0.728	11.8	+05
1200	0.4469	44.90	−06	10.05	−05	1.280	78.00	−03	0.737	80.9	+04
1300	0.4125	47.40	−06	11.49	−05	1.298	82.50	−03	0.746	57.1	+04
1400	0.3831	49.70	−06	12.97	−05	1.313	86.70	−03	0.753	41.6	+04

TABLE A-5
Properties of H_2 at $P = 101.325$ kPa; $\mathfrak{M} = 2.016$

T, K	ρ, kg/m³	μ, Pa·s		ν, m²/s		c_p, kJ/(kg·K)	k, W/(m·K)		Pr	$g\beta/\nu^2$, 1/(K·m³)	
100	0.2457	42.10	−07	17.14	−06	11.16	67.60	−03	0.695	33.4	+07
150	0.1638	55.70	−07	34.01	−06	12.61	98.60	−03	0.712	56.5	+06
200	0.1228	67.80	−07	55.20	−06	13.52	12.80	−02	0.716	16.1	+06
250	0.0983	79.00	−07	80.39	−06	14.03	15.60	−02	0.711	60.7	+05
300	0.0819	89.40	−07	10.92	−05	14.27	18.15	−02	0.703	27.4	+05
350	0.0702	99.40	−07	14.16	−05	14.38	20.33	−02	0.703	14.0	+05
400	0.0614	10.91	−06	17.76	−05	14.49	22.12	−02	0.715	77.6	+04
450	0.0546	11.84	−06	21.69	−05	14.49	23.89	−02	0.718	46.3	+04
500	0.0491	12.74	−06	25.93	−05	14.50	25.64	−02	0.720	29.2	+04
600	0.0409	14.50	−06	35.41	−05	14.53	29.10	−02	0.724	13.0	+04
700	0.0351	16.10	−06	45.87	−05	14.60	32.50	−02	0.723	66.5	+03
800	0.0307	17.70	−06	57.64	−05	14.71	36.00	−02	0.723	36.9	+03
900	0.0273	19.20	−06	70.34	−05	14.83	39.40	−02	0.723	22.0	+03
1000	0.0246	20.70	−06	84.26	−05	14.99	42.80	−02	0.725	13.8	+03
1100	0.0223	22.20	−06	99.40	−05	15.17	46.20	−02	0.729	90.2	+02
1200	0.0205	23.60	−06	11.53	−04	15.36	49.50	−02	0.732	61.5	+02
1300	0.0189	25.00	−06	13.23	−04	15.57	52.80	−02	0.737	43.1	+02
1400	0.0175	26.30	−06	14.99	−04	15.80	56.20	−02	0.739	31.2	+02

TABLE A-6
Properties of He-4 at $P = 101.325$ kPa; $\mathfrak{M} = 4.003$

T, K	ρ, kg/m³	μ, Pa·s		v, m²/s		c_p, kJ/(kg·K)	k, W/(m·K)		Pr	$g\beta/v^2$, 1/(K·m³)	
100	0.4870	97.80	−07	20.05	−06	5.194	73.60	−03	0.690	24.4	+07
150	0.3252	12.50	−06	38.44	−06	5.194	96.90	−03	0.670	44.2	+06
200	0.2439	15.10	−06	61.91	−06	5.193	11.80	−02	0.665	12.8	+06
250	0.1951	17.60	−06	90.20	−06	5.193	13.70	−02	0.667	48.2	+05
300	0.1626	19.90	−06	12.24	−05	5.193	15.50	−02	0.667	21.8	+05
350	0.1394	22.20	−06	15.93	−05	5.193	17.20	−02	0.670	11.0	+05
400	0.1220	24.30	−06	19.93	−05	5.193	18.90	−02	0.668	61.7	+04
450	0.1084	26.40	−06	24.35	−05	5.193	20.50	−02	0.669	36.7	+04
500	0.0976	28.40	−06	29.11	−05	5.193	22.10	−02	0.667	23.1	+04
600	0.0813	32.20	−06	39.61	−05	5.193	25.10	−02	0.666	10.4	+04
700	0.0697	35.90	−06	51.52	−05	5.193	28.00	−02	0.666	52.8	+03
800	0.0610	39.40	−06	64.62	−05	5.193	30.70	−02	0.666	29.3	+03
900	0.0542	42.80	−06	78.97	−05	5.193	33.40	−02	0.665	17.5	+03
1000	0.0488	46.20	−06	94.71	−05	5.193	36.00	−02	0.666	10.9	+03
1100	0.0443	49.40	−06	11.14	−04	5.193	38.50	−02	0.666	71.8	+02
1200	0.0407	52.50	−06	12.91	−04	5.193	41.00	−02	0.665	49.0	+02
1300	0.0375	55.60	−06	14.82	−04	5.193	43.40	−02	0.665	34.3	+02
1400	0.0348	58.60	−06	16.82	−04	5.193	45.70	−02	0.666	24.7	+02

TABLE A-7
Properties of NH_3 at $P = 101.325$ kPa; $\mathfrak{M} = 17.033$

T, K	ρ, kg/m^3	μ, Pa $\cdot$ s		ν, m^2/s		c_p, kJ/(kg $\cdot$ K)	k, W/(m $\cdot$ K)		Pr
250	0.8469	85.29	-07	10.07	-06	2.213	19.68	-03	0.959
300	0.6971	10.27	-06	14.73	-06	2.170	24.55	-03	0.908
350	0.5948	12.06	-06	20.28	-06	2.211	30.21	-03	0.883
400	0.5193	13.90	-06	26.77	-06	2.289	36.48	-03	0.872
450	0.4610	15.76	-06	34.19	-06	2.381	43.24	-03	0.868
500	0.4146	17.63	-06	42.52	-06	2.477	50.42	-03	0.866
550	0.3768	19.51	-06	51.78	-06	2.572	57.97	-03	0.866
600	0.3453	21.38	-06	61.92	-07	2.665	65.63	-03	0.868
650	0.3187	23.24	-06	72.92	-06	2.755	73.24	-03	0.874
700	0.2959	25.09	-06	84.79	-06	2.844	81.20	-03	0.878

transport properties are independent of pressure, and the densities may be computed from $P/\rho = (R/\mathfrak{M})T$, where $R = 8314.34$ J/(kmol $\cdot$ K) is the universal gas constant and $\mathfrak{M}$ is the molecular weight, kg/kmol. The gas data for Tables A-1 to A-5 have been primarily obtained from three volumes of *Thermophysical Properties of Matter*, published by the Thermophysical Properties Research Center (TPRC) at Purdue University.[1-3] The table entries for specific heat and density are ideal-gas values, except near the critical temperature of the gas. For air, oxygen, and nitrogen the first three table entries, and for carbon dioxide the first four table entries, contain real-gas specific heat and density values obtained from the National Bureau of Standards Circular 564.[4] For all gas tables the volume coefficient of thermal expansion for an ideal gas was assumed, $\beta = 1/T$. Table entries for air above 1000 K are from Ref. 5.

The properties of helium gas (Table A-6) were obtained from National Bureau of Standards Technical Note 631.[6]

Tables A-7 and A-10 give gaseous and saturated liquid data for ammonia. Density and specific heat data were obtained from Haar and Gallagher,[7] and the other properties were obtained from the 1976 ASHRAE Handbook.[8]

The properties of steam in the superheated state at two different pressures are given in Table A-8, and the properties of saturated liquid water are given in Table A-9. For these tables the density and specific heat values come from the equations of the 1967 IFC Formulation for Industrial Use, given in the ASME Steam Tables.[9] Viscosity and thermal conductivity values are those issued by the International Association for the Properties of Steam.[10,11]

TABLE A-8
Properties of steam

T, °C	200	250	300	350	400	450	500	550	600
Density, kg/m^3									
$P = 100$ kPa	0.460	0.416	0.379	0.348	0.322	0.300	0.280	0.263	0.248
$P = 7500$ kPa			37.4	30.8	27.1	24.4	22.4	20.7	19.3
Viscosity, μPa · s									
$P = 100$ kPa	16.2	18.2	20.3	22.4	24.5	26.5	28.6	30.6	32.6
$P = 7500$ kPa			19.7	22.1	24.4	26.6	28.8	30.9	32.9
Thermal conductivity, mW/(m · K)									
$P = 100$ kPa	33.4	38.3	43.5	49.0	54.7	60.7	66.9	73.3	79.9
$P = 7500$ kPa			64.1	60.6	62.9	67.5	72.8	78.7	84.8
Specific heat, kJ/(kg · K)									
$P = 100$ kPa	1.979	1.990	2.010	2.037	2.067	2.099	2.132	2.166	2.201
$P = 7500$ kPa			4.686	3.228	2.726	2.512	2.420	2.386	2.379
Prandtl number									
$P = 100$ kPa	0.959	0.947	0.938	0.930	0.924	0.917	0.910	0.905	0.989
$P = 7500$ kPa			1.443	1.179	1.057	0.990	0.956	0.936	0.924

TABLE A-9
Properties of saturated liquid water

T, °C	ρ, kg/m^3	μ, Pa · s	ν, m^2/s	c_p, kJ/(kg · K)	k, W/(m · K)	Pr
0.01	999.8	17.91 −04	17.91 −07	4.217	0.562	13.44
10	999.8	13.08 −04	13.08 −07	4.193	0.582	9.42
20	998.2	10.03 −04	10.05 −07	4.182	0.600	6.99
30	995.6	79.77 −05	80.01 −08	4.179	0.615	5.42
40	992.2	65.31 −05	65.80 −08	4.179	0.629	4.34
50	988.0	54.71 −05	55.37 −08	4.181	0.640	3.57
60	983.2	46.68 −05	47.48 −08	4.185	0.651	3.00
70	977.7	40.44 −05	41.36 −08	4.190	0.659	2.57
80	971.8	35.49 −05	36.52 −08	4.197	0.667	2.23
90	965.3	31.50 −05	32.63 −08	4.205	0.673	1.97
100	958.3	28.22 −05	29.45 −08	4.216	0.677	1.76
140	926.1	19.61 −05	21.17 −08	4.285	0.685	1.23
180	886.9	14.94 −05	16.85 −08	4.408	0.674	0.98
220	840.3	12.10 −05	14.40 −08	4.613	0.648	0.86
260	784.0	10.15 −05	12.95 −08	4.983	0.606	0.83

TABLE A-10
Properties of saturated liquid ammonia

T, °C	ρ, kg/m³	μ, Pa · s		ν, m²/s		c_p, kJ/(kg · K)	k, W/(m · K)	Pr
−30	677	26.3	−05	38.9	−08	4.447	0.609	1.92
−20	665	23.7	−05	35.6	−08	4.501	0.586	1.82
−10	652	21.2	−05	32.4	−08	4.556	0.562	1.72
0	639	19.0	−05	29.7	−08	4.618	0.539	1.63
10	625	17.0	−05	27.2	−08	4.683	0.516	1.54
20	610	15.2	−05	24.9	−08	4.758	0.493	1.47
30	595	13.6	−05	22.9	−08	4.843	0.470	1.40
40	580	12.2	−05	21.0	−08	4.943	0.447	1.35
50	563	11.0	−05	19.5	−08	5.066	0.423	1.32
60	545	98.7	−06	15.7	−08	5.225	0.400	1.29

TABLE A-11
Properties of saturated liquid oxygen

T, K	ρ, kg/m³	μ, Pa · s		ν, m²/s		c_p, kJ/(kg · K)	k, W/(m · K)	Pr
56	1299	58.5	−05	45.0	−08	1.664	0.191	5.10
62	1273	47.5	−05	37.3	−08	1.664	0.186	4.25
68	1246	38.8	−05	31.1	−08	1.666	0.179	3.61
76	1209	30.0	−05	24.8	−08	1.672	0.170	2.95
82	1181	24.9	−05	21.1	−08	1.679	0.162	2.58
90.2	1141	19.6	−05	17.2	−08	1.696	0.151	2.20
100	1091	15.1	−05	13.8	−08	1.732	0.138	1.90
112	1024	11.4	−05	11.1	−08	1.814	0.122	1.70
126	933	87.1	−06	93.4	−09	2.029	0.102	1.73
138	833	72.3	−06	86.8	−09	2.537	0.0838	2.19
150	675	53.7	−06	79.6	−09	5.598	0.0724	4.15

TABLE A-12
Properties of saturated liquid parahydrogen

T, K	ρ, kg/m³	μ, Pa · s		ν, m²/s		c_p, kJ/(kg · K)	k, W/(m · K)	Pr
14	76.9	24.8	−06	32.2	−08	6.47	0.0746	2.15
16	75.1	19.4	−06	25.8	−08	7.36	0.0889	1.61
18	73.2	15.9	−06	21.7	−08	8.42	0.0954	1.40
20.3	70.8	13.2	−06	18.6	−08	9.66	0.0989	1.29
22	68.7	11.6	−06	16.9	−08	10.82	0.1010	1.24
24	66.0	10.1	−06	15.3	−08	12.52	0.1008	1.25
26	62.8	88.0	−07	14.0	−08	14.80	0.0984	1.32
28	59.0	76.2	−07	12.9	−08	18.48	0.0938	1.50
30	53.9	64.6	−07	12.0	−08	26.59	0.0866	1.98
32	46.0	51.3	−07	11.2	−08	65.37	0.0915	3.67

The properties of saturated liquid oxygen (Table A-11), were obtained from NASA SP-3071,[12] and similar properties for liquid parahydrogen (Table A-12) were obtained from NASA SP-3089.[13]

The liquid-metal properties in Table A-13 were obtained primarily from *Liquid Metals Handbook*[14,15] and supplemented from Ref. 16.

The properties of some typical hydrocarbon fuels and oils (Tables A-14 to A-16) were adapted from data presented in *SAE Aerospace Allied Thermodynamics Manual.*[16]

Table A-17 contains properties of the atmosphere, taken from the 1976 U.S. Standard Atmosphere published by NOAA.[17]

Table A-18 gives values of the Schmidt number for a number of dilute binary mixtures, based on data in *International Critical Tables*[18]

TABLE A-13
Properties of mercury, sodium, 22 percent sodium/78 percent potassium, and potassium

T, °C	0	50	100	200	300	400	500	600	700
				Density, kg/m³					
Hg	13,600	13,460	13,350	13,110	12,880				
Na	—	—	927	905	882	858	834	809	783
NaK	—	—	847	823	799	775	751	727	703
K	—	—	819	795	771	747	723	701	676
				Viscosity, mPa · s					
Hg	1.68	1.39	1.21	1.01	0.862				
Na	—	—	0.705	0.450	0.345	0.284	0.243	0.210	0.186
NaK	—	—	0.529	0.354	0.276	0.229	0.195	0.168	0.146
K	—	—	0.463	0.298	0.227	0.190	0.169	0.145	0.134
				Specific heat, kJ/(kg · K)					
Hg	0.140	0.140	0.140	0.136	0.135				
Na	—	—	1.384	1.339	1.305	1.280	1.264	1.255	1.255
NaK	—	—	0.941	0.908	0.887	0.879	0.874	0.874	0.883
K	—	—	0.812	0.791	0.774	0.766	0.761	0.766	0.774
				Thermal conductivity, W/(m · K)					
Hg	8.2	9.3	10.4	12.4	14.0				
Na	—	—	—	81.5	75.7	71.2	66.8	62.7	59.0
NaK	—	—	—	24.7	25.9	26.2	26.2	25.9	25.5
K	—	—	—	44.1	42.0	39.4	37.2	34.9	32.5
				Prandtl number					
Hg	0.029	0.021	0.016	0.0011	0.0083				
Na	—	—	—	0.0074	0.0059	0.0051	0.0046	0.0042	0.0039
NaK	—	—	—	0.013	0.0095	0.0077	0.0065	0.0057	0.0051
K	—	—	—	0.0053	0.0042	0.0037	0.0035	0.0032	0.0036

TABLE A-14
Properties of engine oil (unused)

T, °C	ρ, kg/m³	μ, Pa · s		ν, m²/s		c_p, kJ/(kg · K)	k, W/(m · K)		Pr
0	898	38.3	−01	42.7	−04	1.788	14.7	−02	46,540
15	889	10.6	−01	12.0	−04	1.845	14.5	−02	13,530
30	881	36.9	−02	41.9	−05	1.905	14.3	−02	4,912
45	872	15.3	−02	17.5	−05	1.969	14.1	−02	2,130
60	863	73.2	−03	84.8	−06	2.035	14.0	−02	1,065
75	855	39.4	−03	46.1	−06	2.101	13.8	−02	598
100	840	17.2	−03	20.5	−06	2.214	13.6	−02	280
125	826	91.5	−04	11.1	−06	2.328	13.4	−02	159
150	811	56.4	−04	69.5	−07	2.440	13.2	−02	104

TABLE A-15
Properties of hydraulic fluid (MIL-H-5606, $P = 101.325$ kPa)

T, °C	ρ, kg/m³	μ, Pa · s		ν, m²/s		c_p, kJ/(kg · K)	k, W/(m · K)		Pr
−15	879	70.6	−03	80.3	−06	1.737	12.7	−02	962
0	868	34.2	−03	39.4	−06	1.788	12.6	−02	483
15	858	19.2	−03	22.4	−06	1.845	12.4	−02	286
30	847	12.2	−03	14.4	−06	1.905	12.1	−02	192
45	836	84.4	−04	10.1	−06	1.969	11.7	−02	143
60	826	63.0	−04	76.3	−07	2.035	11.2	−02	114
75	815	49.9	−04	61.2	−07	2.101	10.8	−02	97

TABLE A-16
Properties of JP-4 fuel

T, °C	ρ, kg/m³	μ, Pa · s		ν, m²/s		c_p, kJ/(kg · K)	k, W/(m · K)		Pr
−15	802	15.3	−04	19.1	−07	1.858	14.1	−02	20
0	789	11.2	−04	14.2	−07	1.926	14.0	−02	15
15	776	86.8	−05	11.2	−07	1.995	13.8	−02	13
30	763	70.1	−05	91.9	−08	2.063	13.7	−02	11
45	750	58.6	−05	78.2	−08	2.132	13.6	−02	9
60	737	50.4	−05	68.4	−08	2.201	13.5	−02	8
75	724	44.4	−05	61.3	−08	2.269	13.3	−02	8

TABLE A-17
Properties of the atmosphere

Geometric altitude, m	Temperature, K	Pressure, Pa		Density, kg/m³		Viscosity, μPa · s	Thermal conductivity, mW/(m · K)	Speed of sound, m/s
0	288.2	1.01325	+05	1.225	+00	17.89	25.36	340.3
250	286.5	9.836	+04	1.196	+00	17.82	25.23	339.3
500	284.9	9.546	+04	1.167	+00	17.74	25.11	338.4
750	283.3	9.263	+04	1.139	+00	17.66	24.97	337.4
1,000	281.7	8.988	+04	1.112	+00	17.58	24.85	336.4
1,250	280.0	8.719	+04	1.085	+00	17.50	24.72	335.5
1,500	278.4	8.456	+04	1.058	+00	17.42	24.59	334.5
3,000	268.7	7.012	+04	9.093	−01	16.94	23.81	328.6
4,500	258.9	5.775	+04	7.770	−01	16.45	22.03	322.6
6,000	249.2	4.722	+04	6.601	−01	15.95	22.23	316.5
7,500	239.5	3.830	+04	5.572	−01	15.44	21.44	310.2
10,000	223.3	2.650	+04	4.135	−01	14.58	20.09	299.5
12,500	216.7	1.793	+04	2.884	−01	14.22	19.53	295.1
15,000	216.7	1.211	+04	1.948	−01	14.22	19.53	295.1
17,500	216.7	8.182	+03	1.316	−01	14.22	19.53	295.1
20,000	216.7	5.529	+03	8.891	−02	14.22	19.53	295.1
25,000	221.6	2.549	+03	4.008	−02	14.48	19.95	298.4
30,000	226.5	1.197	+03	1.841	−02	14.75	20.36	301.7
45,000	264.2	1.491	+02	1.966	−03	16.71	23.45	325.8
60,000	247.0	2.196	+01	3.097	−04	15.84	22.06	315.1
75,000	208.4	2.388	+00	3.992	−05	13.76	18.83	289.4

and *Chemical Engineer's Handbook*.[19] From the Schmidt number the binary diffusion coefficient $D_{12} = D_{21}$ may be deduced, since the viscosity and density are the values for the medium of diffusion. The Schmidt number for gases varies only slightly with pressure and does not vary greatly with temperature; it does, however, vary significantly with mixture ratio. With these facts, it is thus possible to estimate D_{12} and γ_{12} over a wide range of pressure, temperature, and mixture ratio. The Schmidt number for liquids varies considerably with temperature, although it is not significantly affected by pressure. Little data appear to be available on the variation of Schmidt number with mixture ratio.

Table A-19 has been prepared to provide an idea of the variation of properties with mixture composition in binary gas mixtures. The molecular weight, mole fraction, mass fraction, and specific heat for a gas mixture are given in the mass-transfer chapters. The mixture density was calculated according to the ideal-gas law. For the table air was the medium of diffusion, and the binary diffusion coefficient D_{12} was calculated using the viscosity of air and the Schmidt number from Table A-18. This allowed calculation of γ_{12} from D_{12} and the mixture density.

TABLE A-18
Some values of the Schmidt number for dilute binary mixtures at approximately normal atmospheric conditions

Diffusing substance	Medium of diffusion	Schmidt number $Sc = \dfrac{\mu}{\gamma_{12}} = \dfrac{\mu}{\rho \mathscr{D}_{12}} = \dfrac{\nu}{\mathscr{D}_{12}}$
Ammonia	Air (gas)	0.61
Ammonia	Water (liquid)	570
Benzene	Air (gas)	1.71
Benzene	Carbon dioxide (gas)	1.37
Carbon dioxide	Air (gas)	0.96
Carbon dioxide	Hydrogen (gas)	1.58
Carbon dioxide	Water (liquid)	559
Carbon dioxide	Ethyl alcohol (liquid)	445
Carbon disulfide	Air (gas)	1.48
Carbon tetrachloride	Air (gas)	2.13
Chlorine	Air (gas)	1.42
Chlorine	Water (liquid)	824
Ethane	Air (gas)	1.22
Ethyl alcohol	Air (gas)	1.30
Ethyl alcohol	Water (liquid)	1005
Ethyl ether	Air (gas)	1.70
Glycerol	Water (liquid)	1630
Helium	Air (gas)	0.22
Hydrochloric acid	Water (liquid)	381
Hydrogen	Air (gas)	0.22
Hydrogen	Oxygen (gas)	0.182
Methane	Air (gas)	0.84
Methyl alcohol	Air (gas)	1.00
Nitrogen	Air (gas)	0.98
Nitrogen	Water (liquid)	613
Oxygen	Air (gas)	0.74
Oxygen	Nitrogen (gas)	0.681
Oxygen	Water (liquid)	558
n-Octane	Air (gas)	2.62
Sodium chloride	Water (liquid)	745
Sodium hydroxide	Water (liquid)	665
Sucrose	Water (liquid)	2230
Sulfuric acid	Water (liquid)	580
Water (vapor)	Air (gas)	0.60

The mixture viscosity was calculated using the Wilke[20] method and the mixture thermal conductivity was calculated using the Lindsay–Bromley[21] method, as reviewed by Gambill.[22] References 1, 3, and 23 contain a review of these and other methods of computing mixture thermophysical properties.

Table A-20 has been prepared as an aid in mass-transfer problems

TABLE A-19
Binary gas mixtures ($P = 101.325 \text{ kPa}$, $T = 290 \text{ K}$)

$\dfrac{m_{H_2O}}{m_{H_2}}$ m_{CO_2}	$\mathfrak{M}$, kg/kmol	c_p, kJ/(kg · K)	γ, kg/(m · s)		Pr	Sc	Le
			Air–H_2O				
0.00	28.97	1.003	0.3008	−04	0.714	0.600	1.189
0.10	27.31	1.089	0.2836	−04	0.729	0.597	1.220
0.20	25.82	1.175	0.2682	−04	0.751	0.595	1.264
0.30	24.50	1.261	0.2544	−04	0.780	0.592	1.318
0.40	23.30	1.347	0.2419	−04	0.813	0.589	1.381
0.50	22.21	1.433	0.2307	−04	0.850	0.586	1.451
0.60	21.22	1.518	0.2204	−04	0.890	0.583	1.528
0.70	20.32	1.604	0.2110	−04	0.934	0.580	1.611
0.80	19.49	1.690	0.2024	−04	0.980	0.576	1.701
0.90	18.72	1.776	0.1944	−04	1.030	0.573	1.797
1.00	18.01	1.862	0.1870	−04	1.083	0.570	1.900
			Air–H_2				
0.00	28.97	1.003	0.8204	−04	0.714	0.220	3.244
0.10	12.40	2.326	0.3511	−04	0.425	0.472	0.900
0.20	7.88	3.648	0.2233	−04	0.448	0.664	0.674
0.30	5.78	4.970	0.1637	−04	0.482	0.819	0.588
0.40	4.56	6.292	0.1292	−04	0.517	0.951	0.544
0.50	3.77	7.614	0.1068	−04	0.550	1.066	0.516
0.60	3.21	8.937	0.0909	−04	0.583	1.171	0.498
0.70	2.80	10.259	0.0792	−04	0.614	1.268	0.484
0.80	2.48	11.581	0.0702	−04	0.644	1.359	0.474
0.90	2.22	12.903	0.0630	−04	0.674	1.446	0.466
1.00	2.02	14.226	0.0571	−04	0.703	1.529	0.460
			Air–CO_2				
0.00	28.97	1.003	0.1880	−04	0.714	0.960	0.743
0.10	29.99	0.986	0.1947	−04	0.711	0.911	0.780
0.20	31.09	0.969	0.2018	−04	0.709	0.862	0.822
0.30	32.28	0.952	0.2095	−04	0.708	0.815	0.869
0.40	33.55	0.935	0.2178	−04	0.708	0.768	0.923
0.50	34.94	0.918	0.2268	−04	0.711	0.722	0.984
0.60	36.44	0.901	0.2365	−04	0.715	0.677	1.056
0.70	38.08	0.884	0.2471	−04	0.722	0.634	1.139
0.80	39.87	0.867	0.2588	−04	0.731	0.591	1.238
0.90	41.84	0.850	0.2715	−04	0.745	0.549	1.356
1.00	44.01	0.833	0.2856	−04	0.763	0.509	1.501

involving the air–water-vapor system. The vapor pressure of H_2O as a function of temperature is taken from Ref. 9, and then the concentration of H_2O in a *saturated* air–water-vapor mixture at 1 atm total pressure is plotted. The latter has been prepared from Eq. (22-4). Similar curves can be prepared for other total pressures, using the same vapor-pressure

TABLE A-20
Vapor pressure of H_2O and mass concentration of H_2O in saturated air at $P = 101.325$ kPa

T_{sat}, °C	Vapor pressure, Pa	m_{H_2O}	T_{sat}, °C	Vapor pressure, Pa	m_{H_2O}
−20	103	0.00064	40	7,378	0.0468
−15	165	0.00102	45	9,585	0.0614
−10	260	0.00160	50	12,339	0.0799
−5	401	0.00248	55	15,745	0.1033
0	611	0.00377	60	19,925	0.1330
5	872	0.00539	65	25,014	0.1705
10	1,227	0.00760	70	31,167	0.2180
15	1,704	0.0106	75	38,554	0.2784
20	2,337	0.0145	80	47,365	0.3556
25	3,167	0.0198	85	57,809	0.4554
30	4,243	0.0266	90	70,112	0.5861
35	5,623	0.0354	100	101,325	1.0000

curve; and, of course, other gas–vapor systems can be handled in the same way if vapor-pressure data are available.

REFERENCES

1. Touloukian, Y. S., P. E. Liley, and S. C. Saxena: *Thermophysical Properties of Matter,* vol. 3: *Thermal Conductivity. Nonmetallic Liquids and Gases.* IFI/Plenum, New York, 1970.
2. Touloukian, Y. S., and T. Makita: *Thermophysical Properties of Matter,* vol. 6: *Specific Heat. Nonmetallic Gases and Liquids,* IFI/Plenum, New York, 1970.
3. Touloukian, Y. S., S. C. Saxena, and P. Hestermans: *Thermophysical Properties of Matter,* vol. 11: *Viscosity,* IFI/Plenum, New York, 1970.
4. Hilsenrath, J., et al.: "Tables of Thermal Properties of Gases," NBS Circular 564, Washington, November 1955.
5. Poferl, D. J., R. A. Svehla, and K. Lewandowski: "Thermodynamic and Transport Properties of Air and the Combustion Products of Natural Gas and of ASTM-A-1 Fuel with Air," NASA Technical Note D-5452, Washington, 1969.
6. McCarty, R. D.: "Thermophysical Properties of Helium-4 from 2 to 1500 K with Pressures to 1000 Atmospheres." NBS Tech. Note 631, Washington, November 1972.
7. Haar, L., and J. S. Gallagher: "Thermodynamic Properties of Ammonia," *J. Phys. Chem. Ref. Data,* vol. 7, p. 635, 1977.
8. *ASHRAE Thermophysical Properties of Refrigerants,* American Society of Heating, Refrigeration, and Air-Conditioning Engineers, New York, 1976.
9. *ASME Steam Tables,* 3d ed., American Society of Mechanical Engineers, New York, 1977.
10. Nagashima, A.: "Viscosity of Water Substance—New International Formulation and Its Background," *J. Phys. Chem. Ref. Data,* vol. 6, p. 1133, 1977.
11. "Release of Thermal Conductivity of Water Substance," International Association for the Properties of Steam, December 1977.

12. Roder, H. M., and L. A. Weber: *Thermophysical Properties*, vol. 1: *ASRDI Oxygen Technology Survey*, NASA SP-3071, Washington, 1972.
13. McCarty, R. D.: *Hydrogen Technological Survey—Thermophysical Properties*, NASA SP-3089, Washington, 1975.
14. Lyon, R. N. (ed.): *Liquid Metals Handbook*, 2d ed., Atomic Energy Commission and Department of the Navy, Washington, 1952.
15. Jackson, C. B. (ed.): *Liquid Metals Handbook*, 3d ed., Atomic Energy Commission and Department of the Navy, Washington, 1955.
16. *SAE Aerospace Applied Thermodynamics Manual*, 2d ed., Society of Automotive Engineers, New York, 1969.
17. *U.S. Standard Atmosphere*, 1976, NOAA, NASA, and USAF, Washington, October 1976.
18. *International Critical Tables*, McGraw-Hill, New York, 1926–1930.
19. Perry, J. H.: *Chemical Engineer's Handbook*, McGraw-Hill, New York, 1950.
20. Wilke, C. R.: *J. Chem. Phys.* vol. 18, p. 157, 1950.
21. Lindsay, A. L., and L. A. Bromley: *Ind. Engng Chem.*, vol. 42, p. 508, 1950.
22. Gambill, W. R.: *Chem. Engng*, vol. 64, p. 277, 1957.
23. Reid, R. C., J. M. Prausnitz, and T. K. Sherwood: *The Properties of Gases and Liquids*, 3d ed., McGraw-Hill, New York, 1977.

APPENDIX
B

DIMENSIONS AND CONVERSION TO SI

BASIC UNITS

Length	meter (m)
Mass	kilogram (kg)
Time	second (s)
Electric current	ampere (A)
Thermodynamic temperature	kelvin (K)
Luminous intensity	candela (cd)
Amount of substance	mole (mol)

DERIVED UNITS

See Table B-2.

PHYSICAL CONSTANTS

Gas constant	$R = 8314.34 \text{ J}/(\text{kmol} \cdot \text{K})$
Stefan–Boltzmann constant	$\sigma = 5.66961 \times 10^{-8} \text{ W}/(\text{m}^2 \cdot \text{K}^4)$
Freefall, standard	$g = 9.806650 \text{ m/s}^2$

TABLE B-1
SI prefixes

Factor	Prefix	Symbol
10^{18}	exa	E
10^{15}	peta	P
10^{12}	tera	T
10^{9}	giga	G
10^{6}	mega	M
10^{3}	kilo	k
10^{2}	hecto	h
10	deka	da
10^{-1}	deci	d
10^{-2}	centi	c
10^{-3}	milli	m
10^{-6}	micro	μ
10^{-9}	nano	n
10^{-12}	pico	p
10^{-15}	femto	f
10^{-18}	atto	a

TABLE B-2
Conversion factors

Physical quantity	Unit	SI equivalent†
Acceleration, m/s^2	foot/second^2	3.048 000‡ −01
Area, m^2	foot^2	9.290 304‡ −02
	inch^2	6.451 600‡ −04
Density, kg/m^3	gram/centimeter^3	1.000 000‡ +03
	pound-mass/inch^3	2.767 990 +04
	pound-mass/foot^3	1.601 846 +01
	slug/foot^3	5.153 788 +02
Energy, J = N · m	Btu$_{IT}$	1.055 056 +03
	Btu$_{th}$ (thermochemical)	1.054 350 +03
	calorie$_{IT}$	4.186 800‡ +00
	calorie$_{th}$ (thermochemical)	4.184 000‡ +00
	erg	1.000 000‡ −07
	foot pound-force	1.355 818 +00
	foot-poundal	4.214 011 −02
	watt-hour	3.600 000‡ +03
Energy flux, W/m^2 = J/(s · m^2)	Btu$_{th}$/(s · ft^2)	1.134 893 +04
	Btu$_{th}$/(h · ft^2)	3.152 481 +00
	cal$_{th}$/(s · cm^2)	4.184 000‡ +04
Enthalpy J/kg	Btu$_{th}$/lb$_m$	2.324 444 +03
Force, N = kg · m/s^2	dyne	1.000 000‡ −05
	kilogram-force	9.806 650‡ +00
	pound-force	4.448 222 +00
	poundal	1.382 550 −01
Heat-transfer coefficient,	Btu$_{th}$/(s · ft^2 · F)	2.042 808 +04

TABLE B-2 Continued

Physical quantity	Unit	SI equivalent†	
$W/(m^2 \cdot K)$	$Btu_{th}/(h \cdot ft^2 \cdot °F)$	5.674 466	+00
	$cal_{th}/(s \cdot cm^2 \cdot °C)$	4.184 000‡	+04
Kinematic viscosity, m^2/s	centistoke	1.000 000‡	−06
	$foot^2/second$	9.290 304‡	−02
	stoke	1.000 000‡	−04
Length, m	foot	3.048 000‡	−01
	inch	2.540 000‡	−02
	mile (US)	1.609 347‡	+03
Mass, kg	pound-mass	4.535 924	−01
	slug	1.459 390	+01
	ton (metric)	1.000 000	+03
Mass flow rate, kg/s	$lb_m/second$	4.535924	−01
	$lb_m/hour$	1.259979	−04
Power, $W = J/s$	$Btu_{IT}/hour$	2.930 711	−01
	$Btu_{th}/hour$	2.928 751	−01
	$Btu_{th}/second$	1.054 350	+03
	$cal_{IT}/second$	4.186 800‡	+00
	$cal_{th}/second$	4.184 000‡	+00
	$ft \cdot lb_f/s$	1.355 818	+00
	horsepower ($550 ft \cdot lb_f/s$)	7.456 999	+02
Pressure, $Pa = N/m^2$	atmosphere (normal = 760 Torr)	1.013 250‡	+05
	atmosphere (technical = $1 kg_f/cm^2$)	9.806 650‡	+04
	bar	1.000 000‡	+05
	$dyne/centimeter^2$	1.000 000‡	−01
	inch of Hg (0°C)	3.386 389	+03
	inch of Hg (60°F)	3.376 85	+03
	inch of H_2O (4°C)	2.490 82	+02
	inch of H_2O (60°F)	2.488 4	+02
	mmHg (0°C) = Torr	1.333 22	+02
	$poundal/foot^2$	1.488 164	+00
	$pound-force/foot^2$	4.788 026	+01
	$pound-force/inch^2$	6.894 757	+03
Specific heat, $kJ/(kg \cdot K)$	$Btu_{IT}/(lb_m \cdot °F)$	4.186 800‡	+00
	$Btu_{th}/(lb_m \cdot °F)$	4.184 000‡	+00
	$cal_{IT}/(g \cdot °C)$	4.186 800‡	+00
	$cal_{th}/(g \cdot °C)$	4.184 000‡	+00
Temperature, K	$T(K) = T(°C) + 273.15$		
	$T(K) = T(°R)/1.8$		
	$T(K) = [T(°F) + 459.67]/1.8$		
	$T(°C) = [T(°F) − 32]/1.8$		
Thermal conductivity,	$Btu_{IT}/(h \cdot ft \cdot °F)$	1.730 735	+00
$W/(m \cdot K) = J/(s \cdot m \cdot K)$	$Btu_{th} \cdot in/(h \cdot ft^2 \cdot °F)$	1.441 314	−01
	$Btu_{th}/(h \cdot ft \cdot °F)$	1.729 577	+00
	$cal_{th}/(s \cdot cm \cdot C)$	4.184 000‡	+02
Thermal diffusivity, m^2/s	$foot^2/second$	9.290 304‡	−02
	$foot^2/hour$	2.580 640 ‡	−05
Velocity, m/s	foot/second	3.048 000‡	−01
	kilometer/hour	2.777 778	−01
	mile (US)/hour	0.447 04‡	+00

TABLE B-2 Continued

Physical quantity	Unit	SI equivalent†
Viscosity, $Pa \cdot s = N \cdot s/m^2$	centipoise	1.000 000‡ −03
	poise	1.000 000‡ −01
	$lb_m/(s \cdot ft)$	1.488 164 +00
	$lb_m/(h \cdot ft)$	4.133 789 −04
	$lb_f \cdot s/ft^2$	4.788 026 +01
	$slug/(s \cdot ft)$	4.788 026 +01
Volume, m^3	$foot^3$	2.831 685 −02
	in^3	1.638 706‡ −05
	gallon (US liquid)	3.785 412 −03
	liter	1.000 000‡ −03
	ounce (US fluid)	2.957 353 −05
Volume flow rate, m^3/s	$foot^3$/minute	4.719 474 −04
	$foot^3$/second	2.831 685 −02
	gallon (US liquid)/minute	6.309 020 −05

† As in App. A, an abbreviated notation is used: $-01 = 10^{-1}$ etc.
‡ An exact definition.

REFERENCES

Mechtly, E. A.: *The International System of Units—Physical Constants and Conversion Factors,* 2d ed., NASA SP-7012, Washington, 1973.
Metric Manual, McDonnell Douglas Corporation, 1976.

APPENDIX
C

SOME TABLES OF FUNCTIONS USEFUL IN BOUNDARY-LAYER ANALYSIS

TABLE C-1
Gamma function

$$\Gamma(n) = \int_0^\infty x^{n-1}e^{-x}\,dx, \qquad \Gamma(n+1) = n\Gamma(n)$$

n	$\Gamma(n)$	n	$\Gamma(n)$	n	$\Gamma(n)$
1.00	1.000	1.35	0.891	1.70	0.909
1.05	0.974	1.40	0.887	1.75	0.919
1.10	0.951	1.45	0.886	1.80	0.931
1.15	0.933	1.50	0.886	1.85	0.946
1.20	0.918	1.55	0.889	1.90	0.962
1.25	0.906	1.60	0.894	1.95	0.980
1.30	0.897	1.65	0.900	2.00	1.00

TABLE C-2
Beta function and incomplete beta function

$$\beta_1(m, n) = \int_0^1 Z^{m-1}(1 - Z)^{n-1}\, dz$$

$$= \beta_1(n, m) = \frac{\Gamma(m)\Gamma(n)}{\Gamma(m + n)}$$

Incomplete beta function

$$\beta_r(m, n) = \int_0^r Z^{m-1}(1 - Z)^{n-1}\, dZ$$

$$= \beta_1(m, n) - \beta_{1-r}(n, m)$$

TABLE C-3
Tabulation of $\beta_1(m, n)$ for some particular values of m and n encountered in Chaps. 10 and 13 and values of β_r/β_1 as a function of r for the same values of m and n[†]

m	$\frac{1}{3}$	$\frac{1}{3}$	$\frac{1}{3}$	$\frac{2}{3}$	$\frac{1}{9}$	$\frac{1}{9}$	$\frac{1}{9}$	$\frac{8}{9}$
n	$\frac{2}{3}$	$\frac{4}{3}$	$\frac{8}{3}$	$\frac{4}{3}$	$\frac{8}{9}$	$\frac{10}{9}$	$\frac{20}{9}$	$\frac{10}{9}$
$\beta_1(m, n)$	3.628	2.650	2.015	1.209	9.185	8.844	7.984	1.021
r				β_r/β_1				
0	0	0	0	0	0	0	0	0
0.02	0.225	0.307	0.402	0.0912	0.635	0.659	0.728	0.0432
0.06	0.325	0.441	0.568	0.189	0.717	0.744	0.819	0.0917
0.10	0.387	0.521	0.663	0.264	0.760	0.787	0.862	0.142
0.20	0.492	0.651	0.801	0.312	0.821	0.849	0.920	0.261
0.40	0.634	0.804	0.928	0.634	0.890	0.915	0.970	0.477
0.60	0.745	0.900	0.980	0.800	0.934	0.954	0.990	0.673
0.80	0.850	0.964	0.997	0.924	0.968	0.981	0.998	0.851
1.00	1.000	1.000	1.000	1.000	1.000	1.000	1.000	1.000

[†] *Taken from Baxter and Reynolds.*[1]

TABLE C-4
Error function

$$\text{erf}(x) = \frac{2}{\sqrt{\pi}} \int_0^x e^{-\eta^2}\, d\eta = \frac{2}{\sqrt{\pi}} \int_{-x}^0 e^{-\eta^2}\, d\eta$$

x	erf (x)	x	erf (x)	x	erf (x)
0.00	0.0000	0.45	0.475	1.10	0.880
0.02	0.0226	0.50	0.521	1.20	0.910
0.04	0.0451	0.55	0.563	1.30	0.934
0.06	0.0676	0.60	0.604	1.40	0.952
0.08	0.0901	0.65	0.642	1.50	0.966
0.10	0.1125	0.70	0.678	1.60	0.976
0.15	0.168	0.75	0.711	1.80	0.989
0.20	0.223	0.80	0.742	2.00	0.995
0.25	0.276	0.85	0.771	3.00	1.000
0.30	0.329	0.90	0.797	∞	1.000
0.35	0.379	0.95	0.821		
0.40	0.428	1.00	0.843		

REFERENCE

1. Baxter, D. C., and W. C. Reynolds: *J. Aero. Sci.*, vol. 25, no. 6, 1958.

APPENDIX
D

OPERATIONS IMPLIED BY THE ∇ OPERATOR

Let $\mathbf{V}$ represent a vector quantity and S a scalar quantity. Subscripts designate the scalar magnitudes of the components of a vector in the indicated direction. Note that

∇S is a vector

$\nabla \cdot \mathbf{V}$ is a scalar

$\nabla^2 S$ is a scalar

$\nabla^2 \mathbf{V}$ is a vector

$\mathbf{V} \cdot \nabla S$ is a scalar

RECTANGULAR COORDINATES (x, y, z)

See Fig. D-1.

$$\nabla \cdot \mathbf{V} = \frac{\partial V_x}{\partial x} + \frac{\partial V_y}{\partial y} + \frac{\partial V_z}{\partial z}$$

$$\nabla^2 S = \frac{\partial^2 S}{\partial x^2} + \frac{\partial^2 S}{\partial y^2} + \frac{\partial^2 S}{\partial z^2}$$

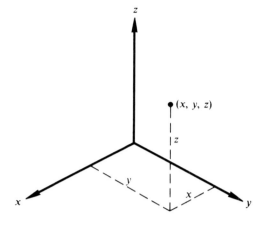

FIGURE D-1

$$(\nabla S)_x = \frac{\partial S}{\partial x}, \quad (\nabla S)_y = \frac{\partial S}{\partial y}, \quad (\nabla S)_z = \frac{\partial S}{\partial z}$$

$$\mathbf{V} \cdot \nabla S = V_x \frac{\partial S}{\partial x} + V_y \frac{\partial S}{\partial y} + V_z \frac{\partial S}{\partial z}$$

$$\nabla \cdot \alpha \nabla S = \frac{\partial}{\partial x}\left(\alpha \frac{\partial S}{\partial x}\right) + \frac{\partial}{\partial y}\left(\alpha \frac{\partial S}{\partial y}\right) + \frac{\partial}{\partial z}\left(\alpha \frac{\partial S}{\partial z}\right)$$

$$(\nabla^2 \mathbf{V})_x = \frac{\partial^2 V_x}{\partial x^2} + \frac{\partial^2 V_x}{\partial y^2} + \frac{\partial^2 V_x}{\partial z^2}$$

$$(\nabla^2 \mathbf{V})_y = \frac{\partial^2 V_y}{\partial x^2} + \frac{\partial^2 V_y}{\partial y^2} + \frac{\partial^2 V_z}{\partial z^2}$$

$$(\nabla^2 \mathbf{V})_z = \frac{\partial^2 V_z}{\partial x^2} + \frac{\partial^2 V_z}{\partial y^2} + \frac{\partial^2 V_z}{\partial z^2}$$

CYLINDRICAL COORDINATES (r, ϕ, x)

See Fig. D-2.

$$\mathbf{V} \cdot \mathbf{V} = \frac{1}{r}\frac{\partial}{\partial r}(rV_r) + \frac{1}{r}\frac{\partial V_\phi}{\partial \phi} + \frac{\partial V_x}{\partial x}$$

$$\nabla^2 S = \frac{1}{r}\frac{\partial}{\partial r}\left(r\frac{\partial S}{\partial r}\right) + \frac{1}{r^2}\frac{\partial^2 S}{\partial \phi^2} + \frac{\partial^2 S}{\partial x^2}$$

$$(\nabla S)_r = \frac{\partial S}{\partial r}, \quad (\nabla S)_\phi = \frac{1}{r}\frac{\partial S}{\partial \phi}, \quad (\nabla S)_x = \frac{\partial S}{\partial x}$$

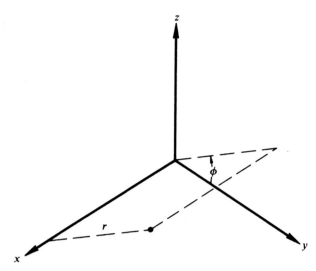

FIGURE D-2

$$\mathbf{V} \cdot \nabla S = V_r \frac{\partial S}{\partial r} + \frac{V_\phi}{r} \frac{\partial S}{\partial \phi} + V_x \frac{\partial S}{\partial x}$$

$$\nabla \cdot \alpha \nabla S = \frac{1}{r} \frac{\partial}{\partial r} \left(r\alpha \frac{\partial S}{\partial r} \right) + \frac{1}{r} \frac{\partial}{\partial \phi} \left(\frac{\alpha}{r} \frac{\partial S}{\partial \phi} \right) + \frac{\partial}{\partial x} \left(\alpha \frac{\partial S}{\partial x} \right)$$

$$(\nabla^2 \mathbf{V})_r = \frac{\partial}{\partial r} \left[\frac{1}{r} \frac{\partial}{\partial r} (rV_r) \right] + \frac{1}{r^2} \frac{\partial^2 V}{\partial \phi^2} - \frac{2}{r^2} \frac{\partial V_\phi}{\partial \phi} + \frac{\partial^2 V_r}{\partial x^2}$$

$$(\nabla^2 \mathbf{V})_\phi = \frac{\partial}{\partial r} \left[\frac{1}{r} \frac{\partial}{\partial r} (rV_\phi) \right] + \frac{1}{r^2} \frac{\partial^2 V_\phi}{\partial \phi^2} + \frac{2}{r^2} \frac{\partial V_r}{\partial \phi} + \frac{\partial^2 V_\phi}{\partial x^2}$$

$$(\nabla^2 \mathbf{V})_x = \frac{1}{r} \frac{\partial}{\partial r} \left(r \frac{\partial V_x}{\partial r} \right) + \frac{1}{r^2} \frac{\partial^2 V_x}{\partial \phi^2} + \frac{\partial^2 V_x}{\partial x^2}$$

SPHERICAL COORDINATES (r, θ, ϕ)

See Fig. D-3.

$$\mathbf{V} \cdot \mathbf{V} = \frac{1}{r^2} \frac{\partial}{\partial r} (r^2 V_r) + \frac{1}{r \sin \theta} \frac{\partial}{\partial \theta} (V_\theta \sin \theta) + \frac{1}{r \sin \theta} \frac{\partial V_\phi}{\partial \phi}$$

$$\nabla^2 S = \frac{1}{r^2} \frac{\partial}{\partial r} \left(r^2 \frac{\partial S}{\partial r} \right) + \frac{1}{r^2 \sin \theta} \frac{\partial}{\partial \theta} \left(\sin \theta \frac{\partial S}{\partial \theta} \right) + \frac{1}{r^2 \sin^2 \theta} \frac{\partial^2 S}{\partial \phi^2}$$

$$(\nabla S)_r = \frac{\partial S}{\partial r}, \qquad (\nabla S)_\theta = \frac{1}{r} \frac{\partial S}{\partial \theta}, \qquad (\nabla S)_\phi = \frac{1}{r \sin \theta} \frac{\partial S}{\partial \phi}$$

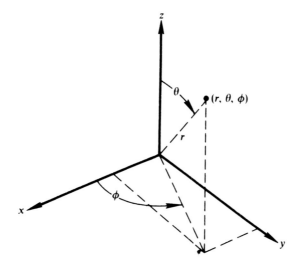

FIGURE D-3

$$\mathbf{V} \cdot \nabla S = V_r \frac{\partial S}{\partial r} + \frac{V_\theta}{r} \frac{\partial S}{\partial \theta} + \frac{V_\phi}{r \sin \theta} \frac{\partial S}{\partial \phi}$$

$$\nabla \cdot \alpha \nabla S = \frac{1}{r^2} \frac{\partial}{\partial r} \left(r^2 \alpha \frac{\partial S}{\partial r} \right) + \frac{1}{r^2 \sin \theta} \frac{\partial}{\partial \theta} \left[(\sin \theta) \alpha \frac{\partial S}{\partial \theta} \right]$$

$$+ \frac{1}{r^2 \sin^2 \theta} \frac{\partial}{\partial \phi} \left(\alpha \frac{\partial S}{\partial \phi} \right)$$

$$(\nabla^2 \mathbf{V})_r = \nabla^2 V_r - \frac{2V_r}{r^2} - \frac{2}{r^2} \frac{\partial V_\theta}{\partial \theta} - \frac{2V_\theta \cot \theta}{r^2} - \frac{2}{r^2 \sin \theta} \frac{\partial V_\phi}{\partial \phi}$$

$$(\nabla^2 \mathbf{V})_\theta = \nabla^2 V_\theta + \frac{2}{r^2} \frac{\partial V_r}{\partial \theta} - \frac{V_\theta}{r^2 \sin \theta} - \frac{2 \cos \theta}{r^2 \sin^2 \theta} \frac{\partial V_\phi}{\partial \phi}$$

$$(\nabla^2 \mathbf{V})_\phi = \nabla^2 V_\phi - \frac{V_\phi}{r^2 \sin^2 \theta} + \frac{2}{r^2 \sin \theta} \frac{\partial V_r}{\partial \phi} + \frac{2 \cos \theta}{r^2 \sin^2 \theta} \frac{\partial V_\theta}{\partial \phi}$$

APPENDIX
E

TURBULENT
BOUNDARY-LAYER
BENCHMARK
DATA

This appendix is provided with two purposes in mind. In developing and testing turbulence models for use in calculating procedures for predicting boundary-layer behavior, it is useful to have a set of heat-transfer test data for a variety of boundary conditions against which the models and procedures can be proven. Furthermore, it is especially useful to have a set of data that is internally consistent and is presented in a consistent manner.

The data presented in the following pages cover a wide variety of boundary conditions, including favorable and adverse pressure gradients, blowing and suction, rough and smooth surfaces, flat and curved surfaces, constant and varying surface temperature, high free-stream turbulence, and various combinations of the above. The source of the data is primarily W. M. Kays' own laboratory (although a few runs come from other sources), because these are data for which the authors have some degree of confidence, and furthermore most of the data in the literature is not available in raw original form, and little in tabular form.

Most of these data have been obtained using relatively small temperature differences, so that fluid properties can be treated as constant. As a matter of fact, there are very few data on the external turbulent boundary layer with large temperature differences available in the literature, although there is a considerable quantity of such data for flow inside tubes.

The experimental uncertainty in the reported values of Stanton number is estimated to be generally about ±5 percent. In all cases the heat-transfer rates leading to the Stanton numbers have been measured directly by wall heat meters of one type or another. The uncertainty in friction coefficient is considerably higher because no direct measurements have been made. The friction coefficients have been determined by a combination of momentum balances, the law of the wall, and velocity measurements close to the wall within the viscous sublayer. Since the raw measurement data is presented, the user is free to employ his or her own procedures to determine friction coefficients if desired.

All of the data are for two-dimensional boundary layers. In any two-dimensional boundary-layer experiment there always exists some three-dimensionality. A fair indication of three-dimensional effects can be obtained for heat-transfer experiments through an *energy balance* if temperature and velocity profiles are measured at two or more points in addition to surface heat-transfer measurements. For most of the data *momentum thickness* Reynolds numbers will be found along with *enthalpy thickness* Reynolds numbers, at locations where velocity and temperature transverses have been made. At or near those locations will be found another *enthalpy thickness* Reynolds number, which will have been determined from wall heat-transfer measurements. A comparison of these Reynolds numbers is an indication of the effects of three-dimensionality. As will be seen, the energy-balance results are generally quite good, with the exception of the one set of data for a very high-turbulence free stream, IDENT G3000.

A second reason for including this appendix is that data of this type, in its unreduced form, provides excellent material for analysis by students. From velocity profiles, for certain cases at least, friction coefficients can be evaluated using the law of the wall, and then the velocity profiles can be expressed in (u^+, y^+) coordinates. The thermal "law of the wall" can be explored. The effect of pressure gradient and transpiration on sublayer thickness can be investigated. The adequacy of the various integral methods of analysis, and the closed-form solutions presented in the text, can be examined. The turbulence models presented in the text, and any other turbulence models, can be critically examined.

IDENT 91871: Blackwell constant u_∞. Fluid: air at 760 mmHg. Smooth surface

Start profile, $x = 0.0508$ m

y, m	u, m/s	T, K
0	0	310.48
0.000099	1.47	308.07
0.00015	2.18	306.63
0.000201	2.79	305.24
0.000251	3.41	304.18
0.000302	3.89	303.29
0.000353	4.39	302.55
0.000429	4.91	301.59
0.000531	5.37	300.62
0.000632	5.75	299.96
0.000836	6.25	299.04
0.00109	6.69	298.33
0.001217	6.84	298.07
0.001344	6.99	297.87
0.001598	7.21	297.52
0.001852	7.42	297.25
0.002233	7.68	296.97
0.002868	8.05	296.67
0.003503	8.38	296.50
0.004646	8.67	296.38
0.005408	9.13	296.29
0.006678	9.44	296.26
0.007948	9.59	296.26
0.009218	9.63	296.26

x, m	u_∞, m/s	T_{wall}, K	$\dot{m}''/G$	St	Re_{Δ_2}	$c_f/2$	Re_{δ_2}
0.25	9.61	310.43	0	0.00299	559	0.00222	816
0.36	9.65	310.37	0	0.00276	724		
0.46	9.60	310.43	0	0.00261	882	0.00202	1204
0.56	9.61	310.37	0	0.00251	1034		
0.66	9.63	310.43	0	0.00242	1181	0.0019	1534
0.76	9.64	310.37	0	0.00235	1324		
0.86	9.64	310.37	0	0.00339	1464		
0.97	9.64	310.37	0	0.00225	1601	0.0019	1862
1.07	9.64	310.37	0	0.0022	1736		
1.17	9.65	310.43	0	0.00216	1869	0.00181	
1.27	9.65	310.37	0	0.00213	1999		2154
1.37	9.65	310.43	0	0.0021	2128	0.00174	
1.47	9.65	310.43	0	0.00207	2256		
1.57	9.64	310.37	0	0.00204	2381		2481
1.68	9.64	310.32	0	0.00202	2506	0.00168	
1.78	9.64	310.37	0	0.002	2629		
1.88	9.65	310.32	0	0.00198	2750		2809
1.98	9.65	310.37	0	0.00196	2871	0.00163	
2.08	9.65	310.37	0	0.00194	2991		2971
2.18	9.65	310.43	0	0.00192	3109	0.00161	
2.29	9.64	310.37	0	0.00191	3227		

Profile at $x = 2.29$ m

y, m	u, m/s	T, K
0	0.00	310.16
0.00014	1.35	308.71
0.00019	1.80	308.05
0.00024	2.18	307.35
0.00029	2.65	306.78
0.00035	2.91	306.29
0.00042	3.37	305.68
0.00052	3.74	304.99
0.00062	4.06	304.39
0.00085	4.63	303.55
0.00123	5.11	302.74
0.00174	5.55	302.15
0.00301	6.03	301.44
0.00428	6.31	300.93
0.00555	6.57	300.51
0.00809	6.99	299.98
0.01063	7.32	299.48
0.01444	7.78	298.91
0.01952	8.24	298.26
0.02714	8.88	297.44
0.0373	9.45	296.59
0.04238	9.58	296.31
0.04746	9.60	296.17
0.05254	9.60	296.16

IDENT 62067: Whitten mixed boundary conditions. Fluid: air at 759.5 mmHg. Smooth surface

Profile at $x = 0.45$ m,
$\quad \mathrm{Re}_{\delta_2} = 1200$

y, m	u, m/s	T, K	x, m	u_∞, m/s	T_{wall}, K	$\dot{m}''/G$	St	Re_{Δ_2}
0	0.00	316.09	0.15	13.17	319.50	0.0003	0.00315	530
0.000147	3.60	311.53	0.25	13.17	319.53	0.0005	0.0027	830
0.000198	4.15	310.19	0.36	13.17	319.53	0.0007	0.00238	1120
0.000249	4.82	308.86	0.46	13.17	319.51	0.0009	0.00219	1400
0.0003	5.42	307.99	0.56	13.17	319.55	0.0011	0.0021	1680
0.000452	6.56	305.60	0.66	13.17	308.21	0.0013	0.00084	1960
0.00063	7.22	303.85	0.76	13.17	308.22	0.0015	0.00138	2250
0.000757	7.55	303.00	0.86	13.17	308.26	0.0017	0.00135	2540
0.000909	7.82	302.37	0.97	13.17	308.36	0.0019	0.00137	2840
0.001163	8.13	301.46	1.07	13.17	308.31	0.0021	0.00124	3150
0.001468	8.46	300.73	1.17	13.17	308.39	0.0023	0.00115	3470
0.001798	8.75	300.11	1.27	13.17	312.08	0.0025	0.00141	3790
0.002052	8.93	299.69	1.37	13.17	316.07	0.0027	0.00134	4130
0.002433	9.21	299.30	1.47	13.17	321.41	0.0029	0.00133	4470
0.003068	9.62	298.51	1.57	13.17	323.46	0.0031	0.00125	4830
0.003957	10.12	297.66	1.68	13.17	327.09	0.0033	0.00111	5200
0.004973	10.61	296.77	1.78	13.17	330.96	0.0035	0.00115	5570
0.006243	11.14	295.86	1.88	13.17	330.98	0.0037	0.00097	5960
0.007513	11.58	295.16	1.98	13.17	330.97	0.0039	0.00085	6370
0.009418	12.11	294.16	2.08	13.17	331.00	0.0041	0.00086	6780
0.011323	12.42	293.57	2.18	13.17	330.24	0.0043	0.00078	7210
0.012593	12.55	293.24	2.29	13.17	329.14	0.0044	0.00073	7650
0.013863	12.61	293.07	2.39	13.17	328.33	0.0047	0.00063	8110
0.015133	12.66	292.95						

IDENT 110871: Blackwell equilibrium adverse pressure gradient. Fluid: air at 762 mmHg. Smooth surface

Start profile, $x = 0.0508$ m

y, m	u, m/s	T, K
0	0	306.37
0.000099	1.72	304.07
0.00014986	2.15	302.57
0.0020066	2.66	301.24
0.0025146	3.17	300.17
0.0030226	3.59	299.29
0.0035306	3.97	298.59
0.0040386	4.31	298.00
0.0050546	4.75	297.13
0.0060706	5.07	296.30
0.0070866	5.31	295.82
0.0093726	5.78	294.97
0.00119126	6.13	294.40
0.00157226	6.45	293.83
0.00182626	6.67	293.55
0.00246126	7.09	293.09
0.00309626	7.44	292.84
0.00373126	7.75	292.71
0.00436626	8.02	292.62
0.00500126	8.26	292.59
0.00563626	8.47	292.57
0.00690626	8.77	292.57
0.00817626	8.91	292.57
0.00908304	8.95	292.57

x, m	u_∞, m/s	T_{wall}, K	$\dot{m}''/G$	St	Re_{Δ_2}	$c_f/2$	Re_{δ_2}
0.25	7.88	306.26	0	0.00032	527	0.0174	1018
0.36	7.47	306.26	0	0.00293	668		
0.46	7.12	306.32	0	0.00277	798	0.00129	1718
0.56	6.85	306.26	0	0.00266	918		
0.66	6.65	306.26	0	0.00258	1033		
0.76	6.46	306.26	0	0.00251	1141		
0.86	6.31	306.26	0	0.00246	1246	0.00119	2316
0.97	6.17	306.32	0	0.00241	1347		
1.07	6.05	306.32	0	0.00237	1444		
1.17	5.95	306.21	0	0.00234	1539	0.00114	2825
1.27	5.86	306.26	0	0.0023	1631		
1.37	5.78	306.26	0	0.00227	1721		
1.47	5.71	306.26	0	0.00225	1809	0.0011	3258
1.57	5.64	306.26	0	0.00222	1896		
1.68	5.59	306.26	0	0.0022	1980		
1.78	5.53	306.26	0	0.00218	2063	0.00107	3659
1.88	5.48	306.26	0	0.00216	2145		
1.98	5.43	306.21	0	0.00214	2225		
2.08	5.37	306.32	0	0.00213	2304	0.00103	4139
2.18	5.30	306.26	0	0.00211	2382		
2.29	5.24	306.32	0	0.00209	2459	0.00101	4533

Profile at $x = 2.29$ m

y, m	u, m/s	T, K
0	0	306.43
0.000168	0.311	305.43
0.000244	0.454	304.91
0.000345	0.637	304.07
0.000447	0.811	303.33
0.000549	0.969	302.69
0.000777	1.271	301.46
0.001031	1.512	300.57
0.001412	1.817	299.54
0.00192	2.042	298.73
0.00319	2.298	297.79
0.00446	2.521	297.41
0.007	2.676	296.78
0.01081	2.822	296.28
0.01843	3.158	295.63
0.02986	3.584	294.86
0.03621	3.831	294.63
0.04256	4.051	294.29
0.04891	4.316	294.08
0.06161	4.691	293.61
0.07431	5.054	293.22
0.08701	5.209	292.97
0.09971	5.243	292.86
0.11241	5.243	292.82

IDENT 72968: Thielbahr strong acceleration. Fluid: air at 757.4 mmHg. Smooth surface

Profile at $x = 0.35\,m$

y, m	u, m/s	T, K
0	0.00	316.22
0.0001143	1.21	312.09
0.0001651	1.75	310.38
0.0002159	2.08	309.38
0.0003175	2.75	307.86
0.0004191	3.26	306.58
0.0005715	3.75	304.78
0.0007239	4.07	303.33
0.0009525	4.47	301.95
0.0012065	4.71	301.03
0.0017145	4.99	299.75
0.0024765	5.30	298.54
0.0031115	5.49	297.85
0.0037465	5.66	297.25
0.0043815	5.81	296.76
0.0056515	6.11	295.90
0.0069215	6.36	295.17
0.0081915	6.60	294.57
0.0094615	6.85	294.10
0.0107315	7.06	293.65
0.0120015	7.22	293.31
0.0145415	7.46	292.82
0.0170815	7.51	292.58
0.0196215	7.53	292.53

x, m	u_∞, m/s	T_{wall}, K	$\dot{m}''/G$	St	Re_{Δ_2}	$c_f/2$	Re_{δ_2}
0.05	7.60	316.03	0	0.0423	169		
0.15	7.60	316.08	0	0.0033	362		
0.25	7.55	316.08	0	0.0031	525		
0.35	7.53	316.12	0	0.0029	678	0.0023	881
0.45	7.60	316.18	0	0.0028	822		
0.55	7.94	316.26	0	0.0026	961		
0.65	8.53	316.27	0	0.0024	1100	0.00245	905
0.76	9.29	316.24	0	0.0022	1240		
0.86	10.28	316.20	0	0.0021	1380	0.00252	796
0.96	11.43	316.20	0	0.0021	1530		
1.06	12.85	316.22	0	0.002	1690	0.00248	747
1.16	14.76	316.20	0	0.002	1880		
1.26	17.36	316.27	0	0.0019	2080		
1.37	20.68	316.24	0	0.0018	2320		
1.47	21.23	316.12	0	0.0019	2580	0.00222	1234
1.57	21.20	316.08	0	0.0018	2850		
1.67	21.22	316.12	0	0.0018	3110	0.00191	1793
1.77	21.25	316.11	0	0.0018	3360		
1.87	21.28	316.07	0	0.0018	3610		
1.98	21.28	316.12	0	0.0017	3860		
2.08	21.33	316.07	0	0.0017	4100		
2.18	21.36	316.08	0	0.0017	4350	0.00175	2760
2.28	21.35	316.03	0	0.0017	4590		
2.38	21.39	316.11	0	0.0017	4830		

Profile at $x = 1.16\,m$

y, m	u, m/s	T, K
0	0.00	316.48
0.0001143	3.83	312.08
0.0001651	5.54	310.43
0.0002159	6.70	309.35
0.0003175	8.41	307.55
0.0004191	9.46	306.02
0.0005715	10.35	304.56
0.0008255	11.12	303.15
0.0010795	11.54	302.26
0.0015875	12.10	301.13
0.0023495	12.62	299.82
0.0029845	12.99	298.93
0.0036195	13.26	298.31
0.0042545	13.48	297.64
0.0048895	13.66	297.14
0.0055245	13.79	296.61
0.0067945	14.01	295.68
0.0080645	14.17	295.03
0.0093345	14.28	294.43
0.0106045	14.40	293.83
0.0118745	14.49	293.47
0.0144145	14.56	292.87
0.0169545	14.58	292.61
0.0194945	14.58	292.54

IDENT 70869: Kearney acceleration close to laminarization. Fluid: air at 759 mmHg

Start profile, $x = 0.351$ m

y, m	u, m/s	T, K	x, m	u_∞, m/s	T_{wall}, K	$\dot{m}''/G$	St	Re_{Δ_2}	$c_f/2$	Re_{δ_2}
0	0.00	310.21	0.15	7.10	310.72	0	0.00348	336		
0.000089	0.75	308.01	0.25	7.09	310.61	0	0.00311	492		
0.0001397	1.17	307.14	0.35	7.10	310.22	0	0.0029	624	0.0025	754
0.0001905	1.60	306.42	0.36	7.07	310.67	0	0.0029	631		
0.0002667	2.05	305.30	0.46	7.28	310.67	0	0.00269	763		
0.0003937	2.82	303.92	0.55	7.62	309.94	0	0.00248	895	0.00255	817
0.0005715	3.55	302.58	0.56	7.78	310.78	0	0.00249	886		
0.0007747	4.01	301.56	0.66	8.66	309.28	0	0.00222	990	0.0026	738
0.0010287	4.37	300.68	0.66	8.76	310.78	0	0.00223	1012		
0.0012319	4.57	300.22	0.76	10.06	308.28	0	0.00198	1120	0.0026	665
0.0014859	4.74	299.76	0.76	10.22	310.72	0	0.00186	1147		
0.0017399	4.89	299.38	0.85	11.89	308.33	0	0.00177	1236	0.00257	595
0.0021209	5.04	298.94	0.86	12.28	310.61	0	0.00174	1287		
0.0025019	5.16	298.58	0.95	14.72	308.72	0	0.00159	1345	0.00248	550
0.0030099	5.32	298.23	0.97	15.42	310.72	0	0.00157	1433		
0.0036449	5.49	297.80	1.07	19.99	310.50	0	0.00135	1627		
0.0042799	5.64	297.45	1.17	20.49	310.33	0	0.00198	1878		
0.0055499	5.92	296.87	1.27	20.47	310.33	0	0.0021	2144		
0.0068199	6.16	296.41	1.37	20.52	310.28	0	0.00193	2425		
0.0087249	6.51	295.85	1.47	20.56	310.22	0	0.00191	2701		
0.0106299	6.78	295.44	1.57	20.59	310.33	0	0.00188	2931		
0.0131699	7.00	295.09	1.68	20.59	310.22	0	0.00185	3209		
0.0157099	7.09	294.93	1.78	20.64	310.28	0	0.00181	3449		
0.0182499	7.10	294.87	1.88	20.62	310.17	0	0.00178	3719		
			1.98	20.60	310.17	0	0.00173	3957		
			2.08	20.63	310.22	0	0.00171	4164		
			2.18	20.64	310.28	0	0.00172	4382		
			2.29	20.63	310.11	0	0.00168	4670		

Profile at $x = 0.951$ m

y, m	u, m/s	T, K
0	0.00	308.71
0.000089	2.62	306.78
0.00014	4.11	305.70
0.000191	5.61	304.87
0.000292	8.02	303.43
0.000394	9.70	302.20
0.000546	11.06	300.98
0.000673	11.70	300.13
0.000755	12.00	299.67
0.000902	12.41	299.16
0.001054	12.65	298.70
0.001232	12.96	298.32
0.001486	13.18	297.75
0.00174	13.35	297.38
0.002121	13.57	296.98
0.002629	13.78	296.53
0.003391	14.00	296.03
0.00428	14.18	295.49
0.00555	14.35	294.89
0.00682	14.49	294.41
0.00809	14.57	294.02
0.009995	14.68	293.62
0.0119	14.71	293.39
0.01444	14.72	293.32

IDENT 50167: Whitten step-blowing. Fluid: air at 765 mmHg. Smooth surface

Profile at x = 0.91 m

y, m	u, m/s	T, K
0	0.00	314.98
0.000155	3.54	310.56
0.000206	4.07	309.01
0.000257	4.80	307.86
0.000333	5.62	306.38
0.000536	6.96	303.77
0.000841	7.72	301.94
0.001298	8.34	300.58
0.00206	8.90	299.38
0.002568	9.19	298.75
0.003457	9.63	297.90
0.004473	10.08	297.28
0.005743	10.53	296.47
0.007521	11.01	295.58
0.009553	11.59	294.69
0.011458	11.97	293.95
0.013363	12.28	293.37
0.015903	12.65	292.75
0.017173	12.79	292.52
0.018443	12.91	292.27
0.020983	13.04	292.01
0.024793	12.99	291.84
0.029873	13.14	291.76

x, m	u_∞, m/s	T_{wall}, K	$\dot{m}''/G$	St	Re_{Δ_2}	Re_{δ_2}
0.0508	13.15	314.86	0	0.00428	280	
0.1524	13.15	315.14	0	0.0031	520	
0.254	13.15	314.74	0	0.00279	810	
0.3556	13.15	314.78	0	0.00261	1060	
0.4572	13.15	314.67	0	0.00248	1300	
0.5588	13.15	314.67	0	0.00238	1530	
0.6604	13.15	314.61	0	0.00228	1750	
0.762	13.15	314.72	0	0.00222	1970	
0.8636	13.15	314.84	0	0.00216	2170	1910
0.9144	13.15	314.92	0.0039	0.00122	2530	2070
1.0668	13.15	314.98	0.0039	0.00111	2980	
1.1684	13.15	314.74	0.004	0.00101	3420	3030
1.27	13.15	314.50	0.0039	0.00098	3860	
1.3716	13.15	314.44	0.004	0.00082	4290	
1.4732	13.15	314.29	0.0039	0.00082	4720	
1.5748	13.15	314.22	0.0039	0.00082	5140	
1.6764	13.15	314.24	0.004	0.00077	5570	5090
1.778	13.15	314.18	0.004	0.00075	5990	
1.8796	13.15	314.01	0.0039	0.0007	6410	
1.9812	13.15	314.07	0.0039	0.00067	6830	
2.0828	13.15	314.32	0.0039	0.00065	7240	
2.1844	13.15	314.52	0.0039		7650	
2.286	13.15	314.49	0.004	0.00065	8060	
2.3876	13.15	314.38	0.0039	0.00059	8470	

Profile at x = 1.67 m

y, m	u, m/s	T, K
0	0.00	313.93
0.0001524	1.57	312.13
0.0002032	1.94	311.27
0.000254	2.25	310.66
0.0003302	2.79	309.79
0.0005842	3.81	307.93
0.0010922	4.78	306.11
0.0019812	5.52	304.63
0.0031242	6.21	303.47
0.0050292	6.95	302.18
0.0082042	7.87	300.89
0.0101092	8.39	300.02
0.0126492	8.99	299.01
0.0151892	9.52	298.16
0.0177292	10.07	297.39
0.0215392	10.78	296.17
0.0253492	11.46	295.12
0.0304292	12.21	293.82
0.0355092	12.71	292.89
0.0405892	12.97	292.17
0.0431292	13.05	291.98
0.0456692	13.09	291.86
0.0507492	13.13	291.75
0.0558292	13.14	291.71

IDENT 30868: Thielbahr strong blowing with acceleration. Fluid: air at 756 mmHg. Smooth surface

Start profile, x = 0.761 m

y, m	u, m/s	T, K
0	0.00	295.59
0.00014	3.47	296.73
0.000191	4.51	296.99
0.000241	5.26	297.21
0.000292	5.77	297.49
0.000445	6.98	297.94
0.000826	8.39	298.58
0.001334	9.44	299.17
0.002477	11.00	299.85
0.003747	12.19	300.46
0.005652	13.40	301.18
0.008192	14.51	302.04
0.012637	15.71	303.18
0.017717	16.47	304.20
0.022797	16.93	305.12
0.027877	17.27	305.80
0.030417	17.39	306.12
0.032957	17.49	306.37
0.035497	17.55	306.57
0.038025	17.60	306.73
0.040577	17.62	306.85
0.043117	17.64	306.91
0.045657	17.64	306.96
0.048197	17.64	306.99

Profile at x = 1.98 m

y, m	u, m/s	T, K
0	0.00	295.59
0.00014	3.47	296.73
0.000191	4.51	296.99
0.000241	5.26	297.21
0.000292	5.77	297.49
0.000445	6.98	297.94
0.000826	8.39	298.58
0.001334	9.44	299.17
0.002477	11.00	299.85
0.003747	12.19	300.46
0.005652	13.40	301.18
0.008192	14.51	302.04
0.012637	15.71	303.18
0.017717	16.47	304.20
0.022797	16.93	305.12
0.027877	17.27	305.80
0.030417	17.39	306.12
0.032957	17.49	306.37
0.035497	17.55	306.57
0.038025	17.60	306.73
0.040577	17.62	306.85
0.043117	17.64	306.91
0.045657	17.64	306.96
0.048197	17.64	306.99

x, m	u_∞, m/s	T_{wall}, K	$\dot{m}''/G$	St	Re_{Δ_2}	$c_f/2$	Re_{δ_2}
0.15	9,53	298.25	0.00624	0.00164	717		
0.25	9,53	297.87	0.00622	0.00125	1170		
0.36	9,53	297.67	0.00624	0.00106	1600		
0.46	9,53	297.53	0.00623	0.00093	2030		
0.56	9,53	297.48	0.0062	0.00085	2440		
0.66	9,53	297.25	0.00623	0.00077	2850		
0.76	9,37	296.29	0.00616	0.00072	3051	0.0006	2793
0.76	9,53	297.23	0.00624	0.00071	3260		
0.86	9,55	297.11	0.00622	0.00066	3670		
0.97	9,71	297.02	0.00624	0.00062	4080		
1.07	10,02	296.96	0.00626	0.00062	4500		
1.17	10,48	296.98	0.00619	0.00069	4930		
1.27	11,03	296.98	0.00619	0.00062	5380		
1.37	11,48	296.15	0.00621	0.00063	5698	0.0009	3386
1.37	11,68	296.84	0.00621	0.00062	5860		
1.47	12,43	296.74	0.00617	0.0006	6370		
1.57	13,30	296.58	0.00626	0.00061	6910		
1.68	14,22	296.49	0.0062	0.00056	7480		
1.70	14,20	296.94	0.00622	0.00058	7107	0.00082	3381
1.78	15,31	296.34	0.00624	0.00056	8100		
1.88	16,56	296.17	0.0062	0.00055	8760		
1.98	17,64	295.59	0.00627	0.00053	8859	0.00086	3368
1.98	18,06	296.01	0.00625	0.00052	9480		
2.08	19,85	295.88	0.00626	0.00051	10300		
2.18	21,81	295.76	0.0063	0.00046	11100		
2.29	24,02	295.83	0.00574	0.00044	12100		
2.39	26,49	295.88	0.00551	0.00043	13000		

IDENT 102271: Blackwell blowing with deceleration. Fluid: air at 764 mmHg. Smooth surface

Start profile, $x = 0.254$ m

y, m	u, m/s	T, K
0	0.00	306.48
0.000117	0.45	305.52
0.000168	0.68	304.97
0.000218	0.91	304.39
0.000269	1.13	303.81
0.000345	1.42	303.07
0.000447	1.73	302.36
0.000549	2.01	301.85
0.000803	2.49	300.93
0.001057	2.83	300.23
0.001565	3.23	299.47
0.002073	3.57	298.98
0.002708	3.87	298.39
0.003978	4.39	297.57
0.005248	4.86	296.90
0.007153	5.55	296.07
0.008423	5.99	295.60
0.009693	6.41	295.17
0.010963	6.84	294.79
0.012233	7.25	294.48
0.013503	7.58	294.23
0.016043	8.04	293.86
0.018583	8.17	293.76
0.021123	8.17	293.76

x, m	u_∞, m/s	T_{wall}, K	$\dot{m}''/G$	St	Re_{Δ_2}	$c_f/2$	Re_{δ_2}
0.25	8.14	306.65	0.00396	0.00165	880	0.00073	1440
0.36	7.85	306.76	0.00395	0.00136	1159		
0.46	7.60	306.65	0.00396	0.00119	1424		
0.56	7.44	306.37	0.00395	0.00108	1672	0.00042	2700
0.56	7.42	306.71	0.00395	0.00108	1680		
0.66	7.27	306.82	0.00394	0.001	1927		
0.76	7.13	306.71	0.00395	0.00094	2168		
0.86	7.01	306.29	0.00394	0.00089	2410	0.00035	3808
0.86	7.01	306.65	0.00394	0.00089	2404		
0.97	6.90	306.76	0.00396	0.00085	2634		
1.07	6.81	306.65	0.00396	0.00081	2861		
1.17	6.74	306.27	0.00398	0.00078	3118	0.00031	4897
1.17	6.72	306.71	0.00398	0.00078	3084		
1.27	6.64	306.76	0.00397	0.00076	3304		
1.37	6.58	306.76	0.00395	0.00073	3520		
1.47	6.52	306.27	0.00396	0.00071	3672	0.00028	5928
1.47	6.51	306.71	0.00396	0.00071	3734		
1.57	6.45	306.76	0.00397	0.00069	3945		
1.68	6.40	306.76	0.00396	0.00068	4154		
1.78	6.37	306.21	0.00392	0.00066	4354	0.00026	6992
1.78	6.36	306.71	0.00392	0.00068	4361		
1.88	6.31	306.76	0.00395	0.00064	4566		
1.98	6.27	306.76	0.00395	0.00063	4769		
2.08	6.22	306.17	0.00394	0.00062	5007	0.00025	7926
2.08	6.22	306.76	0.00394	0.00062	4970		
2.18	6.18	306.71	0.00392	0.00061	5169		
2.29	6.14	306.76	0.00394	0.0006	5365		

Profile at $x = 2.083$ m

y, m	u, m/s	T, K
0	0.00	306.17
0.000244	0.20	305.57
0.000295	0.25	305.43
0.000396	0.35	305.15
0.000599	0.55	304.36
0.000853	0.74	303.64
0.001107	0.95	302.46
0.002123	1.25	301.29
0.003393	1.56	300.52
0.005933	1.88	299.76
0.008473	2.01	299.31
0.013553	2.33	298.50
0.023713	2.85	297.72
0.036413	3.42	296.96
0.042763	3.68	296.66
0.049113	3.89	296.29
0.055463	4.13	295.91
0.061813	4.47	295.69
0.068163	4.72	295.42
0.074513	4.96	295.14
0.080853	5.22	294.89
0.087213	5.47	294.72
0.093563	5.75	294.44
0.099913	5.97	294.20

IDENT 111369: Kearney mixed blowing and acceleration. Fluid: air at 765 mmHg. Smooth surface

Start profile, $x = 0.351$ m

y, m	u, m/s	T, K	x, m	u_∞, m/s	T_{wall}, K	$\dot{m}''/G$	St	Re_{Δ_2}	$c_f/2$	Re_{δ_2}
0.00	0.00	308.15	0.15	7.08	307.93	0.0042	0.00218	479		
0.000089	0.43	306.60	0.25	7.07	308.04	0.0042	0.00182	767		
0.00014	0.67	305.98	0.35	7.10	308.15	0.0042	0.00156	1020	0.0013	1210
0.000191	0.91	305.51	0.36	7.10	308.26	0.0041	0.00157	1032		
0.000292	1.31	304.74	0.46	7.26	308.09	0.0042	0.00136	1316		
0.00047	2.04	303.10	0.56	7.74	308.32	0.0038	0.00126	1571		
0.000724	2.74	301.90	0.66	8.74	308.43	0.0038	0.00118	1846		
0.000851	2.95	301.43	0.76	10.44	308.15	0.0038	0.00107	2200		
0.000978	3.12	301.26	0.86	12.80	307.98	0	0.00172	2473		
0.001232	3.41	300.52	0.97	16.41	307.82	0	0.00148	2668		
0.001613	3.70	299.82	1.07	22.10	307.82	0	0.00125	2872		
0.002121	3.91	299.26	1.17	22.59	307.93	0	0.00193	3084		
0.002756	4.19	298.61	1.27	22.55	307.59	0	0.00183	3443		
0.003518	4.45	298.03	1.37	22.59	307.32	0	0.00174	3787		
0.004153	4.62	297.58	1.47	22.59	307.32	0	0.00174	4041		
0.005423	4.97	296.87	1.57	22.55	307.32	0	0.0017	4308		
0.006693	5.28	296.21	1.68	22.55	307.26	0	0.00169	4564		
0.007963	5.55	295.62	1.78	22.59	307.26	0	0.00165	4829		
0.009868	5.97	294.96	1.88	22.57	307.21	0	0.00165	5100		
0.011773	6.33	294.35	1.98	22.57	307.21	0	0.00163	5325		
0.014313	6.73	293.74	2.08	22.62	307.21	0	0.00161	5602		
0.016853	6.97	293.29	2.18	22.62	307.15	0	0.00162	5839		
0.019393	7.07	293.04	2.29	22.57	307.09	0	0.00158	6109		
0.021933	7.09	292.98								

IDENT 122371: Blackwell adverse pressure gradient with suction. Fluid: air at 766 mmHg. Smooth surface

Start profile, $x = 0.254$ m

y, m	u, m/s	T, K
0.00000	0.00	306.32
0.00018	2.53	301.98
0.00023	3.01	300.57
0.00028	3.45	299.51
0.00033	3.82	298.60
0.00038	4.15	297.82
0.00043	4.46	297.17
0.00048	4.72	296.54
0.00056	5.02	295.84
0.00066	5.32	295.11
0.00078	5.63	294.43
0.00094	5.85	293.93
0.00114	6.04	293.46
0.00147	6.28	292.89
0.00198	6.50	292.39
0.00274	6.77	291.93
0.00388	7.05	291.47
0.00642	7.58	290.99
0.00769	7.80	290.86
0.00896	7.97	290.79
0.01150	8.15	290.74
0.01404	8.16	290.74
0.01587	8.17	290.74

x, m	u_∞, m/s	T_{wall}, K	$\dot{m}''/G$	St	Re_{Δ_2}	Re_{Δ_2} from profiles	$c_f/2$	Re_{δ_2}
0.25	8.19	306.11	−0.00394	0.00509	260	261	0.0038	558
0.36	7.90	306.11	−0.00395	0.00491	310			
0.46	7.64	306.11	−0.00395	0.00479	348			
0.56	7.43	306.22	−0.00396	0.0047	381	374	0.00383	665
0.66	7.28	306.16	−0.00396	0.00463	409			
0.76	7.14	306.16	−0.00396	0.00457	434			
0.86	7.02	306.05	−0.00396	0.00452	457	459	0.00381	714
0.97	6.90	306.11	−0.00397	0.00448	478			
1.07	6.80	306.05	−0.00397	0.00444	498			
1.17	6.71	306.16	−0.00397	0.00441	516	521	0.00379	757
1.27	6.63	306.16	−0.00395	0.00438	534			
1.37	6.57	306.05	−0.00396	0.00435	550			
1.47	6.50	306.16	−0.00396	0.00432	566	570	0.00377	768
1.57	6.45	306.05	−0.00395	0.0043	580			
1.68	6.39	306.11	−0.00398	0.00428	595			
1.78	6.34	306.11	−0.00399	0.00426	608	622	0.00376	789
1.88	6.30	306.11	−0.00396	0.00424	622			
1.98	6.25	306.16	−0.00396	0.00422	634			
2.08	6.20	306.16	−0.00397	0.0042	647	645	0.00375	812
2.18	6.14	306.22	−0.00396	0.00418	658			
2.29	6.09	306.22	−0.00398	0.00417	670	653	0.00375	813

Profile at $x = 2.29$ m

y, m	u, m/s	T, K
0.00000	0.00	306.49
0.00021	1.67	303.20
0.00026	2.01	302.27
0.00031	2.28	301.49
0.00036	2.57	300.72
0.00041	2.81	300.14
0.00052	3.20	299.13
0.00062	3.51	298.16
0.00072	3.72	297.44
0.00082	3.96	296.73
0.00097	4.23	296.03
0.00118	4.44	295.26
0.00118	4.61	294.64
0.00176	4.76	294.14
0.00227	4.94	293.51
0.00328	5.08	293.03
0.00544	5.28	292.46
0.00989	5.50	291.87
0.01814	5.81	291.31
0.02259	5.93	291.10
0.02767	6.03	290.97
0.03275	6.07	290.88
0.03783	6.07	290.87

IDENT 71374: Pimenta fully rough surface. Fluid: air at 759.2 mmHg. Surface of packed spheres, 1.27 mm diameter

Start profile, $x = 0.66$ m											Profile at $x = 2.184$ m		
y, m	u, m/s	T, K	x, m	u_∞, m/s	T_{wall}, K	$\dot{m}''/G$	St	Re_{Δ_2}	$c_f/2$	Re_{δ_2}	y, m	u, m/s	T, K
0	0.00	308.39	0.051	39.82	308.65	0	0.003	458			0	0.00	308.41
0.000178	13.25	301.76	0.152	39.82	308.32	0	0.004	1,455			0.000178	11.84	302.76
0.000229	13.94	301.61	0.254	39.82	308.26	0	0.003	2,434			0.000229	12.38	302.61
0.00033	15.05	301.22	0.356	39.82	308.32	0	0.003	3,276			0.000356	13.67	302.16
0.000483	16.35	300.74	0.457	39.82	308.37	0	0.003	4,054			0.00066	15.79	301.48
0.000686	17.75	300.29	0.559	39.82	308.32	0	0.003	4,791			0.001194	18.16	300.71
0.00106	19.33	299.78	0.66	39.82	308.43	0	0.003	5,508	0.00261	5,935	0.002032	20.45	299.90
0.001473	20.98	299.24	0.762	39.82	308.43	0	0.003	6,203			0.003302	22.67	299.13
0.002134	22.81	298.64	0.864	39.82	308.21	0	0.002	6,869			0.005588	25.03	298.34
0.002997	24.72	298.02	0.965	39.82	308.21	0	0.002	7,522	0.00252	8,093	0.00889	27.44	297.47
0.003556	25.77	297.61	1.067	39.82	308.21	0	0.002	8,165			0.013208	29.69	296.61
0.004191	26.97	297.23	1.168	39.82	308.32	0	0.002	8,800			0.018796	32.27	295.72
0.004953	28.15	296.90	1.27	39.82	308.32	0	0.002	9,430	0.00243	9,974	0.022352	33.61	295.18
0.005969	29.57	296.44	1.372	39.82	308.32	0	0.002	10,052			0.026416	35.21	294.66
0.006985	31.03	295.95	1.473	39.82	308.21	0	0.002	10,661			0.030988	36.56	294.08
0.013081	32.34	295.49	1.575	39.82	308.26	0	0.002	11,625	0.00236	11,582	0.036068	38.07	293.43
0.009144	33.78	295.04	1.676	39.82	308.32	0	0.002	11,865			0.041656	39.06	292.84
0.010414	35.19	294.52	1.778	39.82	308.21	0	0.002	12,468			0.047752	39.76	292.46
0.011684	36.62	294.00	1.88	39.82	308.26	0	0.002	13,071	0.00229	13,344	0.054102	39.81	292.34
0.012954	37.60	293.54	1.981	39.82	308.21	0	0.002	13,668			0.060452	39.86	292.33
0.014224	38.62	293.09	2.083	39.82	308.21	0	0.002	14,256			0.066802	39.85	292.31
0.016129	39.34	292.61	2.184	39.82	308.21	0	0.002	14,838	0.00224	15,142	0.073152	39.83	292.28
0.018034	39.71	292.35	2.286	39.82	308.15	0	0.002	15,423			0.079502	39.81	292.25
0.020574	39.71	292.28	2.388	39.82	308.15	0	0.002	16,007			0.082042	39.79	292.23

IDENT 80874: Pimenta rough surface with blowing. Fluid: air. Surface of packed spheres, 1.27 mm diameter

Start profile, x = 0.66 m

y, m	u, m/s	T, K
0	0.00	313.95
0.000178	5.58	309.69
0.000305	6.46	309.26
0.000483	7.49	308.73
0.000711	8.49	308.14
0.001067	9.67	307.58
0.001524	10.61	306.96
0.002184	11.85	306.36
0.002946	12.98	305.76
0.003937	14.11	305.13
0.005715	15.84	304.27
0.006985	16.95	303.72
0.008255	17.85	303.15
0.01016	19.34	302.42
0.012065	20.60	301.72
0.01397	21.85	301.05
0.015875	22.96	300.41
0.01778	24.03	299.82
0.019685	24.99	299.27
0.02159	25.82	298.81
0.023495	26.33	298.31
0.0254	26.76	298.02
0.02794	26.99	297.83
0.03048	27.08	297.74

x, m	u_∞, m/s	T_{wall}, K	$\dot{m}''/G$	St	Re_{Δ_2}	$c_f/2$	Re_{δ_2}
0.051	27.05	313.59	0.004	0.0019	509		
0.152	27.05	313.54	0.004	0.0031	1,629		
0.254	27.05	313.54	0.004	0.002	2,757		
0.356	27.05	313.48	0.004	0.0017	3,766		
0.457	27.05	313.54	0.004	0.0016	4,737		
0.559	27.05	313.54	0.004	0.0013	5,674		
0.660	27.05	313.43	0.004	0.0012	6,579	0.0012	6,862
0.762	27.05	313.37	0.004	0.0011	7,459		
0.864	27.05	313.37	0.004	0.0011	8,334		
0.965	27.05	313.37	0.004	0.001	9,212	0.0011	9,417
1.067	27.05	313.32	0.004	0.0011	10,076		
1.168	27.05	313.37	0.004	0.001	10,954		
1.27	27.05	313.48	0.004	0.001	11,832	0.0011	12,141
1.372	27.05	313.48	0.004	0.0009	12,674		
1.473	27.05	313.59	0.004	0.0009	13,511		
1.575	27.05	313.59	0.004	0.0009	14,368	0.0011	14,551
1.676	27.05	313.48	0.004	0.0009	15,231		
1.778	27.05	313.48	0.004	0.0008	16,075		
1.88	27.05	313.54	0.004	0.0008	16,904	0.001	17,225
1.981	27.05	313.54	0.004	0.0009	17,740		
2.083	27.05	313.59	0.004	0.0008	18,578		
2.184	27.05	313.59	0.004	0.0007	19,406	0.0009	20,136
2.286	27.05	313.54	0.004	0.0008	20,226		
2.388	27.05	313.59	0.004	0.0007	21,042		

Profile at x = 2.184 m

y, m	u, m/s	T, K
0	0	314.38
0.000178	4.27	310.88
0.000305	5.19	310.47
0.000559	6.27	309.86
0.000914	7.57	309.18
0.001372	8.42	308.67
0.002159	9.68	308.02
0.003302	10.84	307.33
0.005588	12.34	306.44
0.009398	14.16	305.44
0.014478	16.02	304.51
0.021692	17.63	303.58
0.025908	18.67	302.98
0.03048	19.66	302.45
0.03556	20.77	301.83
0.04064	21.81	301.27
0.04572	22.77	300.69
0.05207	23.98	300.06
0.05842	24.97	299.44
0.06477	25.91	298.84
0.07112	26.54	298.34
0.07747	26.90	297.99
0.08382	26.97	297.83
0.09017	27.08	297.74

IDENT 8000: Gibson mild convex surface. Fluid: air at 760 mmHg. Nominal $T_\infty = 289$ K. Nominal temperature difference 14 K. Smooth surface

x, m	Curvature, 1/m	u_∞, m/s	T_{wall}, K	Re_{δ_2}	Re_{Δ_2}	St
-0.705	0	22.25	303	1756	1756	0.00234
-0.248	0	23.14	304.4	2833	3306	0.002
0	0	23.07	304.4	3421		
0.209	0.4098	23.17	304.4	3767	4098	0.00175
0.467	0.4098	23.41	304.19	4268	4873	0.0016
0.743	0.4098	23.61	304.05	4899	5798	0.0015
1.001	0.4098	23.83	304.42	5560	6857	0.00137

Start profile, $x = -0.705$ m

y, m	u, m/s	T, K
0	0	303
0.0003937	12.385	295.02
0.0006477	14.189	294.26
0.0009017	14.882	293.83
0.0011557	15.385	293.42
0.0014097	15.84	293.10
0.0016637	16.252	292.87
0.0019177	16.554	292.66
0.0021717	16.871	292.41
0.0026289	17.477	292.04
0.0031369	18.008	291.75
0.0036449	18.306	291.45
0.0041529	18.974	291.15
0.0046609	19.421	290.95
0.0051689	19.901	290.67
0.0059309	20.36	290.36
0.0069469	21.071	290.05
0.0074549	21.387	289.89
0.0082169	21.795	289.66
0.0089789	22.119	289.46
0.0097409	22.416	289.28
0.0105029	22.49	289.16
0.0112649	22.563	289.08
0.0120269	22.636	289.02

Profile at $x = 1.001$ m

y, m	u, m/s	T, K
0	0	303.42
0.00039	10.26	298.39
0.00065	11.91	297.17
0.0009	12.64	296.63
0.00118	12.96	296.29
0.00141	13.33	295.95
0.00166	13.68	295.73
0.00192	14.03	295.55
0.00217	14.2	295.32
0.00268	14.65	294.97
0.00319	15.03	294.78
0.00364	15.45	294.49
0.00415	15.86	294.33
0.00466	16.06	294.06
0.00517	16.35	293.90
0.00593	16.79	293.64
0.00695	17.35	293.35
0.00822	18.07	292.94
0.0105	19.01	292.39
0.01355	20.02	291.59
0.01685	21.1	290.94
0.02092	22.15	290.27
0.02498	22.95	289.62
0.02904	23.33	289.24
0.03313	23.4	289.00

Note: u_∞ is the potential flow velocity at the wall.

IDENT 8100: Gibson mild concave surface. Fluid: air at 760 mmHg. Nominal $T_\infty = 289$ K. Nominal temperature difference 14 K. Smooth surface

Start profile, $x = -0.705$ m			Curvature,							
y, m	u, m/s	T, K	x, m	1/m	u_∞, m/s	T_{wall}, K	Re_{δ_2}	Re_{Δ_2}	St	$c_f/2$
0	0	303	−0.705	0	22.25	303	1756	1756	0.00234	
0.0004	12.385	295.02	−0.248	0	23.14	304.4	2833	3306	0.002	0.00167
0.0006	14.189	294.26	0	0	23.07	304.4	3421			
0.0009	14.882	293.83	0.229	−0.4098	23.01	304.4	4222	4611	0.00185	0.00158
0.0012	15.385	293.42	0.482	−0.4098	23.3	304.19	4807	5086	0.0019	0.0016
0.0014	15.84	293.10	0.737	−0.4098	23.41	304.05	5193	5677	0.00196	0.00161
0.0017	16.252	292.87	0.989	−0.4098	23.81	304.42	5638	6979	0.00199	0.00167
0.0019	16.554	292.66								
0.0022	16.871	292.41								
0.0026	17.477	292.04								
0.0031	18.008	291.75								
0.0036	18.306	291.45								
0.0042	18.974	291.15								
0.0047	19.421	290.95								
0.0052	19.901	290.67								
0.0059	20.36	290.36								
0.007	21.071	290.05								
0.0075	21.387	289.89								
0.0082	21.795	289.66								
0.009	22.119	289.46								
0.0097	22.416	289.28								
0.0105	22.49	289.16								
0.0113	22.563	289.08								
0.012	22.636	289.02								

Note: u_∞ is the potential flow velocity at the wall.

IDENT 70280: Simon strongly convex surface. Fluid: air at 760 mmHg. Smooth surface

Start profile, $x = 0.0$ m			x, m	Curvature, 1/m	u_∞, m/s	T_{wall}, K	St	Re_{Δ_2}	$c_f/2$	Re_{δ_2}	Profile at $x = 0.613$ m		
y, m	u, m/s	T, K									y, m	u, m/s	T, K
0	0	313.76	-0.013	0	14.65	313.74	0.00233	1755	0.00155	4445	0	0	313.7
0.00014	1.15	310.19	0	0	14.56	313.76	0.0023	1756			0.00014	1.28	311.56
0.000216	1.77	308.5	0.024	2.22	14.88	313.91	0.00199	1814			0.0019	1.74	310.81
0.000241	1.98	308.5	0.073	2.22	14.88	313.88	0.00192	1911			0.000241	2.21	310.19
0.000292	2.40	307.33	0.103	2.22	14.63	313.76	0.002	1827	0.00143	4630	0.000292	2.67	309.69
0.000343	2.82	306.5	0.124	2.22	14.76	313.94	0.00191	1999			0.000394	3.6	308.74
0.000444	3.65	305.38	0.174	2.22	14.71	313.94	0.00179	2089			0.000597	4.9	307.37
0.0008	6.57	303.59	0.225	2.22	14.73	313.86	0.00171	2185			0.001156	6.26	305.58
0.001664	8.5	301.91	0.255	2.22	14.65	313.65	0.00169	2198	0.00121	5024	0.002172	7.08	304.2
0.002934	9.39	300L92	0.276	2.22	14.76	313.86	0.00161	2266			0.003442	7.69	303.19
0.004204	9.93	300.25	0.326	2.22	14.79	313.85	0.00157	2347			0.00522	8.41	302.17
0.0649	10.61	299.45	0.377	2.22	14.78	313.87	0.00155	2421			0.00776	9.25	301.02
0.00776	10.92	299.11	0.405	2.22	14.66	313.82	0.00159	2459	0.00107	5342	0.009792	9.82	300.25
0.009792	11.29	298.67	0.428	2.22	14.76	313.87	0.00152	2496			0.011824	10.3	299.62
0.011824	11.62	298.33	0.478	2.22	14.73	313.91	0.00148	2563			0.014363	10.76	299.02
0.014364	11.99	298	0.529	2.22	14.72	313.95	0.00145	2630			0.016904	11.1	209.59

IDENT 70280 (Continued)

Start profile, x = 0.0 m

y, m	u, m/s	T, K
0.016904	12.31	297.74
0.021984	12.88	297.4
0.027064	13.37	297.24
0.033414	13.85	297.12
0.041034	14.3	297.08
0.048654	14.54	297.06
0.061354	14.54	297.05
0.07054	14.54	297.03

Profile at x = 0.613 m

y, m	u, m/s	T, K
0.021984	11.67	298.05
0.027064	12.15	297.77
0.033414	12.59	297.49
0.041034	13.05	297.31
0.048654	13.22	297.22
0.061354	12.95	297.19
0.074054	12.63	297.18
0.086754	12.33	297.18

Curvature, 1/m	x, m	u_∞, m/s	T_{wall}, K	St	Re_{Δ_2}	$c_f/2$	Re_{δ_2}
2.22	0.58	14.74	313.96	0.00135	2713		
2.22	0.613	14.71	313.7	0.00136	2886	0.001	5647
2.22	0.63	14.88	313.82	0.00131	2785		
2.22	0.681	14.88	313.87	0.0012	2840		
0	0.72	14.64	314.08	0.00147	2848		
0	0.723	14.45	314.23	0.00156	2930	0.00099	5696
0	0.772	14.56	314.36	0.00152	2875		
0	0.824	14.66	314.4	0.00151	2944		
0	0.876	14.72	314.47	0.00145	3005		
0	0.882	14.58	314.73	0.00149	3236	0.00101	6024
0	0.928	14.75	314.21	0.00147	3126		
0	0.981	14.76	314.19	0.0015	3207		
0	1.04	14.56	314.43	0.00162	3539	0.00105	6216
0	1.033	14.76	314.18	0.00157	3286		
0	1.085	14.76	314.22	0.00159	3358		
0	1.137	14.76	313.19	0.00159	3484		
0	1.189	14.76	314.25	0.00156	3511		
0	1.191	14.58	314.41	0.00171	3746	0.00108	6365
0	1.242	14.76	313.92	0.00158	3660		
0	1.294	14.78	313.89	0.00157	3742		

Notes: u_∞ is the potential flow velocity at the wall. $c_f/2$ is based on law of the wall.

IDENT 112188: Hollingsworth concave surface, Fluid: water, $Pr_{average} = 5.90$, average viscosity 8.66×10^{-6} m²/s. Heating starts at $x = 1.14$ m (from trip). Smooth surface

Profile at $x = 3.07$ m (from trip)				Curvature,						Profile at $x = 5.1$ m			
y, m	u, m/s	T, K	x, m	1/m	u_∞, m/s	St	Re_{Δ_2}	$c_f/2$	Re_{δ_2}	y, m	u, m/s	y, m	T, K
0	0	301.42	2.65	0	0.152	0.000845	238	0.00205	1377	0.00000	0.0000	0.00000	301.64
0.00022	0.0118	300.68	3.07	0	0.152	0.000833	287	0.002	1552	0.00022	0.0132	0.00042	299.98
0.00024	0.0126	300.6	3.66	-0.735						0.00030	0.0164	0.00057	299.49
0.00028	0.0147	300.4	4.04	-0.735	0.153	0.000934	441	0.00216	1867	0.00040	0.0216	0.00072	299.08
0.00032	0.0163	300.34	4.39	-0.735	0.149	0.000956	474	0.00234	2102	0.00053	0.0283	0.00088	298.82
0.00036	0.0181	300.17	4.74	-0.735	0.149	0.000964	549	0.00249	2188	0.00070	0.0364	0.00103	298.61
0.00042	0.0208	299.98	5.1	-0.735	0.149	0.000968	599	0.00263	2102	0.00094	0.0458	0.00118	298.51
0.0005	0.0240	299.76								0.00125	0.0570	0.00133	298.36
0.00059	0.0274	299.48								0.00164	0.0680	0.00147	298.28
0.0007	0.0324	299.19								0.00218	0.0761	0.00163	298.25
0.00101	0.0452	298.77								0.00286	0.0871	0.00192	298.13
0.00152	0.0590	298.4								0.00497	0.0995	0.00291	297.99
0.00228	0.0747	298.16								0.00859	0.1060	0.00455	297.86
0.00349	0.0854	298.03								0.01482	0.1125	0.00659	297.81
0.00534	0.0944	297.93								0.02558	0.1162	0.00918	297.77
0.00821	0.1030	297.82								0.04058	0.1207	0.01243	297.74
0.01267	0.1097	297.76								0.05559	0.1245	0.01652	297.72
0.01961	0.1181	297.69								0.07059	0.1294	0.02168	297.70
0.03036	0.1272	297.62								0.08559	0.1355	0.02817	297.65
0.04528	0.1378	297.56								0.10059	0.1416	0.03909	297.62
0.06024	0.1454	297.53								0.11559	0.1457	0.05662	297.58
0.0752	0.1508	297.51								0.13059	0.1507	0.07662	297.59
0.09017	0.1528	297.51								0.14558	0.1531	0.09662	297.51
0.0999	0.1527	297.51								0.15996	0.1546	0.11662	297.51

Notes: u_∞ is the potential flow velocity at the wall. $c_f/2$ is based on the law of the wall.

IDENT G3000: Ames high free-stream turbulence. Fluid: air at 760 mmHg. Smooth surface

Start profile, x = 0.2032 m

y, m	u, m/s	T, K
0	0.0	313.26
0.000414		302.04
0.000502	16.54	301.45
0.000635	17.32	300.64
0.000767	17.99	300.07
0.000952	18.68	299.58
0.001163	19.30	299.16
0.001428	19.84	298.79
0.001745	20.57	298.46
0.002168	21.25	298.13
0.002749	21.67	297.83
0.00341	22.10	297.56
0.004203	22.61	297.31
0.005393	23.08	297.10
0.006714	23.44	296.89
0.008432	23.61	296.76
0.010679	23.83	296.66
0.013323	23.86	296.58
0.016627	23.68	296.54
0.020724	23.85	296.49
0.02654	23.87	296.47
0.033148		296.46

x, m	T_{wall}, K	Tu	u, m/s	St	Re_{Δ_2}	$c_f/2$	Re_{δ_2}
0.042			22.86	0.0039			
0.1524		0.166	22.94	0.00336		0.00249	1107
0.2032	313.26		23.88	0.00315		0.00244	1142
0.252			23.17	0.0031	1035		
0.356			23.5	0.00294			
0.4572		0.137	23.65	0.00284	1985	0.00217	2039
0.508	315.36		23.9	0.00279		0.00221	1836
0.559			23.78	0.00274			
0.66			23.73	0.00269			
0.762			23.69	0.00261			
0.864			23.64	0.0026		0.00204	2567
0.965			23.6	0.00252		0.00207	2464
1.0668	317.71	0.101	23.62	0.00251			
1.1176			23.63	0.00251	3674		
1.168			23.56	0.00245			
1.27			23.58	0.00243			
1.372		0.085	23.59	0.00239			
1.473			23.31	0.00238		0.00197	2841
1.524			23.71	0.00238	4391	0.00199	2877
1.575			23.63	0.00232			
1.676			23.66	0.00231			
1.778			23.69	0.00228			
1.88			23.72	0.00226			
1.981			23.75	0.00221			
2.08	319.73	0.068	23.72	0.00221	5243	0.00189	3607
2.1336			24	0.00223		0.00187	3725
2.184			23.78	0.00219			
2.286			23.78	0.00217			
2.388			23.78	0.00217			

Profile at x = 2.1336 m

y, m	u, m/s	T, K
0	0.00	319.73
0.0005556		306.46
0.0007671	14.65	305.23
0.0010844	15.65	304.00
0.0015604	16.67	303.12
0.0023007	17.62	302.14
0.0035436	18.96	300.82
0.005527	20.35	299.52
0.0068492	20.82	299.20
0.008568	21.43	298.59
0.0108158	21.95	298.29
0.0134601	22.40	297.83
0.0167657	22.85	297.54
0.0208646	23.08	297.17
0.0266822	23.41	296.95
0.0332932	23.60	296.76
0.0425486	23.76	296.59
0.0531263	23.97	296.49
0.0663484	23.97	296.37
0.0795703	24.08	296.34
0.1060143	23.93	296.29
0.1324583	24.10	296.23
0.158903	24.02	296.20
0.1853462	24.05	296.18
0.211791	23.90	296.17

Note: The scale of turbulence can be estimated from the decay of Tu.

IDENT 9000: Boldman supersonic nozzle (NASA TND-3221, March 1966). Inside cylinder to $z = -0.12$ m; then 30° half-angle convergence; then 15° half-angle divergence. The radius of curvature at the throat is equal to the throat diameter (0.0379 m). At $x = -0.334$ m, $Re_{\delta_2} = 3800$. The surface is insulated until $x = -0.137$ m

Axial distance z, m	Distance x, m	Radius, m	P/P_0	T/T_0	ρ/ρ_0	u, m/s	T_w, K	Curvature, 1/m	$\dot{q}''$, W/m^2	St
−0.313	−0.3298	0.165				12.89	532.28	0	0	
−0.12	−0.1367	0.165				12.89	532.28	0	0	
−0.115	−0.1304	0.158	0.999	1.00	1.00	13.54	379.04	0	0.0082	0.0037
−0.089	−0.1011	0.129	0.999	1.00	1.00	21.41		0		
−0.064	−0.0717	0.1	0.996	1.00	1.00	33.64	413.48	0	0.0111	0.00261
−0.055	−0.0613	0.089	0.994	1.00	1.00	43.22	423.54	0	0.0118	0.00235
−0.046	−0.0511	0.079	0.99	1.00	0.99	56.75	433.15	0	0.0123	0.00206
−0.037	−0.0408	0.069	0.981	0.99	0.99	76.05	443.09	0	0.0135	0.00189
−0.028	−0.0305	0.059	0.963	0.99	0.97	106.60	451.93	0	0.0155	0.00175
−0.019	−0.0198	0.048	0.872	0.96	0.91	202.70		26.39		
−0.016	−0.016	0.045	0.825	0.95	0.87	239.20	462.32	26.39	0.016	0.00107
−0.009	−0.0092	0.04	0.71	0.91	0.78	315.80		26.39		
−0.004	−0.0045	0.038	0.58	0.86	0.68	392.85	461.32	26.39	0.0154	0.00086
0	0	0.038	0.459	0.80	0.57	462.58	458.65	26.39	0.0152	0.00086
0.0033	0.00331	0.038	0.373	0.76	0.49	512.98	455.32	26.39	0.0148	0.00087
0.0065	0.00651	0.039	0.29	0.71	0.41	565.25	453.82	26.39	0.0141	0.00092
0.0098	0.00992	0.04	0.23	0.66	0.35	606.59		26.39		
0.01	0.01008	0.041	0.225	0.66	0.34	610.24	443.54	0	0.0141	0.00091
0.0161	0.01644	0.044	0.182	0.62	0.29	643.10	437.65	0	0.0132	0.0009
0.031	0.03188	0.052	0.111	0.54	0.21	708.20	423.21	0	0.0117	0.00091
0.0695	0.07172	0.072	0.037	0.40	0.09	809.83	387.54	0	0.009	0.00101
0.1389	0.14356	0.11	0.009	0.27	0.03	892.59	332.21	0	0.0042	0.00082
0.2083	0.21543	0.147	0.004	0.21	0.02	927.85	308.54	0	0.0027	0.00086
0.2802	0.28988	0.186	0.002	0.18	0.01	945.16	303.65	0	0.0022	0.001

Notes: The Stanton number based on an assumed recovery factor of 0.892. $P_0 = 2064.99$ kPa, $T_0 = 523.3$ K, $\rho_0 = 13.52$ kg/m^3, $R/\mathfrak{M} = 287$ J/(kg · K).

AUTHOR INDEX

Allen, R. W., 369
Ambrok, G. S., 310
Anderson, P. S., 242, 243
Arpaci, V. S., 415
Aung, W., 87, 157, 254, 353

Back, L. H., 43
Bailey, F. J., 415
Barnes, J. F., 369
Baron, J. R., 18, 515, 540
Baxter, D. C., 565
Bell, D., 310
Bergles, A. E., 354
Bhatti, M. S., 157
Bird, R. B., 18
Blackwell, B. F., 309
Blair, M. F., 310
Blasius, H., 107
Blottner, F. G., 191
Boelter, L. M. K., 353, 354
Boerner, C. J., 107
Bradshaw, P., 61, 243, 310
Brim, L. H., 369
Bromley, L. A., 558
Brown, S. N., 191
Brown, W. B., 190, 369

Carlson, L. W., 243
Carslaw, H. S., 158
Cebeci, T., 61, 395
Cess, R. D., 158
Chapman, S., 18

Cheesewright, R., 415
Chen, T. S., 416
Chi, S. W., 369, 395, 516
Chu, H. H. S., 415, 416
Churchill, S. W., 415, 416
Clark, S. H., 87
Clauser, F. H., 243
Crawford, M. E., 310
Cohen, C. B., 190, 369
Cowling, T. G., 18
Curtiss, C. F., 18

Davenport, M. E., 369
DeGroot, S. R., 18
Deissler, R. G., 43, 310, 353, 369, 395
Dennis, S. C. R., 158
Dershin, H., 516, 540
Desmon, L. G., 369
Dewey, C. F., Jr., 107, 190
Dipprey, D. F., 310, 354
Dittus, P. W., 353
Donoughe, P. L., 190, 369
Drake, R. M., Jr., 191
Dyadyakin, B. V., 369

Eckert, E. R. G., 190, 191, 369, 395, 415
Ede, A. J., 415
Eichhorn, R., 415
Elzy, E., 191
Emmons, H. W., 190
Endo, K., 354
Eustis, R. H., 310, 369
Evans, H. L., 515

591

Falkner, V. M., 107
Fand, R. M., 416
Fishenden, M., 416
Flemming, D. P., 87
Fletcher, D. D., 310
Fujii, M., 415
Fujii, T., 415
Futagami, K., 416

Gallagher, J. S., 557
Gambill, W. R., 558
Gartner, D., 353
Gebhart, B., 415
Gibson, M. M., 309
Gnielinski, V., 353
Goldstein, R. J., 310, 416
Gregg, J. L., 415, 416
Gross, J. F., 107, 190

Haar, L., 557
Hallman, T. M., 157, 353
Hancock, P. E., 243, 310
Hansen, A. G., 107
Harlow, F. H., 242
Hartnett, J. P., 254, 353
Hatton, A. P., 353
Heaton, H. S., 87, 158
Hestermans, P., 557
Hilsenrath, J., 557
Hinze, J. O., 61
Hirschfelder, J. O., 18
Hokenson, G. L., 74
Hollands, K. G. T., 415
Hollingsworth, D. K., 309
Hornbeck, R. W., 87, 158
Howe, J. T., 190
Humble, L. V., 360

Irvine, T. F., Jr., 254
Iversen, H. W., 354

Jackson, C. B., 558
Jackson, T. W., 415
Jaeger, J. C., 158
Jenkins, R., 310
Johannsen, K., 353
Johnson, P. L., 242
Johnston, J. P., 242, 309
Hones, D. C., 416
Jones, W. P., 243

Kakac, S., 87, 157, 254, 353
Kays, W. M., 87, 157, 158, 191, 242, 243,
    254, 309, 310, 353, 369, 416, 479

Kearney, D. W., 310
Keenan, J. H., 18
Kim, J., 61, 310
Klein, J., 157, 191
Kline, S. J., 309
Knuth, W. L., 18, 515, 516, 540
Koh, J. C. Y., 254
Kuehn, T. H., 416

Langhaar, H. L., 87
Latzko, H., 254
Launder, B. E., 243
Le Fevre, E. J., 415
Leigh, D., 190
Lel'Chuk, V. D., 369
Leontiev, A. I., 310
Leung, E. Y., 353
Lewandowski, K., 557
Liley, P. E., 557
Lin, S. H., 87
Lindsay, A. L., 558
Livingood, J. N. B., 190
Lloyd, J. R., 415, 416
Loeffler, A. L., Jr., 395
London, A. L., 87, 157, 191, 254, 353, 479
Lowdermilk, W. H., 369
Lum, M.,
Lundberg, R. E., 87, 157
Lundgren, T. S., 87
Lyon, R. N., 558

McCarthy, J. R., 369
McCarty, R. D., 557, 558
McCornas, S. T., 254
McEligot, D. M., 369
Maciejewski, P. K., 310
Makita, T., 557
Mansour, N. N., 61
Mechtly, E. A., 562
Mercer, A. McD., 158
Mersman, W. A., 190
Metzger, D. E., 310
Meyers, G. E., 310
Michelsen, M. L., 157
Mickley, H. S., 190
Mihkowycz, W. J., 415
Moffat, R. J., 242, 243, 309, 310, 416
Moin, P., 61, 310
Moody, L. F., 254
Moore, F. K., 107
Moran, W. R., 415, 416
Morgan, V. T., 191, 416
Mori, Y., 416
Morris, E. W., 416
Mucoglu, A., 416
Myers, G. E., 310

Nagashima, A., 557
Nakayama, P., 242
Nicoll, W. B., 369
Nikuradse, J., 243
Norris, R. H., 354
Notter, R. H., 353

Onsager, L., 18
Orszag, A., 310
Ostrach, S., 413

Pappas, C. C., 395, 516, 540
Patankar, S. V., 157
Perkins, H. C., 416
Perry, J. H., 558
Petukhov, B. S., 243, 369
Pimenta, M. M., 243, 309
Poferl, D. J., 557
Pohlhausen, E., 395
Poots, G., 158
Prausnitz, J. M., 558

Quack, H., 107
Quarmby, A., 353

Raithby, G. D., 191, 415
Ramm, H., 353
Reichardt, H., 243
Reid, R. C., 558
Reshotko, E., 190, 369
Reynolds, A. J., 310
Reynolds, W. C., 87, 157, 158, 309, 310, 353, 565
Roder, H. M., 558
Rodi, W., 242
Roganov, P. S., 310
Rogers, M. M., 310
Rohonczy, G., 369
Rohsenow, W. M., 353
Rose, W. C., 61
Rosenhead, L., 18
Ross, D. C., 415
Ross, R. C., 190
Rouse, M. W., 353
Rubesin, M. W., 61, 395, 516, 540

Sabersky, D. H., 310, 354
Sakakibara, M., 354
Saunders, O. A., 416
Saxena, S. C., 557
Schauer, J. J., 310
Schlichting, H., 43, 107, 190, 243
Schultz-Grunow, F., 242
Schwind, R. G., 242
Seibers, D. L., 416
Seigel, R., 157
Sellars, J. R., 157

Shaffer, E. C., 158
Shah, R. K., 87, 157, 254, 353
Sherwood, T. K., 558
Shihov, E. V., 310
Singh, S. N.,
Simon, T. W., 309
Simpson, R. L., 243
Sisson, R. M., 191
Skan, S. W., 107
Skupinski, W., 353
Slanciauskas, A., 309
Sleicher, C. A., 353, 369
Smith, A. G., 191
Smith, A. M. O., 395
Soehngen, E. E., 415
Spalding, D. B., 3, 107, 191, 242, 369, 395, 502, 515, 516
Sparrow, E. M., 87, 106, 157, 190, 353, 415, 416
Speigel, R., 353
Squyers, A. L., 190
Stewart, W. E., 190
Stewartson, K., 191
Streeter, V. L., 43
Sutherland, W. A., 353
Svehla, R. A., 557

Tewfik, O. E., 18
Thielbahr, W., M., 309
Thwaites, H., 107
Tortel, J., 353
Touloukian, Y. S., 557
Tribus, M., 157, 191, 353

Van Driest, E. R., 242, 395
Vautrey, L., 353
Verriopoulos, C. A., 309
Villadsen, J., 157
Vliet, G. C., 415

Warner, C. Y., 415
Weber, L. A., 558
Whitten, D. G., 310
Wibulswas, P., 158
Wieghardt, K., 310
Wilke, C. R., 558
Williams, J. C., III, 191
Wolf, H., 369
Worsoe-Schmidt, P. M., 369

Yakhot, A., 310
Yakhot, V. S., 310
Yang, K. T., 369, 415
Young, G., 354
Yu, H. S., 190

Zabolotsky, E. V., 310
Zukauskas, A., 191, 309

SUBJECT INDEX

Accelerating flows (*see* Boundary layer, pressure gradient flows)
Adiabatic wall temperature, 372, 377, 380
Adverse pressure gradient (*see* Boundary layer, pressure gradient flows)
Air-water vapor system (*see* Mass transfer problems)
Ambrok solution, 284
Assymptotic accelerating flow, 236
Assymptotic suction layer, 226
Averaging rules, turbulent flow, 49
Axial conduction effects, 120, 128

Beta function, 177, 564
Binary diffusion coefficient, 16
Binary mixture, 16
Blasius equation, 93
    solution, 94
Blowing, effects of (*see* Boundary layer, transpiration flows)
Blowing parameter:
    free convection, 404, 405
    laminar flow, 99
    mass transfer, 449, 501
    turbulent heat transfer, 288
    turbulent momentum transfer, 224, 228, 230
Boundary layer:
    approximations, 3, 20, 21
    Couette flow, 201
    definition:
        forced convection, 20, 21
        free convection, 399
    equations (*see* Equations)
    free convection:
        other geometries, 411
        vertical plate, 402, 409
        flow regimes, 407

laminar integral solutions, 405
laminar similarity solutions, 398, 404
turbulent integral solutions, 409
variable property effects, 406
laminar momentum:
    integral solutions, 101
    similarity solutions, 94, 96, 98
laminar thermal:
    free convection, 401–404
    integral solutions, 172
    similarity solutions, 163, 165, 167, 170, 171
    unheated starting length, 175
    variable t_0 and $\dot{q}_0''$, 175, 179
    variable free-stream velocity, 183
mixed convection, 413
pressure gradient flows:
    laminar momentum, 96, 104
    laminar thermal, 166, 183
    turbulent momentum, 208, 214, 215
    turbulent thermal, 274, 284, 287
stability (*see* Transition)
thickness:
    concentration, 74
    conduction, 71, 180
    displacement, 66, 67
    enthalpy, 71
    free convection, 406
    laminar momentum, 95
    laminar thermal, 174
    momentum, 66, 67
    shear, 103
    turbulent momentum, 208, 209
transpiration flows:
    film cooling, 293, 296
    foreign gas injection, 293, 507, 508, 531
    laminar momentum, 98, 99

595

transpiration flows—(*Continued*)
    laminar thermal, 169, 190
    turbulent momentum, 226, 229
    turbulent thermal, 288
  turbulent momentum:
    eddy diffusivity, 197, 199
    equilibrium, 222, 223
    integral solutions, 207, 209
    $k - \varepsilon$ model, 220
    kinetic energy equation model, 213
    law of the wall, 204
    mixing-length model, 198, 212
    pressure gradient, 206, 208, 214, 215
    roughness, 230, 232
    shear stress, 201, 202
    transition, 193, 194
    transpiration, 226, 229, 230
    van Driest model, 210
  turbulent thermal:
    eddy conductivity, 256
    eddy diffusivity, 256
    film cooling, 293, 296
    flat plate solutions, 277, 278
    integral solutions, 281, 286
    law of the wall, 272, 273
    Reynolds analogy, 257, 259
    roughness, 298, 299
    transpiration, 288
    turbulent Prandtl number, 259, 263
Boussinesq approximation, 398

Channel flow (*see* Tube flow)
Chemical reactions (*see* Mass-transfer
    problems)
Clauser shape factor, 223, 225
Combined entry-length:
  laminar tube flow, 149
  turbulent tube flow, 345
Combustion (*see* Mass-transfer problems)
Compact heat exchanger surfaces, 443,
    452, 453, 461–468
Concentration (*see* Mass concentration)
Conductance (*see* Heat-transfer
    coefficient, Mass-transfer
    conductance, Nusselt/Stanton
    numbers, Overall heat exchanger
    conductance)
Conduction thickness, 71, 180
Conservation of energy principle, 8
Conservation of mass principle, 6
Conserved property of the second kind,
    494
Considered phase, 482
Continuity equation (*see* Equations)
Control surface, 5
Control volume, 5
Convection, definitions:
  forced, 2

  free, 2, 396
  mixed, 413
  natural, 2, 396
Couette flow approximation:
  mass transfer, 512
  transpiration, 228, 288
  turbulent near-wall flows, 269

Declerating flows (*see* Boundary layer,
    pressure gradient flows)
Density, 14, 510
Diffusion coefficient, 15, 16
Diffusion-thermo effect, 14
Displacement thickness:
  definition, 67
  laminar flow, 95
  turbulent flow, 207
Dissipation equation (*see* Equations)
Dissociation (*see* Mass-transfer problems)
Driving force, mass transfer, 499
Drying (*see* Mass-transfer problems)
Dufour effect, 14

Eddy conductivity, 256
Eddy diffusivity:
  heat transfer, 256
  momentum, 197, 199
Eddy viscosity, 197
Energy equation (*see* Equations)
Energy integral equation, 71
Enthalpy:
  conductance, high velocity, 375, 387
  incompressible liquid, 41
  of a mixture, 34, 483, 491
  of a reacting mixture, 491
  partial, 34, 483
  perfect gas, 41
  stagnation, 373
  thickness (*see* Boundary layer,
    thickness)
  transferred-substance state, 484, 485
Entry-length (*see* Tube flow)
Equations:
  continuity:
    general form, 22, 23
    specie, 30
    turbulent flow, 52
  dissipation:
    general, 60
    boundary layer form, 60, 221
  energy:
    boundary layer form, 36
    free convection, 398
    general viscous, 38
    stagnation enthalpy, 54, 55, 374, 375
    stagnation temperature, 376
    turbulent flow, 55, 56
    with mass diffusion, 36, 486

integral:
    energy, 71, 73
    momentum, 65, 68
mass diffusion:
    chemical element, 31, 32
    P equation, 493, 494
    specie, 30, 31
mechanical energy, 56
momentum:
    boundary layer form, 26
    free convection, 398
    Navier–Stokes, 27
    turbulent flow, 53, 54
turbulent kinetic energy:
    boundary layer form, 59
    general, 58
Equilibrium constant, reaction, 529
Error function, 565
Evaporative cooling (*see* Mass-transfer
    problems)

Falkner-Skan flows, 96
Fick's law, 14
Film cooling, 296
Fin effectiveness, 422
Flow work, 8
Foreign gas injection, 531
Fourier's law of heat conduction, 13
Free convection, definition, 2, 396
    (*see also* Boundary layer, free
    convection)
Friction coefficients (friction factors):
    definition, 68, 78
    laminar boundary-layer flow:
        blowing and suction, 98
        body of revolution, 102
        constant u_∞, 89
        variable properties correction, 358–
        362, 365, 367, 368
        variable property, high velocity, 384,
        385
        variable u_∞, 96
    laminar flow in tubes:
        annuli, 81
        circular tube, 79
        equilateral triangular tube, 81
        hydrodynamic entry length, 84, 85
        rectangular tubes, 80
        variable property correction, 358–360
    turbulent boundary layer flow:
        constant u_∞, 208, 209
        rough surface, 232
        transpiration, 229
        variable property correction, 367, 368
        variable property, high velocity, 388,
        389
        variable u_∞, 208

turbulent flow in tubes:
    circular tube, 249
    other shapes, 249
    rough surface, 250–252
    variable property correction, 360–362
Friction velocity, 202

Gamma function, 563
Gibbs–Dalton law, 518
Grashof number, 400, 404, 411

Heat of combustion, 491
Heat exchanger solutions:
    counter-flow, 431
    cross-flow, 433
    counter-flow with cross-flow headers,
        436
    multi-pass, 434
    periodic flow, 437
Heat exchanger effectiveness, 419
Heat-flux vector, 13
Heat-transfer coefficient:
    definitions:
        boundary layer flows, 163
        high velocity flows, 372, 375, 377, 381
        tube flows, 111
        (*see also* Nusselt/Stanton numbers)
Heat-transfer conductance (*see* Heat
    transfer coefficient, definitions)
Heat transfer—friction power calculation,
    445
Hydrodynamic entry length (*see* Tube
    flows)
Hydraulic diameter, 125, 334
Hydraulic radius, 82
Humidity:
    absolute, 518
    relative, 519

Integral equations (*see* Equations)
Integral solutions, laminar flow:
    free convection, 406
    heat transfer:
        body of revolution, 183
        unheated starting length, 175
    momentum transfer:
        body of revolution, 104
        flat surface, 102
Integral solutions, turbulent flow:
    free convection, 409
    heat transfer:
        body of revolution, 286
        unheated starting length, 281
    momentum transfer, 208

Karman–Nikuradse equation, 249
$k - \varepsilon$ model of turbulence, 220

Laminarization, 226
Law of the wall:
　momentum boundary layer:
　　circular pipe, 246
　　continuous, van Driest model, 210
　　logarithmic, 204
　　power-law, 207
　　rough surface, 232
　thermal boundary layer:
　　circular pipe, 314
　　logarithmic, 272, 273
　　rough surface, 299
Lewis number, 17, 488, 496
Liquid metal flows, 320
Local similarity/nonsimilarity:
　heat transfer, 171
　momentum, 99

Mach number:
　definition, 383
　effects of, 384, 389
Mass concentration:
　chemical element, 31
　component, 15, 29
T-state, 484
　water vapor, 518
Mass diffusion coefficient, 15, 16
Mass diffusion equations (see Equations)
Mass flux vector, 22
Mass transfer coefficient (conductance):
　definition, 3, 499
　from heat-transfer data, 513
　laminar boundary layer:
　　constant u_∞, 507
　　variable property correction, 510
　　variable u_∞, 511
　thermal radiation effect, 523
　turbulent boundary layer:
　　constant u_∞, 512
　　variable property correction, 513
Mass-transfer driving force, 499
Mass-transfer problems:
　air-water-vapor system:
　　drying, 524
　　evaporative cooling, 524
　　wet-bulb psychrometer, 519
　combustion, 525, 527
　diffusion-controlled burning rate, 526
　multiple chemical reaction, 529
　simple chemical reaction, 487
　dissociating boundary layer, 531
　foreign gas injection, 531
Mass-transfer rate, 498
Mean velocity, 77
Mixing-length:
　roughness effect, 231
　theory, 198
　theory summary, 212

van Driest model, 210
Mixture velocity vector, 482
Mole fraction, 518
Molecular weight:
　binary mixture, 510
　mixture, 518
Momentum equation (see Equations)
Momentum theorem, 6
Momentum thickness:
　definition, 67
　laminar boundary layer, 95
　turbulent boundary layer, 207
Moody diagram, 251
Natural convection (see Free convection)
Navier–Stokes equations, 27
Neighboring phase, 482
Nusselt/Stanton numbers:
　definitions:
　　boundary layer, 163, 165
　　circular tube, 116
　laminar boundary layer flows:
　　blowing and suction, 170, 171
　　body of revolution, constant t_o, 183
　　circular cylinder, flow normal t_o, 185
　　constant u_∞, constant t_o, 163–166
　　constant u_∞, constant $\dot{q}''$, 179, 180
　　constant u_∞, variable t_o, 176
　　separated flow, 185
　　sphere, flow over, 185
　　stagnation point, 167, 168
　　unheated starting length, 175
　　variable u_∞, constant t_o, 183
　　variable property, high velocity, 391
　　variable property corrections, 365,
　　　367
　laminar combined entry-length, tubes:
　　annuli:
　　　constant $\dot{q}''$, 151
　　circular tubes:
　　　constant $\dot{q}''$, 151
　　　constant t_o, 151
　　parallel planes:
　　　constant $\dot{q}''$, 151
　laminar free-convection:
　　blowing and suction, 404
　　constant $\dot{q}''$, 403
　　constant t_o, 402
　　other geometries, 411
　laminar fully developed flow, tubes:
　　annuli:
　　　asymmetric heating, 123
　　　axial fluid conduction, 120
　　circular:
　　　constant $\dot{q}''$, 116
　　　constant t_o, 117
　　internal heat sources, 120
　　other cross-section shapes, 125, 126
　　parallel planes, 123

peripheral $\dot{q}''$ variation, 119
variable property correction, 358, 359
viscous dissipation, 120
laminar thermal entry-length, tubes:
  annuli:
    constant $\dot{q}''$, 139
    axial variation of $\ddot{q}''$, 147
    axial variation of t_o, 143
  circular:
    constant $\dot{q}''$, 134, 135
    constant t_o, 132, 133
  parallel planes:
    constant $\dot{q}''$, 139
    constant t_o, 137
  square, constant t_o, 137
  rectangular, constant t_o, 137, 138
mass transfer (*see* Mass-transfer
  conductance)
turbulent boundary layer flows:
  boundary of revolution, 286
  constant u_∞, constant t_o, 277, 278
  constant u_∞, variable $\dot{q}''$, 293
  constant u_∞, variable t_o, 283
  transpiration, 288, 289
  unheated starting length, 281
  variable property, high velocity, 389,
    390
  variable property correction, 367, 368
  variable u_∞, variable t_o, 286
turbulent combined entry-length, tubes,
  345
turbulent free-convection:
  flat plate, 409, 410
  other geometries, 411
turbulent fully developed flow, tubes:
  annuli:
    asymmetric heating, 328, 329
    effects of eccentricity, 333
  circular:
    constant $\dot{q}''$, 316, 319, 322
    constant t_o, 324
  high Prandtl number flow, 319
  liquid metal flow, 321
  other flow cross-section shapes, 333
  parallel planes, 332
  perpheral heat flux variation, 325, 326
  rough surface, 348
  bariable properties corrections, 360–
    362
turbulent thermal entry-length, tubes:
  axial variation of t_o and $\dot{q}''$, 344
  circular, 338, 339
  parallel planes, 343, 344

Overall conductance heat exchanger, 421,
  422

Pipe flow (*see* Tube flow)
Prandtl mixing-length theory (*see* Mixing-
  length) Prandtl number, definiton,
  17
Pressure drop calculation, heat
  exchangers, 441
Pressure gradient, boundary layer (*see*
  Boundary layer, pressure gradient
  flows)
Pressure gradient parameter:
  laminar flow, 96
  turbulent flow, 224, 225
Property ratio scheme (*see* Variable
  property effects)
Psychrometer, wet bulb (*see* Mass transfer
  problems)

Rayleigh number, 408
Recovery factor:
  laminar flow, 380
  turbulent flow, 388
Reference temperature scheme (*see*
  Variable property effects)
Reynolds analogy, 259
Reynolds decomposition, 48
Reynolds number:
  diameter, 79
  enthalpy thickness, 166, 277
  free convection, 408
  hydraulic diameter, 125
  momentum thickness, 95
  roughness, 231
  transition, 1094
  x, 95
Reynolds stress tensor, 53
Roughness:
  definition, 231
  momentum boundary layer, 230
  momentum tube flow, 250
  Moody diagram, 251
  thermal boundary layer, 298
  heat transfer in a tube, 348

Sand grain roughness, 230
Schmidt number:
  definition, 17
  turbulent, 512
Separation:
  laminar momentum, 98
  laminar thermal, 185
Shape factor:
  laminar boundary layer, 104
  turbulent boundary layer, 207
  Clauser, 223
Shear stress:
  laminar and turbulent pipe flow, 78
  turbulent, 53, 197
  turbulent boundary layer, Couette flow,
    202
Shear thickness, 103

Shear velocity, 202
Sherwood number, 501
Similarity solutions, laminar flow:
  free convection, 401–405
  heat transfer:
    blowing and suction, 170
    constant u_∞, 163, 164
    variable property, high velocity, 365, 385
    variable u_∞, 167
  mass transfer, 504
  momentum transfer:
    blowing and suction, 100
    constant u_∞, 95
    variable property, high velocity, 365, 384
    variable u_∞, 98
Simultaneously developing flows (*see* Combined entry length)
Soret effect, 17
Specific heat:
  component, 483
  mixture, 483
Stagnation enthalpy, 47, 373
Stagnation enthalpy equation, 47, 375
Stagnation-point heat transfer, 168, 169
Stagnation temperature, 376
Stanton number (*see* Nusselt/Stanton numbers)
Step-function solutions (*see* Boundary layers, unheated starting length)
Stream function:
  compressible flow, 363
  incompressible flow, 93
Stress:
  normal, 2
  shear, 11
  turbulent, 53
Stress tensor:
  laminar, 12
  turbulent, 53
Substantial derivative, 27
Suction, effect of (*see* Boundary layer, transpiration flows)
Suction parameter (*see* Blowing parameter)
Superposition:
  laminar boundary layers, 175, 176
  laminar tube flows, 140–143

Temperature:
  bulk fluid, 111
  fully developed laminar profile, 114, 117
  mixed mean fluid, 111
  mixing-cup, 111
  reference, 366, 386, 391
  stagnation, 376

Tensor notation, 12
Thermal conductivity:
  mixture, 555
  molecular, 13
  turbulent, 256
Thermal diffusivity:
  molecular, 41
  turbulent, 256
Thermal entry-length (*see* Tube flows)
Thermo-diffusion effect, 14
Total derivative (*see* Substantial derivative)
Transferred substance state, 484, 485
Transition, 193, 194
Transpiration cooling (*see* Boundary layer, transpiration flows)
Tube flow:
  equations (*see* Equations)
  laminar momentum:
    circular tube, 75
    hydrodynamic entry length, 82
    other cross-section shapes, 79
  laminar thermal:
    annuli, 121–124
    circular tube, 114–121
    combined entry-length, 149–152
    other cross-section shapes, 124–126
    parallel planes, 125
    thermal entry-length, 126–140
    variable surface boundary conditions, 140–149
  turbulent momentum:
    circular tube, 244–249
    hydrodynamic entry-length, 245
    other cross-section shapes, 249, 250
    rough surface tubes, 250–252
  turbulent thermal:
    annuli, 326–333
    circular tube, 313–326
    combined entry-length, 345–348
    other cross-section shapes, 333–335
    parallel planes, 326–333
    rough surface tubes, 348, 349
    thermal entry-length, 337–344
    variable surface boundary conditions, 344, 345
Turbulence models:
  Prandtl mixing-length, 198
  $k - \varepsilon$ model, 220
  turbulence kinetic energy model, 213
Turbulent kinetic-energy equation (*see* Equations)
Turbulent boundary layer (*see* Boundary layer)
Turbulent heat flux, 55, 256
Turbulent Prandtl number, 257, 259–268
Turbulent shear stress, 53

Unheated starting length (*see* Boundary layer *or* Tube flow)

van Driest damping function, 210
Variable-properties effects:
  free-convection, 406, 407
  laminar boundary layer, 365
  laminar tube flow, 358–360
  mass-transfer conductance, 510, 513
  turbulent boundary layer, 367, 368
  turbulent tube flow, 360, 361
Variable surface boundary conditions (*see* Boundary layer; Tube flow)
Vector operators, 566–569
Velocity:
  fully developed laminar, 76

  mean, 77
Viscosity:
  laminar or molecular, 11
  mixture, 555
  turbulent, 197
Viscous dissipation:
  effects of, 120, 372
  in energy equations, 38
Viscous sublayer, momentum, 196, 203
von Karman constant, 200

Wall coordinates, turbulent flow:
  momentum, 202
  thermal, 270
Wedge flows, 96